Society of
Exploration
Geophysicists

MIGRATION OF SEISMIC DATA

Edited by
Gerald H. F. Gardner

Series Editor
Franklyn K. Levin

Geophysics reprint series
No. 4

ISBN 0-931830-35-4
0-931830-00-1

Library of Congress Catalog Card Number: 85-061313

Society of Exploration Geophysicists
P.O. Box 702740
Tulsa, Oklahoma 74170-2740

Printed in the United States of America

CONTENTS

Chapter 5 Wave equation developments

Preface

Papers selected for this reprint volume sample the literature on migration from 1937 to 1984. They are arranged in chronological order to show the evolution of migration from the initial ruler-and-compass constructions used at the start of seismic exploration to feasible computer algorithms for depth migration in three dimensions. Papers dealing with the direct inversion of seismic data to obtain velocity and density are not included because they are considered a distinct approach to seimsic data analysis. Only a few papers treating velocity analysis are included, although selection of a velocity distribution remains an outstanding unsolved problem for the execution of depth migration algorithms.

The interpretative value of migration was recognized in the early 1920s. An example cited in Dobrin (1976) illustrated a method used by J. C. Karcher. A single-fold reflection profile was used to pick two-way traveltimes for echoes from the interface between the Viola limestone and Sylvan shale formations. Circular arcs were drawn with center at the shot-receiver midpoints

on the ground and with radii equal to half the product of the traveltimes for the picks and the velocity of propagation in the Sylvan shale. The envelope curve for the circular arcs was then drawn and interpreted as the limestone-shale interface in the correct position. Karcher's sketch of the process is shown in Figure 1.

Ideally, the circular arcs would touch the interpreted interface. However, time shifts in the measurements, possibly caused by velocity anomalies in the shale formation, may have caused some radii to be too small and some too large. The interpreter did not force tangency but made a geologically reasonable fit.

The point at which the circular arc and the envelope would ideally touch defines the *reflection point*, the corresponding radius defines the *raypath*, and the circular arc defines the *wavefront*. The wavefront that spreads out from the shot first hits the interface at the reflection point, and the raypath there is perpendicular to the interface. The reflected wave travels back to the receiver with the same raypath if the shot and receiver are close together. When the velocity of propagation

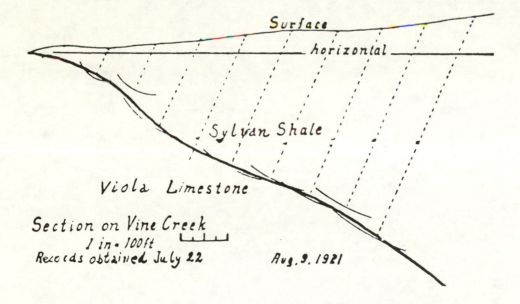
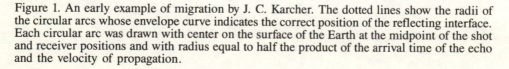

Figure 1. An early example of migration by J. C. Karcher. The dotted lines show the radii of the circular arcs whose envelope curve indicates the correct position of the reflecting interface. Each circular arc was drawn with center on the surface of the Earth at the midpoint of the shot and receiver positions and with radius equal to half the product of the arrival time of the echo and the velocity of propagation.

varies in the Earth, raypaths are no longer straight and wavefronts no longer circular, but the same relationships hold.

This simple construction for constant velocity is the basis for migration in its many presentday forms. Indeed, performance of most migration programs, when scrutinized, reveals the underlying construction. This is done by migrating just one measurement. All input data values are made zero except for one sample for a selected position and time. The migrated section is a picture of the wavefront which traveled from this surface position for this time according to the velocity of propagation assumed for the underlying formations in the Earth. In general, migration is the summation of the wavefronts constructed from each sample value.

Selected reprints are arranged chronologically in chapters and a brief preface for each chapter gives additional references. Chapter titles are: Initial Concepts, Ruler and Compass Methods, Computer Processing, Applications and the Exploding Reflector Model, and Wave Equation Developments.

This progression of concepts and application continues. Problems still almost untouched are: migration before stack, interval velocity analysis in complex structures, true amplitude migration, migration of three-component data, extensions to three dimensions, etc.

Acknowledgments

The editors thank Geophysical Prospecting, Geophysics Journal of the Royal Astronomical Society, Oil and Gas Journal, Shale Shaker, and Soviet Physics-Acoustics for granting permission to reproduce the selected papers and the authors of each paper for their contribution to this work.

The constraints of publication costs have eliminated many excellent papers.

We especially commend the staff of the Society of Exploration Geophysicists, and in particular J. Henry, for making publication possible.

Gerald H. F. Gardner

References

Dobrin, M., 1976, Introduction to geophysical prospecting: McGraw Hill Book Co., 3rd Ed., p. 12.

Chapter 1

Early Concepts

Rieber's approach to migration was to delay and add individual recordings at adjacent ground locations and so create a synthetic antenna which could be steered over a range of directions by varying the time delay per recording. The weakness in the device, called the Geo-sonograph (Rieber 1936a, 1936b; Sawdon 1936), was that analog methods in the 1930s did not allow enough recordings to be combined to bring the full power of the technique to bear on the geologic problems (McDermott, 1936).

Rieber's Company, formed to promote the Geo-sonograph, did not prosper and the concept was not pursued until technological developments made it more practical. In Russia in the 1950s a similar idea was developed under the name controlled directional reception (CDR). Current processing methods which use slant stacks, *p*-tau transforms, or inverse ray tracing are more successful implementations of the same concepts.

Rieber (1937) described his ideas and an experimental verification of how "confused records" could be caused by reflections from several different directions, or by energy scattered by discontinuities such as faults, *Johnson (1938)* gave some field applications in which the Geosonograph was used to resolve such confused records.

References

McDermott, E. E., 1937, The use of multiple seismometers: Petroleum Engineer, **8**, February, 135-136.

Rieber, F. 1936a, A new reflection system with controlled directional sensitivity, Geophysics, **1**, 97-106.

————1936b, Visual presentation of elastic wave patterns under various structural conditions: Geophysics **1**, 196-218.

Sawdon, W. A., 1936, New geophysical method provides reliable data for structure mapping: Petroleum Engineer, **8**, October, 25-29.

Reprinted from GEOPHYSICS, 2, 132-160.

COMPLEX REFLECTION PATTERNS AND THEIR GEOLOGIC SOURCES*

FRANK RIEBER†

ABSTRACT

A technic is shown for producing synthetic records corresponding to assumed ideal structural conditions, the resulting records being reproduced in the usual visual form, and also in the form of directionally analyzed records. Examples of folding, faulting and irregular sedimentation are treated.

The research on which this paper is based was originally undertaken by the writer several years ago. The objective aimed at was the development of a radically new attack on the problem of reflection shooting by which it was hoped that many of the limitations of the usual methods might be overcome.

Exploration by reflection shooting had progressed, at that time, to a point where these limitations had begun to be definitely evident. While it was known that the method as a whole was eminently satisfactory for outlining broad structural features in regions where clear records of reflections could be obtained, users were beginning to recognize that certain areas and certain types of structure presented problems which might be somewhat beyond the capacity of the technic as then employed—and even beyond the reach of such improvements in the apparatus and methods as might reasonably be expected.

It was known that blind spots would occur in otherwise clearly mapped areas and that within these spots only very poor records were obtained. In fact, very extensive regions could be found in which poor records were the rule, and only a few spots existed where passable records could be obtained.

And in some places, reflections could be identified only on certain parts of the records, corresponding to a certain depth zone, above and below which no satisfactory results seemed to be obtainable.

The limitations of the method could, therefore, be expressed in terms of the frequency of occurrence of unsatisfactory records, which, by the way, were usually marked "N. R."—presumably meaning "no reflections."

These poor records were roughly divisible into two types, the first

* Paper read at the Fall Meeting, Houston, Texas, November, 1936.
† Rieber Laboratory, Los Angeles, Calif.

of which certainly contained no reflections. In fact, they could not properly be said to contain anything at all. Beginning with waves of moderate amplitude, the recorded vibrations on such records would die down very rapidly to so small value that, even had reflections been present at the later portions of the record, they could certainly not be seen or identified.

To meet such conditions, it was customary to change the shooting or recording procedure or the location of the set-up, until a record of readable amplitude was obtained.

Poor records of the second class were, however, far more numerous, and presented a much more interesting problem. They contained vibrations of adequate amplitude throughout the useful length of the record, but were characterized by a great dearth or a complete absence of the characteristic pattern by which reflected waves are usually identified.

Such patterns consist of a sharply recognizable group of waves, appearing at approximately equal magnitudes and with similar wave forms on all of the adjacent traces and exhibiting a "line-up" or relative time correspondence. This characteristic appearance is so well-known to every one who has had to do with reflection shooting that it should require no further description here.

Instead of showing any of these patterns, the confused records under discussion would show what appeared to be random wave motion on all traces—frequently throughout the length of the record without even one line-up or correspondence of waves.

If any explanation of such records was attempted, it was usually to the effect that an explosion generated in the earth two classes of phenomena, definable as "reflected waves" and "random vibrations." Both of these were present everywhere to some extent, but in some places, for reasons not clearly understood, the random vibrations became very large and the reflected waves either shrunk to a very small value or were entirely absent.

When regions were encountered where such unpatterned or illegible records were obtained, and where minor modifications in shooting procedure failed to produce clearer ones, the setback was accepted in a philosophical spirit, as merely one of the minor tribulations of a not altogether tranquil profession.

Several recourses were, of course, open to a geophysicist undertaking to work a region where a majority of the records were confused. The first and most logical was to take lots of records, to mark on them anything that might be a reflection, and then to cross-compare all of the results. By this procedure anything that was real and present in the earth should logically give evidence of its presence on a number of different records, where waves of a truly random nature would not necessarily be obtained twice in the same place. This procedure worked well enough in some places, where it was not quite impossible to mark on each record a reasonable minimum number of line-ups faintly resembling reflection patterns.

Another was to refuse to work the area entirely, giving as a reason that "no reflections could be obtained." If this were actually a fact, it could, of course, be a clinching argument against reflection work in certain places.

The above was roughly the state of reflection shooting at the time the writer undertook to develop a new method of attack. Incidentally, while it has been altered slightly by various subsequent developments, it is very largely the condition confronting the most widely used methods of reflection shooting at the present time.

At the outset of the new attack, the problem of the confused and unpatterned record was seen to be at the heart of the whole matter. Such records obviously contained in one form or another practically all of the information sent back by the earth after a shot was fired. If it could be proved that this content was nothing but random waves unrelated to geology, or so broken up that the fragmentary answers could not be translated into terms of structure, there was certainly no point in going further.

On the other hand, the fact that records of this type almost invariably occurred in connection with geological anomalies rather challenged the attention. As a specific example, such records were known to be associated with faulted regions so uniformly that they were customarily used as indirect evidence of the existence of faults.

Geophysicists cannot claim the honor of discovering that it is expedient to draw a fault through an unexplainable inconsistency in the data. Geologists have used this device too long in their own work to object seriously if it is borrowed in its entirety by a young art.

Further, a species of mental parallel makes the connection between faults and confused records seem very plausible. The waves on a confused record are jumbled. The strata in a fault zone are likewise thoroughly scrambled. It is natural to accept the connection between these two facts without inquiring too closely into exactly how the disturbance of the strata might result in disturbance of the recorded waves.

On examining into the confused record, however, it did not seem so certain that reflected waves were absent. True, if there were any reflected waves there, they did not show as patterns in the customary manner. But the presence of vibrations of good amplitude throughout the length of such records had to be accounted for in some way. It scarcely seemed possible that some mysterious random cause, unrelated to geology, and possibly concerned solely with surface conditions, could so invariably produce a confused record in the vicinity of a geological anomaly.

A much more plausible explanation was soon found—namely, that anomalies of many types return waves from the earth from several directions at the same time. While each of these waves—if the others could have been entirely suppressed—would have given a clear pattern on the record, the simultaneous arrival of all of them produced such an overlapping that the resultant combination appeared to be entirely random.

Following out this assumption, which has since been abundantly sustained by experimental and field results, the writer undertook to develop a method by which such a complicated group of returned waves could be recorded and thereafter analyzed into their component elements. In this system, the traditional seismograph form of record was abandoned entirely, and a fresh start made by recording the earth vibrations as sound tracks on a film, in a manner similar to the recording of sound for talking pictures. This film was later passed through an optical analyzer with a photo-electric cell, developed especially for this purpose, and adapted to break down groups of wave trains of mixed direction of arrival into its component wave elements, to each of which a direction of arrival and a time of arrival could be assigned.

This system has been termed the Sonograph, and has been briefly described by the writer on several previous occasions.[1] Unfortunately, however, these descriptions seemed to have been so short, and so lacking in detail, that the purpose and functioning of the new method are not widely and completely understood. For example, an impression seems to exist that if a Sonograph, and a reflection seismograph of the usual pattern, could be set up in the same locality, and used to record the same shot, they should obtain more or less interchangeable results. That is to say, if a reflected wave were recorded on the seismograph

[1] "A New Reflection System with Controlled Directional Sensitivity" by Frank Rieber, published in GEOPHYSICS, Vol. 1, No. 1, January, 1936.

with a clear and unmistakable pattern, it is readily conceded that the Sonograph record, when analyzed, should show this same reflected wave coming from the same depth and the same direction as computed from the seismograph findings.

If, on the other hand, geological conditions were such as to give a confused record on the seismograph, in which no pattern appeared, this would mean that there were really no reflected waves present in the earth. Hence, the Sonograph record when analyzed, should show no reflections.

By a simple extension of this same line of reasoning, it has been argued that if, under such conditions, a seismograph record should show no patterns of reflected waves, and if an analyzed Sonograph record showed reflections to be present, then these reflections could not be real, but must have been somehow manufactured by the analyzing instrument in the process of reproducing the record.

The following demonstration will show that this is not the case, and that an analyzed record from the Sonograph possesses, in fact, a number of advantages over the ordinary visual record.

This demonstration is arranged as an idealized comparison of the visual recording seismograph of the Sonograph methods and consists of the following steps:

First, it will be shown that most common structural anomalies must inevitably return waves from a number of different directions, and overlapping each other with respect to time. These waves may be thought of either in terms of "central rays," commonly used in diagrams of optical reflection, or they may be considered in terms of the shape of their emerging wave fronts at the time when these wave fronts reach the surface of the earth near the receptors. For the purposes of brevity, waves of this class arriving from various directions over the same time range, will be termed "criss cross" waves.

The second step in the demonstration is to show a method by which such criss cross waves can be artificially produced in the laboratory, and thereafter recorded

(a) As seismograph records of the usual form,

(b) As analyzed Sonograph records.

The third step in the demonstration will be to take an assumed and extremely simple geological condition from which criss cross waves will be produced, and to make, by the above technic, equivalent seismograph and Sonograph records of this criss cross pattern.

A comparison of the two methods is then possible, in that an attempt may be made to reconstruct the original geological condition from the seismograph record by the usual visual identification of waves and a similar attempt may be made using the analyzed Sonograph record.

The following diagrams are intended to illustrate a few of the many geologic conditions which can cause criss cross waves. The diagrams have been purposely simplified to their bare essentials, and also the structural irregularities have been exaggerated to give clearer emphasis to the criss crossing of the wave paths.

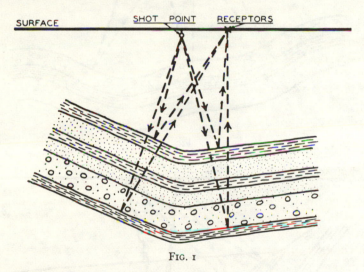

FIG. 1

A simple, sharply folded syncline has been chosen for this demonstration and is shown in Fig. 1. The shot point and the receptors are presumed to be set up over the axis of the fold as shown. Under these conditions, it is quite apparent that waves will be returned by the structure from two directions at once as is indicated clearly by the wave paths in the diagram.

Fig. 2 shows another commonly encountered structural condition causing criss crossed waves which are again indicated by the wave paths on the sketch. Crenulated beds of this type, and other similar irregularities in sedimentation, are recognized as occurring very frequently where deposition has taken place under shore line conditions.

Fig. 3 shows a combination of overlap and truncation, the wave path diagrams clearly illustrating the criss crossing of the waves. In passing, it might be mentioned that all of these conditions have been purposely exaggerated as to dip, in order to clearly separate and indicate the individual wave paths.

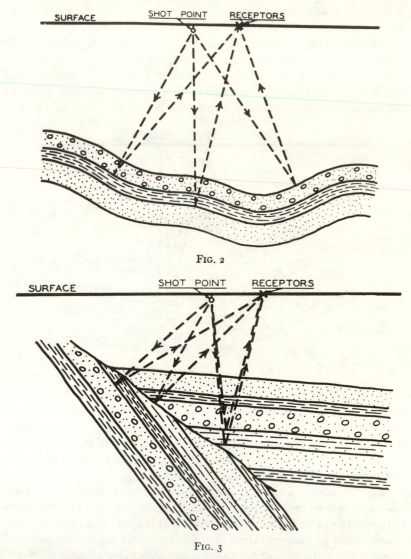

FIG. 2

FIG. 3

Fig. 4 shows a normal fault in a homocline, and illustrates one of the reasons why fault mapping by the ordinary reflection technic must frequently be done in a negative fashion. The heavy wave paths refer to reflected waves of the usual sort coming from successive strata, while the lighter wave paths refer to waves returned from the breaks and discontinuities in the fault region itself.

As will shortly be shown, these waves returning from the fault are not always to be considered as reflections. Diffracted waves are frequently sent back from fault zones, although their presence has not

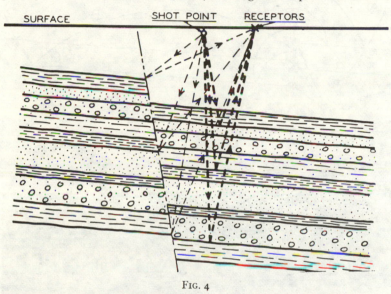

FIG. 4

heretofore been generally recognized due to the impossibility of separately identifying them on the usual form of records.

To illustrate this point more clearly, Fig. 5 shows a similar faulted condition, in which the dotted line method of indicating wave paths has been abandoned in favor of diagramming the actual wave fronts of the downgoing and returning waves.

The writer recently suggested an interesting method by which the shape of returning wave fronts could be visually demonstrated[2] and the existence of diffracted waves from faults could be clearly shown. The apparatus and technic for this purpose were adapted from previous similar work by various acoustical engineers, who employed it for tracing wave patterns in auditoriums and the like. It consists essentially in creating a miniature explosion in air, from which a sharp

[2] "Visual Presentation of Elastic Wave Patterns Under Various Structural Conditions," by Frank Rieber, GEOPHYSICS, Vol. 1, No. 2, July, 1936.

wave front is radiated and thereafter photographing this wave front
at various stages of its progress. If a model of any structural condition

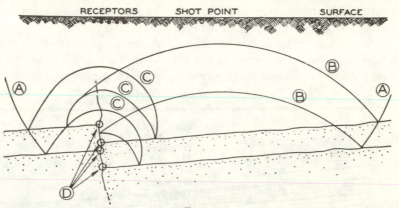

Fig. 5

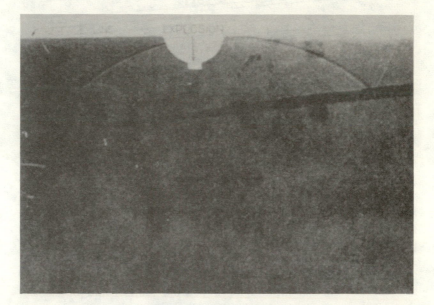

Fig. 6

is interposed in the wave path, the reflected and diffracted waves will
be clearly demonstrated. The form of this apparatus constructed
under the writer's direction has been termed a strobograph.[3]

[3] "Sound Wave Stroboscope," by B. F. McNamee, "Electronics," November, 1936.

Fig. 6 shows an actual group of wave fronts photographed in connection with a model of a single faulted stratum, and will be seen to correspond strikingly with the theoretical diagram in Fig. 5. A criss cross wave pattern from the faulted model is about to emerge at the surface of the illustration. Obviously, if a succession of such strata exist in the earth, and the same fault condition exists in all of them, a succession of these criss cross waves must return.

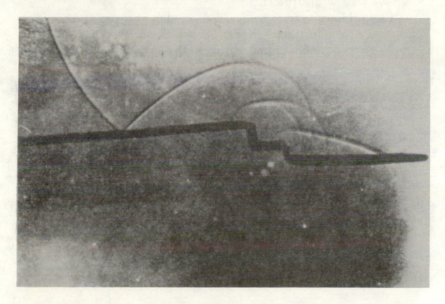

FIG. 7

Fig. 7 shows a model of a double step fault in a stratum as recorded by the strobograph. The criss crossing of reflected and diffracted waves on this picture scarcely needs further comment. If several such strata followed one another at successive depths, the complexity of the emerging patterns can readily be visualized.

The structural conditions shown in the preceding diagrams are, of course, thought of as existing only in the plane of the drawing. Therefore, they indicate only a small fraction of the possibilities for criss crossing of waves under actual field conditions. A three dimensional earth will, of course, return waves to the surface from all possible directions, and far greater confusion will exist than can be shown in any two-dimensional sketch.

A technic will now be shown by which criss cross waves of any desired and predetermined pattern may be produced in the laboratory, and reproduced either in the usual visual form common to seismograph records; or in the analyzed form used in the Sonograph.

As the first step in this process, an electrical source of wave trains was constructed. Transient wave generators of this type are well known to geophysical engineers, and are commonly used to make various tests on instruments.

Using this wave train source, a succession of sound track records are made, as illustrated in Fig. 8, which shows ten such sound tracks on a single film. Similar waves are recorded on all ten tracks,

FIG. 8

but at successively later times, giving a "line-up" corresponding to the arrival of a wave in the earth, as recorded from a succession of receptors.

Such a film record has the advantage that it may thereafter be reproduced either as a visual type seismograph record, or through the Sonograph analyzer, thus permitting a direct comparison of the relative ability of the visual and the analyzed method in recognizing and identifying individual wave trains.

Reproduction in the usual visual form consists in converting each individual sound track into a corresponding oscillograph trace. The record shown in Fig. 8 has been thus reproduced in Fig. 9, which may be thought of as representing a single isolated and idealized reflected wave, as it would appear on a seismograph record if all other vibrations could in some way be eliminated.

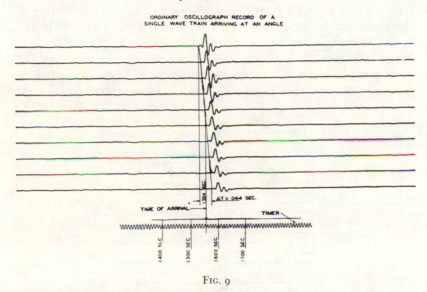

ORDINARY OSCILLOGRAPH RECORD OF A
SINGLE WAVE TRAIN ARRIVING AT AN ANGLE

FIG. 9

Having shown the simple wave train in film form in Fig. 8, and in the usual seismograph record form in Fig. 9, we can now proceed

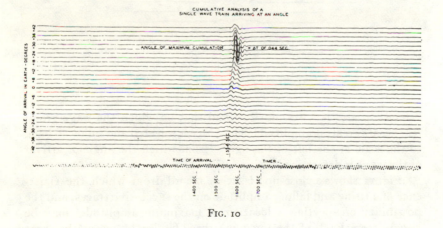

CUMULATIVE ANALYSIS OF A
SINGLE WAVE TRAIN ARRIVING AT AN ANGLE

FIG. 10

to the appearance of the same wave train as reproduced through the Sonograph analyzer as shown in Fig. 10. This form of presenting

the wave is termed an analyzer strip, and will be seen to differ radically in appearance from previous conceptions of wave records. Each
trace on this analyzer strip represents a different angular setting of
the analyzer, corresponding to a known direction of arrival in the
earth. These assumed angles of arrival are shown in the scale at the
left of the figure. To read the record, the presence of an outstanding
wave group is first noted (a simple matter in the figure shown, which
contains only one wave group) and the successive traces on the ana-

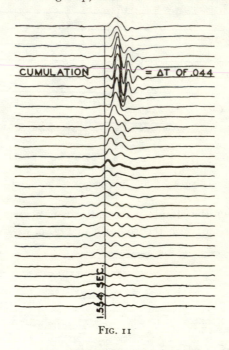

FIG. 11

lyzer strip are then examined to find that on which the wave group
reaches its greatest amplitude. This trace is selected and followed back
to the scale at the left of the figure, from which the angle of arrival
of the wave is determined.

Fig. 11 shows a close-up of such a record, from which the eixist-
ence of a rising and falling amplitude on the successive traces, and the
possibility of selecting a location of maximum amplitude, may be
clearly seen. Each of the foregoing three figures represent the same
event, namely, a single arriving wave train. They may be thought of
as three different languages for expressing the same fact.

Having shown how a single wave train can be recorded in a form similar to a reflected wave pattern, it will be seen that a succession of such wave trains can readily be made by the same method, corre-

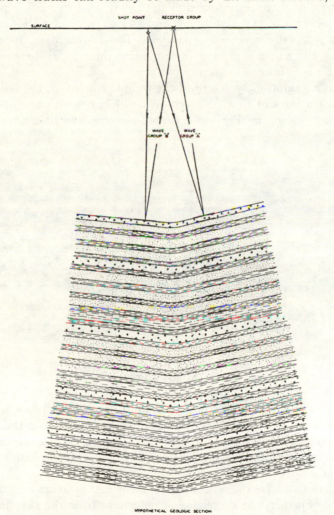

FIG. 12

sponding to successive reflections arising from a series of parallel beds. Such wave trains will hereafter be termed "synthetic reflections."

Having prepared a film containing a series of synthetic reflections from some one bed sequence, a second film may be as readily made, containing a similar set of synthetic reflections from some other group of assumed beds. Both of these films may then be combined, by a process of re-recording similar to that used in motion picture work, where it is termed dubbing, and used to combine various sound effects originally recorded separately. The re-recording apparatus used by the writer is capable of combining three separate sets of reflected waves into one complex pattern.

Having sufficiently described the technic employed, the following demonstration can now be given. We will begin with the assumed structure shown in Fig. 12, a simple chevron syncline, which has

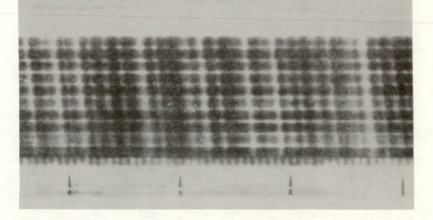

FIG. 13

been idealized for the purpose of simplifying and emphasizing the demonstration. It is assumed that each interface between the strata will give a reflected wave, and the location on each stratum from which the reflections will be returned to the receptor is indicated by a slightly heavier line.

Wave paths for the reflections from the first stratum only are indicated. Further, the succession of reflections from the two limbs of the syncline are divided into one group A and one group B, respectively, for purposes of reference.

Separate film records for wave group A and wave group B are next made in accordance with the technic just described.

Fig. 13 shows a close-up of the film for group A, on which the succession of wave trains corresponding to the various strata may be clearly seen by eye.

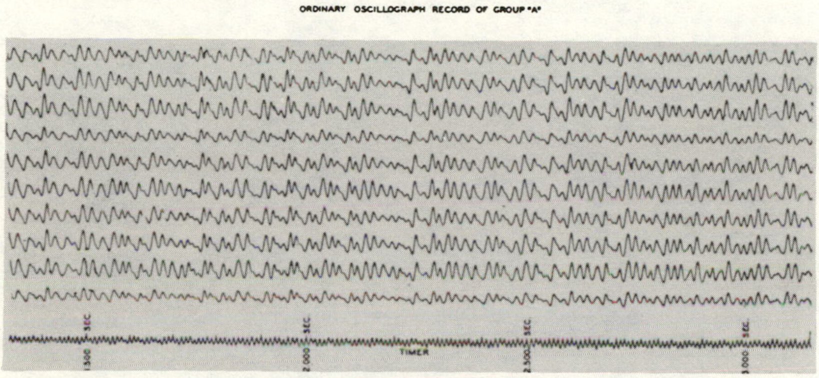

ORDINARY OSCILLOGRAPH RECORD OF GROUP "A"

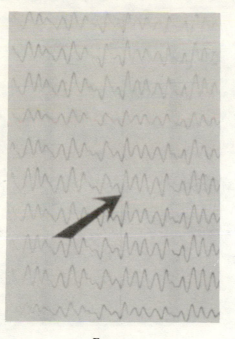

FIG. 14

FIG. 15

Fig. 14 shows this film translated into the usual seismograph form, which corresponds roughly to the type of record which would

be obtained over the syncline if, by some miracle, one side of the structure could have been removed or eliminated and reflections from the

WAVE GROUP 'X CUMULATIVE

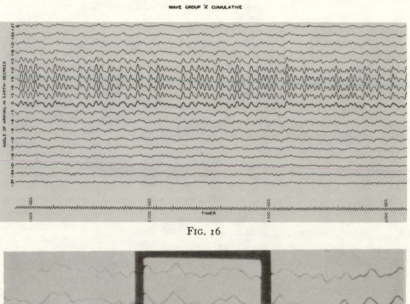

FIG. 16

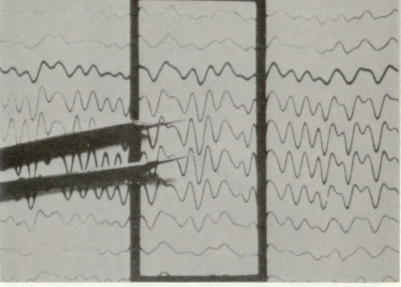

FIG. 17

other side only could be considered. While the wave forms produced
by the transient source are not identical with those appearing on

many seismograph records, especially those in which special filtering and turning circuits are employed, it will be seen from inspecting Fig. 14 that there should be no difficulty in finding any number of line-ups and deducing the presence of the corresponding reflected waves.

Fig. 15 shows a close-up of this same record, in which the sharpness of the line-up may be more clearly evident. Given such a record there would, in practice, be no difficulty in reconstructing from it a close approximation to one limb of the syncline.

Fig. 16 shows the Sonograph analyzer strip corresponding to the seismograph record in Fig. 14.

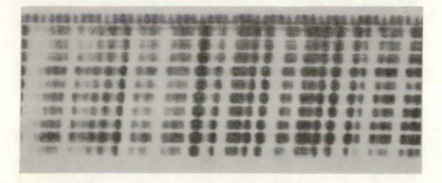

FIG. 18

Fig. 17 is a close-up of the same wave train shown in Fig. 16, to illustrate the positive manner in which the maximum trace may be selected.

The dividers shown in the picture mark one of the maximum waves. It will be seen that by the use of the Sonograph it would likewise be possible to reconstruct one limb only of the syncline, and that so far as the two method are entirely comparable.

We will now assume that the geological miracle just performed can be reversed, and that we can eliminate that side of the syncline containing wave group A, and deal solely with the other side. Fig. 18 shows the film for wave B group alone.

Fig. 19 shows this film reproduced in a manner corresponding to the ordinary seismograph record.

Fig 20 is a close-up of one of the line-ups. Obviously, the seismograph type of record would permit the reconstruction of that part of the structure defined by wave group B, if wave group A were only absent.

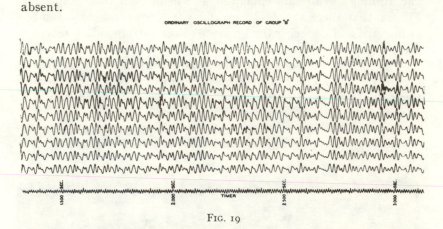

ORDINARY OSCILLOGRAPH RECORD OF GROUP ⅴ

FIG. 19

Fig. 21 shows the analyzer strip for wave group B and Fig. 22 shows a close-up of a portion of this strip—clearly indicating that

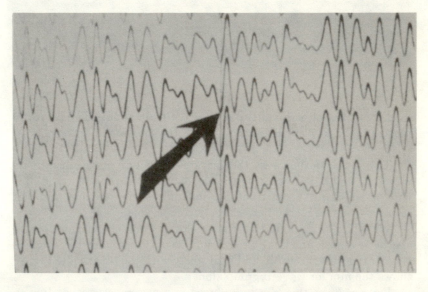

FIG. 20

the maximum waves can again be picked, and their arrival time and directions readily determined from the analyzer strip. Again little choice is left between the seismograph and Sonograph methods. Either could have reproduced the second limb of the syncline.

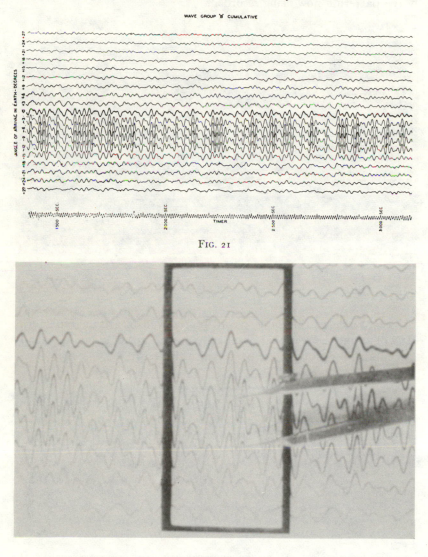

WAVE GROUP 'B' CUMULATIVE

Fig. 21

Fig. 22

Fig. 23 is a record showing the actual condition which would exist in the earth with both limbs of the syncline active in returning

reflections. The miracle has been repealed, and the true criss cross
wave pattern is now being recorded.

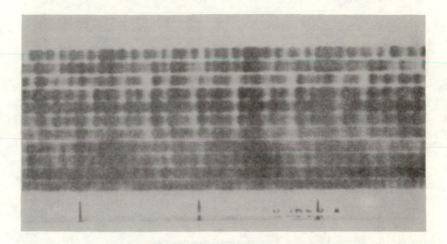

FIG. 23

Fig. 24 shows this same record reproduced in seismograph form.
It will at once be seen that practically no line-ups of a quality usually
acceptable for plotting are present in this record.

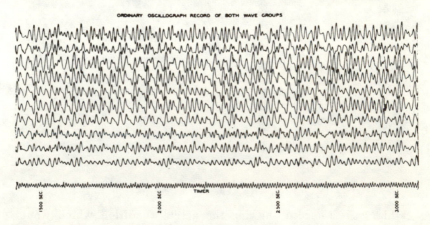

FIG. 24

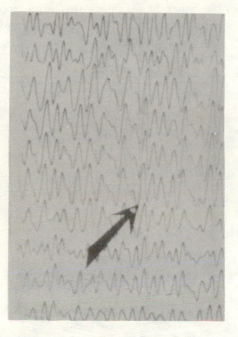

FIG. 25

In Fig. 25 the arrow points to one approximate line-up which is shown in a close-up, and will be seen to be of a very dubious nature, considered from the point of view of the usual critical computer.

SIMPLIFIED CUMULATIVE ANALYSIS SHOWING BOTH GROUPS

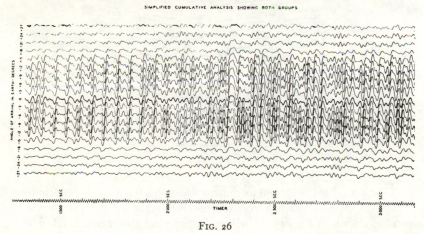

FIG. 26

Fig. 26, on the other hand, shows this same record reproduced through the Sonograph analyzer. Wave group A and wave group B will be clearly seen as separate entities.

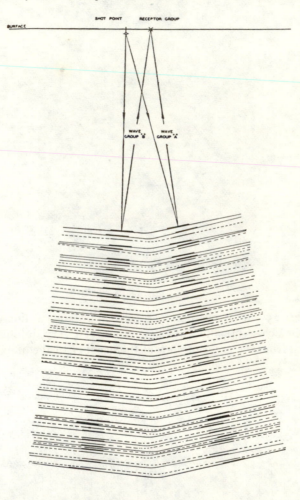

Fig. 27

The actual record given in Fig. 26 was submitted to an experienced computer on Sonograph records, who reconstructed from it the cross section shown in Fig. 27. It will be noted that the reconstructed section does not correspond completely with the original assumed section. Neither the dips of the successive strata nor their positions, are identical with those of the original. This deviation will give some idea of the limits of error of the Sonograph when applied to such a thoroughly interlaced group of criss cross waves.

This test—which, incidentally, has been made for a wide variety of other structural conditions, should demonstrate clearly that the Sonograph analyzer is more than a mere mechanical equivalent for the earlier visual identification of waves on seismograph records. The

ORDINARY OSCILLOGRAPH FIELD RECORD

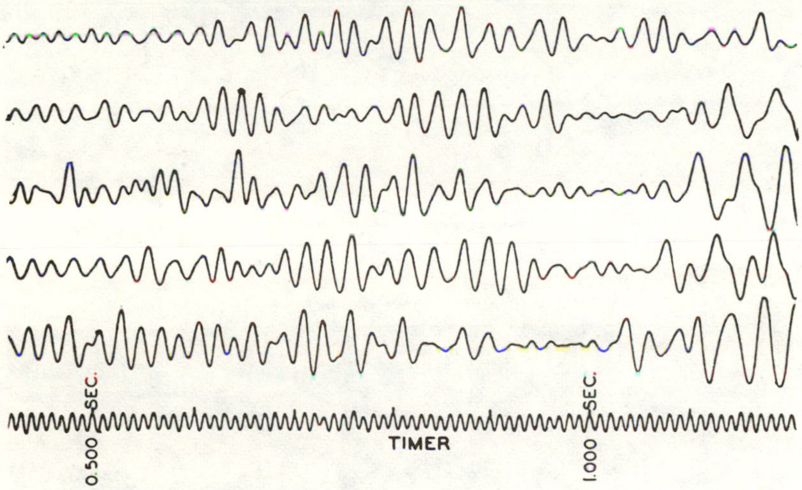

FIG. 28

section just reconstructed could certainly not have been filled in with equal completeness and quality by picking out line-ups and waves on the record in Fig. 24, in the usual manner.

In closing, it may be remarked that the use of the Sonograph in the field has now extended over a sufficient period of time, and covered a sufficient variety of problems, to completely confirm its utility in separating and identifying criss cross waves. Sections have been successfully plotted by the method in any number of locations where visual records showed no usable reflection patterns. And these sections made by the Sonograph have been checked in a very satisfactory way, and in a great majority of instances, against other geological and subsurface information.

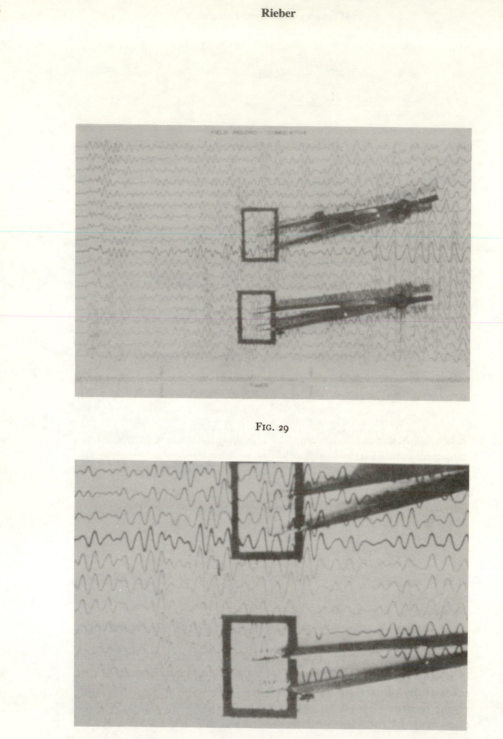

Fig. 29

Fig. 30

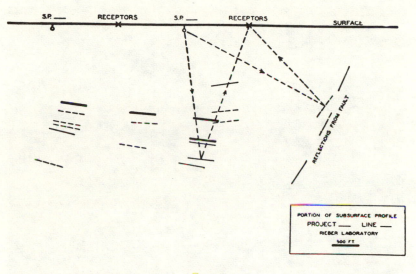

FIG. 31

As a matter of interest, Fig. 28 shows a typical Sonograph film, taken on an actual structure, but reproduced in the customary seismograph form. Little evidence of pattern or line-up will be seen.

Fig. 29 shows the same film reproduced through the Sonograph

ORDINARY OSCILLOGRAPH FIELD RECORD

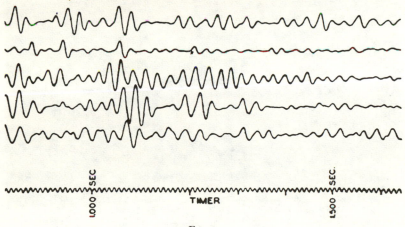

FIG. 32

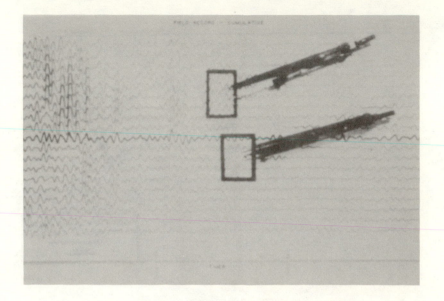

FIG. 33

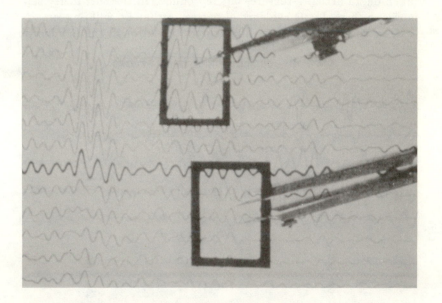

FIG. 34

analyzer. Dividers mark two of the many individual wave trains visible on this record, on their points of maximum amplitude.

Fig. 30 shows the close-up of this divider and the two wave trains to illustrate the nature of the maximum picked.

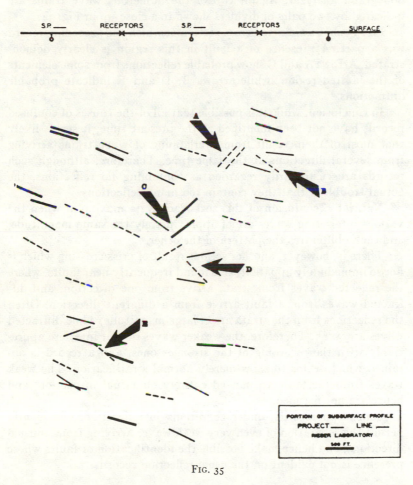

FIG. 35

Fig. 31 shows a reconstructed section in the earth, made from a number of successive shot points, on which the positions of the two individually marked waves from Fig. 30 are indicated by the wave travel paths. Incidentally, the fault shown in this section was amply confirmed by local surface evidence.

Fig. 32 shows another Sonograph record, reproduced from the film in the ordinary seismograph style. Again no striking line-ups will be seen.

Fig. 33 shows the result of passing the same film through the Sonograph analyzer. Again, two of the numerous wave trains are indicated by two pairs of dividers shown in a close-up in Fig. 34.

Fig. 35 shows a plotted cross-section of which the preceding record was a part. A presence of a fault in this region is clearly demonstrated. Arrows A and C show probable reflections from some elements of the faulted region, while arrows B, D and E indicate probable diffractions.

In conclusion, while it is possible that all of the causes of confused records have not been identified at the present time, it seems likely that most of them result from overlapping of wave trains arriving from several directions at the same time. Therefore, although such records are customarily regarded as containing no reflections, the actual trouble is that they contain too many reflections.

Naturally, confusion of this sort reaches its maximum when the various interfering waves are of approximately the same magnitude, and each obliterates the pattern of the other.

There is, however, another serious result of criss crossing which is not so immediately apparent. It is found frequently near faults, where the reflected waves from strata arrive from one direction, and diffracted waves from a fault arrive from a different direction. Often the reflections from the strata are of large magnitude, while diffracted waves are weak. Therefore, the weaker waves do not interfere appreciably with the recording of the stronger ones, and a record is obtained which seems to show merely normal stratification. The weak waves from the fault are missed entirely on visual inspection, and hence are not mapped.

Sonograph records, under conditions just described, frequently permit the separation of even very weak waves arriving from unusual directions, and hence make possible the identification of faults whose presence is not evident on the usual reflection records.

Reprinted from GEOPHYSICS, 3 273-291.

LOCATING AND DETAILING FAULT FORMATIONS BY MEANS OF THE GEO-SONOGRAPH*

CURTIS H. JOHNSON†

ABSTRACT

Mapping of faults has previously presented great difficulties for the reflection method. These difficulties have often arisen from a confused reflection pattern, rather than from entire absence of reflections near the fault. The use of the Geo-sonograph in making a directional analysis of such confused patterns is described, and illustrated with several examples of faults located and mapped by this method.

Reflection seismograph exploration has now arrived at a state of maturity where the benefits derivable from its use are well established. It is, however, a widely conceded fact that the current type of seismo-

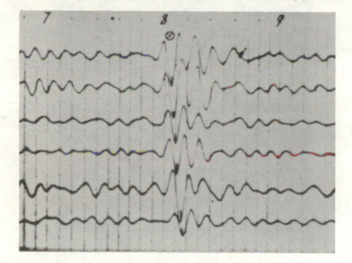

FIG. 1.

graph cannot operate in many regions, and obtains only marginal results in many others, due to the confused quality of the seismograms obtained.

Before discussing the appearance of confused seismograms, their causes, and cure, let us first have a look at a typical good seismogram. Fig. 1 illustrates such a record, and it seems to be characterized by

* Paper read at the Annual Meeting, Los Angeles, California, Mar. 18, 1937.
† Rieber Laboratory, Los Angeles.

a number of lineups, or groups of similar waves traceable continuously across from one side of the record to the other. A typical confused seismogram, however, might appear as in Fig. 2, where there is almost a complete absence of such lineups. The disconcerting feature about the confused seismogram is its tendency to appear whenever an interesting geological situation is approached. Opinions have been privately expressed by many geophysicists that interesting geological regions and confused records invariably go hand in hand.

Whether this be strictly true or not, it is generally observed that confused records and suspected faults very frequently go hand in hand. There are, of course, notable exceptions where, to the delight

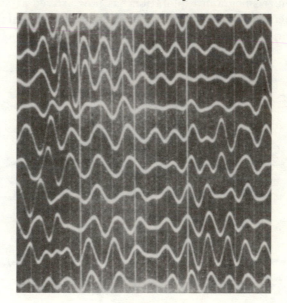

FIG. 2.

of the seismographer, minor displacements may actually be seen on a single seismogram. However, unreadable records and known faulting have been so commonly associated that the presence of confused records, has frequently been assumed, to good effect, as indicating the presence of a fault. Fig. 3 is a somewhat idealized profile, showing the delineation of the stratification up to a certain point—then a blank—then the continuation of the profile. There are many verified cases where the assumption that the "confused record" region, or "blank profile" region, should be interpreted as a fault has been borne out by subsequent drilling. Obviously, the validity of such an assumption increases if several lines are shot in the same region, and

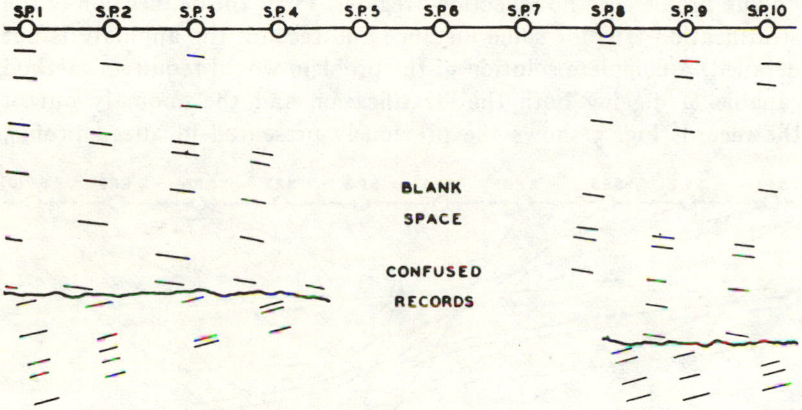

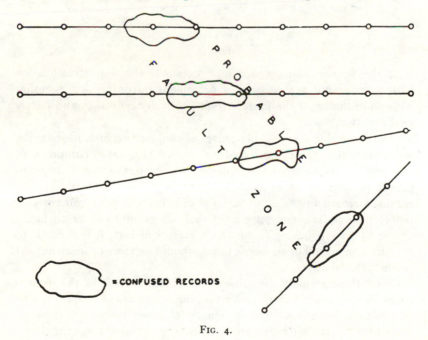

FIG. 3.

there appears a correlatable blank region on all of the series of lines, such as is here shown in Fig. 4. From the lineup of "bad record"

FIG. 4.

regions, or "blank profile" regions, the approximate trend of the fault is sometimes assumed.

It should be observed that there are two inadequacies in the "blank profile" or "no reflection" regions. First, the delineation of the stratification fails for some distance and second, the anomaly is not detailed. A complete solution of the problem would require a method capable of digging both the stratification and the anomaly out of the record. Fig. 5 shows the previously presented idealized profile,

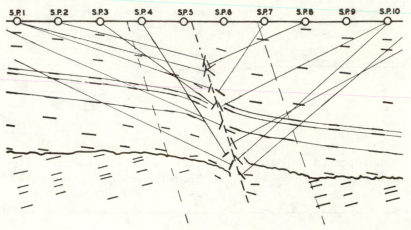

FIG. 5.

with the former blank space now detailed, showing stratification with a prominent key-bed and a prominent unconformity, together with an outlining of the fault surface partly by reflection and partly be diffraction.

In considering the possible causes of confused records near faults, we observe that the typical confused record (Fig. 2) is composed of frequencies characteristic of reflected waves but without any apparent pattern or set of lineups. Assuming, then, that the typical confused record in the neighborhood of a fault is predominantly of reflected wave frequency, we must rule out ground roll or air-borne unrest as a contributing factor. As a matter of fact, it is difficult to see why ground roll or air-borne unrest should be more prominent near faults than elsewhere.

Were these actually the cause of the confused records, the records could be very readily clarified by the simple process of frequency discrimination, for ground roll is usually of lower frequency than reflected waves, and air-borne unrest is usually of higher frequency.

Thus, the probability is that the confusion is caused by a multiplicity of compression waves arriving from the vicinity of the fault. Geometric considerations leading to the conclusion that waves will arrive from many directions over certain geologic structures, including fault regions, were presented by Frank Rieber in the form of an animated diagram at the March, 1937, meeting of the A. A. P. G. In a perhaps less striking but possibly more fundamental manner he has

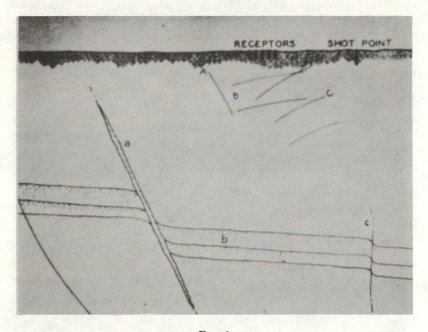

FIG. 6.

demonstrated in a previous paper,[1] the fact that multiple angles of arrival caused by diffractions, as well as reflections, from fault surfaces will produce crisscross arrivals at the surface.

Fig. 6 is one frame of the animated diagram showing geometrically the crisscross arrivals at the surface, "*A*" being a reflection from fault "*a*," "*B*" being reflections from strata "*b*" and '*C*" being diffractions from fault "*c*."

Fig. 7 is a sample strobogram taken from Rieber's previous paper,[1] illustrating the return of diffracted waves from fault corners.

[1] Visual Presentation of Elastic Wave Patterns Under Various Structural Conditions. Frank Rieber, *Geophysics*, Vol. 1. No. 2, July, 1936.

On the assumption that confused records are caused by criss-cross arrivals, it is necessary to design some system for separating such crisscross waves out of a record. Beginnings along this line had been tried prior to the Geo-sonograph. The method of multiple recording,[2] while primarily designed to minimize the effect of ground roll, was useful, in a limited way, to clarify confused records, where the confusion was due to crisscross waves from a depth.

Fig. 8 combines parts of two of the illustrations from an article by E. E. McDermott.[2] It shows the effect of combining eight de-

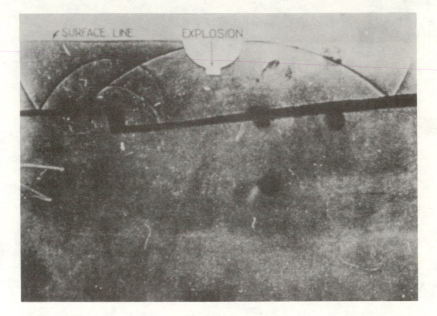

FIG. 7.

tector outputs, when the angle of arrival is respectively 5 degrees, 15 degrees and 25 degrees, with a detector spread of 100 feet, a frequency of 40 cycles per sec. and a velocity of 8,000 ft. per sec.

In Fig. 9 the magnitudes of the cumulations shown by McDermott are plotted as radii vectors pointing in the directions from which the waves reached the surface. The summation for waves arriving from other angles may be calculated analytically. The results of several

[2] The Use of Multiple Seismometers. E. E. McDermott, *Petroleum Engineer*, February, 1937.

such calculations are shown in Fig. 9 as lighter vectors than those obtained from McDermott's graphical solution. If we draw the envelope of all these vectors, it is evident that we have a polar curve representing the effective sensitivity of the detector group to waves of the reflected wave frequency.

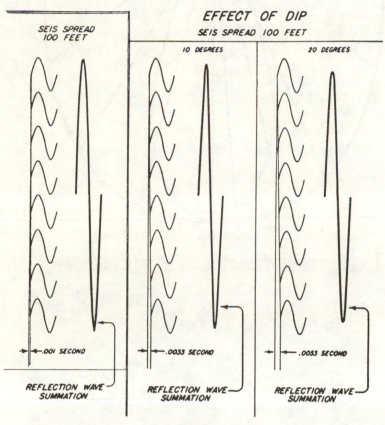

EFFECT OF DIP

SEIS SPREAD 100 FEET

SEIS SPREAD 100 FEET

10 DEGREES   20 DEGREES

.001 SECOND   .0033 SECOND   .0033 SECOND

REFLECTION WAVE SUMMATION   REFLECTION WAVE SUMMATION   REFLECTION WAVE SUMMATION

FIG. 8.

In Fig. 9 is drawn another polar sensitivity curve not designated by vectors. This curve, calculated throughout, illustrates the sharper effect obtained by using a 300 foot detector spread instead of one of 100 feet. Had a 600 foot spread been used, the corresponding sensitivity curve would have been still sharper.

Fig. 10 illustrates how such simple accentuation of vertical arrivals over arrivals from other angles might assist in detailing the

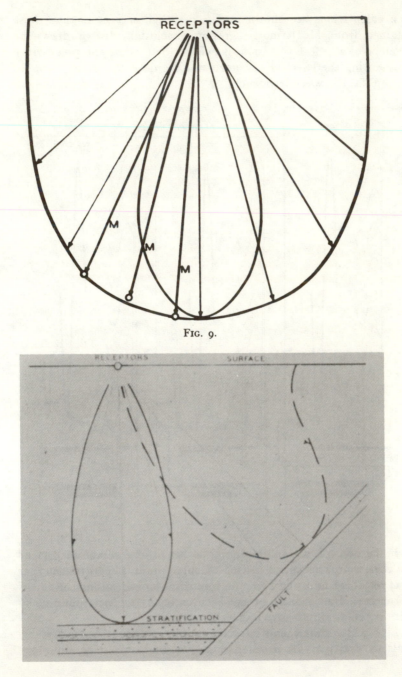

FIG. 9.

FIG. 10.

stratification near a fault, and, if outstanding key-beds were present, in finding the fault by means of its displacement. If the polar sensitivity curve could be pointed in turn at the stratification, and then at the fault surface (as shown by the dotted curve in Fig. 10), it might be possible to detail both of them. The story of the development of the Rieber Geo-sonograph is essentially a story of the development of a method by which this polar sensitivity curve could be rapidly and economically rotated to any desired direction, exploring for any one feature, while eliminating the effect of all others.

The rotatability of the polar sensitivity curve is accomplished in the Geo-sonograph method of directional analysis by cumulating in two steps, recording first a reproducible record of the output of each detector and later combining them not only in their original phase relationship, as is done in ordinary multiple recording, but also in any desired phase relationship. The finally perfected technique of directional analysis will not be described here, for it has been adequately described in various publications.[3,4]

For the present discussion it should be sufficient to show one variable density record with the corresponding visual seismograph record and the directional analysis obtained by means of the Geo-sonograph Analyzer from the variable density record. This is shown in Fig. 11 which illustrates a very simple, uncomplicated type of record. It serves to relate the ordinary seismograph record or seismogram, the sound film record (possibly called sonogram), and the directional analysis of the sonogram. Each of the traces on the seismogram is a reproduction of the corresponding sound track of the sonogram. Each of the traces on the directional analysis strip is, however, the sum (or multiple recording) of the outputs of all of the detectors.

The heavy trace is the combination of all the detector outputs unaltered in phase, and corresponds exactly to the result which would have been obtained had the detectors been connected in series to a common recording element. Each of the other traces of the directional analysis strip also represents the sum of the outputs of all of the detectors, but with a constant phase difference introduced between the outputs from adjacent detectors before cumulation. The traces near the heavy, or zero-phase cumulation, have small phase

[3] New Geophysical Method Provides Reliable Data for Structure Mapping. Wallace A. Sawdon, *Petroleum Engineer*, October, 1936.
[4] A New Reflection System with Controlled Directional Sensitivity. Frank Rieber, *Geophysics*, Vol. 1, No. 1. January, 1936.

differences introduced between the outputs of adjacent detectors
before cumulating, while the traces farther away from the heavy, or
zero, trace represent the cumulation when larger phase differences
were introduced between the outputs of adjacent detectors.

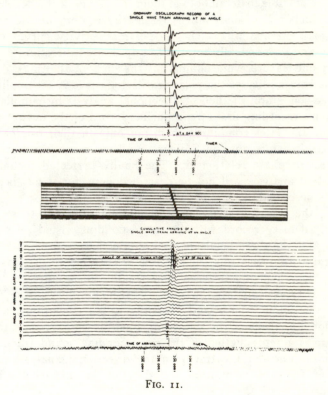

FIG. 11.

Waves arriving at the surface of the earth at an angle reach the detectors at different times—or, in other words, there is a phase difference between the wave as it appears on the record obtained from one
detector and the record obtained from an adjacent detector. This is
the ordinary ΔT value ascertained visually on seismograms obtained
in relatively simple shooting territory. The various traces on the
directional analysis strip, representing, as they do different ΔT values,
may be calibrated in terms of angle of arrival from the earth, as shown
at the left hand side of the analysis strip. It is sufficient here to call
attention to the large range of angles of arrival on the cumulative
analysis strip which is unaffected by the waves recorded here. Clearly,

if another wave had arrived at an angle represented by one of these unaffected traces, it could have been almost as easily observed as if the waves now on the record had not been present.

Laboratory and field tests of the relative abilities of the analyzer and the visual method to extract two waves occurring on the record at nearly the same time and at different angles have been described by Frank Rieber in a previous publication.[5] This present paper will discuss some examples of results of shooting in complicated territory with the Geo-sonograph.

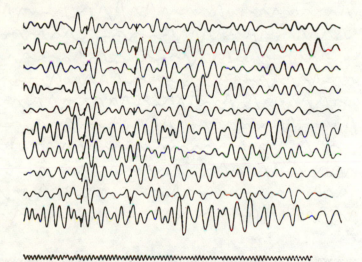

FIG. 12.

The first example of faults located in the field by the Geo-sono-graph has not been verified by any other method to date. However, it is included in the presentation because it is an admirable illustration of a fault found by diffraction, and also because the results are so consistent as to hardly require substantiation by other means. These faults occur in territory characterized by relatively flat dips, where three-dimensional control is not vitally necessary as far as the stratification is concerned, but, as will be pointed out, without three-dimensional control these faults could not have been outlined.

[5] Complex Reflection Patterns and their Geologic Sources. Frank Rieber. *Geophysics* Vol. 2, No. 2, March, 1937.

A short distance away from these faults, the records are of a rather simple type and are suitable for ordinary seismograph recording, as shown in Fig. 12.

However, records obtained in the vicinity of the fault are of the confused type already discussed. One of these records, shown in Fig. 13, was taken only a little over half a mile from the previous record. Figure 14 shows how this confused record has been separated into its component parts by analysis.

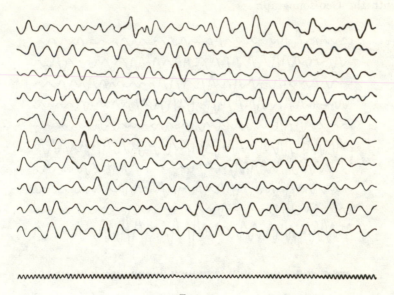

FIG. 13.

The profile obtained from records along this line is shown in Fig. 15. It should be noted that there is a fairly continuous set of flat dips, without any particularly outstanding key-bed. Neither are there any outstanding unconformities on this line from which a fault throw might have been determined. The detailing of the fault surface itself was the only method by which the fault could be located.

It should be noted from this figure that the diffracted waves apparently fall in no set pattern, and if this were assumed to be a true vertical profile of the line, no deductions as to the nature of the fault could be made. However, by a process of continuous three-dimensional control on the line of profile, which has been developed to go hand in hand with the Sonograph system of directional analysis, the location of each of these diffractions to one side or the other of the line of exploration is known.

In a manner similar to plotting a strike and dip from three-dimen-

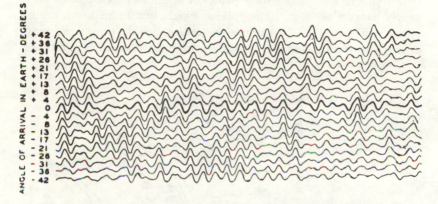

FIG. 14.

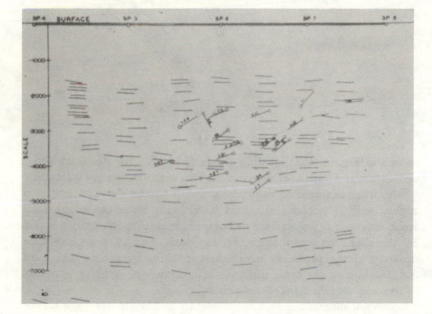

FIG. 15.

sional measurements the origin of the diffracted waves may be located three-dimensionally in space. Fig. 16 shows a plan view of the region, with the diffraction origin points plotted and the depths indicated. All but four of these points of origin fall on a pair of very nearly parallel surfaces, almost plane, which, in all likelihood, represent a pair of faults. Discarding the unexplainable points, we may draw approximate contours on the basis of the remaining points, as

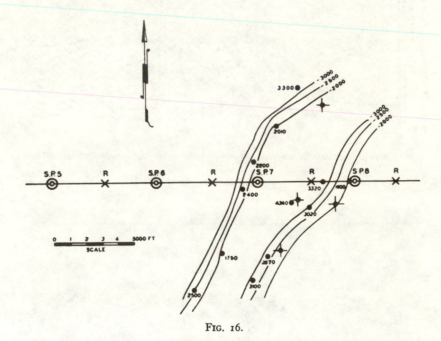

Fig. 16.

shown in Fig. 16, thus locating a pair of faults with a hade of about 40 degrees and a strike of about 50 degrees with the line of exploration.

Another example of field performance of the Geo-sonograph in highly faulted territory is shown in Fig. 17. Since the geophysical work was completed, surface geology confirmation has been released by the client for whom the work was done. Fig. 17 is a small-scale drawing of the profile in question, one of many in the region all contributing to the same ultimate interpretation. A quick scanning of the line reveals many complexities.

The unconformities and faults shown in the figures were all located prior to knowledge of the surface geology. The agreement with surface geology is better than should be expected. Some of the unconformities are required by differences in line component of dip, while others were required by differences in cross-line component of dip. Some of the faults are indicated by dip changes (either line or cross-line components) while some are indicated by reflections or diffractions from fault corners or blocks in fault zones. Both faults and unconformities are supported by evidence from adjacent and cross lines though only this one section was released for publication.

Primarily the number of different directions of arrival at any one point on the surface should be noticed. Evidently any method of

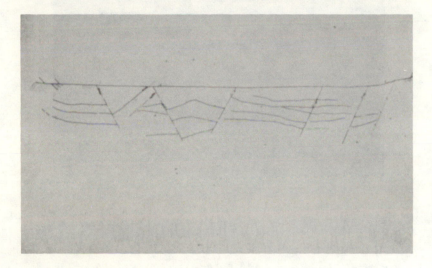

FIG. 17.

exploring this type of territory must feature directional separation to succeed.

For a third example of Geo-sonograph field performance, we are fortunate in having a direct comparison between results obtained by a Geo-sonograph field party and those obtained by a party employing what is conceded to be one of the better visual seismographs, in a typical faulted region, very similar to the one just shown. Typical records obtained by each party are shown in Figs. 18 and 19, where, for purposes of comparison, the Geo-sonograph record has been reproduced track by track in the equivalent visual form. Differences in the general appearance of the two records may be partly ascribed to dif-

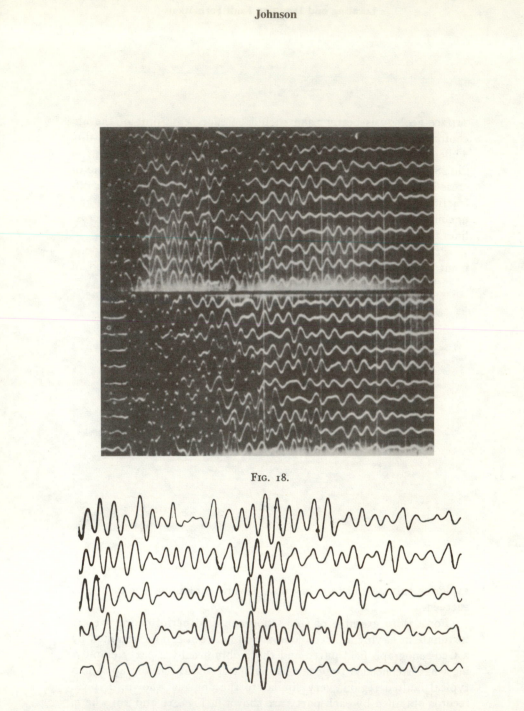

Fig. 18.

Fig. 19.

ferent time scales, but principally to differences in the recording instruments.

Since the Geo-sonograph achieves its results in two steps, the recording circuits may admit a large frequency and amplitude range which may be more narrowly limited in the step of analysis. The seismograph on the other hand, must completely limit frequency and amplitude ranges on the original record.

Overlooking these differences for the moment, it should be evident that neither record lends itself readily to visual interpretation,

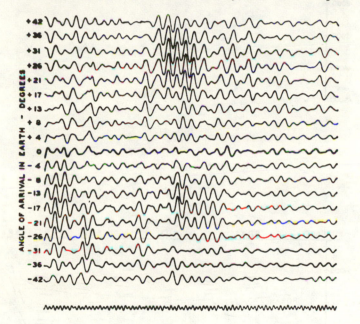

FIG. 20.

each in its own way being a typical confused record. Instead, from the original film record the cumulative analysis shown in Fig. 20 is made. On this record, certain features are strikingly evident, which were not noticeable at all on the visual record. These are, principally, a set of strong waves arriving from high angles together with others arriving from low angles. There cannot be the slightest suspicion that the interpreter of the Geo-sonograph cumulative record was searching for high angle waves because the fault was known to be present. Even the novice at reading these cumulative records can see that the

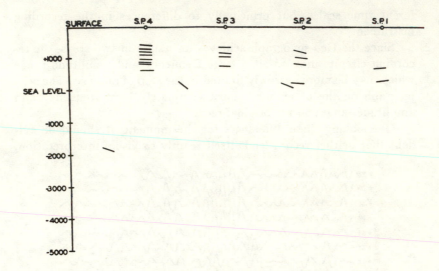

FIG. 21.

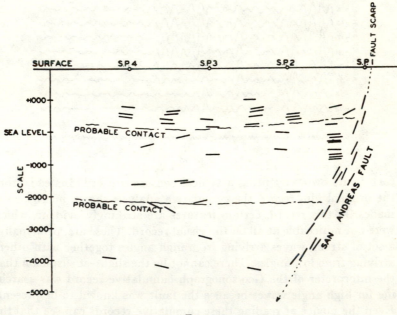

FIG. 22.

high angle waves are as outstanding as any set of waves on the entire record.

Fig. 21 shows the profile which it was possible to obtain from the confused visual records. The complete absence of the fault, and the sparse reflections in the vicinity of the fault should be noticed. Fig. 22 shows the profile plotted from the results obtained from the Geo-sonograph cumulative record. Four advantages of this profile over that obtained by the seismograph are at once evident. First, the fault is very clearly outlined; second, reflections from stratification are practically as numerous near the fault as they are farther away; third, the reflections throughout the length of the line are more numerous on the Geo-sonograph profile than on the profile obtained from the seismograph records; fourth, the depth to which the stratification is mapped in this case is two or three times as great on the Geo-sonograph profile as on the seismograph profile.

In conclusion it may be observed that successful mapping of faulted areas with the Geo-sonograph is as much a vindication of the assumption that multiple angles of arrival cause "confused records" as of the method and apparatus employed to analyze the "confused record" into its components.

SUPPLEMENTARY NOTE (JUNE, 1938)

Since this paper was presented before the Society, much further experience with the Geo-sonograph has been obtained. In the light of this experience, the author believes the paper placed too much emphasis, in both theory and results, on diffractions as a major factor in Geo-sonograph performance. Far more impressive than its use of diffractions is the ability of the method to extract pertinent reflections from masking waves of undetermined origin and to separate two or more sets of pertinent reflections, relating each to its geologic origin. Thus, in filling in details in the "Blank Spaces" described in the paper, the extension of reflecting horizons far into the "Blank Spaces" and the separation of reflections arriving at the same time from strata on both sides of faults are more common occurrences than the detailing of the fault face by diffractions.

Chapter 2

Ruler and compass methods

A symposium titled Seismograph Dip Migration was held in 1957 by the Oklahoma City Geophysical Society and the papers were printed in its journal *The Shale Shaker*. The titles, listed in Table 1, probably indicate the problems and solutions of interest at that time. A variety of mechanical devices were available to assist the interpreter, but data processing in the sense visualized by Rieber had not yet arrived.

In the simple case of constant velocity, the downward traveling wavefront from an impulse applied to the surface of the Earth is a circular arc with center at the point of application. When velocity increases linearly with depth (which is a fair approximation in many areas), the wavefront is still circular but the center is below the point of application and moves downward at a constant rate. Special drafting tools were designed to make wavefront construction and migration easy for this case and for others.

Articles by *Robinson (1957), McGuckin (1957),* and *Hawes (1957)* illustrate the mechanical devices which preceded digital calculations. A detailed discussion of the underlying theory of wavefront propagation as related to migration was given earlier (Hagedoorn, 1954) but is not reproduced here. In fact, Hagedoorn may be the first who defined "migration" as "the procedure of determining the true reflecting surface from a surface determined by a number of vertically plotted points. This true surface can be found, in principle, as the envelope to all surfaces of equal reflection time determined by the vertically plotted points. This is essentially a three-dimensional procedure, setting out from the usually two dimensionally, vertically plotted points."

A later paper (Musgrave, 1961) showed the use of digital computers for generating wavefront charts in place of the mechanical devices.

References

Hagedoorn, J. G., 1954, A process of seismic reflection interpretation: Geophysical Prospecting, **2,** 85-127.

Musgrave, A. W., 1961, Wave-front charts and three-dimensional migrations: Geophysics, **26,** 738-753.

Table 1. contents of a symposium on dip migration held in 1957 (Shale Shaker, 1957, 8, 3-37). Ten years later, digital techniques were being adopted generally.

The need for seismic dip migration	W. B. Robinson
Constant velocity case	A. J. Oden
The McGuckin section plotter	G. M. McGuckin
The application of a linear increase of velocity with depth to seismic dip	J. E. Stones
Chart migrations	F. A. Roberts
An electronic seismic dip plotter	J. A. Westphal
Machine calculation of migration data	E. Usdin
The SDP dip plotter	W. S. Hawes
Slide rule seismic computations	R. H. Mansfield
Three-dimensional control	G. E. Anderson
Projected sections for areas of steep dip	H. J. Fenton
A comparison of the various methods of dip migration	B. Mayo

Reprinted from *Shale Shaker*, 8, 3-6.

SEISMOGRAPH DIP MIGRATION: A SYMPOSIUM

The Need For Seismic Dip Migration

By W. B. ROBINSON.
Gulf Oil Corp.

When we fire a charge of dynamite in a shothole, we generate waves that go outward in all directions, as illustrated in Figure 2. Essentially, there are two useful components of this energy. The one we usually think of is the ray which goes down into the ground, strikes some kind of a reflecting horizon, comes back up to the surface and strikes the detector shown on the surface of the ground. The other useful ray is the one that runs horizontally at a shallow depth and strikes that detector. Those fragments of the total energy are used in our seismic depth calculations. The horizontally moving energy assists us in calculating our weathering corrections or adjustments for the changes in the low speed weathered zone at the surface and, of course, the time of the reflection which travels to some depth is what we use to calculate the depth of the reflecting horizon.

I have used the terms wave front and ray path. I had better define these.

SEISMOGRAPH CROSS SPREADS

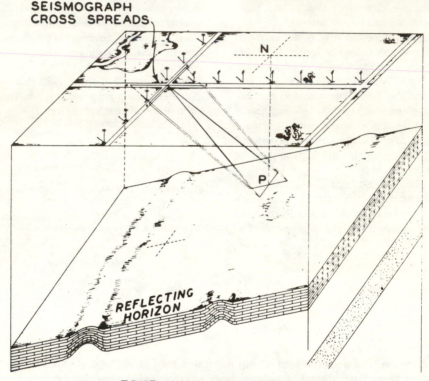

FOUR - WAY DIP SHOOTING

FIGURE 1

The need for dip migration is illustrated in this figure. Seismic cross-spreads are illustrated at the road intersection. The point of reflection for the center of the cross-spreads is at "P" on the reflecting horizon. Proper calculations based on migration principles will yield the drilling depth (NP) required for a well at (N) to reach the reflecting horizon at point (P).

You have all watched a wave caused by the wind blowing over a wheat field. You know that each stock of wheat bows as the wind blows across it but that doesn't mean that one head of wheat goes from one side of the field to the other. It means that, in succession, each little stock of wheat bows down and comes back up into its normal position. Practically the same thing happens in seismic work. When we fire our dynamite, we cause each little partical of the earth to go through that same motion of moving forward and then coming back. Consider this wave in Figure 2 as it moves outward in all directions from the shot; assume that we could stop it,

say we could stop it, or photograph it, at a particular instant of time. If we could map the most forward position of that vibration at a particular time. that position would constitute a wave front and it would be sort of hemispherical. A ray path is, more or less, a figment of our imagination. It is a line drawn always normal to the wave fronts. It represents the path followed by a small amount of our seismic energy as it travels from the shot to the detector. Our calculations using geometrical optics can be readily visualized if we assume that such a path actually exists and that a partical of energy follows it. The depth of a horizontal reflecting horizon is given by

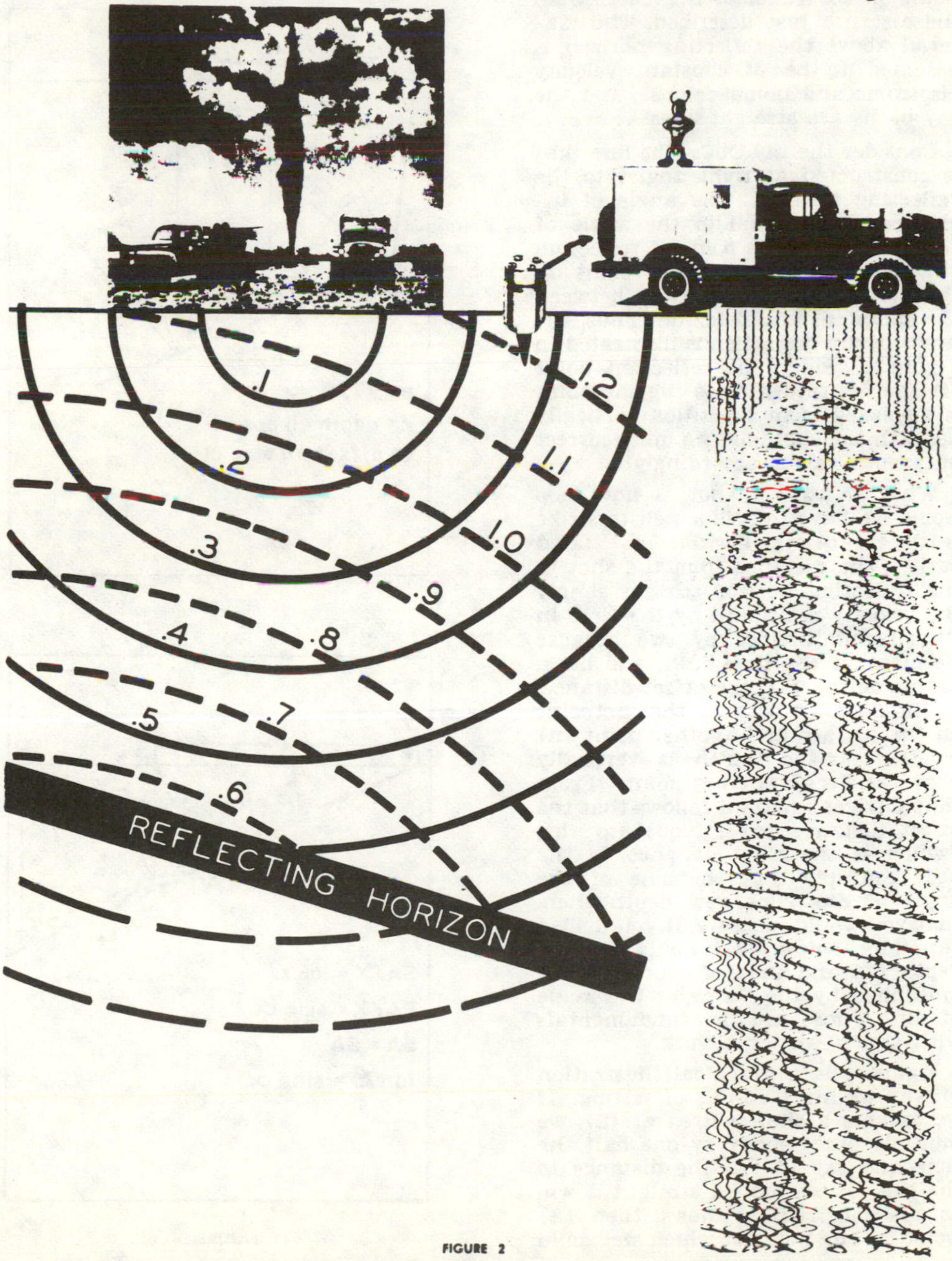

FIGURE 2

multiplying the average velocity by one-half the corrected reflection time for a detector located near the shotpoint.

Consider Figure 3 which illustrates a geologic horizon dipping at the angle a. The shot at "O" will create the advancing wave fronts as before but only three ray paths are illustrated. A detector at "O" would register the time for the ray which travels down and back on a line (Oh) constructed at right angles to the reflecting horizon and through the shotpoint. This is the shortest time path for a reflection from the dipping reflector. Obviously the distance from "O" to any other point on the reflector is greater than the distance just described. The material above the reflecting horizon is assumed to be of constant velocity (isotropic and homogenious), and the ray paths are straight lines.

Consider the ray ObC. The line (kb) is constructed at right angles to the reflecting horizon. The angle of incidence (i) is equal to the angle of reflection (r). The point of reflection (b) is not directly below a point on the surface located midway between the shotpoint and detector. Thus, the need for dip migration is illustrated in Figure 3. Since the reflection point for the ray returning to the shotpoint is displaced from a position vertically beneath the shotpoint we must correct our depth values accordingly.

What we want to know is how deep would we have to drill a well to strike this reflecting horizon. As stated earlier, the distance from the shot to the reflecting horizon, (h), is simply the velocity multiplied by the time in transit and divided by two because we measure the time down and back. Let us define (a), the offset distance, as the distance between the shotpoint on the surface and another point (n) on the surface, which is vertically above the reflecting point. From simple trigonometry it follows that the offset distance (a) is equal to (h), (which is the slant distance to the bed) multiplied by the sine of the angle of dip. The new depth then, that you would measure, if you drilled vertically below (n), would be the depth (Z) and that, of course, is (h) multiplied by the cosine of the angle of dip. Those are the fundamentals with which we must work.

Figure 4 is a graphical illustration of why we must do this offsetting. If we assume a shot is fired at (S), we multiply our velocity by one-half the time and say that is the distance to the bed. If we plotted straight down, as one construction shows, then (A) would be the point at which we would

plot our reflecting horizon, but we know that is wrong because the bed has some dip since it outcrops at (O). The dashed construction shows the actual position on the reflecting horizon from which our energy was turned back. So the difference then between the line (OA) and the line (OA') is the difference between the corrected position of this reflecting horizon and the erroneous or uncorrected position.

In Figure 4 assume that we wish to measure the dip of a bed knowing the reflection times for shotpoints (R) and (S). The bed outcrops at (O). To measure the angle of dip, we start

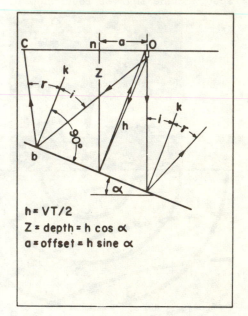

$$h = VT/2$$
$$Z = depth = h \cos \alpha$$
$$a = offset = h \sin \alpha$$

FIGURE 3

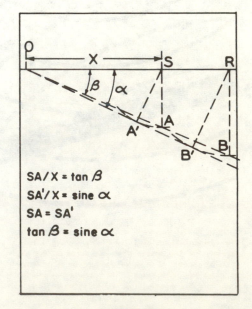

$$SA/X = \tan \beta$$
$$SA'/X = \sin \alpha$$
$$SA = SA'$$
$$\tan \beta = \sin \alpha$$

FIGURE 4

by actually measuring the increase in distance to the bed at two successive shotpoints. On an unmigrated depth profile, they would appear to be vertical distances. Figure 3 says that we may measure "vertically" the increased "depth" and use the tangent to calculate the angle of dip. This is an easy thing to do on a profile. It is perfectly alright because what you measure as the tangent of dip on an unoffset profile, is in truth, the sine of the true angle of dip. You may go ahead and make your migrations on that basis.

Figure 5 is a small section of a seismograph record showing reflections with considerable dip. We shall use this illustration to define the rate of dip which we shall call Delta T. The term will be used frequently in the work that follows. This is a split spread with a shot at the center. Delta T is the difference in time between traces at opposite ends of the record. The normal moveout or angularity, as it is sometimes called, will be cancelled since the spread is located symmetrically across the shotpoint. Delta T, or rate of dip, would have to be defined in another way if the shot was placed at the end of the cable rather than in the center. In this case, Delta T would be the time difference across the spread reduced by the "reflection time difference" introduced by the dimensions of the spread. We could measure the rate of dip in a variety of ways. Usually, it is given as milliseconds per spread length. It could also be given in milliseconds per mile, or if you have a map contoured in feet, you could determine the rate of dip in feet per unit of map distance, whether it be feet per mile, feet per shotpoint spacing, etc. Rate of dip in milliseconds across a split spread is one of the things that is commonly used in our calculations for dip migration. The other thing that we must use is the two-way reflection time. The reflection time, of course, must be adjusted for weathering and surface elevation.

Migration is needed to improve the accuracy of our calculations in order that we may better predict drilling depths. As a general rule, the unmigrated profile will show highs that are too wide and it will show lows that are too narrow. By our migrating processes, we can correct this.

In an extreme case, what appears to be an anticline on a time plot can be migrated properly to become a syncline. See Figure 6 for an interesting display of this condition. The numbers along the reflecting horizon are reflection points which correspond with the numbered observation points along the surface of the ground.

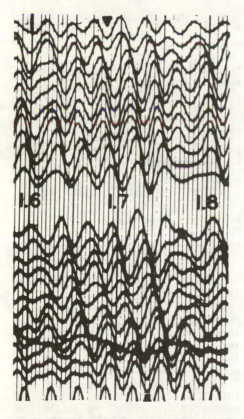

Delta T (sometimes written Δt), as used in this symposium, is the "time" dip across a split spread. For the event near 1.7 seconds, the Δt is 0.052 second. It may be defined in other ways as described in the text. The corrected reflection time for this event (T) is 1.706 seconds with a proper adjustment applied for surface corrections.

FIGURE 5

Types of Devices

The various migrating devices could be divided into several classes as illustrated in the accompanying figures. One group we could call the **HEADS**. They pivot about a calibrated head which is scribed in any suitable measure of "rate of dip." Of course, there is an arm carrying a depth scale on which we mark off the distance from datum to the reflecting horizon. A small triangle is held against the edge of this depth arm at the proper position. A line of proper length drawn along the upper edge of the triangle represents one segment of the reflecting horizon.

An interesting one arm device results when we assume a constant velocity of 10,000 feet per second. No head is required. Dip is "set" at a time of 2 seconds on the depth scale. At that "time" the device is moved 100 feet in the proper direction for

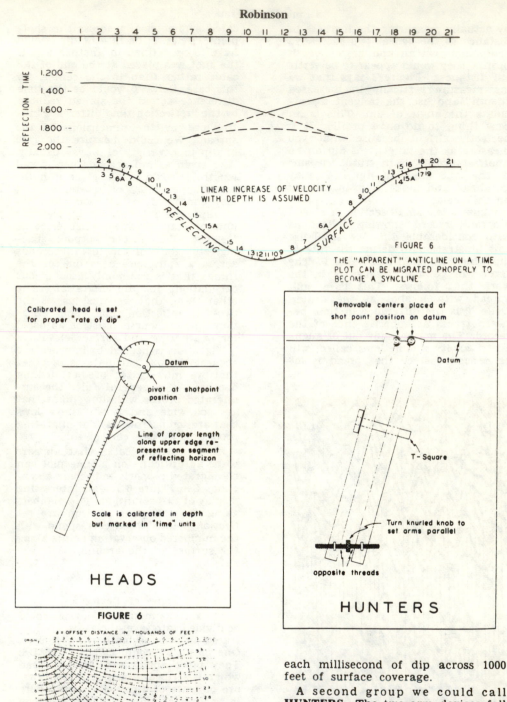

FIGURE 6

THE "APPARENT" ANTICLINE ON A TIME
PLOT CAN BE MIGRATED PROPERLY TO
BECOME A SYNCLINE

LINEAR INCREASE OF VELOCITY
WITH DEPTH IS ASSUMED

HEADS

FIGURE 6

Calibrated head is set
for proper "rate of dip"

pivot at shotpoint
position

Line of proper length
along upper edge re-
presents one segment
of reflecting horizon

Scale is calibrated in depth
but marked in "time" units

Datum

HUNTERS

Removable centers placed at
shot point position on datum

Datum

T - Square

Turn knurled knob to
set arms parallel

opposite threads

CHARTS

each millisecond of dip across 1000
feet of surface coverage.

A second group we could call
HUNTERS. The two arm devices fall
in this class since they merely hunt
for a solution on the basis of two reflection time values, and do not require an evaluation of rate of dip.

A third technique we could call
CHARTS. Using the two families of
curves on the chart, we locate the
proper rate of dip and the proper two-
way time. A line drawn tangent to
the wave front at this point represents
the reflecting horizon at its proper
attitude.

Other tools for migration include
**ANALOG DEVICES, COMPUTING
MACHINES,** etc.

Reprinted from *Shale Shaker*, **8**, 9.

THE McGUCKIN SECTION PLOTTER

By GLENN M. McGUCKIN
Independent Consultant

The McGuckin Seismograph Section Plotter is a semi-automatic mechanical device which solves the image point problems mentioned in the previous article. The device consists of two scale arms, or plotting bars, connected by a head and foot bar so the assembly always constitutes a parallelogram. This parallelogram assembly slides back and forth in a channel in the long fixed base bar above the top edge of the cross-section paper. A T-square scribing bar slides along the plotting bars and is used to locate the reflecting horizon on the cross-section.

Identical time-depth scales, appropriate to the local velocity conditions, are cemented in recesses in both plotting arms. These scales are positioned so that the index line on the T-square scribing bar reads the same value on both scales when the scribing bar is parallel to the datum plane. The datum plane, from which the net reflection times are measured, passes through the centers of the two pivots at the top end of the plotting arms.

In operating the plotter the seismologist puts into the plotter each successive pair of continuous reflection times from adjacent shotpoints in such a way that the plotter automatically determines the depth and offset position of the reflecting horizon. For example; with the left and right plotting arms positioned over Shotpoints 1 and 2, respectively, (See Figure One) the sliding T-square is moved down the left-hand plotting bar to a reflection time of 0.685. The whole assembly is rotated slightly until the index line on the sliding T-square reads 0.680 on the right-hand

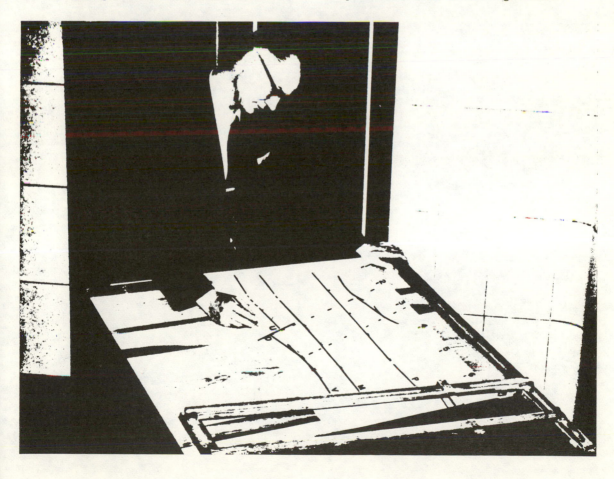

57

plotting arm. With the sliding T-square in this position, a pencil line is drawn the length of the reflecting segment. This line is the migrated position of the reflecting horizon. The above procedure is repeated for each reflecting horizon. The plotter is then shifted to Shotpoints 2 and 3 and the process repeated for the information at these two positions.

Figure One shows the results of plotting three horizons: a Pennsylvanian, a Mississippian, and the Viola, typical of many Oklahoma areas. The Pennsylvanian shows little dip and very little migration. The Mississippian shows a slight syncline under Shotpoint 2 with small migration, and a high under Shotpoints 4 and 5. The Viola shows a more pronounced syncline under Shotpoint 2. The deep, down-thrown Viola segment, found under Shotpoint 4, migrated to that true position from reflection data found on the seismograms recorded at Shotpoints 2 and 3. The Viola time from the up-thrown segment was also recorded at Shotpoint 3. Also shown is the extensive overlapping of the segments from Shotpoints 3-4, 4-5, and 5-6 in going over the anticline.

The McGuckin Seismograph Section Plotter is simple and fast to operate. It is adaptable to split-spread plotting, plotting half-spreads, and the use of curved ray paths. Its construction is simple, sturdy and inexpensive.

Reprinted from *Shale Shaker*, **8**, 20-22.

The SDP Dip Plotter

By W. S. HAWES
Seismic Explorations, Inc.

Mechanical dip plotters of whatever simplicity or complexity are devices for graphically resolving the rate and position of a component of subsurface dip in one plane. The quantities of distance (or time) and dip angle which are applied to the plotter for such resolutions are derived through mathematical equations from observed seismic data.

Many forms of equations have been developed, but in each instance they generally express relatively simple physical and geometrical relationships based upon a particular **set of assumptions**.

The set of assumptions chosen and the resultant equations are valid only to the extent to which they approximate actual subsurface conditions of velocity distribution, etc.

The dip plotter is only one part of a **system** composed of:

1. A set of assumptions.
2. The mathematical equations derived therefrom.
3. Observed data.
4. Mathematical treatment of the observed data.
5. Results introduced to dip plotter.

Since item 4 may (and often does) entail tedious effort, it is of crucial importance, when setting up a procedure for dip plotting, to consider the manner and extent of intended use and to so arrange the procedure that maximum accuracy with minimum time consumption be attained.

In the following paragraphs, the SDP dip plotter and the associated "system" as developed by Seismic Explorations, Inc., is described. This machine, together with the associated computational methods, has been used since 1948 as standard practice on all SEI field parties where its use is indicated as being of possible value to interpretation.

In the development of this plotter, the objective was to derive a system which could be used by a field party to prepare migrated cross sections on **all** data.

To satisfy such a requirement demanded:

1. That all components of the system be simplified to an absolute minimum in order that it could be used by a field party without undue time consumption.
2. That all charts, scales, etc., associated with the system be of a form that could be rapidly prepared (pre-

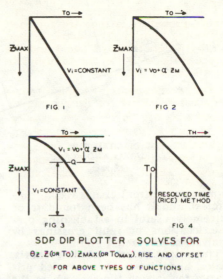

FIG 1   FIG 2   FIG 3   FIG 4

SDP DIP PLOTTER SOLVES FOR
θz, Z (or T_0), Z_{MAX} (or T_{0MAX}), RISE AND OFFSET
FOR ABOVE TYPES OF FUNCTIONS

ferably on 24 hour notice) so there would be no delay in processing data during the early stages of a survey.

3. That the assumptive and mathematical basis of the system be the best approximation to actual velocity distributions over the widest area, and

4. That the mechanical and graphical devices used be of maximum accuracy.

Consideration for these demands led to the development of the SDP plotter and system which is applicable to the four types of functions shown in figures 1 - 4. Figure 1 shows the form of the constant velocity case which may be used with good approximation in some areas where a very low rate of increase of velocity with depth is present. Figure 2 shows the form for the classical case of linear increase of velocity with depth. This, of course, is the type of function probably most widely used in curve ray path migration method.

Figure 3 shows the time-depth relationship for the so-called Type "A" compound function where the velocity increases linearly with depth to some point and below this point the velocity remains constant—being that of the instantaneous velocity at the "Q" point.

Figure 4, the fourth type of "function" which may be used with this plotter, is the Rice "resolved time" method which yields cross sections in the time dimension only.

In addition to the above applications, the plotter may be employed

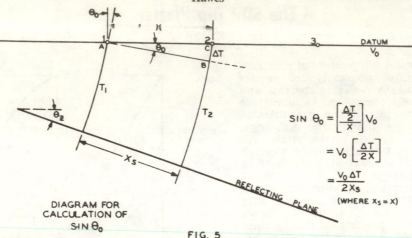

$$\text{SIN } \theta_0 = \left[\dfrac{\frac{\Delta T}{2}}{X} \right] V_0$$

$$= V_0 \left[\dfrac{\Delta T}{2X} \right]$$

$$= \dfrac{V_0 \, \Delta T}{2 X_S}$$

(WHERE $X_S = X$)

DIAGRAM FOR
CALCULATION OF
SIN θ_0

FIG. 5

to make wave front charts (equal time circles) and it has been found to be extremely useful in establishing precise locations of fault crossings by analysis of diffracted events.

The entire system including charts, scales, and plotter was designed primarily for solving the case where the velocity increases linearly with depth. A comparison of the diagram of the geometry of the problem (Figure 6) with the plotter (Plate I) will show the mechanical relationship of the latter to the former.

It will be observed from Figure 6 that three quantities are required for a solution of the dip angle and the position (in the x, y plane) of a subsurface point (P). These quantities are H, R, and θz, or Rise, Offset, and θz. Either set of three values provides a solution. As employed by the SDP plotter, it is the former set (H, R, and θz) which is used for the solution and which automatically yields the Rise and Offset values—as well as the true depth Z.

To obtain a simplified viewpoint for visualizing the problem and its solution, the various quantities involved may be considered as:

1. Quantities given.
2. Quantities observed.
3. Quantities computed.
4. Quantities plotted.

Without going into the equations, we have for any particular area the **given** quantities:

1. Vo (initial velocity).
2. α (constant—rate of increase of velocity with depth).
3. Vi (the instantaneous velocity for all values of depth).

From the field data, we have the **observed** quantities:

1. To (Record time).
2. Xs (Subsurface distance between points of observation).
3. ΔT (Time increment between points of observation).

From the given quantities the fol-

lowing are **computed**:

1. H (The instantaneous depth of the wave front center).
2. R (The instantaneous wave front radius).

From the observed quantities the following is **computed**:

θz (Dip angle of reflecting horizon).

The quantities derived by **plotting** are:

1. θz
2. H
3. R
4. Xs
5. Offset (displacement in x plane).
6. Rise (displacement in y plane).
7. Z (true depth of reflector).

The procedure for solving the quantity **computed** from the observed data (θz) may be demonstrated by reference to Figures 5, 6 and 7. It will be seen from equation V that θz (the angle of dip to the reflector) is a function of θo (the angle of emergence of the reflected ray), and that in making the computation for θz, it is first necessary to solve for θo. The geometry of the problem and the equation for solution of θo are shown in Figure 5. The **graphical** solution for sin θo is illustrated in the lower segment of the chart shown in Figure 7. (The master chart for solution of this quantity is so devised that no computation whatever is required for the construction of specific charts for any value of Vo—all that is necessary is the drafting of the ΔT lines.)

The upper segment of Figure 7 is an example of the chart for solution of θz. It is constructed in terms of sin θo, To, and θz. The construction of this chart for any value of α is also very simple—requiring only a simple calculation and drafting.

The two charts (for θo and θz) are joined on their common parameter (sin θo) and θz is determined from the observed values of ΔT, Xs, and To as shown in the example (Figure 7) by the heavy line and arrows.

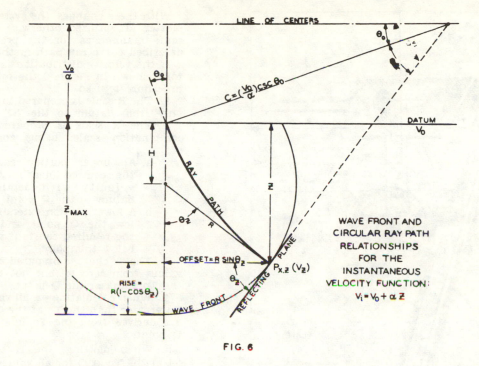

WAVE FRONT AND
CIRCULAR RAY PATH
RELATIONSHIPS
FOR THE
INSTANTANEOUS
VELOCITY FUNCTION:
$V_i = V_0 + \alpha Z$

FIG. 6

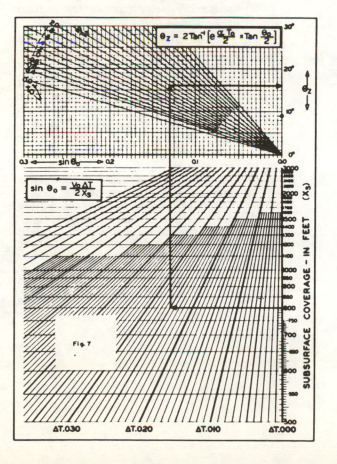

$$\theta_Z = 2\,\text{Tan}^{-1}\left[e^{\frac{\alpha T_0}{2}} \times \text{Tan}\,\frac{\theta_0}{2}\right]$$

$$\sin\theta_0 = \frac{V_0 \Delta T}{2 X_s}$$

Fig. 7

Having solved for θ_z, there remains two "quantities computed" to be solved for. These are the H and R for given values of Vo, α, and To. These H and R quantities are in terms of distance and are in the form of scales (examples show the "H scale strip" and the "R scale strip"—Plate I).

Reference to equations III and IV (Figure 6) will show that when Vo and α are fixed (as for a given area) the H and R become simply a function of To and it is possible to prepare a set of master scales for these quantities. From such a master set any particular set of scales is derived by one simple calculation and photographic reduction (or enlargement) to satisfy any required combinations of Vo and α as well as any desired cross section depth scale.

From the foregoing, it may be seen that the necessary charts and scales for the system are subject to rapid preparation.

The actual plotting of the data is accomplished by use of the plotter in the following manner (refer to Plate I):

With the cross section paper correctly positioned with respect to the plotter, the H and R arms are extended to the desired To value, the angle (θ_z) is turned and the point (or segment) plotted at the R arm head

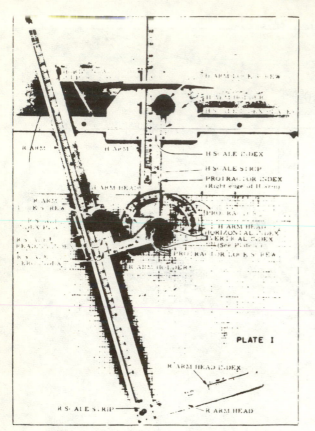

PLATE I

index (or along the R arm head).

The entire assembly is mounted on the horizontal support bar along which it is moved to succeeding observation points.

If it is desired to use the so-called straight line migration method where the velocity distribution is of the form shown in Figure 1 (constant), the following procedural changes are made:

1. The H scale, of course, does not exist.
2. A Zmax scale is mounted on the R arm strip.
3. The sin θ_0 chart is made up using the selected constant velocity as Vo.
4. The θz chart has only the zero To line since under the constant velocity condition $\theta_0 = \theta z$.

With these changes, the system and plotter are utilized otherwise in the same manner as for the previously described curve ray path method.

If the velocity distribution is of the form shown in Figure 3 the following procedure is used.

1. The H scale is prepared to extend only from datum to the "Q" point.
2. The R scale is prepared as a combination scale, being composed of:
   a. An upper portion satisfying the required linear increase of velocity with depth from datum to the Q point.
   b. A lower portion extending below the Q point and being the required constant velocity below this point.
3. The θz chart is prepared with To values from zero to, but not beyond, the To value at the Q point.
4. When calculating θz, all values of To greater than the To at the Q point are derived from the To line at the Q point.
5. When plotting, the H scale is set at the To at Q for all values of To greater than the To at Q.

(These procedures all derive from the fact that the curvature of the wave front does not change in the constant velocity part of the section below the Q point.)

If the system is employed for data analysis by the Rice "resolved time" method, it is necessary to:

1. Construct the lower portion of the chart shown in Figure 7 with Vo equivalent to VH (horizontal velocity).

2. For the upper portion of this chart (Figure 7) only the zero To line is used.

3. The H scale is zero.
4. The R arm scale is simply a convenient linear time scale.
5. The procedures for computing and plotting of the "resolved time" section are the same as for the straight line migration method outlined above with the difference being that in the case of the "resolved time" method, all dimensions are in time rather than footage.

Reprinted from *Shale Shaker*, **8**, 26-30.

Three-Dimensional Control

By G. E. ANDERSON
Pan American Petr. Corp.

No geophysicist would consider interpreting seismograms without obtaining the necessary data to calculate elevation and low-velocity-layer corrections. No geologist would consider interpreting an area without determining where the wells were drilled and what formations were penetrated. Yet both geophysicists and geologists appear to be quite willing to use seismic data to make an interpretation in an area without even making an effort to obtain three-dimensional control. Sometimes, apparently quite by accident, three-dimensional control is obtained, but all too often it appears that no attempt was made to use this control. These statements are based on conversations with many people as well as examination of records from several companies.

The type of three-dimensional control needed for an area will, of course, be determined by the type geology for the area. In areas of relatively gentle dips, crossing seismic profile lines will probably suffice. However, in areas of steep dip and complex subsurface the need for three-dimensional control would demand a four-way spread on each shot point in the seismic profile line, with additional four-way shot points strategically located between the lines. This discussion will be confined to the areas of steep dip and complex geology.

Three-dimensional control can be obtained in two ways. Figure I illustrates one method. The seismometers are laid out in a circle around the shot point. It is apparent that a wave front striking these seismometers will give a reflection from which the direction and amount of dip can easily be determined. As illustrated, the dip would be from the seismometer which has the least amount of time toward the seismometer which has the greatest amount of time, and the amount of dip would be the difference between these two times. If the records are not very good, this method is generally unsatisfactory. The advantage to this method is that in addition to the simplicity of obtaining direction and amount of dip, reflections from two spreads do not need to be correlated.

The other method and the most generally used method is four-way spreads. Figure 2 illustrates the use of four-way spreads. As can be seen from Figure 2, the North-South East-West cables are laid out on the road

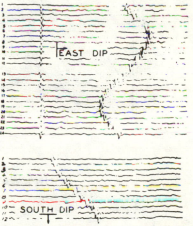

FIG. I

intersection. The point of reflection is up-dip from the cable layout. The problem is to find by proper calculations the depth of Point P, the horizontal distance and direction of Point N from the cable layout, and the true dip at Point P on the reflecting horizon.

Figure 3 illustrates a record which might be obtained from the cable layout in Figure 2. Note that here two components of the dip are obtained. One disadvantage to the four-way method is that reflections from two different cables must be correlated to give the proper three-dimensional control. This is not always easy. The two components must be combined to give the direction and amount of true dip. Figure 4 illustrates how this can be done. As the emerging

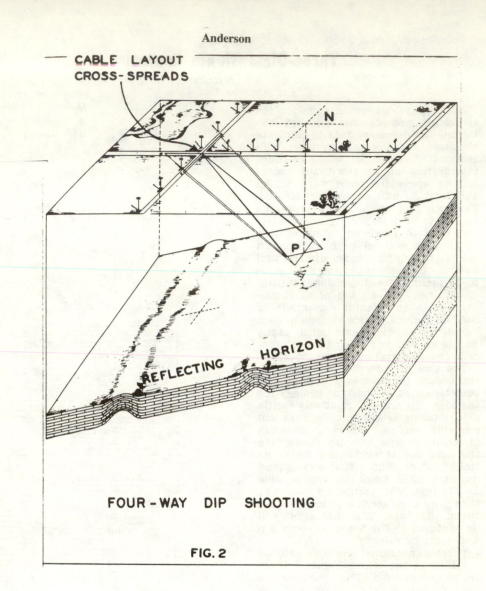

CABLE LAYOUT
CROSS-SPREADS

N

P

REFLECTING   HORIZON

FOUR - WAY DIP SHOOTING

FIG. 2

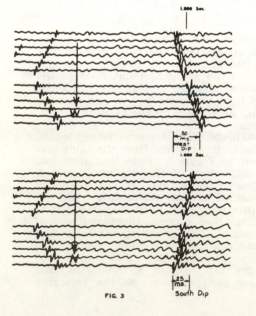

FIG. 3

wave front strikes the cable layout, the ΔT dip moveout can be obtained for the North-South cable and the East-West cable by examination. As illustrated in Figure 4, the sines of the two components can be vectorially added to give the sine of the resultant angle of dip and its direction. It is well to note that if the ΔT dip moveout from each cable is normalized, by making the spread lengths of both cables the same, the resultant ΔT dip moveout is simply the vectorial addition of the ΔT dip moveout from each cable. This ΔT dip moveout would be that dip moveout obtained if one of the cables had been laid out originally in the true direction of dip. The sine of the angle of true dip at the point of reflection is obtained by multiplying the sine of the angle of dip of the emergent wave by the ratio of the velocity at the point of reflection to the emergent velocity. Plates A and B show two charts used

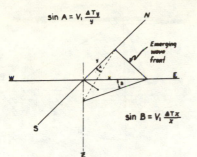

$$\sin A = V_1 \frac{\Delta Ty}{y}$$

$$\sin B = V_1 \frac{\Delta Tx}{x}$$

Let C = angle of emergent wave in direction of true dip

Sin C = a vector whose components are sin A and sin B

$$\sin A = V_1 \frac{\Delta Ty}{y}$$

$$\sin B = V_1 \frac{\Delta Tx}{x}$$

if x = y

$$\text{Sin } C = V_1 \frac{\Delta TA}{A} \qquad S = x = y$$

V = velocity at the point of reflection

θ = angle of true dip at point of reflection

$$\sin \theta = \frac{V}{V_1} \sin C = V \frac{\Delta TA}{S}$$

FIG. 4

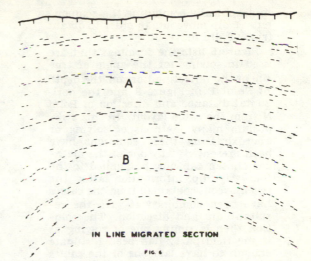

IN LINE MIGRATED SECTION

FIG. 6

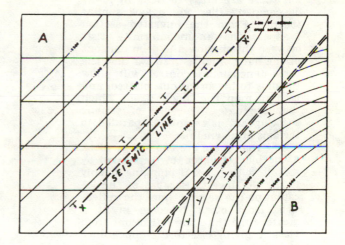

$$T = \int_0^z \frac{dZ}{V\sqrt{1 - P^2 V^2}}$$

$$d = P \int_0^z \frac{V dZ}{\sqrt{1 - P^2 V^2}}$$

Where $P = \dfrac{\sin \alpha}{V_1}$

$$\sin \theta = \frac{V}{V_1} \sin \alpha$$

$$\sin \alpha = \frac{V_1 \Delta T}{X}$$

$$\therefore \sin \theta = \frac{\Delta T}{X} V$$

FIG. 5

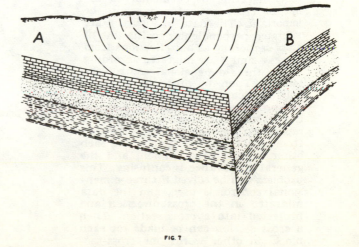

FIG. 7

in graphically combining the component ΔT dip moveouts in order to obtain true ΔT dip moveout.

Figure 5 lists the formulas that are used in solving for the angle of true dip of the reflecting horizon, the depth of Point P in Figure 2, and the horizontal distance and direction of Point N from the cable layout. In this symposium many methods of solving or approximating these equations have been presented. Any one of the methods chosen can now be applied to the four-way dip information after the two moveouts obtained from the cables have been combined to give the resultant dip and direction. The solution then is simply one of solving as if the field party had been fortunate enough to have laid one of the cables out in the direction of true dip.

Figure 6 illustrates the need for three-dimensional control. This is not an observed cross section, of course, but there are cross sections in many oil company files which look similar to this. From the migrated section it appears that an unconformity exists between the shallow and deep portions of the data. Figure 7 shows how the three-dimensional control will allow the dips to be sorted out, plotted, and used to map the blocks from which they originate. The line of seismic shooting XX' is essentially parallel to the axis of a syncline.

Note that there are two blocks, block A and block B, probably separated by a fault, with block B plunging steeply to the Northwest. Because of the location of the seismic line, the reflections from block B come in at a greater time than the reflections from block A, giving rise to the appearance of an unconformity on the seismic cross-section.

Much care is generally used in migrating data in the line of cross-sectioning. It is well to remember that migrating perpendicular to the line of cross-sectioning can also be important. The data in Figure 8 are migrated perpendicular to the line of cross-sectioning. The reflections from blocks A and B are migrated according to the three-dimensional control obtained. Then in order to obtain a true vertical cross-section, the data are projected back into the plane of the vertical cross-section. Figure 9 represents the results. The reflections from blocks A and B are crossing each other on the cross-section, and the general appearance is confusing. This problem can be solved if three-dimensional control is used, and the data migrated in the cross direction and projected into a cross-section. Then a cross-section can be made for each block, in other words one cross-sec-

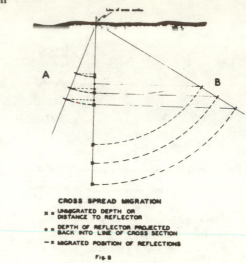

CROSS SPREAD MIGRATION

X = UNMIGRATED DEPTH OR DISTANCE TO REFLECTOR

• = DEPTH OF REFLECTOR PROJECTED BACK INTO LINE OF CROSS SECTION

- = MIGRATED POSITION OF REFLECTIONS

Fig 8

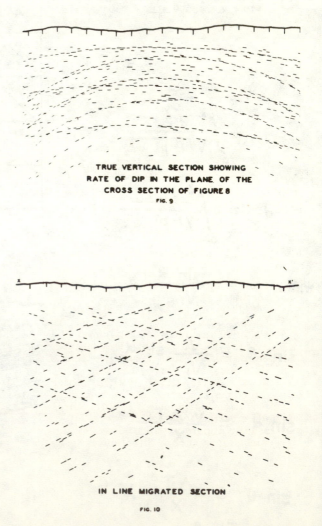

TRUE VERTICAL SECTION SHOWING RATE OF DIP IN THE PLANE OF THE CROSS SECTION OF FIGURE 8

FIG. 9

IN LINE MIGRATED SECTION

FIG. 10

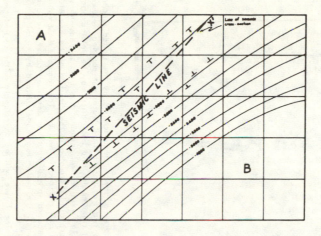

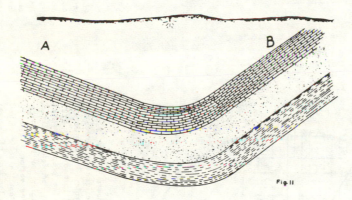

Fig. 11

tion for block A, and one cross-section for block B.

Figure 10 probably illustrates the usual confused seismic cross-section obtained in southern Oklahoma. It is evident, of course, that this situation cannot exist, that these two sets of data must not occupy the same space at the same time. Without three-dimensional control, the correct interpretation is hopeless. With three-dimensional control, however, it is readily apparent that the seismic line was shot in such a direction as to ob-

liquely cross a syncline, one flank of which is dipping more steeply than the other, as seen in Figure 11.

Three-dimensional control in seismic work is not difficult to obtain nor does it appreciably increase the cost of the seismic survey. It is highly recommended that when any seismic program is planned, very serious thought be given to obtaining the proper amount and type of three-dimensional control, which will at least make it possible to approach the correct interpretation.

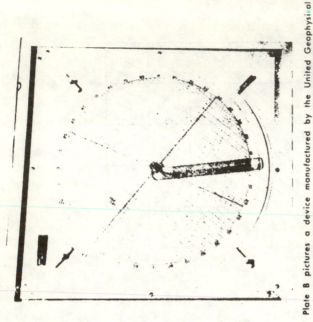

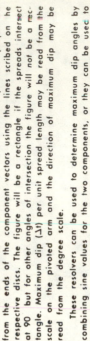

Plate B pictures a device manufactured by the United Geophysical Co. It comprises two transparent discs, each disc carrying its own grid of lines ruled at right angles and both discs may be rotated about the common center. This resolver will combine dips from spreads intersecting at any angle. Each disc is oriented in the direction of the line of shooting it represents and clamped in position. The component Δt values are adjusted to a standard spread length as described above and layed out on the proper axes. The vector sum is determined by visual projections from the ends of the component vectors using the lines scribed on the respective discs. The figure will be a rectangle if the spreads intersect at 90 but for other angles of intersection the figure will not be a rectangle. Maximum dip (Δt) per unit spread length may be read from the scale on the pivoted arm and the direction of maximum dip may be read from the degree scale.

These resolvers can be used to determine maximum dip angles by combining sine values for the two components, or they can be used to combine offset distances to get direction of maximum dip, etc.

Plate B. Courtesy of United Geophysical Co., Pasadena, Calif.

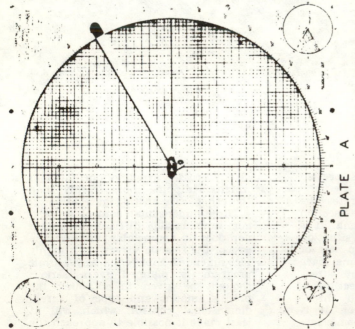

PLATE A

Plates A and B illustrate two types of Dip Resolvers. Plate A is used by (Stanolind) Pan American Petroleum Co. It is designed primarily for use with cross spreads that are oriented at right angles. The component dips (Δt) in milliseconds are adjusted to a standard spread length by a proportional multiplier and layed off as vectors on the proper direction axes of the resolver. The vector sum is determined by completing the rectangle visually using the lines scribed at right angles, on the device. The pivoted arm is set in alignment with the resolved vector. The maximum dip for the standard spread length is read from the scale etched on the arm. The direction of maximum dip is read on the scale of degrees scribed around the disc. The maximum Δt so determined may be converted to a dip angle by methods described elsewhere. This dip resolver can also be used to combine non-perpendicular vectors as shown in the upper left corner of the plate.

Chapter 3

Computer Processing

With the advent of digital recording of seismic data and digital computers for processing, automatic migration of seismic data displaced all mechanical and analog devices. The new technology also allowed several new approaches to be implemented.

The articles by *Lindsey (1970)* and *Rockwell (1971)* typify what is sometimes called the "diffraction-stack" method in contrast to Karcher's wavefront method. In the wavefront method each sampled value of a recording is spread along a wavefront. This is done for every sample of every recording and the superimposition of the wavefronts is the migrated result. In the diffraction-stack method each sample of the migrated result is computed by collecting and adding the values that are superimposed by the overlapping wavefronts. In the case of constant velocity, the values lie along a hyperbola or diffraction curve in the space-time measurements. The two methods get the same answer but add the numbers in a different order.

A weakness in the diffraction-stack method is knowing the curve along which the values should be added. When the velocity of propagation varies laterally and vertically, the curve can be obtained by ray tracing, but the procedure is time-consuming.

Papers by *Timoshin (1970), Claerbout (1970), Claerbout and Johnson (1970),* and *Claerbout and Doherty (1970)* overcame this weakness by formulating the problem of imaging reflectors in terms of a source wave field and a measured reflected wave field. The measurements made at closely and regularly spaced geophone stations on the Earth's surface can be regarded as measurements of an upward traveling wave field. To reconstruct the wave field from the measurements, imagine fictitious sources placed at the points of measurement which radiate vibrations into the Earth equal to what was measured but in the opposite time order. At any point in the Earth, the disturbance generated by this surface distribution of fictitious sources equals the upward traveling wave field, except that the time order is reversed. Similarly, the real sources, which can be any combination of point sources, generate a downward traveling wave field. At a reflector the upward reflected wave field should be equal to the downward source wave field multiplied by the reflection coefficient. Hence, by computing two wave fields, one from the source and one from the measurements, the reflectors can be defined as their intersection.

Claerbout describes practical methods for computing the two wave fields by solving differential equations, whereas Timoshin only formulates the method in principle using terms of integrals. To understand Claerbout's numerical solution of the wave equation, it may be helpful first to study the standard finite-difference techniques described in Alford et al. (1974).

An important concept, introduced by Claerbout, is *one-way* wave propagation through inhomogeneous media. The downward traveling source wave field in reality is complicated by internal multiple reflections. For practical calculations, in many cases only the transmitted field is important. An equation (also derived by Claerbout) computes a transmitted wave, not exactly but closely enough for many practical applications. Similarly, a one-way upward traveling wave field can be constructed, equal to the measured data on the Earth's surface, by omitting all reflections which generate downward traveling waves. The intersection of these two one-way propagating wave fields gives an image of the reflectors with less interference from multiples.

In the early 1970s there was considerable interest in optical holography and the possiblity of applying the concepts to seismic exploration (Milder and Wells, 1970; Hoover, 1972; Marcoux, 1973). After concluding that holographic techniques do not add anything to the seismic methods, these papers were omitted.

References

Alford, R. M., Kelly, K. R., and Boore, D. M., 1974, Accuracy of finite-difference modeling: Geophysics, **39,** 834-842.

Hoover, G. M., 1982, Acoustical holography using digital processing: Geophysics, **37,** 1-19.

Marcoux, M., Matzuk, T., and French, W. S., 1973, Technical limitations of seismic holography: Presented at the 43rd Ann. Internat. Mtg. and Expos., Soc. Explor. Geophys., Mexico City.

Milder, D. M., and Wells, W. H., 1970, Acoustic holography with crossed linear arrays: IBM J. Res. and Dev., **14,** 492-500.

Reprinted from *Oil and Gas Journal*, **38**, 112-115.

DATA PROCESSING
IN EXPLORATION

Digital migration

J. P. LINDSEY
Chief Research Geophysicist
Geocom Inc.
New Orleans, La.

AL HERMAN
Manager of Geophysical Programming
Geocom Inc.
Houston, Tex.

AN ENTIRELY new aid to accurate interpretation of seismic data is now in use.

The new aid is digital migration— a technique which migrates seismic signals and displays the data in time or in depth and, for the first time, reveals seismic reflections in their proper attitude on seismic sections. It was devised by Geocom Inc. New Orleans and Houston based geophysical processing company.

The new process makes it possible to migrate time sections directly from the seismic data, while eliminating the necessity of first picking the events. The result is that all of the important properties of the time section are preserved in the migrated section. These include wavelet character, data band-width, resolution, and signal/noise ratio.

In the process, diffractions are also migrated increasing their clarity over the unmigrated display. The accuracy of migration is limited by the quality of the velocity function used and the extent to which events on the section arrive from out of the plant of profile.

Geophysicists have recognized the need for migration of seismic data for many years. The tedious hand processes they have used for migration can now be replaced by a direct computer technique.

A distorted view. The conventional method of viewing subsurface geology in seismic data is by plotting the reflection arrivals along a vertical time scale and directly under a position midway between the energy source and the receiver.

Subsurface reflection horizons are recognized as lineups of events across this time section and the general degree of dip and depth is apparent.

Although this method of plotting is convenient, it presents a distorted view of the true subsurface since the reflections from dipping beds actually come from the updip side and not from directly under the halfway position as plotted.

Furthermore, diffractions f r o m faults, pinchouts, facies changes, etc. appear as curved lineups rather than a single subsurface source.

The removal of this distortion by moving the seismic events to their p r o p e r positions in the subsurface space is called migration.

It can be accomplished in two dimensions by overlaying wave-front charts on the time section as illustrated in Fig. 1. The wave-front chart is a plot of e q u a l travel-time curves from the surface downward into the subsurface strata.

The reflection time corresponding to the event to be migrated is found on the wave-front chart. The amount of movement along this path to the reflector's proper position is determined by the tangent line drawn to two or more such curves located on laterally displaced wave-front charts.

The ability of any migration procedure to correctly portray the subsurface structure is a function of the accuracy of the subsurface velocity information and the extent to which three-dimensional s e i s m i c data are available.

Velocity is required to relate time to depth, and three-dimensional control is required to resolve the true position of events that come from the side rather than in a direction along the profile line.

The need for lateral control is satisfied by running crossed seismic lines. Velocity information is obtained from field data in the case of common-depth-point field procedure.

Migration using wavefront charts

Fig. 1

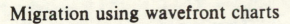
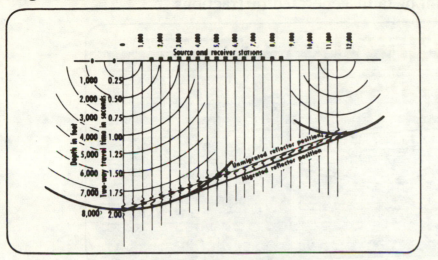

Diffraction from a bed termination

Fig. 2

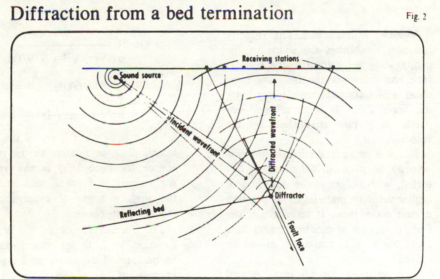

Diffraction and reflection

Fig. 3

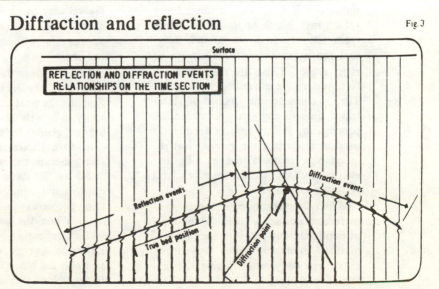

REFLECTION AND DIFFRACTION EVENTS
RELATIONSHIPS ON THE TIME SECTION

Migration of fault-generated diffractions   Fig. 4

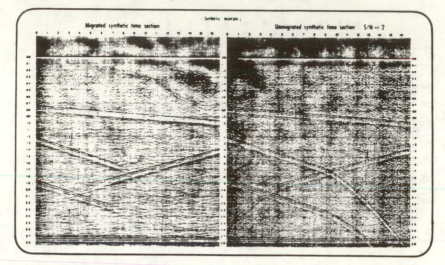

Velocity is important. The velocity function establishes the shape of the propagating wave front in the subsurface and consequently the wave-front chart. For a constant velocity medium, the wave front is spherical and expands at a rate established by the velocity.

For the more realistic nonuniform velocity distribution, the sphere is distorted, with bulges in the direction of higher-velocity material. The position of the wave front at any instant defines a surface of constant travel time from the source, sometimes called an aplanatic surface.

The surfaces for all equal travel times are just as well defined by a set of rays which appear to emanate from the source location and intersect the aplanatic surfaces always at right angles. These are the ray paths followed by the expanding wave front. They represent the paths of least travel time to any given point for the propagating energy and are frequently used in discussing the geometrical relationships between the propagating energy and the source, receiver, and reflection surfaces.

When the velocity of sound in subsurface materials is solely a function of depth, the ray paths or their mathematical equivalent, the wave-front chart, can be generated using the following equations.[2]

$$t = \int_o^z \frac{dz}{V(z)\sqrt{1-\alpha^2 V^2(z)}}$$

$$x = \int_o^z \frac{V(z)\,dz}{\sqrt{1-\alpha^2 V^2(z)}}$$

where z is depth, t is travel time, x is lateral distance relative to the surface source location, $V(z)$ is the velocity for a strata at depth z, and α defines the surface angle of emergence for the ray path chosen.

For certain forms of the velocity function $V(z)$, these integrations can be performed explicitly and algebraic expressions for t and x obtained. If the velocity has a linear increase with depth, the constant travel-time loci are circles with vertically displaced centers.[1]

For a linear increase of velocity with time, the loci become cycloids.[2] For the general case where the velocity is a table of values corresponding to given depths, a computer is used to evaluate the above formulas and generate the wave-front chart.

3D vs. 2D data. The lack of three-dimensional control gives rise to certain problems in migrating seismic data. When the attitude of the subsurface reflecting bed is such that the reflection energy comes from some point to the side rather than from directly below the source-receiver line, the indicated dip is in error.

Such an error, of course, is detected when data from an intersecting seismic line is compared. In the absence of such lateral control, these dip errors are detected when attempts at migration fail to separate the events. Inaccurate velocity information can also prevent the migration process from separating conflicting events even when the points of reflection are directly below the seismic line. The need for accurate velocity data is especially acute when only one seismic line of profile is available for migration.

Diffractions play an important role. The diffraction of sound from small subsurface acoustic discontinuities, a phenomenon illustrated in Fig. 2, has become of more importance since the widespread adoption of the common depth point technique.

Since diffraction occurs in a small region of the subsurface, it is virtually indistinguishable from a reflection event when the field data have been selected for common d e p t h point stacking.

Since the major diffraction events are associated with bed terminations of one kind or another, they are of definite assistance in interpretation when they can be distinguished from reflections. Proper migration of diffraction events is one key to their detection.

Diffractions that are important to seismic prospecting are generated by line discontinuities rather than points. A bed termination at a fault face is a prime example and the line formed by the intersection of the bed with the fault surface can be called a line diffractor.

If the profile crosses the diffraction line at right angles, the diffraction point is essentially stationary on the diffraction line and the classic diffraction curve, or the curve of maximum convexity, will be observed on the record section as in Fig. 3 and the unmitigated section of Fig. 4.

However, if the line of profile parallels the line diffractor, it will exhibit no relief on the time section and will therefore be indistinguishable from a flat bed. It is therefore important to know the angle between the line of profile as projected on the terminating bed and the line diffractor.

For diffracting l i n e s crossed at angles less than 90°, the diffraction event on the time section will appear to be a diffraction associated with an abnormally high velocity.

Migration of a curve of maximum convexity results in a point in space— the point representing the diffractor.[3] Failure to use the correct velocity function for migration of a diffraction exhibiting maximum convexity will either over or under migrate it, resulting in smearing the point in space. The degree of sensitivity to velocity for migrating a diffraction event can be expressed by the algebraic relation between two ratios:

$$M_R = \frac{1}{1 - R^2}$$

The quantity M_R is called the migration ratio and expresses the ratio of lateral extent of a diffraction event before migration as it appears on the time section to its lateral extent after migration. The quantity R is the ratio of the correct velocity to the velocity used in migration.

If the angle between the line diffractor and the line of profile is not known, it may be estimated from the following formula:

$$\sin \alpha = \frac{V'}{V'_m}$$

The velocity V' is that determined for performing the optimum stack and V'_m is the velocity required to migrate the diffractor to a point. The velocity V'_m can be determined directly from the time section if a diffraction event can be identified:

$$V = q \sqrt{\frac{2}{T^2_a - 2T^2_b + T^2_c}}$$

The three normal times T_a, T_b, and T_c are picked from the time section across the diffraction event from traces having a horizontal separation of q feet. Maximum accuracy is provided when q is as large as possible; however, care must be taken not to inadvertently pick the reflection event time to which the diffraction event ties.

What the Geocom process does. The Geocom migration process examines the time section, recognizes reflections, and m o v e s them into their proper structural positions. The technique is able to discriminate between portions of the section which have noise and portions containing signal. The result is that noisy portions tend to be obscured relative to the signal.

The process uses a specified velocity as a function of depth. It will begin with unstacked data after static corrections have been applied and the velocity function determined.

Since the same data and parameters are used that are required for making the final time section, the migrated section is made available at the same time. It bears a close resemblance to the time section, and can be plotted in the same format as the time section. Since subsurface velocities are known, the vertical scale may be linear in e i t h e r time or depth. Wavelet characteristics, bandwidth, and data quality are preserved in the migrated section, and all the data are migrated rather than selected events.

In its present form, the migration program works in two dimensions using data from a single line of profile. However, the concept is readily expanded to perform three-dimensional migration when three-dimensional input data are available.

Benefits. The benefits of a direct-migration process with the characteristics presented are twofold: (1) better interpretation, and (2) more accurate measurements of the subsurface. Interpretation is improved because:

1. The migrated section is available along with the time section. Visualization is easier in true space than in the distorted space of the time section.

2. The wavelet characteristics are preserved. This is more than merely a useful correlation feature. It serves as a reminder of the actual quality of the basic data since no neatly drafted lines are arbitrarily substituted for events recorded in the field.

3. Diffractions are migrated along with reflections. The distinction be-

tween these two types of events is much more pronounced in a migrated section than in a time section as illustrated in the synthetically generated data of Fig. 4. The structural significance of diffraction events should not be ignored.

More accurate subsurface measurements result from the migration process since:

1. Diffractions migrate to a point. Bed terminations frequently provide a blurred indication of their exact location in the conventional time section as illustrated in Fig. 3 and the unmigrated section of Fig. 4.

A diffraction event forms a tangential time tie with the reflection event from the terminating bed surface. Only a small change in curvature for the event is noticed on the time section.

The migration of the diffraction to a point at the termination of the reflection event provides a clear mark of identification. This is particularly useful in defining the correct attitude of a fault face.

2. Migration of diffractors is velocity sensitive. An error in migration velocity does not change the appearance of the reflection event but merely its placement and attitude in the migrated section.

The same velocity error makes a significant difference in the final appearance of a diffractor since it is expected to collapse to a point and does not.

Migrated seismic data can now be made available along with the output time section. This provides data that are more easily interpreted than ever before. Diffractions are migrated as well as reflections resulting in better definition of bed terminations. Migrated data are now available at only a small increase in processing costs.

References

1. Van Melle, F. A., "Wave-front circles for a linear increase of velocity with depth," Geophysics, XIII, 2, Apr. 1948.

2. Musgrave, A. W., "Wave-front charts and ray path plotters," Quarterly of the Colorado School of Mines, Vol. 47, No. 4, Oct. 1952.

3. Hagedoorn, J. G., "A process of seismic reflection interpretation," Geophysical Prospecting, Vol. II, No. 2, June 1954.

More on Data Processing in Exploration on Page 123.

Reprinted from *Oil and Gas Journal*, **69**, 202-218.

Migration Stack aids interpretation

D. W. ROCKWELL
Geophysical Service Inc.
Dallas

A NEW technique—Migration Stack (U.S. patent 3353151)—is now being used on difficult-to-interpret seismic data and in most cases is making possible more meaningful geologic sections for interpretation.

Although the technique is now being applied to seismic data already processed by Common Reflection Point stacking techniques, it promises even greater improvement when input is unstacked multiple coverage data; that is, the data are first migrated and then summed or stacked in their displaced spatial positions.

The "text-book" Gulf Coast salt dome in Figs. 1a, 1b, and 1c illustrates dramatically the confused appearance of seismic reflection time data recorded over complex geologic structures, even when the data quality per se is good. Geologic meaning of the shallow data in Fig 1a is obvious

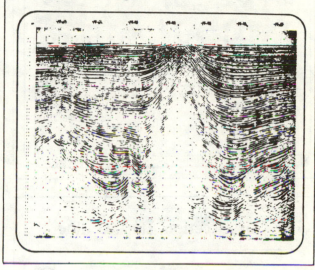

Salt dome, migrated time section

Fig. 1b

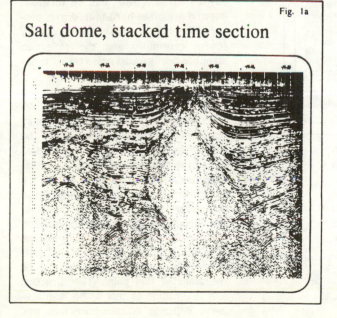

Salt dome, stacked time section

Fig. 1a

enough, but below 2-sec, the picture becomes progressively less clear.

In any case, as can be seen by comparing Figs. 1a and 1b, the complex time data have now been rearranged by migration especially below 2-sec. The diapiric intrusive zone is seen to be slimmer; the actual position of the salt flank is suggested by a concentration of energy within this zone. The typical rim synclines have been created out of the conflicting mass of interfering reflections and diffractions on the lower flanks of the dome.

In Fig. 1c, the same velocities used to displace the data have been reapplied to give a depth-converted version, a geologic section on which the geologist can work directly.

Also, the signal-to-noise ratio is improved by the Migration Stack technique even when applied to already stacked data and it is anticipated that this feature of the process may result in its use even where structures are not complex or are only marginally complex.

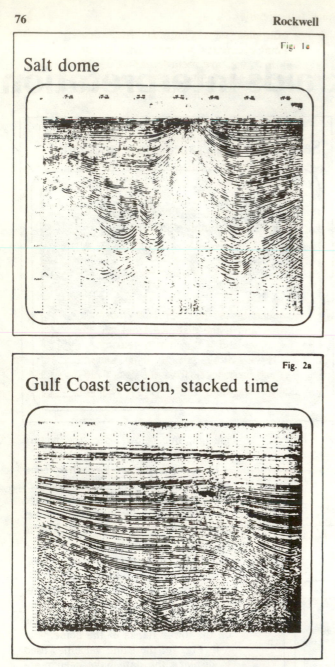

Fig. 1a

Salt dome

Fig. 2a

Gulf Coast section, stacked time

The retention in a geologic section of the appearance of a seismic section is certain to please both geophysicists and geologists and to reduce the credibility gap that has sometimes existed in the past between these two branches of the exploration profession. The mysterious effects of velocity variations on depth conversion and migration, plus the use of a middleman geophysicist to "make" an interpretation, have tended to keep the geologist one step away from the original data. Now perhaps he, as well as his exploration manager and executive managers, can appreciate directly the quality of data

on which their drilling locations are based and the effect that migrating and depth converting these data with different velocity functions can have.

Another example of the results of applying Migration Stack can be seen in Figs. 2a and 2b. Note in Fig. 2b that the deep broad syncline emerges from a relatively restricted pattern of interfering data which experienced interpreters might recognize as a "buried focus" (discussed later on). Also, note the retention and even the clarification of the small faulting detail and the displacement of the deep steep (fault plane?) event. There is an overall marked increase in signal-to-noise ratio, especially at depth.

Unquestionably, Migration S t a c k technology will undergo further development culminating in its economic use under a wide variety of conditions. The goal is the economic stacking of two and three-dimensional data from the most complex as well as the simpler structural configurations and velocity fields.

Past practice in migration technology. Although seismic reflection travel time profiles or sections often give a remarkably direct indication of the structural configuration of the layers of sedimentary rock lying beneath the earth's surface, they can be misleading and are often hard to interpret geologically even when record quality (event continuity and the reflection signal-to-background noise ratio) is excellent.

For years it was the task of the trained and experienced exploration seismologist to interpret the data, sort out and correlate the signal—"pick the record"—and plot a series of points or a line, representative of his correlations, on a separate section. When the structure is complex, showing large deviations from horizontal layering, the subsurface position of the reflecting point (P1 or P61, on Fig. 3) on the interface does not lie under the shot-receiver but is displaced to one side or the other of this point. The direction and magnitude of this updip displacement depends on the direction and magnitude of the inclination or dip of the strata, bedding plane, or seismic interface.

Therefore, besides picking and plotting the reflection "events," the seismologist often calculated the magnitude of the dip and displacement as well as the depth of the event and plotted the event in a displaced or migrated position on the section. Various techniques existed for doing this — plotting arms and pivoted rulers, light tables with overlay charts of the "wave-fronts" and "ray-paths" corresponding to the measured or assumed v e l o c i t y distribution. Dips were "swung" or displaced in accordance with reflection travel times and with Δt's—difference in travel times—to recording points symetrically disposed about the shot. The technique often worked well and many oil-bearing structural traps (folds, faults and domes) were mapped in this way.

The digital computer greatly improved reflection data quality, Figs. 1a and 2a, by stacking of multiple coverage data (Common Reflection Point data) and eased the task of the interpreter by permitting automatic correlation (picking) of events and by migrating these picked data and "plotting" them on film.

The Migration Stack process allows the data to be migrated prior to or without picking and the stacking and migration can be carried out together. Depth conversion can be effected at the same time if desired.

Ray path geometries and associated reflection times. From left to right on Fig. 3, we see a number of ray-path configurations for an interface with various dips, depths, folds and a fault.

The travel times associated with these travel paths are plotted at their surface recording positions in the time profile above.

The various normal times (marked "N") shown, multiplied by the velocity obtaining above the reflecting horizon, yield directly the distance to the reflector and can be used to draw circles to which the reflector is tangent (as at P1 and P5), thus effecting depth conversion and simple "migration" of the reflecting segment.

The various oblique travel time paths also represent useful information for interpretation but require analysis and additional processing.

For example, P10 and P50 are both examples of "Common Reflection Point geometry" wherein shots say at 5, 6, 7, etc. are recorded at 13, 14, 15 symetrically across a center point assumed to be below the surface point 10.

The oblique travel times can be corrected to an equivalent vertical time, 10V, by applying a so-called Normal Moveout Correction based on the image geometry and in the simplest case of constant velocity, on Pythagoras' theorem. This correction, Δt, is given by a simple formula

$$\Delta t = \sqrt{ to^2 + \frac{x^2}{V^2} } - to$$

where to is the corrected (normal) time and x the distance from shot to receiver and V the known velocity.

If the average velocity is not known, it can be deduced from the data themselves by applying the simple formula

$$V = \sqrt{ \frac{X}{tx^2 - to^2} }$$

where tx is the "long trace" oblique time such as the time from surface point 5 to 15 and to is the vertical time at surface point 10.

Once corrected, the multiple coverage or Common Reflection Point data of the set can be summed or "stacked" together for improved signal-to-noise ratio (as was done for Figs. 1a and 2a).

Thus multiple coverage data serve two purposes: data enhancement and velocity determination. In effect they provide the time-difference data used to calculate the velocities used to correct the data for differences in travel path. The by-product velocities are used in depth conversion and for their geologic significance (correlation to rock type).

Modern practice is to "scan" the data with corrections based on a number of assumed velocities selecting the velocity which yields the "best" stack (i.e., most in-phase stack, see Fig. 4 and text).

Diffractions. Diffractions differ from true reflections in that the energy from a shot, say at surface point 15,

Gulf Coast section, migrated time

Fig. 2b

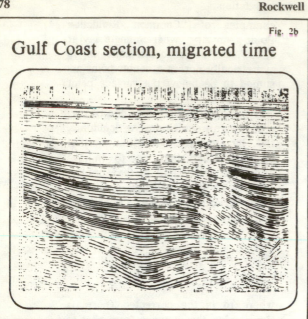

Interface and typical reflection times

Fig. 3

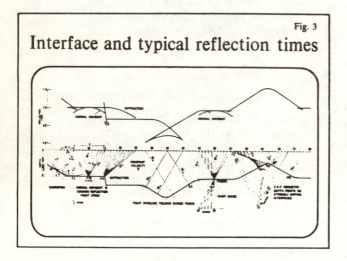

Corrected data*

Fig. 4

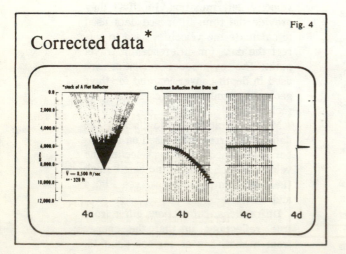

4a   4b   4c   4d

returns from or near the same point without appearing to obey the reflection law (incidence angle equals reflection angle).

P15 on Fig. 3 is an example of a sharp diffracting edge at a fault. These diffraction events can be seen for example on Fig. 1a along the edges of the salt intrusion.

Time differences for diffraction events can also be used to calculate velocity and this velocity, when applied to arrival times, should permit their migration or focusing at the diffraction point, as at P15.

Dipping beds. As the interface or horizon deviates more and more from the strictly horizontal, the C.R.P. geometry shown becomes less and less correct as can be appreciated by comparing the ray-paths associated with P50 and near P61.

There is no longer any true Common Reflection Point. (The reflection points can be found on the interface by drawing a line from the image of the shot normally across the reflector to the receiver such as a "shot" at 55 (shot image 55′) to receiver at 67.)

The result of this fact is that a set of correction times, "N.M.O." or normal moveout times, calculated for and applied to the arrival times from the dipping layer will not correct these times to equal the center or normal ray-time along ray 61N, if the true velocity is used.

Therefore, the miscorrected events, being out of phase, Fig. 5, will not stack or add in optimum fashion. In fact they can cancel out altogether, losing the advantage of stacking.

However, a "false" stacking velocity can be found by "scanning" which will "force" the events to be in phase for best stacking. Scanning is in fact a trial and error method. The so-called "C.R.P." points from a dipping reflector are not truly "Common Reflection Points" even though they are stacked together. Dips have to be quite steep, 15° or more, before these errors become serious and usable stacked results can be obtained at even greater dips employing standard C.R.P. techniques. The stacking velocities derived cannot be used for picked segment migration and depth conversion without at first being corrected

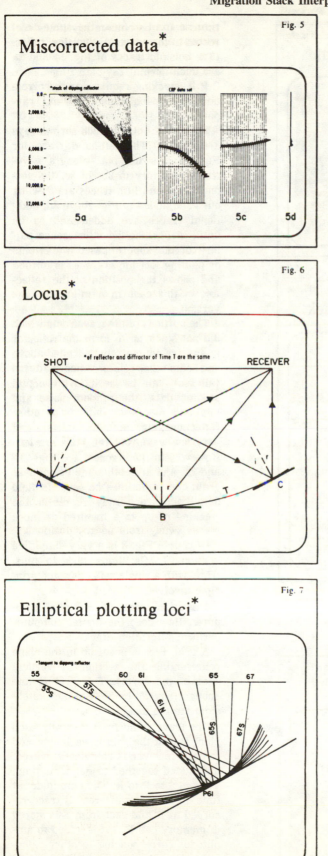

Miscorrected data*

5a 5b 5c 5d

Fig. 5

Locus*

*of reflector and diffractor of Time T are the same

SHOT   RECEIVER

A   B   C   T

Fig. 6

Elliptical plotting loci*

*Tangent to dipping reflector

55   60 61   65   67

P61

Fig. 7

for these dip effects.

The C.R.P. stacking of a flat bed such as at P50 of Fig. 3, is illustrated in Fig. 4 by a synthetic seismogram (4b). The corrected events (4c) are observed to stack in phase resulting in a large signal amplitude in Fig. 4d. In Fig. 5, the same N.M.O. corrections of Fig. 4, for the same center trace time, are applied to the dipping data, 5b, (a case similar to P61, Fig. 3) and the overcorrected data of Fig. 5c result, which stack into a distorted low amplitude event as shown in 5d. True corrections, based on the actual ray paths of Fig. 5a, would yield the same high amplitude trace as in 4d.

Basis of migration stack technology. In its simplest form Migration Stack theory can best be understood by looking at Fig. 6.

Let us assume the shot on the left resulted in a reflection event of time T at the receiver on the right. As far as this one-trace record can tell us, this event could have come from any point on the curve marked T because the travel time to points on this curve along the various paths shown (and in fact along all such paths) is always T whether the reflection occurs from a dipping bed at A, a flat bed at B, or a dipping bed at C. The reflection law is satisfied (angle i = angle r). But even if the source is a diffraction source at A, B, or C, the travel time T still locates that source along the curve for T, which has the form of an ellipse for a constant velocity.

As far as travel times are concerned, there appears to be no difference between reflection and diffraction events. Reflection is a special case of diffraction. If shot and receiver are coincident, the ellipse becomes a circle coincident with the physical wavefront at time T.

Now look at Fig. 7, which is an enlargement of the dipping bed ray geometry below surface position 61 of Fig. 3. The curves shown are all ellipses corresponding to the various times along the oblique ray paths (i.e., shot at 55 recorded at 67, etc.) except one circle corresponding to the normal ray 61N. Each ellipse represents all the possible positions of a reflecting interface satisfying the observed oblique reflection time. Together they deline-

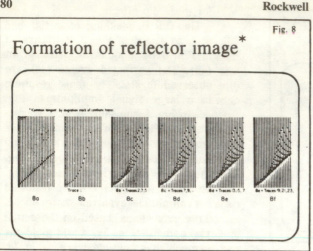

Formation of reflector image[*]

Fig. 8

*(Common tangent by migration stack of synthetic traces)

8a   Trace :   8b   8b + Traces 2,3,5   8c + Traces 7,9, .   8d + Traces 13,15, 7   8e + Traces 19, 21, 23.
8a   8b   8c   8d   8e   8f

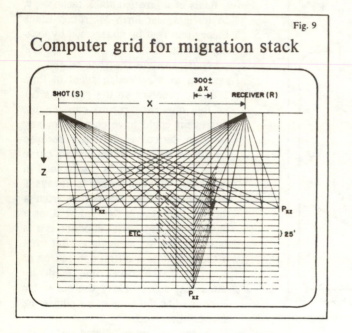

Computer grid for migration stack

Fig. 9

SHOT (S)   300±   RECEIVER (R)
            ΔX
        X

Z

P_{xz} P_{xz}

ETC.   ) 25'

P_{xz}

ate, through their common tangent, the migrated (or depth converted) reflector position.

Synthetic seismograms illustrating migration stack. If an elliptical curve of Fig. 7 were "modulated" with a seismic wavelet, the interference pattern of the resulting curves would show the position of the reflector (or the reflection) to be along the common tangent.

Fig. 8 illustrates by means of a synthetic how this interference pattern is built up. In this instance the reflection profile is shown at the left and can be considered to represent a series of normal incidence travel times (i.e., with a shot at each receiver) such as would be measured along the normal rays, from 55 to 61, in Fig. 3 or the times could be oblique ray times corrected to the equivalent vertical times. (The seismic traces of Fig. 1a and 2a are such normal ray-path times since C.R.P. stacking requires prior correction to the equivalent normal ray-path.)

The wavefronts for such normal rays are circles. The left-hand trace for Fig. 8a is "swung out" along a wavefront trajectory in 8b. In 8c, the second, third and fifth traces are "swung out" and added to Fig. 8b. In 8d, additional traces are added and so on until, in Fig. 8f, the interference pattern of the superimposed wavefronts or plotting loci for 23 traces is formed and shows the position of the reflector, in this case in a migrated time version.

The portions of the wavefronts which did not stack up to form the reflector image are not entirely cancelled above and some other reinforcement patterns (aliases) can be seen. With actual seismic data, the random noise and bits and pieces of data from other reflections, etc. mutually interfere and very few wavefront residuals are left. However, in quiet zones, see Figs. 1b and 2b, and at depth some curvilinear grain can sometimes be seen. In field data each trace may represent the stacking of up to a hundred or more traces swung from nearby positions.

Of course Fig. 8 is a very simplified version of Migration Stack to illustrate, with a single reflection, the principle involved.

Migration stack on a digital computer. How does the digital computer perform Migration Stack?

A grid, Fig. 9, is set up in the plane representing the section (a vertical slice of the earth). This grid is fine enough (say 25' by 300') to adequately simulate the loci.

The travel time from virtually each point P_{xz} on the grid to each shot and to each receiver is calculated, paired, and added together to get total time, T, from S to P to R. The amplitude, at time T, on a seismic record trace recorded at R and shot from S is stored in memory cell P_{xz}. It is added to amplitudes corresponding to other calculated times for other combinations of shot and receiver but to the same grid point since every grid point is a potential source of a diffraction or a reflection. Each point P is on an infinite number of potential elliptical (or circular) loci of which only the ones including actual shot and receiver positions are calculated.

The summed interference patterns formed are the Migration Stack sections and show the reflections or diffractions where the amplitudes "stack up" or add in phase. Random noise is attenuated and unused "random" fragments of loci, being approximately half positive and half negative numbers, on the average also tend to sum to zero.

Field examples reexamined—folds and buried foci. Two other phenomena should be examined diagrammatically before reexamining again the actual field examples discussed above.

On Fig. 3, rays from both flanks of the sharp syncline labeled "buried focus" under point 35 actually cross as from a reflecting mirror focused below ground level. As can be seen from the diagrammatic reflection time-section, the time data are multibranched and look somewhat like a diffraction or a sharp fold (compare with times over point 67). These reflection times are hard to correlate continuously on a time section and must be unraveled by migration. (Compare with Figs. 1 and 2.)

Also note the diagrammatic section, Fig. 3, from point 55 to 75. The time data are broader than the anticlinal fold that produced the anomaly. All dipping data migrate up dip; the anticlines narrow and synclines broaden; the buried foci unfold into sharp synclines. Careful study of the actual field example of the dome will show these same complexities on the time section and their corresponding solution in two dimensions on the migrated sections. Of course their true 3-dimensional solution will require the recording of 3-dimensional data; the recording and processing of such data will undoubtedly be the next major advance in seismic exploration technology.

Conclusions. Migration Stack, as can be seen from the field data examples, has real potential for both data enhancement and as an interpretational aid.

All of the reflection data can be preserved in a format familiar to the geophysicist and yet a geological depth section can be made available for geologists and management. The separate routines of data correction, stacking, correlating, migrating and depth conversion can all be combined in one processing sequence, yielding superior and often dramatically different results. END

NEW POSSIBILITIES FOR IMAGERY
Yu. V. Timoshin

Kiev Expedition, Ukrainian Geological Prospecting Scientific Research Institute
Translated from Akusticheskii Zhurnal, Vol. 15, No. 3,
pp. 421-429, July-September, 1969
Original article submitted October 2, 1967

A new technique is discussed for the conversion of wave fields into images of a medium, and the fundamental characteristics of the technique are investigated. A set of equations is given for the generation of images for homogeneous, inhomogeneous, and layered media over a wide range of frequencies and wavelengths of pulsed and harmonic radiation for continuous and discrete allocation of the points of detection of the oscillations.

The wave methods of investigation widely used in the imagery of various media [1-5] fall into two categories according to the form of representation of the results of investigation. The first category includes methods whose purpose is to acquire an image of the medium by means of lens [6-8] and holographic [9, 10] systems. The second category includes methods for the analysis of the structure of the medium on the basis of the detection of the wave fields at a set of discrete points distributed over some surface or along the line of observation. In both cases it is possible to use either harmonic or pulsed oscillations.

The conversion of wave fields into images of the medium is a highly sophisticated data processing technique and makes it possible to gain information in an extremely visible form. However, the methods in the first category are effective only for the solution of a limited class of problems for homogeneous media and wavelengths considerably smaller than the dimensions of the lens. The more complicated problems encountered in studies of inhomogeneous and layered media containing, in addition to diffracting elements, smooth reflective boundaries of various configurations and large size in comparison with the wavelength are solved by the methods of the second category with the use of radiation having large wavelengths and discrete allocation of the oscillation detectors, along with other special characteristics. For the imagery of a medium under these conditions new techniques are needed for the conversion of the observed wave fields; we consider some of these techniques below.

The most common method of obtaining images is clearly the method of reconstruction of the wave field in the medium. We consider the imagery problem for an inhomogeneous medium in the case of pulsed radiation. Let us assume that a point source at the point O generates a wave pulse $f_0(t)$ in the medium at time $t = 0$ (Fig. 1). As it propagates through the medium the incident wave is reflected by extended boundaries and is scattered by inhomogeneities, forming diffracted waves. The propagation time of the wave pulse from the source to a point M in the medium and from M to the surface of observation Σ_0 in the form of a diffracted wave is determined by the expression

$$t_S = \frac{r_0}{c_0} + \frac{r_S}{c_S} = \frac{1}{c_S}\left(r_0 \frac{c_S}{c_0} + r_S\right), \qquad (1)$$

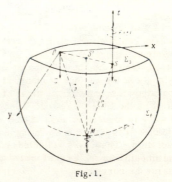

Fig. 1.

where $r_0 = (x_m^2 + y_m^2 + z_m^2)^{1/2}$; $r_C r_S = [(x_m - x_S)^2 + (y_m - y_S)^2 + (z_m - z_S)^2]^{1/2}$; c_i are the propagation velocities of the incident and diffracted waves along the rays r_0 and r_S and are not the same in general (for example, for body elastic waves). For inhomogeneous media we introduce the effective ray velocity, which is equal to the ratio of the length of the corresponding ray to the propagation time of the wave between its end points [2]. The set of reflected and diffracted waves detected on the surface of observation Σ_0 forms a diffraction pattern [6-10], which we refer to henceforth as a wave diagram, interpreting the latter as the time trace of the oscillations on the surface Σ_0. Clearly, in the informational sense the wave diagram is a more general concept than the hologram, which corresponds to coherent radiation.

The conversion of the wave diagram into an image of the medium can be described by means of the Huygens-Kirchhoff diffraction integral for nonstationary processes [8, 11-14]:

$$f(M, t) = \frac{1}{4\pi} \int_{\Sigma} \left\{ \frac{1}{r_s} \left[\frac{\partial f_s}{\partial n} \right] - [f_s] \frac{\partial}{\partial n} \left(\frac{1}{r_s} \right) \right. $$
$$\left. + \frac{1}{c_s r_s} \left[\frac{\partial f_s}{\partial t} \right] \frac{\partial r_s}{\partial n} \right\} d\Sigma, \qquad (2)$$

where $[f_S] = f(S, t - r_S/c_S)$ are the values of the space-time distribution function for the wave field on the surface Σ at time $t - r_S/c_S$ and n is the unit normal to the surface Σ. The integration in Eq. (2) is carried out over the surface Σ, which includes the surface of observation Σ_0 and the part of the spherical surface Σ_1 (Fig. 1) on which the integral tends to zero by virtue of the radiation extinction condition [8, 11, 12].

For coherent radiation the Kirchhoff integral reduces to the form [8, 13, 14]

$$f(M) = -\frac{jk}{2\pi} \int_{\Sigma} \frac{1}{r_s} e^{-j\omega \left(t - \frac{r_s}{c_s} \right)} \cos(n, r_s) [f_s] d\Sigma, \qquad (3)$$

where $k = 2\pi/\lambda$, λ is the wavelength, and ω is the circular frequency.

Integrals (2) and (3) describe the superposition of spherical waves radiated by fictitious sources continuously distributed over the surface Σ_0. The oscillation amplitudes of the sources are proportional to the signals at the surface of observation, which are separated from the arrival times of the waves at point M by the time $t = r_S/c_S$. Therefore, the diffracted wave is focused at point M only at a certain time in pulsed radiation and continuously in coherent radiation. The waves reflected from extensive plane smooth boundaries are focused at the virtual sources for the corresponding boundaries.

In accordance with Eqs. (2) and (3) the reconstruction of the wave field in image space (at any point M thereof) can be realized by the excitation of spherical waves at all points of the surface Σ_0 with amplitudes proportional to the values of the modulus of the function $f(S, t)$ and opposite to the observed phases of the oscillations (the latter operation transforms the divergent waves in the medium into convergent waves for the reconstruction of the wave field, i.e., it corresponds to inversion of the process). In the reconstruction of the wave field, for example by means of acoustic oscillations, the signals recorded on the wave diagram must be transmitted into the image space by means of equivalent radiators, which operate in the opposite direction and have the same allocation as the detectors, i.e., with the rear edges coming first and observance of the diffraction similarity principle [8], which under our conditions assumes the form

$$K(M) = \frac{t_r}{t_i} = \frac{f_i}{f_r} = \frac{1}{M} \cdot \frac{c_i}{c_r(M)}, \qquad (4)$$

where K is the acceleration ratio of the process of reconstruction of the oscillations from the wave diagram; t_r and t_i, f_r and f_i, and c_r and c_i are the times, frequencies, and velocities in the real medium and in the image space, respectively; M is the geometric conversion scale, which is equal to the ratio of the wavelengths in the image space and the real medium. It follows from (4) that for inhomogeneous media with a variable velocity c_r the value of K is variable. For coherent radiation K determines the ratio of the frequencies of the oscillations used for irradiation of the medium and reconstruction of the wave field. In the reconstruction of the wave field for inhomogeneous media, for example by laser irradiation of a hologram, the laser frequency must vary with the image depth in accordance with (4).

As already indicated, the diffraction equations (2) and (3) do not ensure the generation of images of smooth reflecting boundaries of large extent. For the solution of this problem we use a more general integral transformation of the form

$$f(M, 0) = \frac{1}{4\pi} \int_{\Sigma_0} \left\{ \frac{1}{r_{eff}} \left[\frac{\partial f_s}{\partial n} \right] - [f_s] \frac{\partial}{\partial n} \left(\frac{1}{r_{eff}} \right) \right. $$
$$\left. + \frac{1}{c_s r_{eff}} \left[\frac{\partial f_s}{\partial t} \right] \frac{\partial r_{eff}}{\partial n} \right\} d\Sigma_0, \qquad (5)$$

where r_{eff} is the effective radius, which in correspondence with [1] is equal to $r_{eff} = c_s t = r_0(c_S/c_0) + r_S$.

The transformation (5) realizes the transfer of a pulsed source to point M of the image space and estimates the contribution of the corresponding point (element) of the real medium to the wave field detected on the wave diagram. In the case of an inhomogeneity the integration is taken over the diffracted wave front on the wave diagram, and in the image space we obtain an image of the inhomogeneity in the form of a wave pulse whose amplitude is proportional to the energy of the diffracted wave. With a smooth reflecting boundary at M Eq. (6) also gives the image of the inhomogeneity at the corresponding point of the image space due to integration of the energy of the reflected wave on the wave diagram in the contact region of the fronts of the waves diffracted and reflected at point M. The set of such images obtained at different points of the reflecting boundary forms an image of a smooth reflecting boundary of any size in the form of a wave front.

We note that the transformation (5), which we refer to from now on as D-conversion, describes the superposition of waves with nonspherical fronts. Thus, replacing r_S in the transformation (3) by r_{eff}, we have

$$f(M) = \frac{1}{j\lambda} \int_{\Sigma_0} \frac{1}{r_{eff}} e^{-i\omega \left[t - \left(r_0 \frac{c_S}{c_0} + r_S\right) c_S\right]} \cos(\mathbf{n}, \mathbf{r}'_S) [f_S] d\Sigma_0, \tag{6}$$

where $\mathbf{r}'_S$ is the radius vector connecting points M and S' (Fig. 1).

It follows at once from the expression for the exponent of the exponential function in Eq. (6) that equal-phase surfaces in a homogeneous medium have the configuration of ellipsoids with focuses at points O and S. The centers of the isochrones are situated at points S'. In an inhomogeneous medium the isochrones have a more complicated configuration.

On any isochrone through M the time is equal to the propagation time of a wave pulse over the path OMS, i.e., it corresponds to the time $t = 0$ on the wave diagram. Consequently, for the reconstruction of a wave field in the image space by means of D-conversion the wave fronts at $t = 0$ coincide with the reflecting boundaries and diffracting elements in the medium. In the formation of the images by means of equivalent sources at points S' the dimensions of the surface Σ_0 are half as large. At the same time the wavelength of the oscillations in the image space is cut in half, and the right-hand side of Eq. (4) is doubled. The resolving power of the D-conversion in the image space is determined by the same relations as for lens systems.

A very interesting case arises with discrete allocation of the detection points on the surface Σ_0.

The wave field on Σ_0 in this case is given by a lattice function with a finite number of readings [14, 15]. The uniqueness of the representation of a continuous function by a lattice function with a finite number of readings holds for a function with a finite spectrum and quadratically-integrable functions if all the readings outside the surface of observation Σ_0 are zero [14, 16]. A lattice function of the type in question is a function with a finite spectrum and finite extent [16], and the corresponding conversions are limited by the spatial frequencies, whose periods are equal to $2\Delta x$ and $2\Delta y$ [14]. Consequently, for observations the distances between detection points on Σ_0 must be chosen to meet the condition $\Sigma_x = \Sigma_y \leq \lambda/2$. Then the continuous function $f(S, t)$ in the diffraction equations (2, 3), and (5, 6) can be replaced by the appropriate lattice function. Here the diffraction integrals assume the form

$$f(M, t) = \frac{1}{4\pi} \sum_m \sum_n f_S\left(m\Delta x, n\Delta y, t - \frac{r}{c_S}\right) A(S_{mn}, M), \tag{7}$$

where f_S is the lattice function describing the wave field recorded on the wave diagram with discrete allocation of the detectors on the surface Σ_0, $A(S_{mm}, M) = (1/r)\Delta x\Delta y \cdot B(S_{mn}) \cos(\mathbf{n}, \mathbf{r})$, and $B(S_{mn})$ is an amplitude factor introduced for reconstruction of the wave diagram $r = r_{eff}$ or $r = r_S$, depending on the type of diffraction transformation. The possibility of using discrete wave diagrams for imagery by means of the D-conversion or the ordinary Huygens-Kirchhoff transformation considerably facilitates the problem of the practical implementation of the given diffraction transformation either on computers or by the technique of physical reconstruction of the wave field as, for example, in an acoustical model.

We now consider the problem of the uniqueness and validity of the solutions obtained by Eqs. (5-7). These equations are of the same class as the Kirchhoff diffraction integral and therefore satisfy the wave equation. The solutions are unique if the integrands $f(S, t)$ are quadratically integrable and have finite extent (finite spectrum) [14]. These conditions are met for all functions that decrease as $|x| \to \infty$ and $|y| \to \infty$ for the description of the actual observed space-time distributions of the wave fields on the surface Σ_0. With discrete allocation of the detection points on the surface of observation the distances between them must comply with the auxiliary condition $\Delta x = \Delta y \leq \lambda_h/2$ (where λ_h is the wavelength of the oscillations at the highest frequency of the pulse signal).

The validity of the given diffraction transformations follows directly from the fact that with a variation of the wave propagation velocity in the medium or the time values on the wave diagram by an amount ε the image of a diffracting element or reflecting boundary in the image space also changes its configuration by just a small amount. As ε tends to zero, the original image is completely restored.

It is important to note that the integral transformations (5-7) yield images corresponding to the actual structure of the medium under the condition that it is admissible to neglect multiple wave reflections in parallel-layered media, repeated wave scattering by inhomogeneities of the medium, and other interference effects. The presence of wave noise on the wave diagrams produces fictitious boundaries on the images. Without delving at length into the properties of the fictitious boundaries, we point out that in conjugate images obtained for the same domains of the medium with the radiator in different positions the fictitious boundaries will occupy different positions in the image space, whereas the actual reflecting boundaries must be completely congruent. Then in the joint processing of the conjugate images elements can be distinguished that coincide in all the images and correspond to the actual reflecting boundaries and diffracting objects in the medium, while the noncoinciding elements that form the fictitious boundaries in the images can be completely eliminated or significantly attenuated [17]. Various algorithms can be used for the elementwise comparison of the conjugate images. The simplest of these are point-by-point summation or multiplication of the conjugate images:

$$f(M,t) = \sum_p f_p(M,t), \qquad (8)$$

$$f(M,t) = \prod_p f_p(M,t). \qquad (9)$$

where f_p are the conjugate images.

The same objective can also be attained by using a relative difference function of the form

$$\frac{|f_{p+1} - f_p|}{|f_{p+1}| + |f_p|} \leqslant \varepsilon, \qquad (10)$$

where ε is a small quantity (e.g., $\varepsilon \leqslant 0.1$). Better results should be obtained by complex criteria for the discrimination of coinciding elements in the conjugate images, such that the character of the images is taken into account, not only at a given point, but also in its surrounding neighborhood.

In coherent radiation the wave field on the surface Σ_0 is characterized at every point by just two parameters, the amplitude and phase, and there is no information on the total propagation time of the waves. This is enough for construction of the images of diffracting elements and the images of virtual sources for extended boundaries. In the construction of an image by the D-conversion (6-7) every wave reflected from a smooth boundary in the medium will provide a set of boundaries of various configurations, which fill the entire image space, but only one of them is the true one, the rest being fictitious. For elimination of the fictitious boundaries it is necessary to use conjugate images obtained for different positions of the oscillation source, as well as with the use of coherent radiation at a series of discrete frequencies. Under these conditions the number of conjugate images, which is equal to the product of the number of source positions by the number of discrete frequencies, can be made large enough to ensure discrimination of the true boundaries in the images.

The practical realization of D-conversion is possible by at least two routes. On the one hand, digital and analog computers can be used, which implement relation (7) in one form or another for $r = r_{eff}$. Then the time argument of the lattice function has the form $t - r_{eff}/c_S = t - t_{MS}$, where t_{MS} is the diffracted wave propagation time computed by Eq. (1) at every point M of the image space for all detection points on the surface Σ_0. From the computed times on the wave diagram, which is stored in the computer memory, the signals whose sum refers to point M of the image are read out. For observations along a line (profile) only one sum is computed in Eq. (7). It is also possible on computers, particularly those of the analog type, to simulate the superposition of elliptic waves emanating from points S'. The addition of these waves at every point of the image space must be carried out in an accumulator device. It is important to note that algorithms involving the computation of the summation signal separately for every point of the image are more versatile, permitting additional processing of the summation signal, for example, in the comparison of conjugate images with the use of expressions (8-10), as well as in the solution of more complex problems in the construction of images with selection of the optimum value for the wave propagation velocity in an inhomogeneous medium.

General-purpose digital computers used for the construction of images must have a large-capacity operational memory, particularly for the processing of conjugate images, and they must have high-speed operation. The wave diagrams must be represented in digital form with time quantization, and it is advisable at the computer output to use a two-coordinate plotting device. The main difficulty here is associated with the insufficient operational storage capacity of most computers. In order to diminish the influence of this factor one can use peripheral drum storage or data compaction in the memory cells (four to six signals per cell), taking advantage of the fact that the dynamic range of the signals detected in imagery is typically 40–60 dB. Under these conditions it becomes possible to construct single and conjugate images for the majority of problems encountered in practice. In addition, for the construction of the images it is possible to use special-purpose analog and analog-digital hybrid computers of various types with higher operating speeds and lower accuracy than digital computers.

The second approach is physical reconstruction of the wave field in a model by means of equivalent sources allocated in the model as the detectors used for recording the oscillations. For the implementation of this technique the elliptic waves may be approximately replaced by spherical waves of radius equal to the following (from the triangle OMS' in Fig. 1 with the stipulation that point M is below point S'):

$$r_{s'} = \frac{1}{2}\sqrt{r^2_{eff} - r_s^2} = \frac{1}{2}\sqrt{c_s^2 t^2 - r_s^2}, \qquad (11)$$

where $r_S = \sqrt{x_S^2 + y_S^2}$. It is obvious that this is equivalent to the introduction of time delay in the signals to be reconstructed from the wave diagram. For the reconstruction of the wave field in the model the signals are reconstructed simultaneously for all the equivalent radiators with the required acceleration for observance of the similarity criterion (4) and delays according to Eqs. (11). The waves generated by the individual sources are superimposed on one another and form a wave field, which at time t = 0 is fixed as the image. For inhomogeneous media the time scale in the model varies in correspondence with (4), so that the time t = 0 is shifted for different intervals of the recording on the wave diagram. Under these conditions it is convenient to construct the image in sections with gradual transition from one section to the next by repetitive reconstruction of an interval of the recording of definite length with stroboscopic illumination of the model. This makes it possible to obtain a slowly drifting or stationary image of a definite volume of the medium. If acoustic radiators

of dimensions comparable with the wavelength are used in the model, the waves excited in the model will have approximately elliptic fronts, thus improving the approximation. For readout of the signals from the wave diagram with variable speed and time delays it is possible to use buffer storage devices, for example in the form of recirculating delay lines, charged-storage tubes, etc., or to read out the information stored on parallel tracks of a transparent carrier by means of an electron beam across the tracks according to a definite curve taking account of the time delays (11), and along the recording at the speed determined from Eq. (4). The signals reproduced in this case are distributed by means of electronic gates among their channels and are transmitted to the corresponding radiators. The main difficulties here are associated with the construction of a high-speed high-accuracy computer device for the solution of Eqs. (4) and (11). In order to decrease the computation speed significantly it is required insofar as possible to decrease the frequency of the oscillations in the model and to use special materials in it with low propagation velocities for longitudinal waves. With some complication of the reconstruction section the signals from a series of conjugate wave diagrams can be transmitted into the model. Both two- and three-dimensional models can be used to produce the images.

For the investigation of layered media it is customary to use images representing cross sections of the medium in a vertical plane perpendicular to the interfaces. The resulting images of reflecting boundaries appear as a series of parallel bands of elevated intensity (according to the number of total oscillations in the pulse), the first of which coincides with the boundary [17]. The image of an inhomogeneity has the same form, but its size is limited by the resolving power of the D-conversion.

In order to test the possibilities of discrete D-conversion we investigated the conversion of wave diagrams into images of a medium by computer techniques. One of the wave diagrams obtained in the model contained data from observations at 41 points which were spaced discretely at distances of 100 m (in the scale of the model) from one another along a straight line and symmetrically about the pulsed source. An image of the medium was plotted from the data of observations according to Eq. (7) in the plane XOZ (Fig. 2), in which the images of two inhomogeneities located symmetrically about the source at a depth z = 2000 m and about 600 m apart are distinctly seen. De-

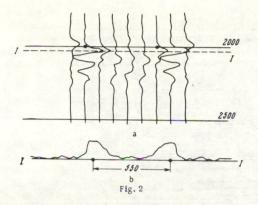

a

b

Fig. 2

spite the small number of detection points and their arrangement along a straight line, the images of the inhomogeneities are well pronounced and very accurately reflect the structure of the model (the actual positions of the inhomogeneities are represented in Fig. 2a by points spaced a distance of 540 m, corresponding to three wavelengths of the visible frequency of the pulse). It follows from the graph in Fig. 2b characterizing the amplitude distribution over the cross section I-I of the image that the D-conversion has a high resolving power. For an aperture angle of 45° and direct irradiation the size of the inhomogeneity in the image does not exceed 0.7λ at the 0.5 level, a result that is consistent with the resolving power of superior microscopes [6-8, 14].

A second example (Fig. 3a) corresponds to the wave diagram (Fig. 3b) obtained in the detection of seismic waves under field conditions for a layered-inhomogeneous medium. The source of oscillations was a detonation in a shallow pocket near the bottom surface at the time t = 0 on the wave diagram. Twenty-three detectors were placed on the ground surface symmetrically about the point of excitation at distances of 40 m from one another along the profile on a baseline of 880 m. For the observations the vertical component of the oscillations was recorded, primarily for longitudinal modes. From the data of the observations the image shown in Fig. 3a was plotted for a section of the medium in the vertical plane XOZ through the line of observations. In the given depth interval (900-4500 m) the investigated medium, according to deep-drilling data, consists of several distinct strata of sedimentary rock with diverse elastic wave propagation velocities and almost horizontal orientation on the pre-Cambrian crystalline substrate at a depth of 2850 m. The interfaces between the beds and their corresponding reflected waves

on the wave diagram are indicated by like indices (Figs. 3a and 3b). In the image of the medium cross section numerous extensive nearly-parallel reflecting boundaries are clearly traced. Below the surface of the substrate (Prt) a series of fictitious concave boundaries are observed, corresponding to waves multiply reflected in the upper part of the medium. We note that in the detection of the oscillations a programmed amplification regulator was used to offset the influence of spreading of the waves with depth. This fact was disregarded in the construction of the image, resulting in growth of the amplitude of the image signals with depth. This effect is easily eradicated if necessary.

The recording obtained in the observations spanned the frequency band from 25 to 60 cps (with the peak of the spectrum occurring at 35-40 cps) and was quantized for input to the computer in 4-msec intervals with a dynamic range of 36 dB. More than 20,000 readings of the signal amplitudes from the wave diagram are entered into the computer memory, which has 4096 cells; this is sufficient for the plotting of the image representing a cross section of the medium over an interval of 3500 m in depth.

It may be concluded on the basis of the foregoing that the construction of an image by D-conversion is an effective technique for the analysis of inhomogeneous media or media with smooth reflecting boundaries of considerable extent in any predetermined cross sections of the image space. Of special interest is the possibility of plotting images from the data of observations at a set of points situated discretely on some surface and even along a line. In all of these typical imagery problems the new information-processing technique has significantly greater possibilities than lens (holographic) systems or direct methods for processing of the observational data.

The D-conversion can be used for wave processes of a diverse physical nature over a very wide range of frequencies and wavelengths under the condition of detection of the corresponding oscillations on the surface of observation in reconstructable form. The detection of the oscillations is realized by the customary technical means for each type of process with the points of detection spaced at distances not exceeding half the wavelength ($\Delta x = \Delta y \leq \lambda_h /2$). The best-suited areas of application of the D-conversion so far are acoustic and radio-wave imagery in gases, liquids, and solids (ultrasonic and radio flaw detection, acoustic microscopy, seismic prospecting, underwater acoustics, radar, etc.).

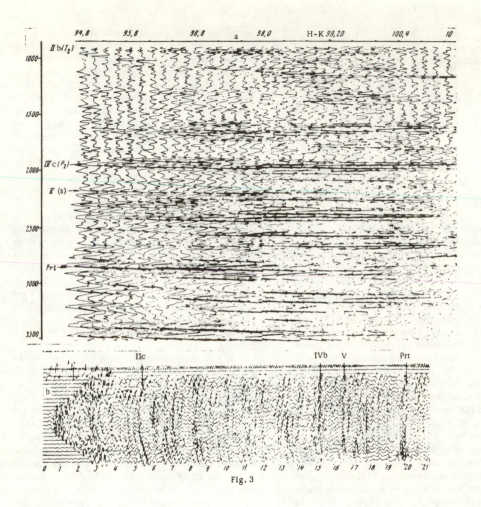

Fig. 3

LITERATURE CITED

1. V. S. Sokolov, Flaw Detection in Materials [in Russian], Gosénergoizdat, Moscow–Leningrad (1957).
2. I. I. Gurvich, Seismic Prospecting [in Russian], Gostoptekhizdat, Moscow (1960).
3. L. Bergmann, Ultrasonics [Russian translation], IL, Moscow (1957).
4. Underwater Acoustics [in Russian], Mir, Moscow (1965).
5. Physics of Sound in the Sea [Russian translation], Sovetskoe Radio, Moscow (1955).
6. G. S. Landsberg, Optics [in Russian], GTTI, Moscow (1954).
7. K. Michel, Fundamentals of Microscope Theory [Russian translation], GTTI, Moscow (1955).
8. A. Sommerfeld, Optics, Academic Press (1954).
9. E. Leit and J. Upatnieks, "Reconstructed wave fronts and communication theory," J. Opt. Soc. Am., 52, No. 10, 1123–1130 (1962).
10. G. W. Stroke, An Introduction to Coherent Optics and Holography, Academic Press, New York (1969).
11. E. Skudrzyk, Fundamentals of Acoustics, Vol. 1 [Russian translation], IL, Moscow (1958).
12. E. F. Savarenskii and D. P. Kirnos, Elements of Seismology and Seismometry [in Russian], GTTI, Moscow (1955).
13. E. L. O'Neill, Introduction to Statistical Optics, Addison–Wesley, Reading, Mass (1963).
14. Ya. I. Khurgin and V. P. Yakovlev, Methods of the Theory of Integer Functions in Radiophys-

ics, Communication Theory, and Optics [in Russian], Fizmatgiz, Moscow (1962).

15. Ya. Z. Tsypkin, Theory of Linear Pulsed Systems [in Russian], Fizmatgiz, Moscow (1963).

16. C. Shannon, "Communication in the presence of noise," Information Theory and Its Applications [Russian translation], Fizmatgiz, Moscow

(1959), pp. 82-112.

17. Yu. V. Timoshin, "Interference analysis of seismic records," Automatic Processing and Conversion of Geophysical Information, Trans. Ukrainian Geological Prospecting Scientific-Research Institute (UkrNIGRI) [in Russian], No. 11, Nedra, Moscow (1965), pp. 13-32.

Reprinted from GEOPHYSICS, 35, 407-418.

COARSE GRID CALCULATIONS OF WAVES IN INHOMOGENEOUS MEDIA WITH APPLICATION TO DELINEATION OF COMPLICATED SEISMIC STRUCTURE†

JON F. CLAERBOUT*

The multidimensional scalar wave equation at a single frequency is split into two equations. One controls the downgoing transmitted wave; the other controls the upcoming reflected wave. The equations are coupled, but in many reflection seismology situations the transmitted wave may be calculated without consideration of the reflected wave. The reflected wave is then calculated from the transmitted wave and the assumed velocity field. The waves are described by a modulation on up- or downgoing plane waves. This modulation function is calculated by difference equations on a grid. Despite complicated velocity models (steep faults, buried focus, etc.), the grid may be quite coarse if waves of interest do not propagate at large angles from the vertical. A one-dimensional grid may be used for a two-dimensional velocity model. With approximations, a point source emitting waves spreading in three dimensions may be included on the one-dimensional grid. Calculation time for representative models is a few seconds. Phenomena displayed are interference, spherical spreading, propagation through focus, refraction, and diffraction. Converted waves are neglected.

A procedure is suggested for the construction of a depth map of reflectors from observations at the surface. Assuming a velocity model, we may integrate the downgoing wave away from a surface source. Likewise, the upcoming wave may be approximately integrated back down into the earth. Since reflection coefficients are real, the ratio of upcoming to downgoing waves tends to be real at a reflector. An example is given in two dimensions which shows that this ratio over a dipping bed gives the dip correctly independent of source/receiver-group offset.

INTRODUCTION

In this paper, we shall describe a rapid means of numerically integrating the wave equation. Our integration technique leads to a seismic reflection data processing scheme which is a radical departure from existing schemes based on ray theory. Being very different in concept, the method is expected to avoid certain computational and theoretical difficulties that occur when ray theory is applied. The frequent occurrence of geological structures which show steeply dipping, faults, buried foci, diffractions, and the like on record sections give impetus to this study. Analysis of these structures by ray theory leads to complicated programs whose results may not even be correct in principle. In a finite difference approach to the wave equation, most of the problems that occur in the application of ray theory do not arise. We often associate finite difference schemes with large computer memories and computer costs. By means of an apparently novel approach to the finite difference formulation of the scalar-wave equation, we need only a one-dimensional computer memory vector in order to compute the amplitude and phase of waves spreading in a two-dimensional inhomogeneous medium. In fact, none of the figures presented in this paper required more than 10 sec of computer time.

In the first four sections, assuming space vari-

† Presented at the 39th Annual International SEG Meeting, Calgary, Alberta, Canada. Manuscript received by the Editor October 13, 1969; revised manuscript received January 23, 1970.

* Geophysics Department, Stanford University, Stanford, California.

able material properties of common geologic occurrence, we try to get approximate solutions to the wave equation. We could solve an initial value problem for the full vector wave field in a complicated model by difference approximations to the full wave equations (Boore, 1969). The accuracy would become arbitrarily good if the step size were reduced and the grid were enlarged. Computer technology, however, limits the accuracy to conform with our budget, i.e., very small step size implies high computing costs. Rather than do the exact problem on a very crude grid, we make some approximations in principle which enable us to go to a much finer grid for the same computation effort. Compromises in principle enable us to treat a more complicated inhomogeneous structure.

Assumptions made in what follows will be distressing to layered media theorists. However, because of our assumptions, we can deal with complicated inhomogeneity which the theorist cannot handle. When we make restrictive assumptions, it is always with an eye to computational practicality which, in effect, enables us to handle a more complicated structure. Though perhaps outrageous from the layered media viewpoint, what we assume should be compared with what is assumed in currently popular programs for handling inhomogeneous structure. Unlike calculations based on ray theory, our methods will handle diffractions correctly. Unlike the situation in classical diffraction theory, our rays are not all straight lines. In contrast to what a straightforward application of difference equations requires, our computational requirments are modest. Examination of Figures 1 through 4 will give an idea of the kinds of calculations which can be made in 5 to 10 sec. The figures were all made with the

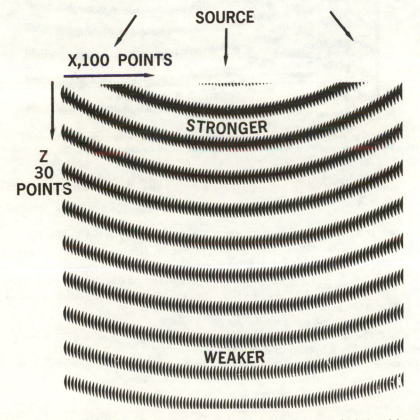

FIG. 1. An expanding monochromatic cylindrical wave. The wavefronts are concentric circles of decreasing amplitude. The computation begins with an analytic solution at the top of the figure in a 100 point linear grid. Using difference equations, we stepped the grid downward, 30 steps making up the whole figure. About 6 complex multipliers are required per point; this amounts to about 5 sec of time on our computer. The display is the x-z plane, although a multichannel seismogram plotter has been used.

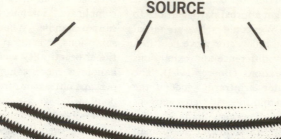

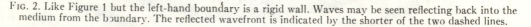

FIG. 2. Like Figure 1 but the left-hand boundary is a rigid wall. Waves may be seen reflecting back into the medium from the boundary. The reflected wavefront is indicated by the shorter of the two dashed lines.

same computer program. It is, of course, an easy task to input much more complicated velocity models.

The sections on computer algorithms, three-dimensional problems-on-a-one-dimensional grid, and reflections contain important though perhaps tedious details, but the section on the inverse problem introduces some novel concepts in data interpretation. The process of normal moveout correction may be done without reference to Dix hyperbolas. In fact, the beds may be dipping or warped, so that the idea of a hyperbola does not even apply. In concept a completely migrated map of reflectors may be made automatically with data from only one source with many receivers. Other sources serve as a check on the velocity model or, alternatively, provide the means for iterative improvement of the velocity model. It will be a challenge to see whether these ideas can be successfully applied to field data.

THE COARSE GRID FORMULATION

Let us begin by neglecting the conversions of pressure waves to shear waves. In reflection seismology such converted waves are observed infrequently. Hence, we shall start with a development of the scalar wave equation. Pressure p is equal to the product of the incompressibility K, and the convergence of displacement $\mathbf{u}$ (negative of the divergence); i.e.,

$$p = -K\nabla \cdot \mathbf{u}. \qquad (1)$$

By Newton's law the product of density ρ and the acceleration $\partial^2 \mathbf{u}/\partial t^2$ equals the negative of the pressure gradient:

$$\frac{\rho \partial^2 \mathbf{u}}{\partial t^2} = -\nabla p. \qquad (2)$$

We take the second time derivative of (1),

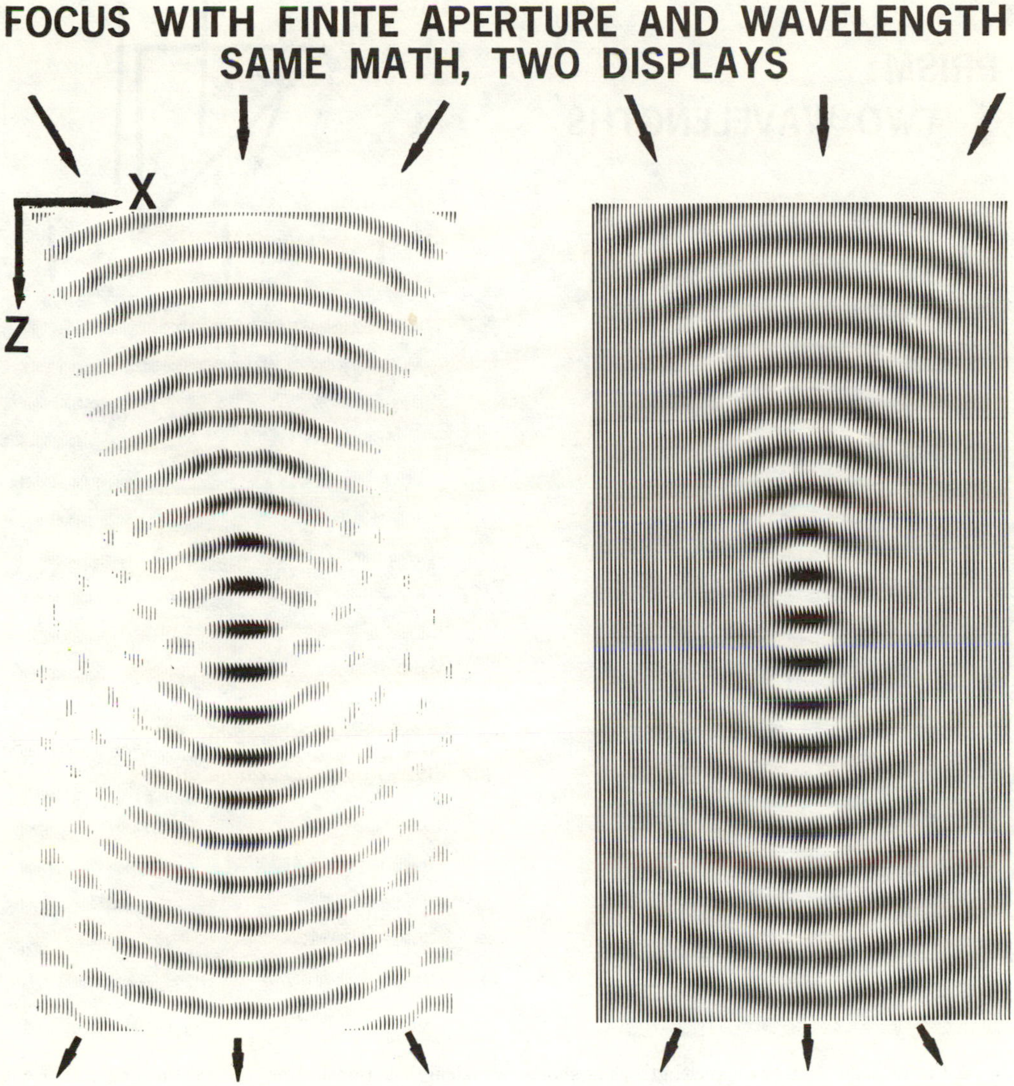

FOCUS WITH FINITE APERTURE AND WAVELENGTH
SAME MATH, TWO DISPLAYS

FIG. 3. Focus with a finite aperature and a finite wavelength: two different displays of the same mathematics. A plane wave coming from above goes through a converging cylindrical lens at the top boundary of the figure. The waves come through a focus at the center of the figure. The $\pi/4$ wavelength stretch on either side of the focus given by the asymptotic relation

$$H_0^{(1)}(r) \sim (2/\pi r)^{1/2} e^{i(r - \pi/4)}$$

may be verified with a measuring device. Some diffracted energy may be seen reflecting off the rigid side walls.

$$\frac{\partial^2 p}{\partial t^2} = - K \nabla \cdot \frac{\partial^2 \mathbf{u}}{\partial t^2} . \qquad (3)$$

Next we divide (2) by ρ and introduce (2) into the right side of (3) to get

$$\frac{\partial^2 p}{\partial t^2} = K \nabla \cdot \frac{\nabla p}{\rho} \qquad (4)$$

We temporarily assume that the density is space independent. In fact, space variation of density is usually much less important than space variation of velocity. Denoting the wave velocity by $c = \sqrt{K/\rho}$, we get

$$0 = \nabla^2 p - \frac{1}{c^2} \frac{\partial^2 p}{\partial t^2} . \qquad (5)$$

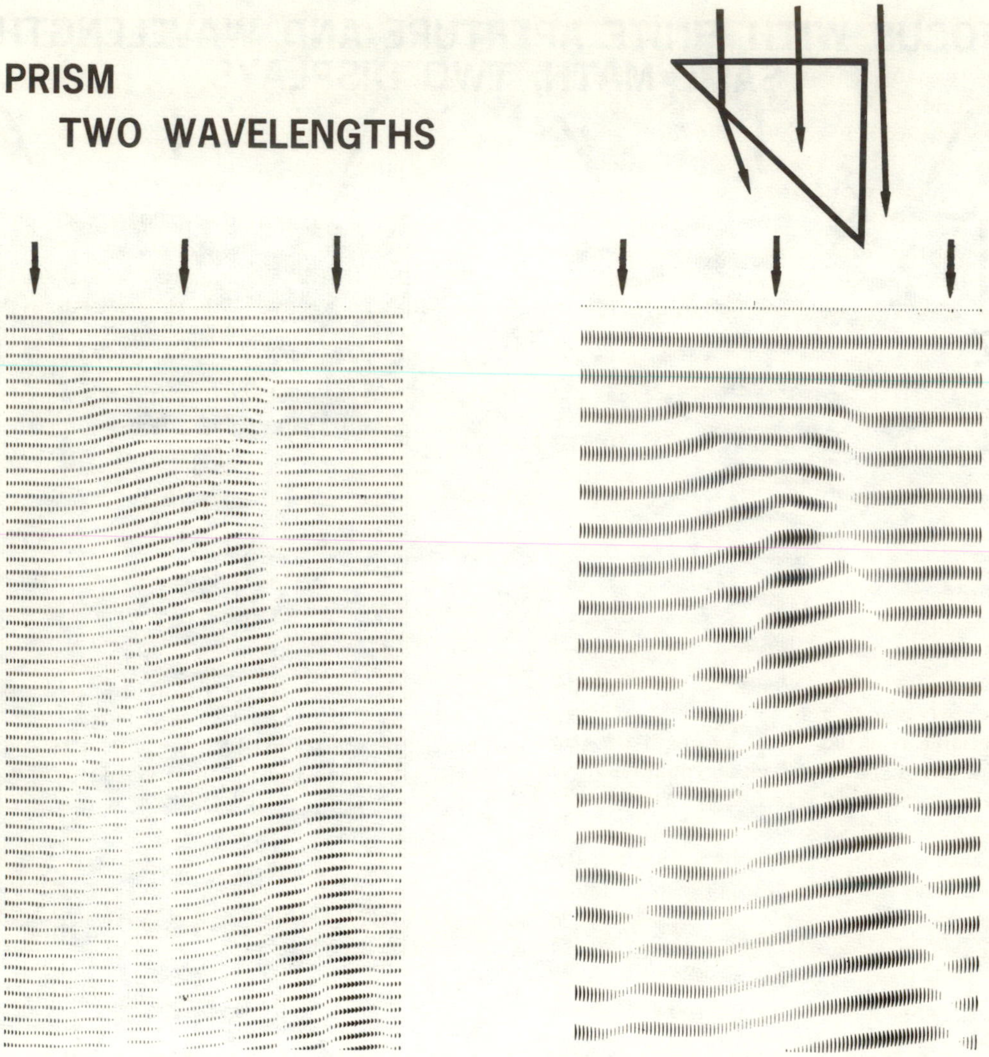

PRISM
TWO WAVELENGTHS

FIG. 4. Embedded prism, two wavelengths. The shorter wavelength emphasizes ray theory; the longer wavelength emphasizes diffraction. A plane wave is incident from above into a 45 degree right prism. The wavelength is shorter inside the prism. The bent ray emerging from the prism is made clearly visible by the angle of the emergent wavefront. There is a geometrical shadow under the left side of the prism. At the bottom right, two rays interfere, causing amplitude modulation. At the vertical boundary on the right side of the right prism (where ray theory is particularly inappropriate), a wave is scattered into the prism; and the vertical boundary becomes depleted of energy.

Now the velocity c may be an arbitrary complicated function of space, but we intend to deal in situations like those encountered in reflection shooting where the waves of interest do not propagate at great angles from the vertical (or any chosen preferred direction). For simplicity, we consider the two-dimensional case. If c is constant, equation (5) admits plane wave solutions of the form

$$p = p_0 e^{imz+ikx-i\omega t}. \qquad (6)$$

By substituting (6) into (5), we find that if $\omega^2/(k^2+m^2) = c^2$, then p_0 is a constant. The case $k=0$, $m=+\omega/c$ is that of a plane wave traveling in the plus z direction. Let us represent the downgoing wave from a surface explosion by

$$p = p^+(x, z)e^{imz-i\omega t}. \qquad (7)$$

At some distance underneath the explosion, the wave is approximately planar, so that $p^+(x, z)$ is slowly variable and may be represented on a coarse grid. Off to the sides, a finer grid is required; but we may consider situations where we do not need to look very far off to the side. Subsequent approximations will restrict the validity of our computations to waves traveling within 30 degrees of the z axis. Since we suppose that there may be velocity inhomogeneities, reflected waves will arise; and a more general solution to (5) looks like

$$p = (p^+(x, z)e^{imz} + p^-(x, z)e^{-imz})e^{-i\omega t}, \quad (8)$$

where p^+ and p^- are slowly variable. However, (8) may be rewritten as

$$p = (p^+(x, z) + p^-(x, z)e^{-i2mz})e^{imz-i\omega t}$$
$$= \tilde{p}(x, z)e^{imz-i\omega t}. \quad (9)$$

A problem now arises that, unless p^- is very small, $\tilde{p}(x, z)$ becomes a very rapidly varying function. Hence we will not be able to use the scalar wave equation on a coarse grid if there are appreciable reflected waves added to the transmitted waves. Thus we seek an equation governing the transmitted wave only. Our equation should not accept the reflected wave as a solution, since that would force us to use a fine grid. Later we can solve for the reflected wave on another coarse grid. We just don't wish to treat the sum of the two waves on the same grid. Our refusal to treat both waves together is somewhat contradictory, since, when a reflected wave is again reflected it becomes a transmitted wave: the calculation of the transmitted wave can't really be separated from the calculation of the reflected wave. However, a common geophysical situation is one where the transmitted wave is of much greater strength than the reflected wave. The reflection of the reflection is often completely negligible. We will neglect it now and later show how it may be picked up if necessary.

Let us write down (7), the transmitted wave, and some of its z derivatives (denoted by subscripts).

$$p = p^+e^{imz-i\omega t},$$
$$p_z = (p_z^+ + imp^+)e^{imz-i\omega t}, \quad (10)$$
$$p_{zz} = (p_{zz}^+ + 2imp_z^+ - m^2p^+)e^{imz-i\omega t}.$$

The wave equation (5) in two-dimensional Cartesian coordinates is

$$p_{xx} + p_{zz} - c^{-2}p_{tt} = 0. \quad (11)$$

Introducing (10) into (11) and cancelling $e^{imz-i\omega t}$ from each term, we get

$$p_{xx}^+ + p_{zz}^+ + 2imp_z^+ - m^2p^+ + p^+\omega^2/c^2 = 0. \quad (12)$$

It is clear that, if the amplitude modulation is slow, the p_{zz}^+ term will be much smaller than the $2imp_z^+$ term, so that we might consider dropping the p_{zz}^+ term for convenience. In fact, as we will see, dropping the p_{zz}^+ term is just what we need to do in order to suppress reflected waves altogether. Define

$$\epsilon(x, z) = \omega^2/m^2c^2(x, z) - 1. \quad (13)$$

In a homogeneous medium, we choose m so that ϵ vanishes. Otherwise m is chosen so that the spatial average of ϵ is small. Keeping ϵ small is necessary to keeping the grid coarse.

$$p_{xx}^+ + 2imp_z^+ + \epsilon(x, z)m^2p^+ = 0. \quad (14)$$

This is the equation which we intend to solve numerically on a grid for the transmitted wave. The procedure is to solve for p_z^+

$$p_z^+ = (i/2m)(p_{xx}^+ + \epsilon m^2p^+). \quad (15)$$

We then approximate the derivative p_z^+ as the difference

$$(p^+(x, z + \Delta z) - p^+(x, z))/\Delta z \simeq p_z^+$$
$$= (i/2m)[p_{xx}^+(x, z) + \epsilon m^2p^+(x, z)]. \quad (16)$$

Given p^+ at some fixed z_0 for all x, we may use (16) to extrapolate downward to any z. An analytic solution may be used near a source. Figures 1–4 show some examples of solutions based on our procedure. To see that we have indeed suppressed all upcoming reflected waves, we only need note the surprising fact that (14) is the Schroedinger equation where z plays the usual role of time t. The Schroedinger equation in the usual quantum mechanical notation is

$$\frac{h^2}{8\pi^2m_0}\psi_{xx} + \frac{hi}{2\pi}\psi_t - V(x, t)\psi = 0, \quad (17)$$

where h is Plank's constant, $\psi^*\psi$ is the probability function, m_0 is mass, and $V(x, t)$ is the potential. Physically it is clear that a future change in the potential $V(x, t)$ cannot affect the present value of the probability function. Likewise, a perturbation in the seismic velocity $c(x, z)$ at a depth z_1 cannot affect the downgoing wave at $z_0 < z_1$ if reflected waves are excluded.

Now supposing that p^+ has been calculated at all points in space, let us calculate p^-. We insert the trial solution (8) into the wave equation (5), cancel the exponential, and subtract equation (14). Omitting the p_{zz}^- term, we get

$$p_{xx}^- - 2imp_z^- + \epsilon(x, z)m^2 p^- = -p_{zz}^+ e^{i2mz}. \quad (18)$$

Equation (18), which may be used to find p^-, is like equation (14) used to determine p^+ except that the term involving i has taken a minus sign (which disappears if we change the sign of z), and there is a source term dependent on p_{zz}^+. The upcoming wave p^- is assumed to be zero at great depth for all x. Solving (18) for p_z^-, we find

$$(p^-(x, z - \Delta z) - p^-(x, z))/\Delta z$$
$$\simeq -p_z^- = i/2m(p_{xx}^- + \epsilon m^2 p^- + p_{zz}^+ e^{i2mz}). \quad (19)$$

As we come up from depth with equation (19), p^- remains zero until we hit the first reflector; that is, the first place where p_{zz}^+ is nonzero. Equation (19) then brings p^- up to the surface. We can obtain higher order reflections in the same way by noting that the dropped p_{zz}^- term and the p_{zz}^+ term give rise to the infinite number of multiples.

Finally, we remark that it is not difficult to include the effect of space variable density as well as space variable velocity. For space variable density, the formula for numerical integration analogous to (15) is found to be

$$p_z = i/2m$$
$$\cdot (p_{xx} + (\epsilon m^2 + im\rho_z/\rho)p - \rho_x/\rho p_x). \quad (20)$$

Formulas for the reflected waves are similar.

COMPUTER ALGORITHMS

Although the difference approximations (16) or (19) approach the correct limiting differential equation, a more careful analysis leading to a slightly more complicated procedure really pays off by letting us use quite large steps of Δx and Δz. We begin by defining the difference notation,

$$p(x, z) = P_{k\Delta x}^{n\Delta z}, \qquad (n, k \text{ integers}).$$

The obvious, so-called explicit scheme, for solving (15) is

$$P_k^{n+1} - P_k^n = (im\epsilon\Delta z/2) P_k^n$$
$$+ (i\Delta z/2m\Delta x^2)(P_{k+1}^n \quad (21)$$
$$- 2P_k^n + P_{k-1}^n).$$

If we consider a plane wave going in the z-direction, we may omit the last term. Let $a = m\epsilon\Delta z/2$. Equation (21) becomes

$$P_k^{n+1} - P_k^n = iaP_k^n, \quad (22)$$

or

$$P_k^{n+1} = (1 + ia) P_k^n. \quad (23)$$

If we fail to go all the way to the limit $\Delta Z \to 0$, this equation incorrectly states that the magnitude of P increases by the factor $|1 + ia|$ for each step in the z-direction. An alternative approach is to put P_k^{n+1} instead of P_k^n on the right side of (22). Then we get the so-called implicit scheme,

$$P_k^{n+1} - P_k^n = iaP_k^{n+1}, \quad (24)$$

or

$$P_k^{n+1} = P_k^n/(1 - ia). \quad (25)$$

The implicit scheme (25) gives nearly the same phase change going from P_k^n to P_k^{n+1} as does the explicit scheme, but it predicts a decreasing magnitude instead of an increasing magnitude. The truth is that the magnitude does not change. This is what we get with the mixed scheme,

$$P_k^{n+1} - P_k^n = (P_k^{n+1} + P_k^n)ia/2, \quad (26)$$

or

$$P_k^{n+1} = [(1 + ia/2)/(1 - ia/2)]P_k^n. \quad (27)$$

Now let us suppose $\epsilon = 0$, so that $a = 0$; and let us focus attention on the second x derivative term in (21). We have the explicit scheme,

$$P_k^{n+1} - P_k^n = (i\Delta z/2m\Delta x^2)(P_{k+1}^n$$
$$- 2P_k^n + P_{k-1}^n). \quad (28)$$

To discover its defect, we introduce the trial solution of a plane wave propagating at a small angle from the z-axis,

$$P_k^n = P^n e^{i\eta k}, \quad (29)$$

into (28), getting

$$P^{n+1} = P^n \left[1 + \{ i\Delta z/(2m\Delta x^2) \}(e^{i\eta} - 2 + e^{-i\eta}) \right]$$

$$= P^n \left[1 - 2i\Delta z \sin^2(\eta/2)/(m\Delta x^2) \right]. \tag{30}$$

Since the magnitude of the factor on the right is always greater than unity, we have the unfortunate result that off-axis waves are amplified unless Δz is arbitrarily small. Furthermore, the farther off axis the waves get, the larger they become. We are led to consider an implicit scheme formed by replacing n by $n+1$ on the right side of (28). Letting $b = \delta z/(2m\delta x^2)$, we get

$$-ibP_{k+1}^{n+1} + (1 + 2ib)P_k^{n+1}$$
$$-ibP_{k-1}^{n+1} = P_k^n. \tag{31}$$

Expression (31) is a set of tridiagonal simultaneous equations for the unknowns P_k^{n+1}. We may see by means of the plane wave trial solution (29) that the implicit scheme (31) attenuates the off-axis waves. Attenuating the waves is certainly better than amplifying them. We introduce a mixed scheme with a parameter θ which takes us continuously from the explicit to the implicit scheme.

$$P_k^{n+1} - P_k^n = ib(P_{k+1}^{n+1} - 2P_k^{n+1}$$
$$+ P_{k-1}^{n+1})\theta + ib(P_{k+1}^n$$
$$- 2P_k^n + P_{k-1}^n)(1 - \theta). \tag{32}$$

If $\theta = 1$, we have the implicit scheme; if $\theta = 0$, we have the explicit scheme; and if $\theta = 0.5$, we have a mixed scheme which neither amplifies nor attenuates off-axis waves. We now show a computing scheme of Gauss, (Richtmyer and Morton) which enables us to use the mixed scheme at very little more cost than the explicit scheme. Dropping the superscript $n+1$, we note that (32) with (26) included is of the form

$$A_{k+1}P_{k+1} + B_kP_k + C_{k-1}P_{k-1} = D_k. \tag{33}$$

Gauss's method proceeds by our writing down another equation with the same solution P_k as (33):

$$P_k = E_kP_{k+1} + F_k, \tag{34}$$

where E_k and F_k are yet unknown.

We write (34) with shifted index as

$$P_{k-1} = E_{k-1}P_k + F_{k-1}, \tag{35}$$

insert (35) into (33), and rearrange the terms to get

$$A_{k+1}P_{k+1} + B_kP_k + C_{k-1}(E_{k-1}P_k + F_{k-1}) = D_k;$$

or

$$P_k = \frac{-A_{k+1}}{B_k + C_{k-1}E_{k-1}}P_{k+1}$$
$$+ \frac{D_k - C_{k-1}F_{k-1}}{B_k + C_{k-1}E_{k-1}}. \tag{36}$$

If we compare (36) to (34), we see that they are the same and that we may develop E_k and F_k recursively from E_{k-1} and F_{k-1}.

Now let us consider boundary conditions. Suppose P_0 is prescribed. Then we may satisfy (34) with $E_0 = 0$, $F_0 = P_0$ and can compute all E_k and F_k. If P_N is prescribed, we may use (34) to calculate successively P_{N-1}, P_{N-2}, ... P_0. Another useful set of boundary conditions is to prescribe the ratios $r_1 = P_0/P_1$ and $r_2 = P_N/P_{N-1}$. If the the ratio has unit magnitude, this prescription effectively chooses the directions of plane waves entering or leaving at the boundaries. We begin by choosing $E_0 = r_1$, $F_0 = 0$. Next we compute E_k and F_k, and then solve the following set of equations for P_N. From (35), we have

$$P_{N-1} = E_{N-1}P_N + F_{N-1},$$
$$P_N/r_2 = E_{N-1}P_N + F_{N-1},$$
$$P_N = F_{N-1}/(1/r_2 - E_{N-1}).$$

From this point, we compute P_{N-1}, P_{N-2}, ... as before.

It may be noted that, although we have presented these computing schemes in terms of equation (15), there is no extra difficulty in handling reflected waves (18) or density variations (20). In practice, we often want to terminate our model with a boundary at which there are no reflections. We have not been able to express such an ideal boundary in a simple workable formula.

In higher dimensions, implicit schemes appear to be somewhat more costly than explicit schemes. Fortunately, due to the great interest in partial differential equations, investigations of implicit and explicit schemes comprise an area with an extensive literature.

A THREE-DIMENSIONAL PROBLEM ON A ONE-DIMENSIONAL GRID

In the field we always use point sources of seismic energy. The energy from these sources spreads out in three dimensions. The effective system often isn't fully three-dimensional though, because we usually try to shoot perpendicular to strike. If the velocity $c(x,z)$ is essentially two-dimensional, we may modify our two-dimensional calculation so that for a point source at $y=0$, we calculate pressure $p'(x,o,z)$ in that plane perpendicular to strike which contains the source.

Since the medium does not vary in the y direction, we may use a generalization of Snell's law which says that the y component of the slowness vector is constant on a ray. Specifically, if $\theta(x,z)$ is the angle between the z axis and the projection of a ray onto the y-z plane,

$$\frac{\sin \theta(x, z)}{c(x, z)} = \text{constant on a ray.}$$

If we take any ray for which θ is small, this expression simplifies to $\theta(x,z) \sim c(x,z)$. For the shot at the origin, the area over which energy spreads is proportional to the y coordinate of the ray. We have

$$\frac{d}{dz} y(x, z) = \theta \sim c(x, z), \quad (x, z \text{ on a ray}),$$

or   (37)

$$y(x, z) \sim \int_{\text{ray}} c(x, z)dz.$$

We would like to scale the pressure $p(x,z)$ by the inverse square root of y, so that the energy decrease due to spreading in the y direction would go as y. This scaling would not be exactly correct, since $y(x,z)$ is defined for (x,z) on a a ray; but x and z are independent variables in $p(x,z)$. Furthermore, $p(x,z)$ may consist of diffractions or many superposed rays at any point. If we compute $y(x,z)$ by integration at constant x rather than along a ray, we have a useful approximation if

$$\int_{\text{ray}}^{z} c(x, z)dz \simeq \int_{x=\text{const}}^{z} c(x, z)dz. \quad (38)$$

In other words, the approximation is good if either the ray is near vertical or $c(x,z)$ changes

slowly with x. Thus, if we let p' denote the pressure as reduced in amplitude by spreading in the third dimension we have

$$p' = p/\sqrt{y},$$

and the z derivative

$$p'_z = p_z/\sqrt{y} - (p/\sqrt{y})(y_z/2y).$$

Using equation (15), we get

$$p'_z = i/(2m\sqrt{y})(\epsilon m^2 p + p_{xx}) - p' y_z/2y,$$

or

$$p'_z = i/2m\left[(\epsilon m^2 + imy_z/y)p' + p'_{xx}\right], \quad (39)$$

which is an equation we may deal with by the methods described under Computer Algorithms. We merely introduce a complex index of refraction to account for energy loss into the third dimension.

REFLECTIONS

In calculating the reflected wave by means of equation (19), we must estimate p_{zz}. A number of problems arise. First of all, dropping the p_{zz} term not only suppresses reflections, but it also sets the transmission coefficient equal to unity. Furthermore, justification for dropping p_{zz} depends on ϵ being small. Actually, p_{zz} is of order ϵ^2 even if there are no reflectors at all. Also, in any attempt to estimate p_{zz}^+ on a coarse grid by a second difference operator, numerical problems arise since p^+ is much greater than p^-. Finally, there are severe problems if a reflector has curved surfaces which do not go neatly through the grid points, especially if the two-way traveltime phase between grid points is comparable to half a period.

Let us start with a one-dimensional analysis. From (12) and (13), we have

$$p_z = (i/2m)(\epsilon m^2 p + p_{zz}). \quad (40)$$

If we assume that to a first approximation p_{zz} may be neglected, we have

$$p_z = (im/2)\epsilon p. \quad (41)$$

Differentiating with respect to z, we have

$$p_{zz} = (im/2)(p\epsilon_z + \epsilon p_z). \quad (42)$$

Insertion of (42) into (40) gives

$$(1 + \epsilon/4)p_z$$
$$= (i/2m)(m^2(\epsilon + (i/2m)\epsilon_z))p. \qquad (43)$$

We define

$$\epsilon' = (\epsilon + i\epsilon_z/2m)/(1 + \epsilon/4). \qquad (44)$$

Recalling the p_{xx} term, we get

$$p_z = (i/2m)(\epsilon' m^2 p + p_{xx}). \qquad (45)$$

Thus we have an equation like (15) but which includes a transmission coefficient that may be different from unity. As was true for the approximation to spherical spreading, we need only incorporate a complex index of refraction. The foregoing analysis likewise suggests using (42) in equation (19) for the reflected waves. In the examples, we have indeed used

$$p_z^- = (-i/2m)(\epsilon m^2 p^- + p_{xx}^-$$
$$+ (im/2)\epsilon_z p^+ e^{i2mz}). \qquad (46)$$

Since p^+ and p^- are computed on coarse grids, certain problems may arise because $\exp(i2mz)$ usually will not be densely sampled. These problems do not arise if the reflection surfaces parallel the grid coordinates, but at any other reflection angle, something like Bragg scattering occurs off the grid itself. A partially satisfactory result is obtained if we treat a reflector, a delta function in ϵ_z at z_1, not as an impulse but as a sampled $\sin \pi(z_1 + n\Delta z)/[\pi(z_1 + n\Delta z)]$ function. Folding frequency problems still occur if $2m\Delta z > 1$; however the most satisfactory solution is the following: If a delta function in ϵ_z should occur at z_1 which is not on a grid point, the delta function is multiplied by $\exp[i2m(z_1 - n\Delta z)]$ and put at $n\Delta z$, the closest grid point.

Considering the usual uncertainty in the magnitude of the reflection coefficients, the slow variation of magnitude with angle from vertical, and the existence of converted waves, we do not believe it worthwhile at the present time to attempt to get the correct angular dependence of reflected wave magnitude.

THE INVERSE PROBLEM

By the forward problem, we mean that given the sources and the medium, we wish to find the waves. In this section, we discuss the inverse problem: namely, given the sources and the waves seen at the surface, we want to determine the structure

of the medium. In recent years in GEOPHYSICS, a considerable number of authors have given a reasonably complete picture of the one-dimensional inverse problem. Although our remarks on the multidimensional inverse problem will be considerably less complete than those past discussions, they are suggestive of data processing procedures which could become important. The key to the approach is that we may integrate the transmitted wave p^+e^{imz} and the reflected wave p^-e^{-imz} downward from the surface where they are known. If there is only one reflector, the transmitted and reflected waves must be in phase with one another at the reflector. Thus the reflector may exist only where the ratio $[p^-(x,z) e^{-imz}/p^+(x,z)e^{imz}]$ is real. The magnitude of the ratio equals the magnitude of the reflection coefficient. The ratio will turn out to be real at a good many places where there is no reflector, but this is merely an ambiguity which cannot be resolved at a single frequency. It is resolvable with more frequencies.

Let us consider the scheme in more operational detail. We begin with a single shot, many receivers, and an assumed crude velocity model. The data and the shot (an impulse) are Fourier transformed so that we may work with a single frequency at a time. The shot wave is integrated downward through the assumed velocity structure. Integration is started from the analytic solution e^{ikr}/r. In the event of a buried shot, an image source may be added for a negligible computation cost. The received wave may likewise be integrated downward from the geophone array. Here a problem arises in that we cannot do the integration for p^- exactly unless we assume we know dc/dz. Although knowledge of c is mathematically equivalent to knowledge of dc/dz, the two behave much like independent functions in this particular application. Essentially, c determines wave arrival times, and dc/dz determines reflected wave amplitudes. A crude model for c is a model in which dc/dz is effectively unknown. Thus we define $\hat{p}^-$ by dropping the dc/dz term in the equation for p^-, i.e.,

$$\hat{p}^-(z = 0) = p^-(z = 0),$$

$$\frac{d\hat{p}^-}{dz} = (-i/2m)(\epsilon m^2 \hat{p}^- + \hat{p}_{xx}^-).$$

Then, in a computer we sum over frequency

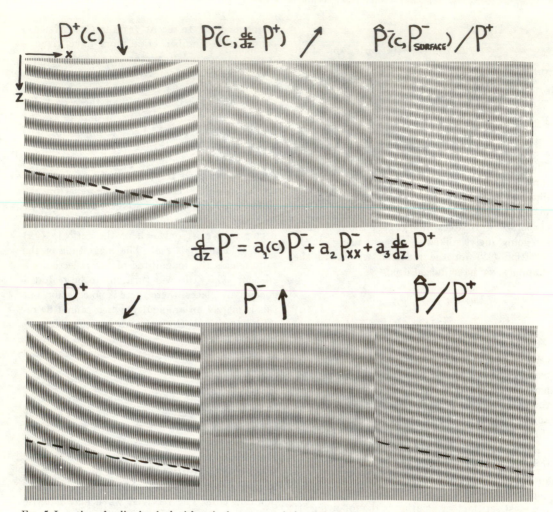

FIG. 5. Location of a dipping bed with a single source emitting a single frequency. Top and bottom represent two possible source locations. The left panel shows the downgoing wave p^+. The dashed line represents a transition from slow to fast velocity as can be seen by the greater wavelength below. The center panel represents the upcoming reflected wave p^-, which originates at the velocity jump. On the right panel is plotted the real part of the ratio $\hat{p}^-/p^+$. The right panel tells where downgoing and upcoming waves are in-phase. It gives the correct dip of the bed. With only a single frequency, the depth cannot be determined except to within multiples of a half wavelength. The estimated upcoming wave $\hat{p}^-$ is computed from the true upcoming wave p^-, observed at the surface, and the velocity c of the medium; dc/dz is not used.

some quantity like the ratio $[\hat{p}^-(x,z)/p^+(x,z)]$ e^{-i2mz}. Thus, in concept, we produce a depth section from the time section of a single shot. When several shots at different locations are used, they give independent estimates of the depth section. The depth sections will not be exactly the same, of course, and the ways in which they differ may be used to make an improved velocity model. The fact that the source location more or less cancels out in the ratio of upgoing to downgoing waves is somewhat surprising and is illustrated for the case of two different source locations over a dipping bed in Figure 5.

If some such scheme is to be applied to real data, it is probably wiser from a noise point of view to stack the product of p^-e^{-imz} and the complex conjugate of p^+e^{imz}. This product has the same phase as the forementioned ratio but has the advantage that it is never indeterminate and is small in regions where waves are small. When a stack is made over all frequencies, the process may be viewed in the time domain as follows: At every point in the depth section consider the upgoing and the downgoing waveforms. (They are never actually computed.) The important thing is the zero lag value of the crosscorrelation function. If it is nonzero, there is a reflector just below. The strength of the zero lag crosscorrelation function increases linearly with the reflection coefficient. Depth resolution improves with the shortness of the transmitted wave.

We should observe that the data processing scheme envisioned here eliminates the need of the conventional processes of normal moveout, common-depth-point stacking, and migration. Also, if we envision statics as a random timing error from trace to trace, the error may be smoothed out by a choice of $\theta > 1/2$ in the integration. Of course, any systematic static correction should be made in the usual way. It must be repeated that the data processing scheme we have described has been tested only on artificial data. Success on real data no doubt will involve an extra measure of shrewdness required by the quasi-statistical nature of real data. However we are optimistic because the process is a completely

linear operation on the data. Even the last step, stacking the cross product of the upgoing and downgoing waves, is linear because the down going wave is not a function of the observations. In our experience, linear stacks are statistically robust.

REFERENCES

Boore, D. M., 1969, Ph.D. thesis at Massachusetts Institute of Technology: Geophysics Department, MIT.
———— 1970, Love waves in nonuniform wave guides: Finite difference calculations: J. Geophys. Res., v. 75, p. 1512–27.
Richtmyer and Morton, 1967, Difference methods for initial value problems: New York, Interscience.

ACKNOWLEDGMENT

I was led to study the seismic use of numerical solutions to partial differential equations by the excellent results being obtained by David M. Boore in his Ph.D. thesis work. My work was principally funded by a matching grant of the National Science Foundation (GA 4495) to Stanford University. I wish to express my thanks to the Chevron Oil Field Research Company for the use of their seismic section plotter.

Reprinted from *Geophysics*, Journal of Royal Astronomical
Society, **26**, 285-293.

Extrapolation of Time-Dependent Waveforms
along their Path of Propagation

Jon F. Claerbout and Ansel G. Johnson

(Received 1971 July 1)

Summary

Time-dependent waveforms are commonly extrapolated in space by means
of rays and occasionally by means of diffraction integrals. It is possible to
extrapolate time-dependent waves in space with a partial differential
equation derived from the wave equation. There are stable numerical
approximations. An example illustrates a mechanism for ' signal-generated
noise ' which is consistent with observations.

1. Introduction

When a wave propagates in an inhomogeneous medium the waveform changes.
Given that the wave has been observed at a suitable number of points in space we
may attempt to solve two types of problems. First we may attempt to ascertain the
nature of material inhomogeneity along the wave paths, and second, we may attempt
to extrapolate the disturbance back to the source in an attempt to discover the
nature of the source. With few exceptions, the methods used during the past decade
for doing this kind of geophysical work may be summarized as follows: when wave
equations are to be used, separability is achieved by considering cases in which the
material inhomogeneity is a function of only one spatial co-ordinate. When higher-
dimensional inhomogeneity is so severe that it cannot be ignored, then the wave
equations are almost always specialized to ray theory. Ray theory is especially useful
when only the travel time is required. Although the amplitude may also be obtained
by ray theory it is often of marginal utility because amplitude measurement is made
ambiguous by changing waveforms. What we develop in this paper is a finite
difference approach to the wave equation which tracks the time dependent waveform
of a travelling wave in two-dimensionally inhomogeneous material. This is an
extension of earlier work done by one of the authors in the frequency domain.
Although the Fourier transform relates time-domain solutions to frequency-domain
solutions, there are several compelling practical factors which give impetus to this
study. When a waveform is small at certain times of interest and large at times which
are not of interest, then a satisfactory approximation to the Fourier integral may be
difficult to obtain even if values are obtained with good accuracy at many frequencies:
an example is the head wave. Another example occurs in reflection seismology where
the most interesting part of the waveform is the late-arriving weak echoes. Another
example is when the time function is of long duration but only a small portion of it is
of interest; this is usually the case with short-period earthquake seismograms where
there is never any hope of interpreting more than the wave packets which come
from identifiable phases.

Waves move quickly into a large volume of space. Given that it seems to require about 10 sample points per wavelength to achieve even modest computational accuracy and that a typical computer memory contains 100 000 memory cells, it is evident that some kind of practical limit is attained when a wave emitted by a point source in two dimensions has expanded to a radius of 15 wavelengths. This is grossly inadequate for most geophysical examples of air waves, water waves, and seismic waves. Simplification can often be achieved by not attempting to describe the entire volume V, but only a reduced volume V_1 which surrounds the path from the source to the receiver. If a reduced volume V_1 is to be used, care must be taken to avoid artificial reflections from the sides of the volume V_1. Further economy may be achieved if an even smaller volume, say V_2, moves with a wave packet along a wave path.

2. The differential equation

Let us begin the analytical discussion by transforming the scalar wave equation

$$0 = \frac{\partial^2 P}{\partial x^2} + \frac{\partial^2 P}{\partial z^2} - \frac{1}{c^2} \frac{\partial^2 P}{\partial t^2} \tag{1}$$

into a co-ordinate frame which translates along the z-axis at the wave speed c. For simplicity in discussion we will make various restricting assumptions. Specialists will recognize that most of these assumptions can be relaxed at the cost of a more complicated development. First, we have obviously chosen the z-axis as being along the path of interest. Second, we will neglect gradients of velocity c while retaining the space variation of c. This is a high-frequency approximation.

Energy which propagates with a component in the positive z-direction in a fixed frame will remain stationary or fall backwards with respect to a frame which translates along z at speed c. Let us choose for the co-ordinate transformation

$$\left.\begin{array}{l} x' = x \\ z' = ct - z \\ t' = t \end{array}\right\} \tag{2}$$

(see Fig. 1). Since we have chosen z' to be directed opposite to z we will have energy moving with a positive velocity component in either co-ordinate frame. Let P' denote the disturbance in the moving frame. We have

$$P(x, z, t) = P'(x', z', t'). \tag{3}$$

It will be convenient to use subscripts to denote partial derivatives. Obviously $P_x = P'_{x'}$ and

$$P_{xx} = P'_{x'x'}. \tag{4}$$

Also

$$P_z = P'_{x'} x'_z + P'_{z'} z'_z + P'_{t'} t'_z = -P'_{z'}$$

so

$$P_{zz} = P'_{z'z'} \tag{5}$$

and

$$P_t = P'_{x'} X'_t + P'_{z'} z'_t + P'_{t'} t'_t = cP'_{z'} + P'_{t'}$$

so

$$P_{tt} = c(cP'_{z'z'} + P'_{t'z'}) + cP'_{z't'} + P'_{t't'}$$

$$= c^2 P'_{z'z'} + 2c \, P'_{z't'} + P'_{t't'}. \tag{6}$$

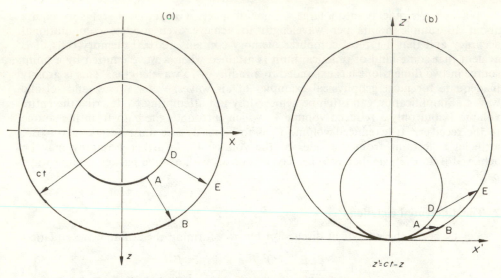

FIG. 1. Expanding spherical wave in fixed co-ordinates (left) and in co-ordinates which translate in the z-direction with the velocity of the wave (right).

Now we may insert (4), (5) and (6) into (1) to obtain

$$0 = P'_{x'x'} - 2c^{-1} P'_{z't'} - c^{-2} P'_{t't'}. \qquad (7)$$

Our main interest is with those waves which propagate with approximately the velocity of the new co-ordinate frame. In the moving frame such waves are Doppler-shifted near to zero frequency. This suggests omitting the $P'_{t't'}$ term from (7). Thus (7) becomes

$$P'_{t'z'} = (c/2)P'_{x'x'}. \qquad (8)$$

If $\partial/\partial_{t'}$ is replaced by $-i\omega$ then (8) is said to be in the frequency domain. This equation has been studied extensively in the frequency domain by one of the authors (JFC) in a number of earlier papers. Claerbout (1970a) considers the important details relating to computer implementation; Claerbout (1970b) shows the validity of the approximation (8) to be in the range of 20 degrees from the z-axis, and gives a method to extend the range to 90° in concept or 45° at the same computational cost as (8). Claerbout (1971) pays closer attention to space variation of the velocity $c(x,z)$. In this paper we develop a solution directly in the time domain.

To solve equation (8) in a computer the disturbance is defined initially in the $x'-z'$ plane and then time is augmented in steps of $\Delta t'$. A formulation more closely related to observations would be to assume the initial disturbance was measured as a function of x and t and then extrapolate the result along the wave path in the z direction. To achieve this we introduce the change of variables

$$x'' = x \qquad (9a)$$

$$z'' = z \qquad (9b)$$

$$t'' = t - z/c. \qquad (9c)$$

(see Fig. 2).

The new co-ordinate frame stays fixed in space relative to the old one, but time is function of position in the new frame (not unlike the difference between universal time G.M.T. and local solar time). When an observer moves in the $+z = +z''$ direction (west) time will seem to go slower. If he moves at velocity c then time stands

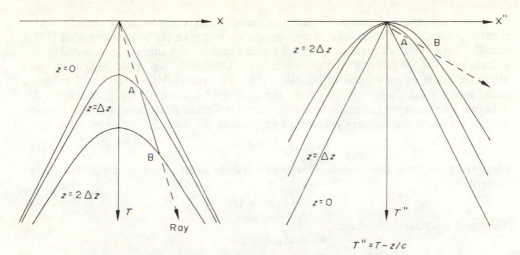

FIG. 2. A point source at $x = 0$, $z = 0$, $t = 0$. Hyperbolas at left indicate arrival times t at $z = 0$, Δz, and $2\Delta z$. When time is a function of position given by $t'' = t - z/c$ the arrival times t'' are as indicated on the right. Energy moves in the direction $+t''$, since on a wave front $z = ct \cos \theta$ and we have

$$t'' = t - z/c = t(1 - \cos \theta).$$

still. Referencing time with respect to the time of the earliest possible ray is a great computational convenience because it means that a wave at the source with onset at $t_0 = 0$ will at some distance $z_1 = z_1''$ have its onset at $t_1 = z_1/c$ in the old frame but in the new frame the onset is still at $t_0'' = 0$. This means the wave onset does not move off the finite, perhaps short, computational grid on which the wave packet has been defined. Define the disturbance in the new frame by P'', where

$$P(x, z, t) = P''(x'', z'', t''). \tag{10}$$

Proceeding as before we obtain

$$P_{xx} = P''_{x''x''} \tag{11}$$

$$P_{zz} = P''_{z''z''} - 2c^{-1} P''_{t''z''} + c^{-2} P''_{t''t''} \tag{12}$$

$$P_{tt} = P''_{t''t''}. \tag{13}$$

Inserting these into the wave equation (1) we obtain

$$P''_{t''z''} = (c/2)(P''_{x''x''} + P''_{z''z''}). \tag{14}$$

To see that the last term of (14) is small of higher order for waves travelling at small angles from the z-axis, recall that the solution to the wave equation for waves in the $+z$ direction is an arbitrary function $f(t - z/c) = f''(t'')$. Thus $\partial f''/\partial z''$ vanishes for a wave along the z'' axis. Neglecting $P''_{z''z''}$ we find that (14) reduces to

$$P''_{t''z''} = (c/2)P''_{x''x''} \tag{15}$$

which is the same as equation (8).

3. The difference approximation and a method of solution

Omitting primes from (8) or (15) we have

$$P_{tz} = (c/2)P_{xx}. \tag{16}$$

From a mathematical point of view this equation is completely symmetric with regard to t and z. For the sake of definiteness we will take the point of view of (15), namely, that P was defined at z_0 for all x and t and we are intending to use (16) to extrapolate the waveforms in the z-direction. Obviously we could also take the alternate point of view, that of equation (8). To express (16) in terms of sample time, as we must do for computation, the notion of a Z-transform is essential. A valuable introductory reference to Z-transforms is Treitel & Robinson (1964). First we put (16) into the frequency domain by replacing the time derivative by $-i\omega$.

$$-i\omega P_z = (c/2)P_{xx}. \tag{17}$$

Next we re-express the angular frequency variable ω in terms of the Z-transform variable

$$Z = \exp(i\omega\Delta t).$$

Taking the logarithm we have

$$i\omega\Delta t = \ln Z.$$

Using a well-known expansion for the logarithm we have

$$i\omega\Delta t = 2\left[\frac{Z-1}{Z+1} + \frac{1}{3}\left(\frac{Z-1}{Z+1}\right)^3 + \frac{1}{5}\left(\frac{Z-1}{Z+1}\right)^5 \cdots\right].$$

By retaining only the first term we restrict the validity of our results to waveforms sampled moderately densely (say eight points/wavelength; see Fig. 3). Thus we take

$$-i\omega\Delta t = 2(1-Z)/(1+Z). \tag{18}$$

If P is taken to be a function sampled in the time domain then its Z-transform (which on the unit circle is its Fourier transform) takes the form

$$P = \ldots p_{-1}Z^{-1} + p_0 + p_1 Z + p_2 Z^2 + \ldots. \tag{19}$$

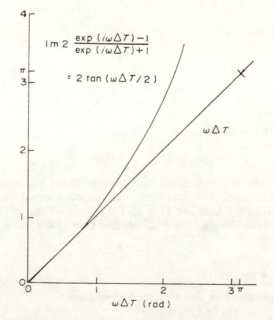

FIG. 3. This shows how the Z-transform approximation forces the data sampling rate to be at least four times the Nyquist sampling rate.

Without loss of generality for the application in mind we may ask P to vanish before $t = 0$, which means that (19) specializes to

$$P = p_0 + p_1 Z + p_2 Z^2 + \ldots = \sum_j p_j Z^j. \tag{20}$$

Putting (18) and (20) into (17) we have

$$\frac{2}{\Delta t} \frac{1-Z}{1+Z} \frac{\partial}{\partial z} P = \frac{c}{2} \frac{\partial^2}{\partial x^2} P. \tag{21}$$

As with all Z-transform equations one has an equation in the frequency domain if one regards Z as taking on all numerical values on the unit circle, and one has an equation at each point in the time domain if one identifies coefficients of various powers of Z. Next let us make the z co-ordinate discrete in equation (21); this follows the approach of earlier papers. Let $P(z)$ be denoted by $P^{n\Delta z}$ or more simply by P^n. Utilizing central differences (21) becomes

$$(1-Z)(P^{n+1} - P^n) = \frac{c\Delta t \Delta z}{4} (1+Z) \frac{\partial^2}{\partial x^2} \frac{P^{n+1}+P^n}{2}. \tag{22}$$

The reader will observe that to avoid defining $P^{n+1/2}$ we have used the average $(P^{n+1}+P^n)/2$. It is shown by Claerbout (1970a) that the more general form $\theta P^{n+1} + (1-\theta) P^n$ (where θ is the implicit/explicit parameter whose value lies between one-half and one) offers some advantage in attenuation of off-axis waves. This is an important parameter is practice but we set $\theta = \frac{1}{2}$ for simplicity since a satisfactory discussion of its use already was given by Claerbout (1970a).

Finally it remains to make (22) discrete with respect to the x-co-ordinate. Let p_j^n for fixed n (depth) and j (time) be called a vector because it contains an unwritten subscript which refers to variation in the x-direction. Then $\partial^2/\partial x^2$ is like a tri-diagonal matrix with the second-difference operator $(1, -2, 1)/\Delta x^2$ on the main diagonal. Denoting by T a tri-diagonal matrix with $(-1, 2, -1)$ on the main diagonal, (22) may be written

$$(1-Z)(P^{n+1} - P^n) = -\frac{c\Delta t \Delta z}{8\Delta x^2} (1+Z) T(P^{n+1}+P^n) \tag{23}$$

which we abbreviate by

$$(1-Z)(P^{n+1} - P^n) = -a(1+Z) T(P^{n+1}+P^n). \tag{24}$$

We bring terms depending on the disturbance at $(n+1)\Delta z$ to the left and the others to the right to get

$$[(1+aT) - Z(1-aT)] P^{n+1} = [(1-aT) - Z(1+aT)] P^n. \tag{25}$$

Recognizing that P^n and P^{n+1} represent Z-transform polynomials (20) of the disturbance at z and $z+\Delta z$, we may identify in (25) the coefficient of Z^{j+1}. For any $j > 0$ the coefficient is

$$(1+aT) p_{j+1}^{n+1} - (1-aT) p_j^{n+1} = (1-aT) p_{j+1}^n - (1+aT) p_j^n \tag{26}$$

which we may rearrange to

$$(1+aT) p_{j+1}^{n+1} = (1-aT) p_j^{n+1} + (1-aT) p_{j+1}^n - (1+aT) p_j^n. \tag{27}$$

Suppose for some particular n and j that everything on the right-hand side is known. Computationally the right side then represents a known vector of NX components where NX refers to the number of points which have been sampled on the x-axis. The factor $(1+aT)$ represents a tri-diagonal matrix of size NX. Thus we have a very sparse set of simultaneous equations which may be solved (by the method of Richtmyer & Morton, 1967, 198–201, for example) for the vector p_{j+1}^{n+1}. Of course boundary conditions are required at the ends of the vector. The authors have thus far used zero-slope conditions at the extremes on the x-axis. This is satisfactory when either the function or its x-derivative is sufficiently small at the boundaries.

It is clear that a satisfactory initial condition for the recursion (27) is knowledge of both p_j^0 for all j and p_0^n for all n, both of course for all values of the x-co-ordinate ($k\Delta x$). Let us consider an example of a point source at $t = 0$ and $z = -10\Delta z$. Clearly $p_0^n = 0$ for all $n \geqslant 0$. Also p_j^0 vanishes for enough values of j for the wave to have time to get to $z = 0$. After that p_j^0 is an arbitrary source waveform. By elementary geometry the x-dependence of p_j^0 must be worked out to conform to the well-known hyperbola in the $x - t$ plane.

4. Stability

In this section we will show that stability is assured for all values of Δx, Δz, and Δt. Stability is lost if the calculation is set up in an unnatural direction. For example, we know that waveforms move in the $+z'$ direction in the x'-z' plane. In other words information at $z' - \Delta z'$ will later be at z'. Thus it is reasonable and stable to calculate $P'(z')$ from present and past values of $P'(z' - \Delta z')$ but it is unreasonable and unstable to try to calculate $P'(z')$ from the present and past values of $P'(z' + \Delta z')$. To get $P'(z')$ from $P'(z' + \Delta z')$ it would be necessary to use present and future values.

First of all the stability of the recursion on x is assured because a is positive and the matrix $(1 + aT)$ has $1 + 2a$ on the main diagonal and $-a$ off the main diagonal, so the matrix is diagonally dominant.

Next let us consider the recursion on time. Eigenvalues for the T matrix may be shown to lie between zero and $+4$. Thus we may consider T in (25) to be replaced by an arbitrary number between zero and $+4$. To determine P^{n+1} from P^n it is necessary to divide (25) by the left-side-polynomial $[(1 + aT) - Z(1 - aT)]$. This polynomial will be minimum-phase (Treitel & Robinson 1964) because a is positive, T is positive, and the coefficient of Z^0 always dominates the coefficient of Z^1. Notice that the polynomial on the right-hand side of (25) will be definitely not be minimum phase, so that one cannot determine P^n from P^{n+1}.

Finally we show stability with regard to stepping in the z-direction. For this it is satisfactory to show that the transfer function of (25), namely

$$\frac{(1 - aT) - Z(1 + aT)}{(1 + aT) - Z(1 - aT)}$$

has unit magnitude for all frequencies (Z on the unit circle) and all horizontal wavelengths ($0 \leqslant T \leqslant 4$). The transfer function is of the form $(b - cZ)/(c - bZ)$. Since $Z = \exp(i\omega\Delta t)$ we have $Z^{-1} = \exp(-i\omega\Delta t)$. Therefore multiplying the transfer function by its conjugate we have

$$\frac{b - cZ}{c - bZ} \cdot \frac{b - c/Z}{c - b/Z} = \frac{b^2 + c^2 - bc(Z + 1/Z)}{b^2 + c^2 - bc(Z + 1/Z)} = 1.$$

From the point of view of stability with respect to extrapolation in the z direction it is irrelevant whether we go from P^n to P^{n+1} or the reverse. It is the recurrence on time which must go in one particular direction. Finally we remind the reader that z and t, n and j, and ω and k_z (vertical wave number) may be interchanged if they are interchanged *everywhere* throughout the discussion subsequent to equation (16).

5. Example—disturbed plane wave

While most theoretical solutions in wave propagation deal with highly symmetric waves (plane, cylindrical, etc.), nature is seldom so regular. The example shown in Fig. 4 represents a disturbed plane wave which might have been produced as shown in Fig. 5. At the bottom of Fig. 5 a plane atmospheric pressure wave is incident upon a series of circulating cells which tend to advance the wave in some regions while retarding it in other regions, thus producing the waveform shown at the top of Fig. 5. A somewhat similar situation is the phase grating in optics (Goodman 1968, p. 69), where the monochromatic solution is usually obtained at infinity.

The seven frames in Fig. 4 illustrate the subsequent development of the disturbed wave shown in frame No. 1. Note the frames may be thought of interchangeably in either steps of $\Delta t'$ or $\Delta z''$, as has been discussed previously. The most obvious development is that the energy spreads out as one moves down the figure. The single pulse of the top frame has become an extended oscillatory arrival by the last frame. As time goes on, less and less energy is in the first pulse and more and more is in the oscillatory tail. Note: in the figure the gain is adjusted with each frame to give better contrast. Also, as might be expected, the wave onset time, which is a dramatic function of x in the first frame, is nearly independent of x by the last frame. A surprising feature is that although energy moves back from the first arrival (see Figs 1 and 2) a point of constant phase in the wave tail moves forward toward the first arrival (a point of constant phase is marked by an X on the right side of each frame). Another clear feature of the wave tails is that dip (arrival time dependence on x) increases going down a single frame. The phase shift of the two-dimensional focus, which causes doublets to form, is clearly visible at the point in frame No. 2 marked with an A.

In order to represent a disturbance of infinite extent in x on a finite computer grid, we initialized the problem with a periodic disturbance having zero slope at the side boundaries. Zero-slope boundary conditions are then equivalent to infinite periodic extension in x. A value $a = \frac{1}{4}$ was chosen to given an appropriate variation in progressive frames, with each frame in Fig. 4 representing 5 computational iterations. The solution may be rescaled in several ways due to the interdependence in the constant a (see equation (24)) of $c\Delta t$, Δx and Δz. The calculation in Fig. 4 required about 0·5 min on the IBM 360/67.

It might be valuable to consider various data enhancement processes in the light of Fig. 4. In the process called 'beam-steering', observations as in Fig. 4 would be

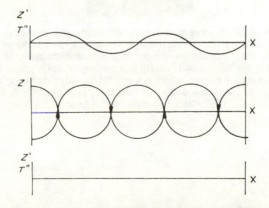

FIG. 5. One means of producing a disturbed plane wave. Incident plane wave at bottom is altered by a material inhomogeneity (centre), resulting in the disturbed wave front at top.

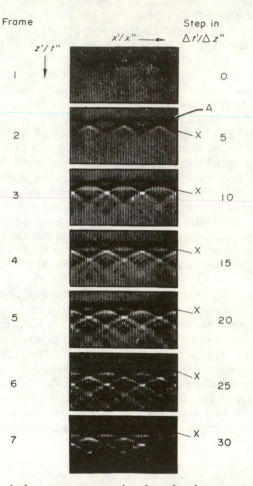

FIG. 4. Disturbed plane wave propagating through a homogeneous space in a moving co-ordinate system. Features of note are: (1) energy moves backwards and toward the sides; (2) the wave onset is a dramatic function of x near the beginning, but by the last frame is nearly independent of x; (3) phase of the wave tail (marked by x in each frame) moves forward; (4) the dip of the wave tail increases in a given frame as one scans down the frame; (5) in frame No. 2, the letter A indicates a two-dimensional focus.

summed over the x-co-ordinate in an effort to enhance signal and reject noise. Clearly beam-steering will enhance the first arrival while rejecting random noise. What it will also do is to tend to cancel the signal energy which resides in the oscillatory wave tails. If one is really interested in enhancing signal-to-noise ratio it would hardly seem desirable to use a processing scheme which cancels signal energy. As z' or t'' is increased the situation becomes increasingly severe since signal energy moves from the initial pulse toward the oscillatory wave tails. The central practical conclusion of this section is that what has often been regarded as 'signal-generated-noise' may turn out to be signal in a potentially valuable form. We can, indeed, expect dramatic results if we are able to learn how to design data enhancement techniques on entire waveforms rather than on the initial pulse alone.

Acknowledgment

This work was supported in part by the Advanced Research Projects Agency of the Department of Defense and was monitored by the Air Force Office of Scientific Research under Contract Number F44620-69-C-0073. The work was also supported in part by a matching grant of the National Science Foundation to Stanford University.

Geophysics Department,
Stanford University,
Stanford, California 94305

References

Claerbout, J. F., 1970a. Coarse grid calculations of waves in inhomogeneous media with application to delineation of complicated seismic structure, *Geophysics*, **35**, 407–418.

Claerbout, J. F., 1970b. Numerical holography in *Acoustical Holography*, Vol 3, ed. A. F. Metherell, Plenum Publishing Corp., New York.

Claerbout, J. F., 1971. Toward a unified theory of reflector mapping, *Geophysics*, **36.** 467–481.

Goodman, J. W., 1968. *Introduction to Fourier Optics*, McGraw-Hill Book Co., Inc., New York.

Richtmyer, R. D. & Morton, K. W., 1967. *Difference Methods for Initial Value Problems*, Interscience, New York.

Treitel, S. & Robinson, E. A., 1964. The stability of digital filters, *IEEE Trans. Geoscience Electronics*, **2**, 6–18.

Reprinted from GEOPHYSICS, 37, 741-768.

GEOPHYSICS

DOWNWARD CONTINUATION OF MOVEOUT-CORRECTED SEISMOGRAMS†

JON F. CLAERBOUT* AND STEPHEN M. DOHERTY*

Earlier work developed a method of migration of seismic data based on numerical solutions of partial differential equations. The method was designed for the geometry of a single source with a line of surface receivers. Here the method is extended to the geometry of stacked sections, or what is nearly the same thing, to the geometry where a source and receiver move together along the surface as in marine profiling. The basic idea simply stated is that the best receiver line for any reflector is just at (or above) the reflector. Data received at a surface line of receivers may be extrapolated by computer to data at a hypothetical receiver line at any depth. By considering migration before stacking over offset, it is found that certain ambiguities in velocity analysis may be avoided.

THE SIMPLEST CASE

For the sake of clarity we begin with the simplest case for which migration is a useful concept. Then realistic complications can be included in order of their practical importance. First, consider a two-dimensional model of the earth, y horizontal, z downward, in which the seismic velocity c is constant but the reflectors have arbitrary dips and curvatures. Here we neglect shear waves, although they can be treated by the method of Landers and Claerbout (1972). We also neglect multiples and energy spreading into the third dimension. Multiples will be treated in a later paper. Let there be sources and receivers uniformly spaced over the y axis at intervals of Δy. We can suppose that all of our shots are set off in unison. (Even if they are not we may synthesize it in a computer by adding seismograms together.) At some depth which is very great compared to Δy the semicircular wavefronts will combine together in the fashion of Huygens secondary sources to make a downgoing wave which is essentially a plane wave. In other words, at sufficient depth, point sources along the surface are indistinguishable from a surface line source. It turns out that in practice there are usually not enough shots and deep enough reflectors for the plane-wave approximation to be very good. Nevertheless, this is a useful starting point and we will return later to consider the fact that the downgoing wave is not a plane wave.

Considering the downgoing wave to be merely an impulsive plane wave simplifies the task of migration because we then may turn our attention entirely to the upgoing wave; since, according to the basic principle (Claerbout, 1971b)

---

† Paper presented at the 41st Annual International SEG Meeting, November 9, 1972, Houston, Texas. Manuscript received by the Editor November 19, 1971; revised manuscript received April 6, 1972.

* Stanford University, Stanford, California 94305.

"reflectors exist at points in the earth where the first arrival of the downgoing wave is time coincident with an upgoing wave."

A useful departure from our earlier work is that we are now migrating data from many shotpoints at a time, whereas previously each shotpoint was migrated separately before summation. The major effort in migration is to take the observed upgoing wave at the surface and project it back downward. A frequency-domain technique for this downward projection is given in Claerbout (1970 and 1971b). Here we will give a time-domain method because it leads to our central topic of downward continuation of moveout-corrected seismograms.

We are basically interested in projecting upgoing waves back into the earth. As pointed out in earlier papers, great economy and stability can be achieved by specializing the wave equation, which is second order and has both upgoing and downgoing solutions, to a first-order equation with only upgoing solutions. The first step is to re-express the scalar wave equation in a coordinate frame which translates upward with the speed c. In such a moving coordinate frame, the upcoming waves will be Doppler shifted to lower frequencies and the downgoing waves will be Doppler shifted to higher frequencies. Then a low-pass filtering type of operation can separate up from downgoing waves.

We have the scalar wave equation

$$0 = \frac{\partial^2 P}{\partial y^2} + \frac{\partial^2 P}{\partial z^2} - \frac{1}{c^2}\frac{\partial^2 P}{\partial t^2}, \qquad (1)$$

which we abbreviate as

$$0 = P_{yy} + P_{zz} - c^{-2}P_{tt}. \qquad (2)$$

We have the transformation to a coordinate frame translating upward with velocity c

$$y' = y, \qquad (3a)$$

$$z' = z + ct, \quad \text{and} \qquad (3b)$$

$$t' = t, \qquad (3c)$$

and we have the statement that the new coordinate frame contains the same wave disturbance as the old frame.

$$P(y, z, t) = P'(y', z', t'). \qquad (4)$$

To express the wave equation (2) in terms of the translating coordinate system, we may use the

chain rule for partial differentiation. Obviously, $P_y = P'_{y'}$ and

$$P_{yy} = P'_{y'y'}. \qquad (5)$$

Also

$$P_z = P'_{y'}y'_z + P'_{z'}z'_z + P'_{t'}t'_z = P'_{z'},$$

so

$$P_{zz} = P'_{z'z'}. \qquad (6)$$

The nontrivial differentiation is

$$P_t = P'_{y'}y'_t + P'_{z'}z'_t + P'_{t'}t'_t = cP'_{z'} + P'_{t'},$$

which on a second differentiation leads to

$$\begin{aligned} P_{tt} &= c(cP'_{z'z'} + P'_{t'z'}) + (cP'_{z't'} + P'_{t't'}) \\ &= c^2 P'_{z'z'} + 2cP'_{t'z'} + P'_{t't'}. \end{aligned} \qquad (7)$$

Now we insert (5), (6), and (7) into the scalar wave equation (2) and obtain the scalar wave equation in a translating frame,

$$0 = P'_{y'y'} - \frac{2}{c}P'_{t'z'} - c^{-2}P'_{t't'}. \qquad (8)$$

The rightmost term $P'_{t't'}$ is proportional to the square of the Doppler-shifted frequency of a wave. Thus, dropping this term may be expected to have little effect on upgoing waves but a drastic effect on downgoing waves. In fact, dropping the last term of (8) has the desired effect of eliminating the downgoing waves altogether as can be seen by comparing these results to earlier work (Claerbout 1970, 1971a, b). Simply dropping the term does have the undesired effect of limiting velocity accuracy to about a percent at 15 degrees. Economical procedures for obtaining better than a percent accuracy at 45 degrees also may be found in the earlier work.

Dropping the last term from (8), we have

$$P'_{t'z'} = \frac{c}{2}P'_{y'y'}. \qquad (9)$$

A computer algorithm for solving (9) is described in Claerbout and Johnson (1972) along with a more detailed discussion including accuracy, stability, and an air-wave example.

Knowing the form of the upcoming wave at the

surface of the earth, we can use equation (9) to project the upgoing wave backwards in time and, thus, find the upgoing wave at greater and greater depths. Using the basic principle "reflectors exist at points in the earth where the first arrival of the downgoing wave is time coincident with an upgoing wave," and considering the downgoing wave to be a delta function at time $t = z/c$, we get for the earth structure $S(y, z)$

$$S(y, z) = \int P(y, z, t)\delta(t - z/c)dt, \quad (10)$$

where $P(y, z, t)$ is found by solving (9) for $P'(y', z', t')$ and then transforming $P'(y', z', t')$ to $P(y, z, t)$ with the inverse to equation (3).

MOVEOUT-CORRECTED SEISMOGRAMS

In the previous section we considered the data which would be recorded if all surface shots were set off at the same time. If the shots are not set off at the same time but data is recorded from each shot separately, simultaneous shooting can by synthesized in a computer by stacking the data. For each receiver all seismograms of the different shots are aligned by shot time and added together. In practice, moveout correction would be applied before stacking. This correction is intended to remove source-receiver geometrical effects. Since it was ignored in the previous section, the results were limited to very great depths where the correction is small.

Figure 1 defines a moveout-corrected profile of seismic data. Great improvement in the partial differential-equation migration method results from the idea that an NMO profile and often a zero-offset section can be governed by a differential equation. Figure 1 illustrates how an NMO profile M may be constructed from an upcoming wave U by transcribing data values from the (y, t) plane to the (x, d) plane. This operation is actually a coordinate transformation of the data. Since the moveout-corrected profile is just the upcoming wave with coordinate axes deformed, it is not surprising that moveout-corrected profiles can be governed by an equation derived through a coordinate stretching transformation of the wave equation. The idea that there should be any advantage to using differential equations on synthetic things like moveout-corrected profiles (as compared to natural things like waves) arose out of the following observations: It is comparatively inefficient to let a wave packet propagate across a grid in a computer. It is much more efficient to describe a wave in a coordinate frame which moves along with the wave, as was done for the surface line source case in the beginning of this paper. In such a frame, things happen slowly and larger time increments may be used. There is a similar situation in a nearly layered medium where seismic arrivals (upcoming waves) fit nearly hyperbolic traveltime curves, and the object of a migration program is to deform the hyperboloids into lines which represent the layer-

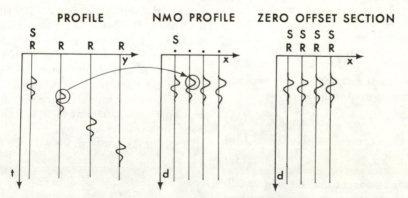

FIG. 1. Definition of profile and section. The left frame is indicative of observed upcoming waves recorded with one shot and a surface receiver line. This is called a *profile*. Data points are moved from this profile $(y, t$ plane) to the central frame $(x, d$ plane) under control of transformation equations like (13), with $z' = 0$. The central frame will be called an *NMO profile*. The third frame depicts upcoming waves recorded with a different shot-receiver geometry. Each recording is made with the shot and receiver at the same location. This will be called a (zero-offset) *section*.

Although the methods of this paper assume the geometry of the *NMO profile*, they will often be applicable to data recorded as *sections*.

ing. A migration partial differential equation does a lot of work just moving energy around on a grid from one more or less predictable place to another. Considerable effort can be saved by doing moveout correction (including application of a time-variable velocity if necessary) to get energy in approximately the right place before requiring a differential equation to migrate the data to its final proper position. If reflecting layers are perfectly flat and level, migration makes no change to the moveout-corrected data. The greater the structural dips and curvatures, the greater will be the task for the migrating differential equation. Grid spacing can be chosen according to the maximum anticipated dip.

Although efficiency was the motivation in the search for a differential equation to control migration from moveout-corrected data, there are two concomitant benefits which are far more important than efficiency. First, the data from spatially separated shotpoints may be stacked before migration, thereby enhancing signal-to-noise ratio. Second, we often may dispense altogether with the line of surface receivers and migrate data recorded in the geometry, where shotpoint and receiver point move together across the earth's surface as in the simplest type of marine profiling.

By "downward continuation of moveout-corrected seismograms" we mean that beginning with moveout-corrected data observed at the surface, we will synthesize moveout-corrected seismograms corresponding to hypothetical receivers at successively increasing depths.

One reason for wanting the moveout-corrected data for buried receivers is that it is related, through geometry, to the upcoming wave which is needed to make a migrated profile. Another reason, which is really the same, relates to the nature of seismic diffraction. Figure 2 (after Hilterman, 1970) illustrates that a data section can be expected to resemble a cross-section through the reflector if the radius of curvature of the reflector is much greater than the distance from the reflector to the receiver. Otherwise, one has a buried focus or diffraction. (Technically, a diffraction is a limiting case where some radius of curvature of a structure goes to zero.) In other words, moveout-corrected data gives a better representation of a structure if the receivers are near the structure, than if they are far away. In fact, when the receivers are at the depth of the structure the buried

focus problem disappears altogether. Diffractions from point scatterers also collapse to points when the receivers are at the same depth as the scatter. The method of migration proposed here is that as data are projected to successively greater depths, that part of the data corresponding to the receiver depth is set aside as belonging to the migrated data at that depth. Thus, various depths on the depth section are developed in succession as the moveout-corrected data are projected downward.

To be precise about the meaning of moveout-corrected data for buried receivers, we refer to Figure 3. Since the moveout-corrected profile M is created by a coordinate stretching of the upcoming wave U, we have

$$M(x, d, z') = U(y, t, z). \quad (11)$$

Observe the conceptual similarity of the relationship between $U(y, t, z)$ and $M(x, d, z')$ to the relationship between $P(y, z, t)$ and $P'(y', z', t')$. The fact that P and P' are the same thing expressed in different coordinates is analogous to the fact that U and M are the same thing expressed in different coordinates. At the surface $z = 0$ we record the upcoming wave $U(y, t, 0)$, and using a presumed velocity (which need not be precisely correct), we transform axes to the NMO profile $M(x, d, 0)$. The downward (z) continuation of receivers of $U(y, t, z)$ with the wave equation will be equivalent to downward (z') continuation of $M(x, d, z')$ with an equation we are about to derive. Although these two different downward continuations would be expected to give the same results, i.e., one could transform from U to M or M to U at any depth, there are several reasons to prefer downward continuation with M: 1) profiles from various shotpoints may be summed before downward continuation, 2) a coarser grid mesh may be used, and 3) sections may be downward continued.

To obtain the differential equation for U, first we must define the coordinate transformation from (y, t, z) to (x, d, z') and then use the chain rule to compute the required partial derivatives. From Figure 3 by means of elementary geometry, one may deduce the transformation

$$y(x, d, z') = x \cdot (2 - z'/d), \quad (12a)$$

$$t(x, d, z') = (2d - z')(1 + x^2/d^2)^{1/2}/c, \quad (12b)$$

and

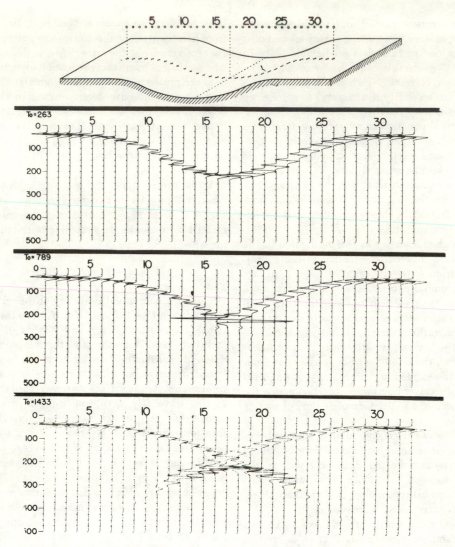

FIG. 2. (After Hilterman, 1970) Time sections recorded at various heights above a model. The top section (T_0 = 263) has shotpoint below the curvature axis, center section at the curvature axis, and bottom section above the curvature axis.

$$z(x, d, z') = z', \qquad\qquad (12c)$$

and, by means of tedious algebra, the inverse transformation is found to be

$$d(y, t, z) = (z + (c^2t^2 - y^2)^{1/2})/2, \qquad (13a)$$

$$x(y, t, z) = (y/2)(1 + z/(c^2t^2 - y^2)^{1/2}), \qquad (13b)$$

and

$$z'(y, t, z) = z. \qquad\qquad (13c)$$

In constant velocity material one could find an equation for M which is valid for all offsets x. The algebra would be overwhelming, so we make the simplifying practical assumption that $x/d \ll 1$. The authors were surprised to discover that even if offset terms like x/d or y/t are completely neglected one still obtains a result which is a big improvement over equation (9) of the introductory section. The reason is that even as offset goes to zero, the ratio of x to y remains important. Although (12) and (13) require a somewhat careful and detailed deduction, the zero offset relations are much easier and will be shown in detail. By

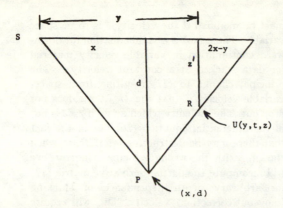

FIG. 3. Moveout-correction geometry for buried receivers. There is a surface shot S and buried receiver R. The wave is assumed to reflect at the point P. The receiver R at (y, z') measures the upcoming wave $U(y, t, z)$. From the upcoming wave a moveout-corrected profile $M(x, d, z')$ is constructed by axis stretching according to the geometry of the raypath SPR.

similar right triangles, the ratio of x to d is the same as the ratio $2x-y$ to z'. Thus $x/d = (2x-y)/z'$ or

$$y(x, d, z') = x(2 - z'/d). \qquad (14a)$$

The traveltime along SP will be d/c and along PR it will be $(d-z')/c$. Thus,

$$t(x, d, z') = (2d - z')/c. \qquad (14b)$$

The zero offset limit of (12) is (14). Now for the inverse relations solve (14b) for d and use $z'(y, t, z) = z$,

$$d(y, t, z) = (ct + z)/2. \qquad (15a)$$

Solve (14a) for x eliminating d with (15a) and using $z' = z$,

$$x = y/(2 - z/d)$$

and

$$x(y, t, z) = y/[2 - 2z/(ct + z)]. \qquad (15b)$$

From the coordinate transformation (14) and its inverse (15), it will be an easy matter now to compute the partial derivatives required for the transformation of the upgoing-wave equation. One derivative of particular interest, x_y, is computed from (15b) to be

$$x_y = 1/(2 - 2z/(ct + z)),$$

which in terms of the other variables is

$$x_y = 1/(2 - z'/d) = d/(2d - z'). \qquad (16)$$

Computing all the partial derivatives at the zero offset limit and arranging into a matrix, we have

$$
\begin{bmatrix} x_y & x_t & x_z \\ d_y & d_t & d_z \\ z'_y & z'_t & z'_z \end{bmatrix}
$$

$$
= \begin{bmatrix} d/(2d - z') & 0 & 0 \\ 0 & c/2 & 1/2 \\ 0 & 0 & 1 \end{bmatrix}. \qquad (17)
$$

Now there are two possible ways to proceed. The simplest way is to use the chain rule and insert partial derivatives into the wave equation. Another way is to insert the partial derivatives into the upcoming-wave equation. Let us insert the partial derivatives into the wave equation (2).

$$U_{yy} + U_{zz} - c(y, z)^{-2} U_{tt} = 0. \qquad (18)$$

We must be aware that (18) also has downgoing solutions which we will later eliminate by dropping a second order z' derivative. Recalling (11) that $U(y, t, z) = M(x, d, z')$, we compute the

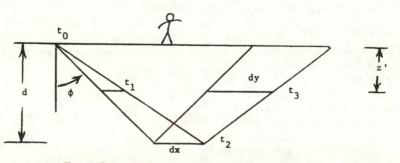

FIG. 4. Geometry for calculation of $(\partial x/\partial y)_t = 1/(\partial y/\partial x)_d$.

necessary terms for insertion into (18). Before insertion we simplify with the zeros in (17) and with the high-frequency approximation (gradients of waves are taken to be much greater than gradients of the coordinate transformation coefficients).

$$U_t = M_x x_t + M_d d_t = M_d d_t, \quad (19)$$

$$U_{tt} = M_{dd} d_t^2, \quad (20)$$

$$U_y = M_x x_y + M_d d_y = M_x x_y, \quad (21)$$

$$U_{yy} = M_{xx} x_y^2, \quad (22)$$

$$U_z = M_{z'} + M_x x_z + M_d d_z$$
$$= M_{z'} + M_d d_z, \quad (23)$$

$$U_{zz} = M_{z'z'} + 2d_z M_{dz'}$$
$$+ M_{dd} d_z^2, \quad (24)$$

and

$$c(y, z) = c[y(x, d, z'), z(x, d, z')]. \quad (25)$$

Inserting (20), (22), (24), and (25) into (18) we obtain for the downward projection of the moveout-corrected data

$$M_{xx} x_y^2 + M_{z'z'} + 2M_{dz'} d_z$$
$$+ M_{dd}(d_z^2 - c^{-2} d_t^2) = 0, \quad (26)$$

which, utilizing (17) and the assumption that the true (wave-equation) velocity equals the constant stacking velocity (in d_t), simplifies to

$$M_{dz'} = -\left(\frac{d}{2d - z'}\right)^2 M_{xx} - M_{z'z'}. \quad (27)$$

The last term arises mainly because the wave equation we started from also has downgoing waves. The omission of this kind of term is discussed in the first section and in earlier papers. There is a formal similarity between (9) and (27), so the same computer algorithm may be used.

VARIABLE VELOCITY

The result of the last section, equation (27), is restricted to material of constant velocity. It will be useful in practice to generalize the result to space-variable velocity. First note that *the velocity $\bar{c}$ in the wave equation (18) need not be the same as the velocity, say $\bar{c}$, in the moveout-correction equations* (11) to (17). It can be shown that the wave equation is still valid at high frequencies $(\omega/c \gg |\nabla c|/c)$ when the velocity $\bar{c}$ is space variable, although the moveout-correction equations

must be modified if the velocity $\bar{c}$ is to be space variable. Thus, (26) is a valid equation for downward continuation in variable-velocity material for data stacked at a constant velocity $\bar{c}$. The principal change to (27) resulting from space-variable velocity is that the M_{dd} term does not drop out. The difficulty which arises if $\bar{c} \neq \bar{c}$ is not in the differential equation (26) but in the fact that there may be destructive interference when stacking with the wrong velocity. Therefore we will recompute the partial-derivative matrix (17) for arbitrary depth dependence of stacking (moveout-correction) velocity. This will require some care. The matrix of (17) is really the Jacobian matrix of the transformation from (y, t, z) variables to (x, d, z') variables, that is:

$$\begin{bmatrix} dx \\ dd \\ dz' \end{bmatrix} = \begin{bmatrix} x_y & x_t & x_z \\ d_y & d_t & d_z \\ 0 & 0 & 1 \end{bmatrix} \begin{bmatrix} dy \\ dt \\ dz \end{bmatrix}. \quad (28)$$

The inverse transformation may be defined as

$$\begin{bmatrix} dy \\ dt \\ dz \end{bmatrix} = \begin{bmatrix} y_x & y_d & y_{z'} \\ t_x & t_d & t_{z'} \\ 0 & 0 & 1 \end{bmatrix} \begin{bmatrix} dx \\ dd \\ dz' \end{bmatrix}. \quad (29)$$

The statement that (29) is indeed inverse to (28) is

$$\begin{bmatrix} x_y & x_t & x_z \\ d_y & d_t & d_z \\ 0 & 0 & 1 \end{bmatrix} \begin{bmatrix} y_x & y_d & y_{z'} \\ t_x & t_d & t_{z'} \\ 0 & 0 & 1 \end{bmatrix}$$
$$= \begin{bmatrix} 1 & 0 & 0 \\ 0 & 1 & 0 \\ 0 & 0 & 1 \end{bmatrix}. \quad (30)$$

Near the zero offset limit x and y tend to zero and are, therefore, independent of t, d, and z, so that $x_t = x_z = y_d = y_{z'} = 0$. Also, a small change in x or y has a second-order effect on traveltime so $t_x = d_y = 0$. Thus, at small enough offsets (30) reduces to

$$\begin{bmatrix} x_y & 0 & 0 \\ 0 & d_t & d_z \\ 0 & 0 & 1 \end{bmatrix} \begin{bmatrix} y_x & 0 & 0 \\ 0 & t_d & t_{z'} \\ 0 & 0 & 1 \end{bmatrix}$$
$$= \begin{bmatrix} 1 & 0 & 0 \\ 0 & 1 & 0 \\ 0 & 0 & 1 \end{bmatrix}. \quad (31)$$

The interesting parts of (31) are

$$x_y = 1/y_x, \qquad (32)$$

$$d_t t_d = 1, \qquad (33)$$

and

$$d_t t_{z'} + d_z = 0. \qquad (34)$$

To compute the partial derivative x_y, refer to Figure 4 which, for clarity, exaggerates the offset. For rays which are essentially vertical

$$dz = c \, dt. \qquad (35)$$

We have Snell's law,

$$(\sin \phi)/c = \text{const} = p. \qquad (36)$$

Tracing rays in time gives

$$x = \int_0^{t_2} c \sin \phi \, dt = p \int_0^{t_2} c^2 dt,$$

and using (35),

$$x = p \int_0^d c \, dz \qquad (37)$$

and

$$y = p \int_0^{t_3} c^2 dt,$$

which, from Figure 4, may be written as

$$y = p\left(2 \int_0^d c \, dz - \int_0^{z'} c \, dz \right). \qquad (38)$$

It is convenient to use d and z' as superscripts to indicate integration in the same way we use them as subscripts to denote differentiation. Thus, we define

$$c^d = \int_0^d c(z) \, dz, \quad c^{z'} = \int_0^{z'} c(z) \, dz,$$

and use (37) to eliminate p from (38);

$$y = x(2c^d - c^{z'})/c^d. \qquad (39)$$

Differentiating with respect to x holding d and z' constant, gives

$$y_x = (2c^d - c^{z'})/c^d. \qquad (40)$$

Thus, using (32),

$$x_y = c^d/(2c^d - c^{z'}). \qquad (41)$$

In the constant velocity limit (41) obviously reduces to (16).

The partial derivatives of traveltime t with respect to receiver depth and with respect to recording position are found in a similar manner. The traveltime dt through a layer of thickness dz is given by

$$dt = dz/c(z) \cos \phi,$$

where ϕ is the angle from the ray to the vertical. Replacing $1/\cos \phi$ by $1 + \sin^2 \phi/2$, we have

$$dt = \left(\frac{1}{c(z)} + \frac{\sin^2 \phi}{2c(z)} \right) dz.$$

Using (36) and (37) to second order in x, we get

$$dt = \left(\frac{1}{c(z)} + \frac{x^2 c(z)}{2(c^d)^2} \right) dz.$$

Referring to Figure 4 we note that the traveltime is twice the time from 0 to d, less the time from 0 to z'. Thus,

$$t = 2 \int_0^d \left[\frac{1}{c(z)} + \frac{x^2 c(z)}{2(c^d)^2} \right] dz$$

$$- \int_0^{z'} \left[\frac{1}{c(z)} + \frac{x^2 c(z)}{2(c^d)^2} \right] dz.$$

This equation may be used to find the partial derivatives of traveltime to first-order accuracy in x. However, in this section we are interested in finding the derivatives to zero order in x, so the terms proportional to x^2 may be omitted, giving

$$t = 2 \int_0^d \frac{1}{c(z)} \, dz - \int_0^{z'} \frac{1}{c(z)} \, dz,$$

which we abbreviate as

$$t = 2(c^{-1})^d - (c^{-1})^{z'}. \qquad (42)$$

Differentiating (42) with respect to d holding x and z' fixed, gives

$$t_d = + 2/c(d), \qquad (43)$$

where $c(d)$ is velocity $c(z)$ evaluated at $z = d$.

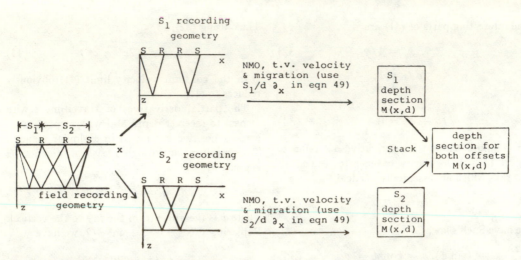

FIG. 5. Procedure for migrating data with several offsets. Field data are recorded using two receiver offsets, S_1 and S_2. The field data are separated into two sets: data recorded with offset S_1 and data recorded with offset S_2. Moveout corrections are done for both sets of data. Then the S_1 data are migrated with equation (49) with $(S_1/d)\partial_z$ substituted for $(x/d)\partial_z$, giving the S_1 depth section, $M(x,d)$. Next the S_2 data are migrated with equation (49) with $(S_2/d)\partial_z$ substituted for $(x/d)\partial_z$, giving the S_2 depth section, $M(x, d)$. Finally, the S_1 depth section and S_2 depth section are stacked to give the composite depth section.

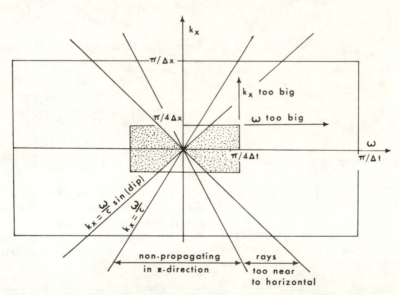

FIG. 6. The (ω, k) plane. Field data may be expected to have some energy everywhere in the (ω, k) plane. Only in the speckled region will our difference equations properly simulate the wave equation. Energy with $|k| > |\omega/c|$ does not represent free waves; it represents either surface waves or errors in data collection (often statics, random noise, or gain not smoothly variable from trace to trace). Such energy can mean nothing in a migration program, hence it should be rejected by filtering. This may be done by fan-filtering (as in Treitel et al, 1967) or as was done here by means of numerical viscosity (Claerbout, 1970). Actually, for practical reasons one frequently may wish to reject rays outside a certain dip angle. This gives the larger fan-filter reject region $|k| > |\omega/c \sin (\mathrm{dip})|$. In fact our present implementation is inadequate for dips greater than about 45 degrees (Claerbout, 1971). Although information can be carried up to the folding frequency in both ω and k, in practice the use of operators of finite length narrows the useful bandwidth. As shown in Claerbout and Johnson (1972) the use of simple difference operators results in a practical bandwidth restriction to about a quarter of the folding frequency. This presents no problem in principle; data may be interpolated before processing, or more elaborate (i.e., longer) difference operators may be used. Finally, the figure was drawn with $\Delta x > c\Delta t$ because it represents the usual case in practice where extra points in time are more cheaply obtained than extra points in space.

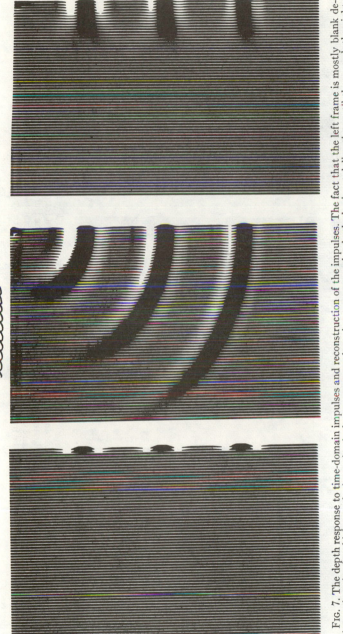

Fig. 7. The depth response to time-domain impulses and reconstruction of the impulses. The fact that the left frame is mostly blank depicts a situation in which no echo is received when a source and receiver move together in the horizontal direction until they reach the right-hand edge of the frame where the three blips indicate that there are three echoes at successively increasing times. With this as observed data, the logical conclusion is that the reflection structure of the earth is three concentric circles with centers on the right margin. The central frame shows the circles. (For economy the right edge of the frame is a plane of symmetry.) It will be noticed that the bottom of the circles is darker than the top. This is indicative of the 45-degree phase shift of bringing two-dimensional waves from a focus away from the focus. Waves with dips greater than about 45 degrees have been filtered away by numerical viscosity. The loss of this energy plus the loss of the energy of waves which propagate at complex angles results in a reconstruction (right-hand frame) in which the impulses are somewhat spread out in the horizontal direction.

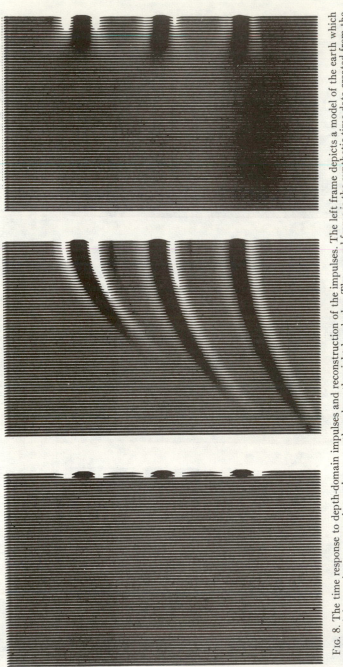

FIG. 8. The time response to depth-domain impulses and reconstruction of the impulses. The left frame depicts a model of the earth which consists of three point scatterers beneath one another along the right-hand edge. The second frame is the synthetic time data created from the model. Basically one observes the hyperbolic traveltime curves to the reflecting points. The third frame represents migration of the synthetic data back to the point scatterers. As in Figure 6 there is a reduced resolution because, in principle, horizontal resolution cannot be better than vertical resolution (which is controlled by the frequency content of the waves) and in practice we have included only rays up to angles of about 40 degrees.

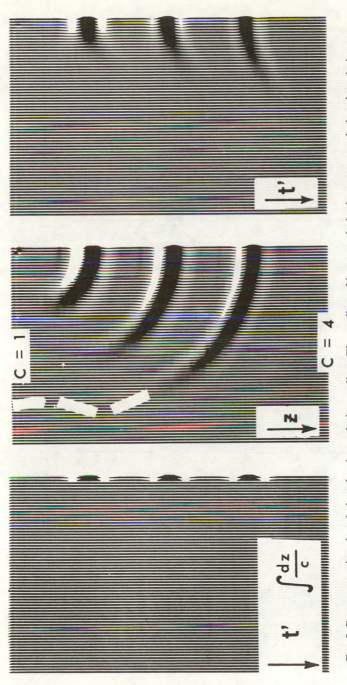

FIG. 9. Response to time-domain impulses in a velocity gradient. The gradient of inverse velocity is constant such that the velocity ranges over a factor of 4 from top to bottom of the frames. The first frame appears like the uniformly spaced (say 1, 2, 3) pulses of Figure 7. However, here the vertical coordinate is not time t but

$$t' = \int c^{-1}dz = \int (az + b)dz.$$

Thus the pulses are really spaced at times proportional to (.9, 1.6, and 2.0). The second frame represents the depth section. The circles of Figure 7 have changed to a narrower shape. If the rays were not cut off around 45 degrees, the shapes would resemble roughly that of a light bulb as indicated by the dashes. The third frame shows the reconstruction of the time data. It clearly deteriorates with depth. This is probably a result of the fact that a 40-degree beam halfwidth at the bottom reflector collapses to a 10-degree beam halfwidth at the surface because of ray curvature. There is also some erroneous bias in the reconstruction (the pulses "bend" a little). This is probably a result of inadequate dip filtering. At the time of writing we are uncertain how much the resolution loss in a velocity gradient resuls from our method and how much is fundamental.

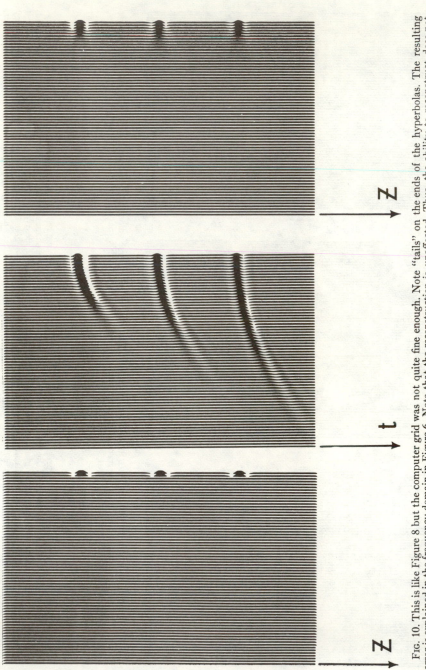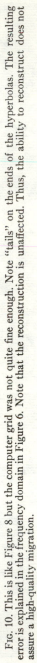

Fig. 10. This is like Figure 8 but the computer grid was not quite fine enough. Note "tails" on the ends of the hyperbolas. The resulting error is explained in the frequency domain in Figure 6. Note that the reconstruction is unaffected. Thus, the ability to reconstruct does not assure a high-quality migration.

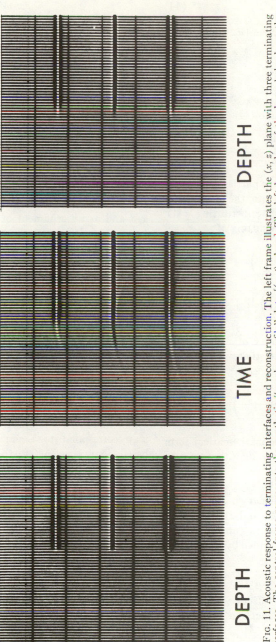

DEPTH

TIME

DEPTH

FIG. 11. Acoustic response to terminating interfaces and reconstruction. The left frame illustrates the (x, z) plane with three terminating interfaces. The central frame represents the synthetic "sparker-profile" data $((x, t)$ plane]. The left branches of the hyperbolas have the polarity of the interface, but the right branches have the opposite polarity. This phenomenon was predicted by Trorey (1970). The rightmost frame is the attempted reconstruction of the model in the first frame.

Differentiating (42) with respect to z' holding x and d fixed, gives

$$t_{z'} = -1/c(z'). \qquad (44)$$

Now we can proceed to find d_t and d_z. Substituting (43) into (33) we find

$$d_t = \frac{c(d)}{2}. \qquad (45)$$

Substituting (45) and (43) into (34) gives

$$d_z = \frac{c(d)}{2c(z')}. \qquad (46)$$

Finally, we may substitute the partial derivatives from (46), (45), and (41) into the downward-continuation equation (26). Substituting (41), (45), and (46) into (26) and omitting $M_{z'z'}$, we get

$$\frac{c(d)}{c(z')} M_{dz'} + M_{xx} \left(\frac{c^d}{2c^d - c^{z'}} \right)^2 + \frac{c(d)^2}{4}$$
$$\cdot \left(\frac{1}{c(z')^2} - \frac{1}{\bar{c}(x, d, z')^2} \right) M_{dd} = 0. \qquad (47)$$

If we assume that the velocity of waves in the earth depends only on depth, then we may take $c(z') = c(x, d, z')$ and (47) reduces to

$$M_{dz'} = -\frac{c(z')}{c(d)} \left(\frac{c^d}{2c^d - c^{z'}} \right)^2 M_{xx}. \qquad (48)$$

This equation has the same mathematical form as (9) and the same computer algorithm may be used again.

Equation (48) is limited to rays of moderate angle from the z axis. This equation was used for Figures 11, 13, 14, 18, 19, 20, and 21. In the remaining figures where steeper ray angles were of interest, the M_{zz} term was included by the method of Claerbout (1971a).

SHOT-GEOPHONE OFFSET

If first-order terms in the offset parameter x/d had been retained in the deduction of (27), we would have instead

$$\left(\partial_d + \frac{x}{d} \partial_x \right) \partial_{z'} M$$
$$= -\left(\frac{d}{2d - z'} \right)^2 \partial_{xx} M. \qquad (49)$$

The new term in (49) becomes important if the inequality

$$1 \gg \frac{x}{d} \frac{M_x}{M_d} \qquad (50)$$

is not sufficiently strong.

The parameter x/d is a measure of shot-geophone offset. The expression M_x/M_d is the tangent of the local angle of dip. Thus, the new term in (49) is not really necessary if the offset, or the dip, or the product of offset with dip is sufficiently small. Thus, we may expect that the simpler equation without the offset term will be valid for migrating data over a fairly large range of offsets if the dip is not too large. Now consider the occurrence of the symbol x in the coefficient x/d and as an argument of $M(x, d, z')$. On the basis of our derivation, these two x's are the same x. However, since x/d can usually be neglected (i.e., reset to an arbitrary small value), we may regard the x in x/d as independent of the x in $M(x, d, z')$. In $M(x, d, z')$ the argument x refers to the horizontal coordinate (depth point) on the depth profile. In x/d the x refers to the offset of the shot from the depth point. We have shown that this offset is not very important. This is a mathematical justification for stacking moveout-corrected data from many shotpoints before migration in cases where the product of refractor dip and offset is small. Problems may be expected to arise when the dip or offset is great enough. This may correspond to Levin's statement (1971), based on ray geometry, that stacking velocity depends on dip. A procedure (refer to Figure 5) will be described which is intended for use with data which has several offsets or dips (hence several stacking velocities) present in the same region of y and t.

First, moveout-corrected time sections S_0, S_1, S_2, $\cdots$ can be constructed such that S_0 is made up of all data of zero and small offset S_1 is data with a somewhat larger offset, etc. Then the different S_j may be separately migrated through the use of (49) or its generalization to depth-variable velocity. Since the earth itself is invariant to changes in shot-receiver offset, the migrated S_j should also be invariant to offset. Thus, regardless of dip, the migrated S_j should stack without destructive interference. In fact, the velocity which should be used in construction of the S_j is that for which the migrated S_j stack best.

DEPTH

TIME

DEPTH

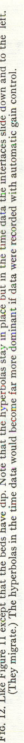

FIG. 12. Like Figure 11 except that the beds have dip. Note that the hyperbolas stay in place but in the time data the interfaces slide down and to the left. (They migrate.) The hyperbolas in the time data would become far more dominant if data were recorded with automatic gain control.

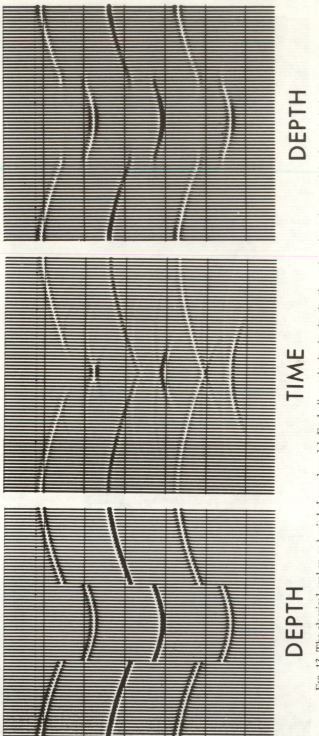

DEPTH   TIME   DEPTH

FIG. 13. The classical graben or buried-channel model. Each discontinuity in the depth section gives rise to a faint hyperbola in time.

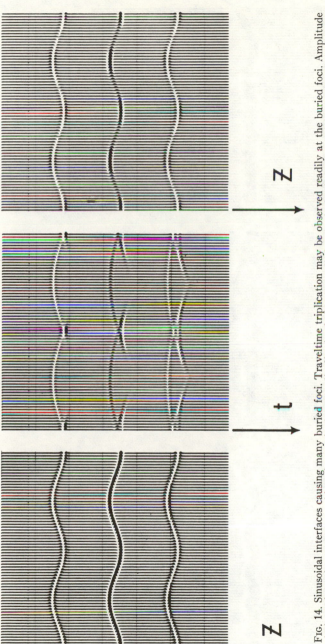

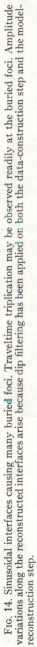

FIG. 14. Sinusoidal interfaces causing many buried foci. Traveltime triplication may be observed readily at the buried foci. Amplitude variations along the reconstructed interfaces arise because dip filtering has been applied on both the data-construction step and the model-reconstruction step.

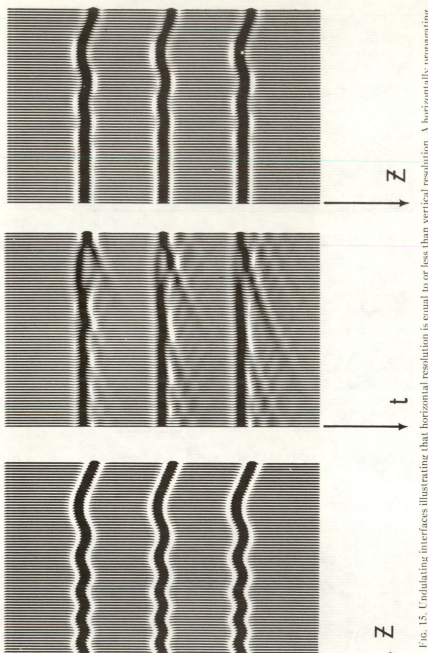

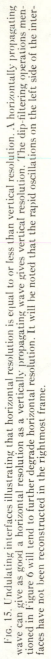

FIG. 15. Undulating interfaces illustrating that horizontal resolution is equal to or less than vertical resolution. A horizontally propagating wave can give as good a horizontal resolution as a vertically propagating wave gives vertical resolution. The dip-filtering operations mentioned in Figure 6 will tend to further degrade horizontal resolution. It will be noted that the rapid oscillations on the left side of the interfaces have not been reconstructed in the rightmost frame.

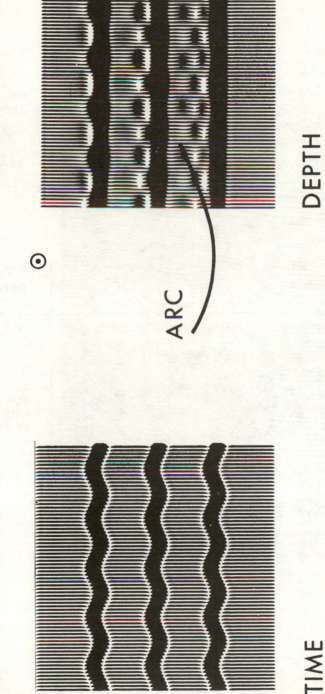

MIGRATION WITHOUT
STATICS REMOVED

DEPTH

ARC

FLAT LAYERS
WITH STATICS

TIME

Fig. 16. The effect of unremoved static errors on migration. The left frame depicts reflections from flat layers plus a static timing error which varies sinusoidally from trace to trace. The result of migration is shown on the right. Note that maxima may be aligned into circular arcs.

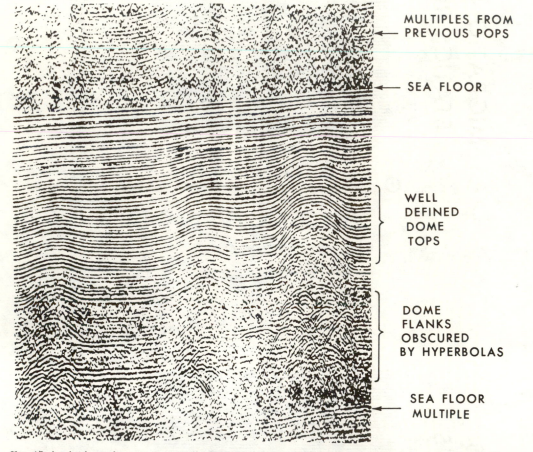

MULTIPLES FROM
PREVIOUS POPS

SEA FLOOR

WELL
DEFINED
DOME
TOPS

DOME
FLANKS
OBSCURED
BY HYPERBOLAS

SEA FLOOR
MULTIPLE

FIG. 17. A seismic section across some diapir-like structures. This unprocessed data is from the Umnak Plateau in the southeastern Bering Sea (Scholl and Marlow, 1970). The water depth is about a mile. The vertical exaggeration is about 8. This is single-channel, constant-offset data. For processing purposes it was considered to be a zero-offset section.

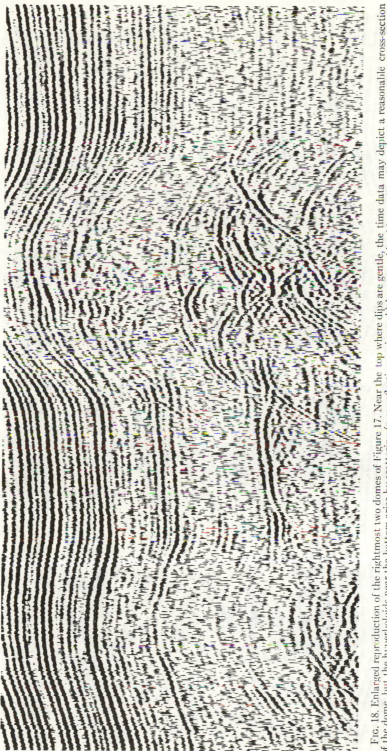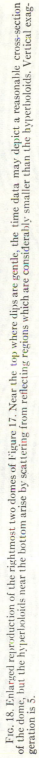

Fig. 18. Enlarged reproduction of the rightmost two domes of Figure 17. Near the top where dips are gentle, the time data may depict a reasonable cross-section of the dome, but the hyperboloids near the bottom arise by scattering from reflecting regions which are considerably smaller than the hyperboloids. Vertical exaggeration is 5.

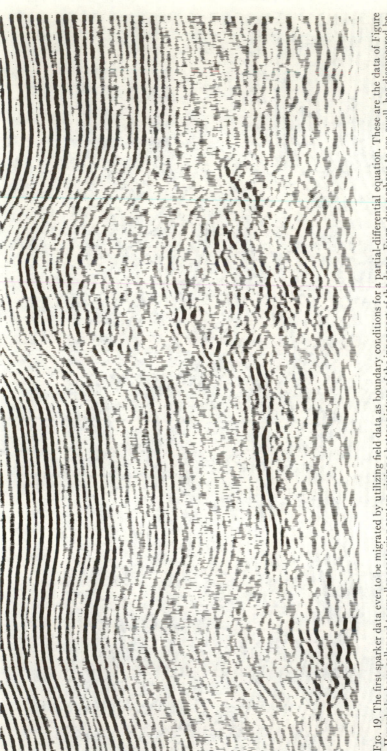

FIG. 19. The first sparker data ever to be migrated by utilizing field data as boundary conditions for a partial-differential equation. These are the data of Figure 18. Hyperbolas have collapsed to smaller scattering centers giving a better picture of the inner part of the dome. Fuzz, due perhaps to sea swell, has disappeared because impossibly steep dips have been attenuated with numerical viscosity. Since all interference effects have not been eliminated (even in the top of the dome) the necessary conclusion is that echoes simultaneously arrive from both sides of the ship. (By the nature of the process, there cannot be simultaneous front-back echoes in the processed data.) Thus, the structure is more complicated than a dome of rotation about the vertical axis. Hence more profiles are required to uniquely delineate the features. The vertical exaggeration is about 5. The data have been processed to dip tangents of about 1/10.

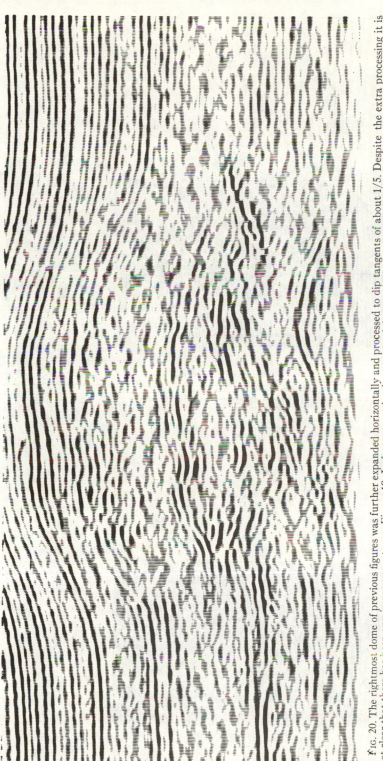

FIG. 20. The rightmost dome of previous figures was further expanded horizontally and processed to dip tangents of about 1/5. Despite the extra processing it is not clear that there has been any improvement over Figure 19. In fact, some circular arcs reminiscent of Figures 7 or 16 are beginning to appear. Such arcs could result from spikes in the data. Actually, we do not believe there are spikes in the data but suspect that these arcs result from some data collection problem as the shipboard AGC changing from channel to channel.

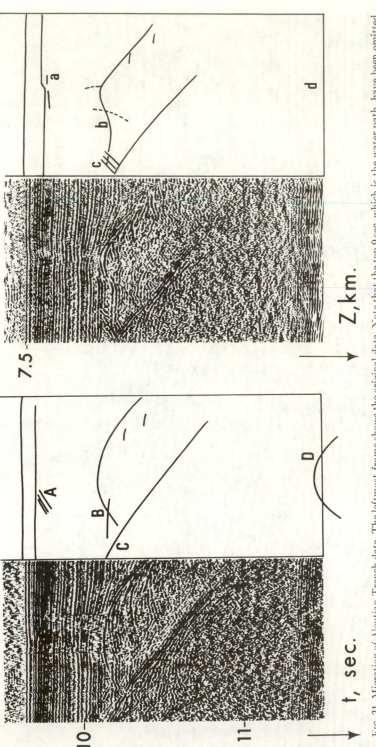

F$_{\mathrm{IG}}$. 21. Migration of Aleutian Trench data. The leftmost frame shows the original data. Note that the top 9 sec, which is the water path, have been omitted. The vertical exaggeration with respect to water velocity is 5.3. Hyperbolic branches at A migrate to a small zone a where flat-lying sediments are abruptly offset. Traveltime triplication seen at B migrates to the concavity at b which caused the buried focus. Some undesirable side effects of the process occur around the edges of the frame because of the limited computer memory available. For example the reflector labeled C should migrate leftward and upward and partly out of the figure (= the computer memory). In fact, what happens is that a Snell's law of reflection takes place at the side boundary so that the energy appears at C. There is also a boundary effect at the bottom of the page. If a hyperbola were to appear as indicated at D, its bottom flanks would be cut off by the page boundary. This limits the amount of dip which may be seen at d, hence the smoothness in the bottom part of the migrated section.

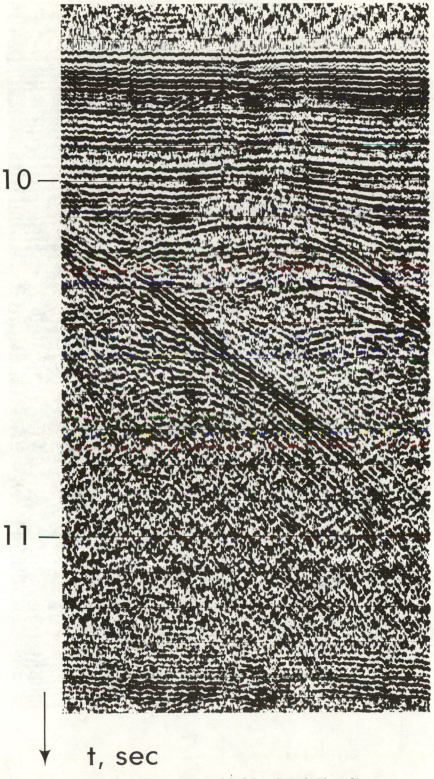

10 —

11 —

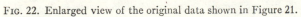

t, sec

FIG. 22. Enlarged view of the original data shown in Figure 21.

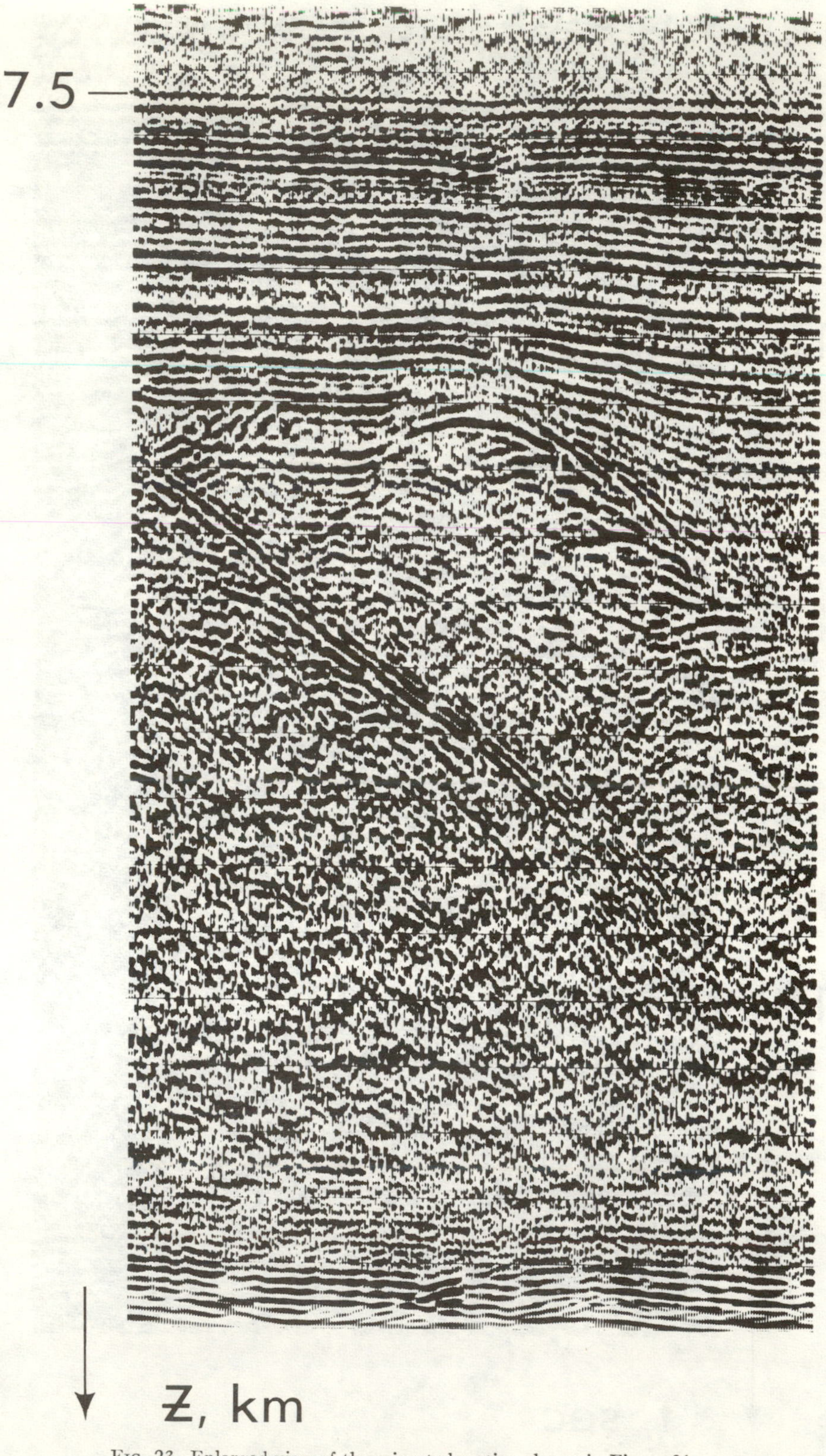

7.5 —

Z, km

Fig. 23. Enlarged view of the migrated section shown in Figure 21.

This velocity, because it is based on migrated data, should not be subject to the diffraction problem discussed by Dinstel (1971). If the data is "in plane" it should be the true material velocity despite bedding curvature or diffraction.

ACKNOWLEDGMENTS

We wish to express our thanks to the National Science Foundation who supported this work in the early stages, to the donors to the Petroleum Research Fund of the American Chemical Society, and to the Chevron Oil Field Research Company.

REFERENCES

Claerbout, J. F., 1970, Coarse grid calculations of waves in inhomogeneous media with application to delineation of complicated seismic structure: Geophysics, v. 35, no. 3, p. 407–418.
——— 1971a, Numerical holography, *in* Acoustical holography, vol. 3: New York, Plenum Press, p. 273–283.
——— 1971b, Toward a unified theory of reflector mapping: Geophysics, v. 36, no. 3, p. 467–481.
Claerbout, J. F., and Johnson, A. G., 1971, Extrapolation of time dependent waveforms along their path of propagation: Geophys. J. of the Roy. Astr. Soc., v. 26, no. 1–4, p. 285–294.
Dinstel, W. L., 1971, Velocity spectra and diffraction patterns: Geophysics, v. 36, no. 2, p. 415–417.
Hilterman, F. J., 1970, Three-dimensional seismic modeling: Geophysics, v. 35, no. 6, p. 1020–1037.
Landers, T. E., and Claerbout, J. F., 1972, Elastic waves in laterally inhomogeneous media: J. Geophys. Res., v. 77, p. 1476–1482.
Levin, F. K., 1971, Apparent velocity from dripping interface reflections: Geophysics, v. 36, no. 3, p. 510–516.
Scholl, D. W., and Marlow, M. S., 1970, Diapirlike structures in the southeastern Bering Sea, Bull. AAPG, v. 54, no. 9, p. 1644–1650.
Treitel, S., Shanks, J. L., Frasier, C. W., 1967, Some aspects of fan filtering: Geophysics, v. 32, no. 5, p. 789–800.
Trorey, A. W., 1970, A simple theory for seismic diffractions: Geophysics, v. 35, no. 5, p. 762–784.

Chapter 4

Applications and the Exploding Reflector Model

The paper by *French (1974)* demonstrated the practicality and importance of three-dimensional (3-D) data acquisition and migration. He also introduced a 3-D scaled seismic physical modeling technique which allows a realistic laboratory test of migration. The papers by *Gardner et al. (1974)* and *Sattlegger et al. (1976)* showed how velocity analysis can be based on migrated data. Problems with migrating two-dimensional seismic lines over 3-D structures are illustrated with physical model data in *French (1975)*. The use of image rays to correct for lateral velocity variations was discussed in *Hubral (1977)*.

Loewenthal et al. (1976) introduced the concept of the "exploding reflector model" which opened the way for a flood of elegant and efficient migration schemes. Up to this point wave equation techniques dealt with the source wave field and the measured reflected wave field but did not clearly explain how to handle a stacked section. The exploding reflector model solved this problem by treating the measured data as a boundary condition, the geologic model as an initial condition, and the wave equation as the connecting link. In mathematical symbolism, if $B(x,t)$ are the measured data (with x horizontal distance and t time), then the wave equation is used to generate a solution $p(x,y,t)$, where y is depth, such that $p(x,0,t) = B(x,t)$. Having constructed $p(x,y,t)$, the values at $t = 0$ are calculated. Then $p(x,y,0)$ is interpreted as the distribution of reflectivity in the Earth. Symbolically,

$$B(x,t) \rightarrow p(x,y,t) \rightarrow p(x,y,0)$$

boundary values　　wave field　　initial condition.

Conversely, if a geologic section is given, $p(x,y,0)$ is constructed by placing wavelets at all points along the interfaces with amplitude proportional to the local reflection coefficient. From this initial condition, $p(x,y,t)$ is calculated by solving the wave equation. Then the measured data are $p(x,0,t)$.

In solving the wave equation to migrate or model stacked or zero-offset seismic data, the velocity of propagation is taken equal to half of the real local velocities to ensure that the one-way traveltime from a reflector to the Earth's surface in the model will agree with the two-way traveltime measured in practice.

The power of the exploding reflector model lies in treating a stacked seismic section as a measurement of a wave field. Although only an approximation, it is such a useful model that methods based on it still dominate migration.

Reprinted from GEOPHYSICS, 39, 265-277.

GEOPHYSICS

TWO-DIMENSIONAL AND THREE-DIMENSIONAL MIGRATION OF MODEL-EXPERIMENT REFLECTION PROFILES

WILLIAM S. FRENCH*

A reflection profile represents an unfocused picture of the subsurface. In areas of rapid structural change, this unfocused picture may not reveal directly the true geometry of subsurface structures. Computer processing techniques, collectively called migration, have been used by many companies to focus 2-D reflection data. A description of the migration process can be given which allows immediate generalization to three-dimensions with arbitrary source and receiver positions.

Reflection profiles digitally recorded in the laboratory over known acoustically semitransparent structural models establish the effectiveness of migration. Processed reflection data over 3-D models demonstrate that 3-D migration eliminates many of the lateral correlation ambiguities caused by "sideswipes" and "blind structures."

Structure maps developed from the results of 3-D migration of reflection data give a true and precise picture of 3-D models. When the same data are processed using 2-D migration, the mapped structures are distorted.

In structurally complex areas it is desirable to collect 3-D reflection data. Single profiles cannot, and conventional grids may not, reveal adequate cross-dip information.

INTRODUCTION

In areas of rapid structural change, unprocessed seismic reflection profiles may not reveal directly the true geometry of subsurface structures. In some cases, it is possible to identify an isolated structure on the basis of its characteristic reflection pattern (e.g., a fault or syncline). In general, however, it may be necessary to employ some technique to "focus" the data from complex areas.

Two calculational procedures to focus seismic reflection data have been discussed in the literature: downward continuation of moveout corrected seismograms (Claerbout and Doherty, 1972) and digital migration (Schneider, 1971). In essence, the methods utilize a surface-recorded reflection profile to calculate the hypothetical profile that would be obtained if the geophones could be placed in a plane deep in the earth. Their usefulness results from the fact that the closer the profile is to a structure, the more the reflection patterns take on the geometry of the structure. Both techniques are founded on the mathematics of wave propagation and should provide identical results if properly executed. The first technique is a numerical solution to the differential equations of wave motion. The second technique is based upon numerical solution of integral equations describing the wave motion (Maginness, 1972). The integral equations admit a construct which proves to be of great value to the intuition,

Presented at the 43rd Annual International SEG Meeting October 25, 1973, Mexico City. Manuscript received by the editor December 6, 1973.

* Gulf Research & Development Co., Pittsburgh, Penn. 15230.

and such an approach will be adopted in this paper; the term migration will be used as a synonym for the construct.

In this paper the assumptions inherent to migration will be stated, the technique will be extended to handle the case of 3-D objects with arbitrary source-receiver arrays, and the results of both 2-D and 3-D migration of seismic model data will be presented.

ASSUMPTIONS OF DIGITAL MIGRATION

The typical seismic record is the result of recording at several locations the reflected waves generated by an impulsive source. A profile consists of a collection of such records for several source locations. For convenience of discussion we will deal with the case of marine profiles. Thus the detectors are hydrophones and the profile consists of the pressure-time records $P(s_j, r_j^i, t)$, where

$s_j = (x_j, y_j, z_j)$ is the location of the jth shot-point,

$r_j^i = (x_j^i, y_j^i, z_j^i)$ is the location of the ith hydrophone recording the jth shot, and

t is a time variable which is reset to zero at the instant of each successive shot.

At this point, no particular shooting geometry has been assumed (i.e., the shotpoints and receiver locations do not necessarily lie in the same line or even in the same plane). The task is to interpret $P(s_j, r_j^i, t)$ in terms of subsurface structural geometry. Migration will be used to accomplish this end.

Computer migration schemes represent a search for scattering centers (i.e., diffraction or reflection points). The process involves assigning to each subsurface point a number which is a measure of the probability that scattered energy emanated from that point. The number is determined by summing the recorded data for all shotpoints and receiver locations at times where energy from that subsurface point could arrive. A number with large absolute value (positive or negative depending upon the wavelet shape) indicates that scattered (reflected or diffracted) energy probably did come from that particular subsurface point. In other words, the coherence of the recorded data along a traveltime versus distance surface appropriate to the source-receiver geometry and subsurface point deter-

mines the assigned probability that a reflector or diffractor exists at that subsurface point.

The assumptions of the migration method used in this paper are as follows:

(a) Shear waves can be ignored.
(b) Each subsurface point represents a possible scattering center.
(c) Reflecting surfaces can be considered a continuum of scattering centers.
(d) The pulse shape for the scattered waves is the same for all directions. The pulse is short enough so that the delay in arrival time of later portions can be neglected.
(e) Traveltime versus distance surfaces can be calculated to sufficient accuracy by averaging horizontal velocity variations. That is, a vertical path is considered the least-time travel path from surface to subsurface point, and moveout times for other surface points can be calculated on the basis of surface distance and a root-mean-square velocity-depth function (see Taner and Koehler, 1969).
(f) A coherent signal summed along the appropriate time-distance surface leads to an average with large absolute value. Any noise summed along the time-distance surface will lead to a small average value due to the equal probability of positive and negative numbers in the noise field.

It is known that some of the above assumptions are not strictly true [e.g., the shape of a pulse scattered by a small sphere depends upon the scattering angle (Morse and Ingard, 1968)]. However, they are consistent with the usual assumptions of seismic data processing.

GENERAL EQUATIONS FOR MIGRATION

The above description of migration can be stated mathematically. We assume that conditions (a) through (f) hold and that we have available the pressure-time data $P(s_j, r_j^i, t)$. Let the time required for a pulse to travel the dashed line path of Figure 1 [from the source at s_j to the subsurface point $r = (x, y, z)$ and subsequently to a receiver at r_j^i] be given by $t_j^i(r)$. As before, we establish a measure of the probability that a scattering center exists at r by forming the sum

$$M(r) = \sum_j \sum_i P[s_j, r_j^i, t_j^i(r)]. \quad (1)$$

Equation (1) represents 3-D migration for arbitrary source and receiver arrays. If the source and receiver lie in the xy-plane, then [according to condition (e)]:

$$t_j^i(r) = \left\{ T_{1/2}^2(r) + \frac{(x_j - x)^2 + (y_j - y)^2}{V^2(z)} \right\}^{1/2}$$
$$+ \left\{ T_{1/2}^2(r) + \frac{(x_j^i - x)^2 + (y_j^i - y)^2}{V^2(z)} \right\}^{1/2}, \tag{2}$$

where $T_{1/2}(r)$ is the one-way vertical traveltime to r and $V(z)$ is the rms velocity function as described by Taner and Koehler (1969).

If we consider the special case where there is but one receiver for each source and both source and receiver are at the same location, then equation (2) becomes

$$\overset{1}{t_j}(r) = t_j(r) \tag{2a}$$
$$= \left\{ T_0^2(r) + \frac{4\left[(x_j - x)^2 + (y_j - y)^2\right]}{V^2(z)} \right\}^{1/2},$$

where T_0 is the two-way vertical traveltime to r.

Equation (1) involves an implicit time-to-depth conversion since the rms velocity is given as a function of depth. In practice, we know the rms velocity more accurately as a function of time. Therefore, in the following applications of equation (1) we replace z by T_0 and plot migrated time sections rather than depth section. Equations (1) and (2a) have been used to focus data collected with both 2-D and 3-D models.

Two-D models have thickness variations only in the direction of the profile—cross-sections perpendicular to the profile line are of constant

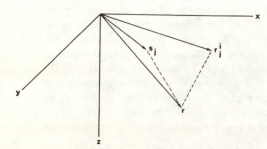

FIG. 1. Coordinate system for migration equation.

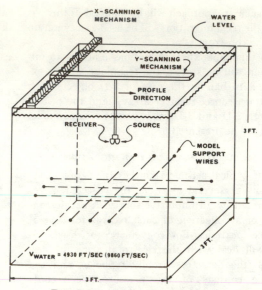

FIG. 2. Water tank arrangement.

thickness. For such models, only a single profile of data is required. Three-D models have thickness variations both in-line with and perpendicular to the profile line; these models require an areal coverage of reflection data.

A 3-D model was studied, and in order to set a standard for the quality of results desired from the 3-D processing, an experiment was run on a 2-D model whose cross-section displayed the same structures and dips found in certain cross-sections of the 3-D model. Results of these experiments are given in the following sections along with others for an additional 2-D experiment which displays a pertinent phenomenon.

EXPERIMENTAL ARRANGEMENT

Ultrasonic pulse-echo electronics with digital recording was used to obtain the reflection data from a model supported in a water tank by fine wires (Figure 2). Two-D or 3-D single coverage reflection data were recorded at some predetermined height above the model. The 2-D data consisted of the usual single coverage profile while the 3-D data consisted of numerous parallel 2-D profiles resulting in a square grid of source-receiver locations. The model dimensions are such that the data represent simulated field data. The scaling is indicated by the following equivalent dimensions and velocities:

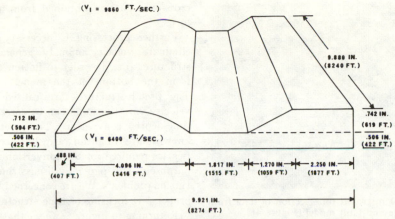

FIG. 3. Ridge and fault model with equivalent field dimensions. The equivalent field velocities are shown.

	Model	Field
Time	$.2\ \mu sec$	1 msec
Length	6.336 inches	1 mile
Velocity	v	$2v$

Hereafter, equivalent field values will be given in parentheses. In all cases the source-receiver separation was .827 inches (675 ft). The data were

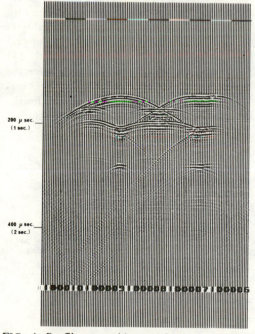

FIG. 4. Profile over ridge-and-fault model. Shotpoint separation = .133 inch (111 ft).

processed, however, as if the source and receiver were coincident. The actual sound speed in the material of which the models were constructed was less than that of water. Thus, the models represent structures with a velocity less than that of the overburden.

TWO-DIMENSIONAL RESULTS

The 2-D model consisted of a rounded ridge and an angular step (or normal fault) whose dimensions and corresponding field values are given in Figure 3. A single coverage reflection profile taken 4.11 inches (3430 ft) above the top of the step is shown in Figure 4. This profile was focused by 2-D migration according to equations (1) and (2a). The migration results are shown in Figure 5. The simple process described by equations (1) and (2a) generates some background noise seen in the figure. This migration was carried out using a constant field-equivalent rms velocity for all depths [i.e., $V(z) = 9860$ ft/sec in equation (2a)]. As a result, the top surface of the model is properly in focus but the velocity-anomaly generated structures of the bottom surface are slightly out of focus. Furthermore, due to the fact that the same rms velocity-depth function was used for all traces (i.e., horizontal velocity variations were ignored), the migrated profiles generated are what we have called "true vertical time sections" rather than "true structural geometry sections." Thus, velocity-anomaly generated pseudostructures are not eliminated by the process. They are, in fact, brought into approximate focus.

A clear example of this pseudostructural focus-

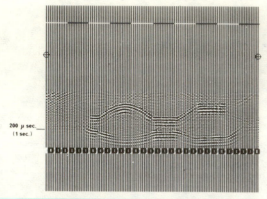

200 μ sec.
(1 sec.)

FIG. 5. 2-D migrated time section of data from ridge-and-fault model (Figure 4).

ing is presented in Figures 6, 7, and 8. Figure 6 gives dimensions for the 2-D model of a low-velocity layer with a high curvature ridge. Note that the bottom of the model is flat. A reflection profile taken at a height of 8.9 inches (7440 ft) above the center of this model is shown in Figure 7, and the migrated result using the same $V(z)$ for all traces is shown in Figure 8. A velocity-anomaly generated pseudosyncline in the reflection from the base of the model is brought into focus by the migration just as is the true ridge on the surface of the model. There is a three-to-one exaggeration of the vertical scale in these profiles which explains the stubby appearance of the migrated result when compared to the model shown in Figure 6. An incidental result (apparent in Figure 7) is the fact that the velocity-anomaly generated syncline produces the same reflection

cross-over pattern obtained from a real buried focus.

Further processing is necessary in order to eliminate velocity-anomaly generated pseudo-structures (i.e., in order to flatten the reflection from the bottom of the model in this case). The pseudostructures are caused by interval velocity variations and can be corrected through time-depth conversion. The methodology is known and straightforward, but accurate velocities are required for the conversion. It is not the purpose of this paper to discuss time-depth conversion problems. We are concerned here with the proper 3-D focusing of time structures.

Returning to Figure 5, we see that the migrated results give a true picture of the upper surface of the model. This was verified by comparison with measurements taken directly from the model. (From here on we will be concerned only with the upper surface of the models as the pseudostructures from the lower surfaces have been improperly focused.)

Figure 5 represents a 48 trace-scan migration (24 to each side), i.e., data from 48 traces were summed to give each migrated trace. A field equivalent 30–60 hz, 24 db/octave band-pass filter was applied to the results (the equivalent bandwidth of the somewhat "ringing" source used was 30–60 hz). The figure indicates the quality of focusing obtainable when equations (1) and (2a) are applied to simple 2-D model data. This 2-D experiment was carried out as a control for the 3-D experiment (to be described in the next section), and the processing parameters used for the 2-D migration (trace-scan distance, velocity,

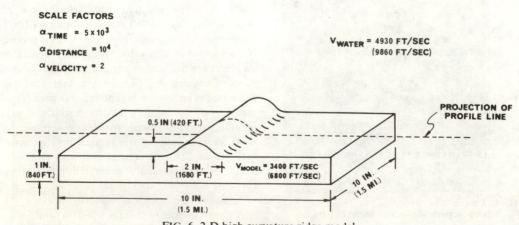

SCALE FACTORS

$\alpha_{TIME} = 5 \times 10^3$

$\alpha_{DISTANCE} = 10^4$

$\alpha_{VELOCITY} = 2$

$V_{WATER} = 4930$ FT/SEC
(9860 FT/SEC)

PROJECTION OF PROFILE LINE

0.5 IN (420 FT.)

1 IN.
(840 FT.)

2 IN.
(1680 FT.)

$V_{MODEL} = 3400$ FT/SEC
(6800 FT/SEC)

10 IN.
(1.5 MI.)

10 IN.
(1.5 MI.)

FIG. 6. 2-D high curvature ridge model.

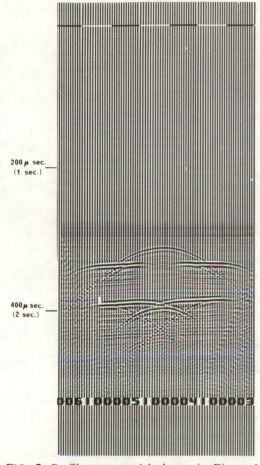

FIG. 7. Profile over model shown in Figure 6.
Shotpoint separation = .2 inch (167 ft).

shotpoint distances, final filter parameters) were also used for the 3-D migration.

THREE-DIMENSIONAL RESULTS

The model used for the 3-D experiment is shown in Figure 9 (see Figure 11 for a perspective view). The purposes of the experiment were to study 3-D complications in reflection profiles and to provide data from a known structure for testing 3-D migration as defined by equations (1) and (2a). Reflection data were recorded over the model and processed to produce 13 different 3-D migrated cross-sections of the model.

Figure 10 represents a plan view of the data collection scheme with the model shown schematically in solid lines. The numbers on the left of Figure 10 represent the successive source-receiver

locations for the first of 96 profiles shot over the model. Numbers along the top of Figure 10 represent the locations of the 96 profiles. The dashed lines which are numbered at the bottom of Figure 10 show the locations of the cross-sections which were calculated for the purpose of constructing a structure map. Our choice of these thirteen 3-D migrated cross-sections was quite arbitrary. The total number of 3-D migrations performed on field data will depend only upon the number of true cross-sections required to delineate the structures. Figure 11 is a photograph of the model showing relief along the cross-sections of interest.

Results for cross-sections 3, 7, and 11 are shown in Figures 12 to 14. Each figure represents a different cross-section, and the display labeled A is the 2-D raw data profile over that cross-section. The display labeled B is a conventional 2-D migration of the raw data. Display C of each figure is the 3-D migration result for that cross-section. The single raw data profile shown is used in the 2-D migration. The 3-D migration, on the other hand, uses data from a number of profiles on either side of the reconstruction plane of interest. We chose to collapse the data from 48 consecutive profiles (24 on each side) into each of the 3-D migrated cross-sections shown in Figures 12–14.

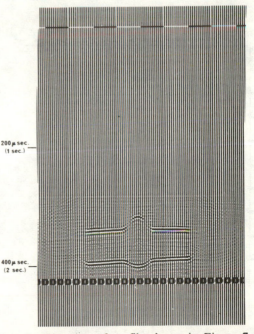

FIG. 8. Migration of profile shown in Figure 7.

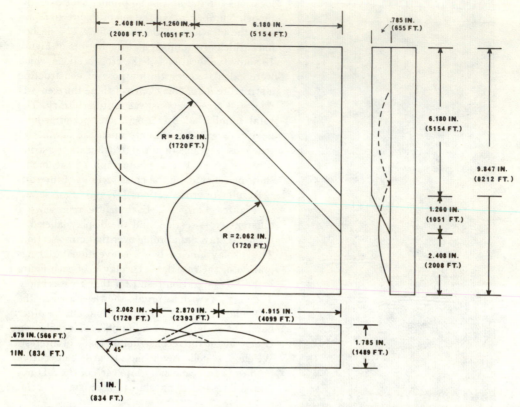

FIG. 9. 3-D model with equivalent field dimensions. See Figure 11 for perspective view.

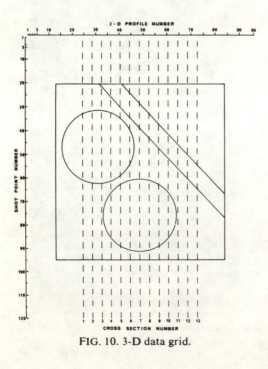

FIG. 10. 3-D data grid.

The quality of the results indicates that less data probably could have been used without serious deterioration of the results. The 96 parallel raw data profiles were taken .133 inches (111 ft) apart while the 3-D migrated profiles were reconstructed at intervals of .532 inches (444 ft).

It is most instructive to examine each of Figures 12 to 14, comparing the raw data, 2-D migrated data, and 3-D migrated data with the true cross-section (Figure 11). Here, the central cross-section (no. 7 in Figure 11) and the corresponding data of Figure 13 are singled out for specific discussion.

Several statements can be made concerning the results shown in Figure 13.

(1) The raw data are extremely difficult to interpret due to diffraction and sideswipe events. Of course, when one starts with a knowledge of the model geometry, it is easy to account for all the arrivals in the raw data.

(2) True structures in the profile plane can be

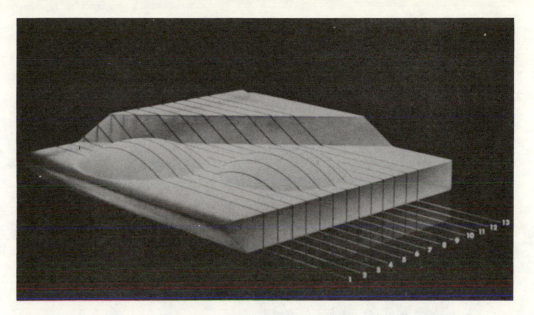

FIG. 11. Photograph of 3-D model.

absent in the raw data due to a component of dip perpendicular to the profile plane (blind structures). For example, no reflected energy from the portion of the fault which lies in the plane of the profile is recorded in the raw data profile.

(3) Conventional 2-D migration of the raw data profile does not eliminate sideswipe events or enhance blind structures. For example, the central hump in Figure 13b is a sideswipe event and has not been removed by conventional migration. This sideswipe causes an ambiguity in lateral correlation. (One could even go so far in this example as to assume the horizontal feature under the hump to represent a gas-fluid contact under an anticlinal trap.) Notice also that arrivals from the correct fault plane position are absent. Correlation across this blind zone is difficult.

(4) Two-D migration produced an increase in the background noise. This is probably a result of using the simplified integration scheme represented by the sum in equation (1).

(5) Three-D migration eliminated sideswipes and brought out blind structures. The resultant profile, shown in Figure 13c, is devoid of any interpretational ambiguities caused by these phenomena.

(6) The background noise created by the 3-D migration is similar to that of the 2-D migration and can probably be reduced by use of more sophisticated numerical techniques.

Similar statements could be made concerning the cross-sections displayed in Figures 12 and 14.

STRUCTURE MAPS FROM 2-D AND 3-D RESULTS

The upper surface of the model was mapped by first using a more conventional 2-D shooting and migration plan and then using the 3-D migration results. These maps are shown, respectively, in Figures 15 and 16. Contour values used to construct the maps were taken from every fourth shot point as this interval is equal to the distance between the processed profiles. The location of layer terminations and the fault-plane boundaries were read (to the nearest shotpoint) from the profiles.

Several statements can be made concerning the mapping of these experimental data:

(1) Structure could not be determined from the raw data profiles; migration was necessary.

(2) The structural map constructed from the 3-D migrated profiles resulted in a true and precise picture of the upper surface of the model (compare Figure 16 with Figure 10).

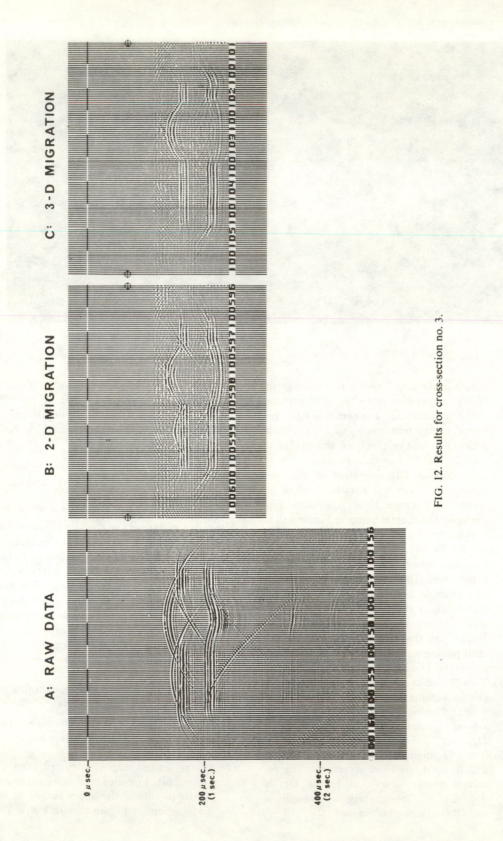

FIG. 12. Results for cross-section no. 3.

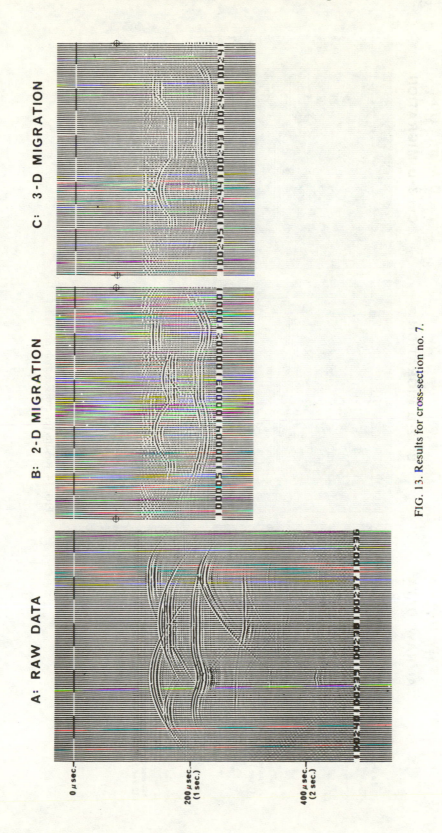

A: RAW DATA

B: 2-D MIGRATION

C: 3-D MIGRATION

0 μsec.

200 μsec.
(1 sec.)

400 μsec.
(2 sec.)

FIG. 13. Results for cross-section no. 7.

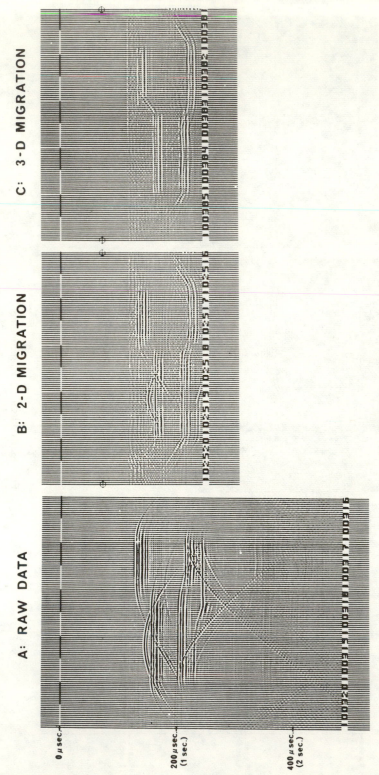

FIG. 14. Results for cross-section no. 11.

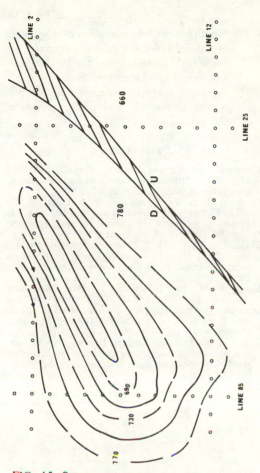

FIG. 15. Structure map constructed from migration of conventional 2-D profiles (contours in msec).

would have been inadequate. The advantages of 3-D processing will probably be even more evident in layered cases where greater correlation ambiguities exist. The 3-D process will also allow us to image features which are some distance removed from our data-gathering area.

CONCLUSIONS

The following conclusions can be drawn from the results and discussion presented above:

(1) The migration method of processing seismic-reflection data can be extended to handle 3-D exploration problems with arbitrary source and receiver arrays. A mathematical description is given by equations (1) and (2) above.

(2) When an average velocity-depth function is used over an entire section, false structures caused by lateral velocity variations in the overburden are not eliminated but, rather, are brought into focus by migration.

(3) The structural map made from the 2-D migrated profiles is distorted in shape (compare Figure 15 with Figure 10).

With dense enough 2-D coverage and 2-D migration, structural distortions in the 2-D data can be corrected through careful interpretation. As the interpreter develops his map, he can detect which events are sideswipes on individual 2-D profiles and apply appropriate areal displacements. In other words, this experiment does not demonstrate that 3-D migration is essential for delineating the features of this model, although it clearly shows that at least 2-D migration is indispensable. Furthermore, the gathering of dense areal data is essential to an accurate resolution of the model; conventional 2-D recording

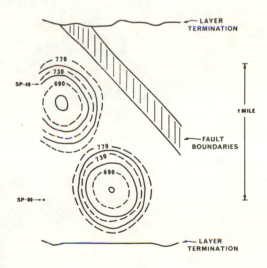

FIG. 16. Structure map constructed from 3-D migrated profiles (contours in msec).

Subsequent trace-by-trace, time-to-depth conversion will eliminate these pseudo-structures but a rather detailed knowledge of the section velocity is required.

(3) When single coverage reflection profiles over 2-D models are processed by conventional migration, an exact reproduction of a true vertical-time section of the model results. The background noise is, however, increased. It has been shown that these noise problems can be eliminated by better numerical techniques (Gardner et al, 1973).

(4) Single coverage profiles recorded over 3-D models are almost totally uninterpretable unless a migration process is applied.

(5) When single coverage reflection profiles over 3-D models are processed by conventional 2-D migration, correlation ambiguities may arise due to sideswipes and blind structures.

(6) Structure maps constructed from a conventional survey with profiles independently processed by 2-D migration show distortions of the structures. However, the distortions can be reduced by increased coverage and careful interpretive techniques.

(7) Simultaneous 3-D migration of a sequence of parallel reflection profiles eliminates the correlation ambiguities caused by sideswipes and blind structures.

(8) Structure maps constructed from the 3-D results give a true and precise picture of the model.

REFERENCES

Claerbout, Jon F., and Doherty, Stephen M., 1972, Downward continuation of moveout corrected seismograms: Geophysics, v. 37, p. 741–768.

Gardner, G. H. F., French, W. S., and Matzuk, T., 1973, Elements of migration and velocity analysis: Presented at the 43rd Annual International SEG Meeting, October 25, 1973, Mexico City.

Maginness, M. G., 1972, The reconstruction of elastic wave fields from measurements over a transducer array: J. of Sound and Vibration, v. 20, no. 2, p. 219–240.

Morse, P. M., and Ingard, K. U., 1968, Theoretical acoustics: New York, McGraw-Hill Book Co., Inc.

Schneider, William A., 1971, Developments in seismic data processing and analysis (1968–1970): Geophysics, v. 36, p. 1043–1073.

Taner, M. Turhan, and Koehler, Fulton, 1969, Velocity spectra—digital computer derivation and applications of velocity functions: Geophysics, v. 34, p. 859–881.

Reprinted from GEOPHYSICS, **39**, 811-825.

ELEMENTS OF MIGRATION AND VELOCITY ANALYSIS

G. H. F. GARDNER,* W. S. FRENCH,* AND T. MATZUK‡

This paper describes a theory for combining migration and velocity analysis of seismic data and uses experimental model data to illustrate some of the practical problems. It is shown that adaptive weighting of the data, based on a measure of coherence, can be effective in suppressing false images and noisy backgrounds caused by finite migration apertures, finite sampling in space and time, and random noise in the data. Velocity analysis is presented as a "range finding" method in which images for two different constant offset profiles are brought into coincidence by using the correct velocity. The method is valid even in the presence of diffraction, dipping interfaces, or curved reflectors.

INTRODUCTION

Early applications of migration theory appeared in the sonograph method (Rieber, 1937) to unscramble "random" or confused seismic records and in the "aplanatic surfaces" method (Gardner, 1949) to outline boundaries of salt domes. In recent years, such seismic processing has been largely automated, and digital programs are available for migrating standard seismic profiles, provided a suitable velocity-depth relationship is known (Schneider, 1971).

Velocity-depth relationships are often derived from CDP data by using either hand methods (Green, 1938; Dix, 1952) or semiautomatic digital methods (Schneider, 1971). However, these methods are least certain in regions of steep dip, faults, and other complex geologic features, which are also the very regions in which migration might be most useful.

In this paper, we discuss the possibility of combining velocity analysis with migration so that both may be improved and usefully applied in regions with complex structures.

MIGRATION OF CONSTANT OFFSET PROFILES

For brevity, the following discussion is limited to two-dimensional problems. The extension of migration to three-dimensional problems has been discussed and illustrated by French (1974).

The subsurface of the earth is often modeled as a layered medium, each layer having uniform acoustic properties. The reflected signals for such a model arise at the interfaces between media. In this paper we are assuming that for purposes of migration the layers can be replaced by scattering surfaces at each interface, and the reflections can be treated as the sum of the energy scattered from a large number of closely spaced points located on the interface. The scattering amplitude at each point is taken as proportional to the reflection coefficient of the interface. This conception forms the basis of an elementary theory of migration. The wave field is processed so as to display point-by-point the interface sheets which would give rise to the observed field.

The practical utility of approximating a continuous reflecting sheet by discrete scattering points was illustrated experimentally in a two-dimensional scaled seismic test tank. A horizontal plane reflecting sheet was approximated by horizontal cylindrical threads as illustrated in Figure 1. The distance between the parallel threads was

Paper presented at the 43rd Annual International SEG Meeting, October 25, 1973, Mexico City. Manuscript received by the Editor January 30, 1974; revised manuscript received May 7, 1974.

* Gulf Research & Development Co., Pittsburgh, Penn. 15230.

‡ Consultant, Pittsburgh, Penn.; formerly, Gulf Research & Development Co.

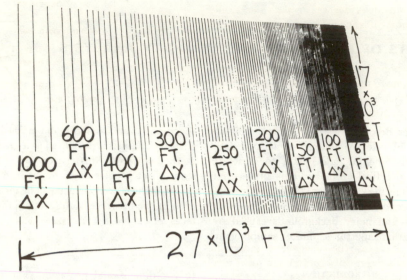

FIG. 1. Model used to illustrate the validity of approximating a plane reflecting sheet by discrete scattering line elements. One-fourth inch in the model represents 250 ft in field dimensions. The model was placed in a water tank at a depth of 8 inches.

reduced in steps, as illustrated, from a separation of many wavelengths at the left to a fraction of a wavelength at the right.

In terms of equivalent scaled field dimensions, the reflector was at a depth of 8000 ft in a uniform medium in which the velocity of sound was 10,000 ft/sec. The principal wavelength of the pulse was 250 ft; the separation of the cylinders varied from 1000 ft maximum to 67 ft minimum; and the diameter of the cylinders was 22 ft. The source and receiver had a constant offset horizontal separation of 476 ft, and the profile data were collected along a surface line perpendicular to the axes of the cylinders.

Figure 2 is a variable density photograph of the signals returning from the threads. It was ob-

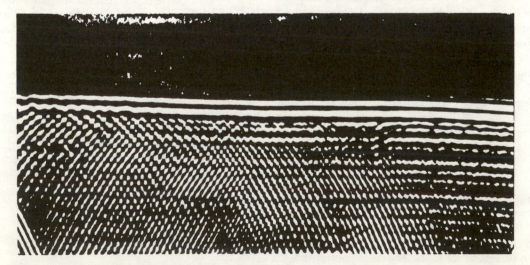

FIG. 2. Variable density display of data scattered by line elements shown in Figure 1. Elements are 1000 ft apart at the left and 67 ft apart at the right.

tained by time-exposing a cathode ray tube on which each recorded trace was displayed vertically at a horizontal distance proportional to the horizontal distance moved by the source-receiver pair. The first arrival of the pulse is overmodulated in the photograph because of the large gain required to make the later events visible.

At the left of the photograph, where the scattering points are 1000 ft (4 wavelengths) apart, the signal from each thread can be traced along a hyperbolic arc extending downward from the thread. The shoulders and tails of these curves form an overlapping pattern which obscures the signature of the pulse following the first strong arrivals. At the right of the photograph, where the scattering points are 67 ft (0.27 wavelengths) apart, the signals have developed lateral continuity. The apexes of the curves from individual threads form a practically continuous event. The signals along the shoulders and tails of the curves interfere destructively, and the signature of the pulse following the first arrival is visible as horizontal continuous bands. By following this signature horizontally to the left one can see the gradual development of a discrete overlapping pattern as the spacing between the scattering points increases in steps. The loss of continuity appears to occur when the spacing exceeds about 150 ft (0.6 wavelengths).

The data displayed in Figure 2 give intuitive support for the argument that the wave field from closely spaced scattering points approaches the

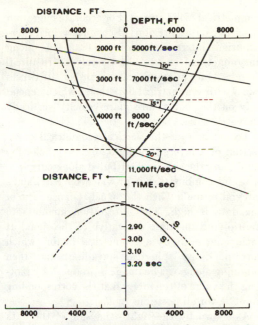

FIG. 4. The two-way traveltime from P to the surface (A and B coincident) is given by S (dashed) when the interfaces (dashed) are flat and by S' when the interfaces are dipping; the velocities, thicknesses, and dips are as indicated.

wave field reflected by a continuous sheet in the limit of zero spacing. A mathematical supporting argument is given in Appendix A.

A basic principle in migration theory is the imaging of each point in the subsurface by detecting the field scattered by that point. Hence, the traveltime from the shot to each subsurface point and back to the receiver, for each position of shot and receiver, is of fundamental importance. For a constant offset profile (AB constant, in Figure 3) this gives rise to a correspondence between points P in (x, z) space and corresponding curves S in (x, t) space. In Figure 3, the traveltime from A to P to B is plotted as CD under the shot position. As A moves along the profile, point D sweeps out curve S.

Although each point in the subsurface corresponds to a curve S, the location and shape of curve S is also dependent upon the distribution of the velocity of sound above point P. Figure 4 illustrates the nature of this dependence for a layered subsurface. If the interfaces are horizontal, the traveltime curve is symmetrical (dashed curve), but if the interfaces dip, the curve is un-

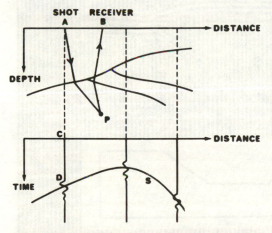

FIG. 3. CD is the time taken by a pulse to travel from A to P to B. As AB, the fixed offset separation, moves along the profile, D sweeps out the curve S.

symmetrical (solid curve). For accurate migration of seismic data in regions of complex velocity distribution, interrogation of the data field along unsymmetrical curves like S' may be required. Standard migration programs approximate curves like S with symmetrical hyperbolas which depend only on the root-mean-square velocity vertically above P.

To reconstruct the vertical geometrical cross-section of a layered medium, one may begin by combining the signals distributed along curves S. A useful combination is provided by Kirchhoff's integral formula when the signal-to-noise ratio of the data is high. This integration amounts to multiplying the time derivative of the signal at each point on S by a geometrical factor (which corresponds to spherical spreading) and then summing along S. The image is produced by plotting the value of the integral at the corresponding point P, for all positions of P.

As shown by the right-hand side of the data field in Figure 2, the signal from points on a continuous reflecting interface is very small outside a relatively short segment on curve S because of destructive interference. If the integration in Kirchhoff's formula is extended much beyond this segment, the addition of noise starts to offset the addition of signal. To avoid adding noise in excess of signal, Kirchhoff's formula may be modified by restricting the range of integration to a limited arc of S which contains the signal.

A second modification of Kirchhoff's formula is often made by integrating the amplitude rather than the time derivative of the amplitude. This is done as a convenience and can be corrected by filtering the migrated data field.

The end result of these modifications is that in practice migration consists of weighting amplitudes along S and summing along S; zero weights may be assigned outside some limited arc of S. Just how the weights should be picked depends primarily on the noise field, and the difference between good and fair migration programs often depends upon how cleverly the weights are picked.

Figure 5 shows (in variable area format) a constant offset profile over a two-dimensional scaled model of an anticline and fault. The model was a silicone rubber molding immersed in a tank of

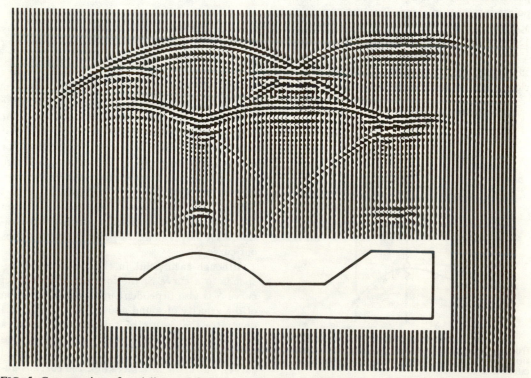

FIG. 5. Cross-section of anticline and fault model and the corresponding constant offset seismic profile.

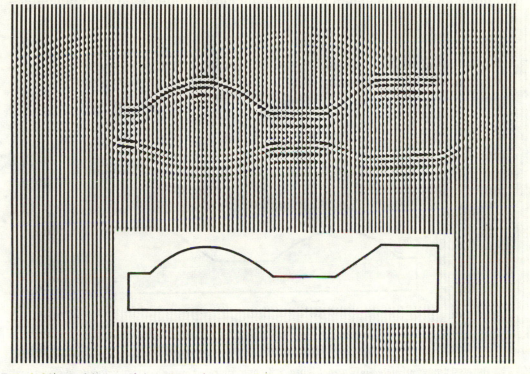

FIG. 6. Migrated image of data shown in Figure 5. Image migrated using correct velocity, symmetrical unit weights on 48 traces, and zero weights elsewhere.

water. The distance between shotpoints in the experiment was 0.14 inches, and the traces were digitized every 0.8 μsec; corresponding field magnitudes were 115 ft and 4 msec.

A migrated image of this data field is shown in Figure 6. The traveltime curves for the migration were determined by assuming that above each point P there is a constant velocity equal to the velocity of sound in water. The weights were unity for 24 traces on either side of P and zero for all further traces. A filter was applied to the migrated traces to restore the bandwidth of the original data.

Defects in the image in Figure 6 can be traced to the combined effects of (a) integration along finite intervals of S, (b) integration with unit weights, (c) random noise, and (d) finite sampling in space and time. These defects can be reduced in various ways. The false images above the first interface are caused partly by the abrupt truncation of the integration along S and partly by the finite sampling in space. The effect of abrupt truncation of the integration can be reduced by

tapering the weights from unity at the apex of S to zero at the ends of the range of integration, as can be seen by comparing Figure 6 and Figure 7 for which tapered weights were used. However, the dip resolution in Figure 7 is poorer than in Figure 6 because of the tapered weights. Dip resolution could be restored by increasing the range of integration.

The benefits derived by increasing the range of integration are offset by the increased addition of random noise. For this reason it is desirable to examine the data along S and limit integration only to that portion of S that contains the main contribution to the image.

Finding the portion of S that contains the main contribution to the migrated image is facilitated by studying the amplitude values along several migration curves as shown in Figure 8.

The amplitudes of the signals along curves A and B are plotted as histograms labeled A and B in Figure 8. Coherence occurs where the curves are tangent to events in the data. The coherence length L can be estimated for smooth flat reflec-

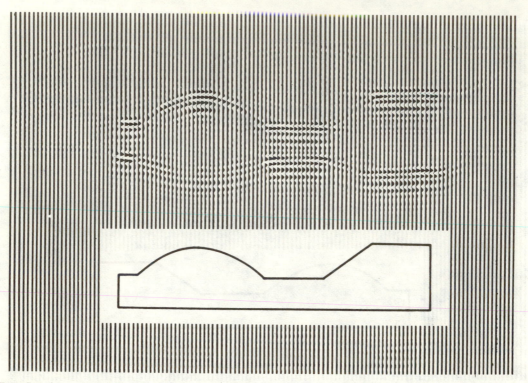

FIG. 7. Migrated image of data shown in Figure 5. Image migrated using correct velocity, symmetrical weights linearly tapered to zero over 48 traces, and zero weights elsewhere.

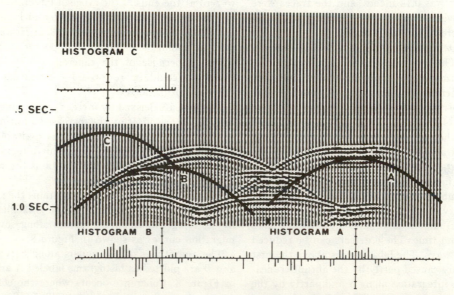

FIG. 8. Amplitude distribution along several migration curves.

tors. The equation for S in a medium with sound velocity V can be approximated for distance x small compared with depth z by

$$t = 2z/V + x^2/zV.$$

Reasonable coherence can occur if the increase in t does not exceed one-half a period for the pulse. Hence the maximum x over which coherence can occur is given by

$$x^2/zV = (1/2)(\lambda/V).$$

From this the distance $L(=2x)$ is given by

$$L = \sqrt{2\lambda z}.$$

In the example shown in Figure 8, $z \doteq 3500$ ft and $\lambda \doteq 250$ ft. Hence $L \doteq 1300$ ft. Since the spacing between recordings is 115 ft, we would expect coherence over approximately $1300/115 = 11$ consecutive traces. Examination of the amplitudes along curves A and B in Figure 8 will show that this formula is a useful estimate of the coherence length. Outside the coherence interval the summed amplitudes have positive and negative signs and their sum tends to be small.

By correlating neighboring traces for this coherence length along curves S and over a short time window, one can automatically weight the integration process for migration. These weights are similar to the semblance calculated for velocity analyses (Taner and Koehler, 1969, appendix 3). The effectiveness of this procedure is illustrated in Figure 9 where a comparison is shown for (a) raw data with random noise, (b) migration with unit weights, and (c) migration with an an adaptive taper based on a semblance measure.

Finally, adequate sampling of the reflection field in space and time is important in obtaining a migrated image with a minimum of false images. Curve C in Figure 8 shows two large positive amplitudes at the right end. Finer spatial sampling would result in negative values in between and, in consequence, a reduced sum. The migrated image shown in Figure 6 was based on sampling with a scaled space interval of 115 ft and a time interval of 4 msec. The effect of increasing the spatial sampling, while maintaining the same time sampling, is illustrated in Figure 10. The three images, based on spacings of 230 ft, 345 ft,

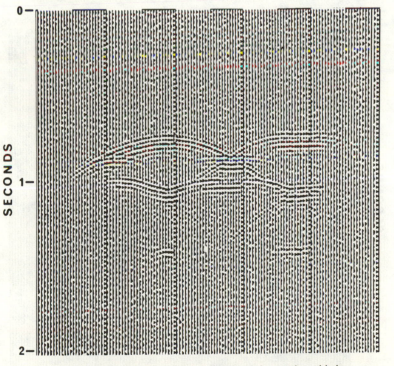

FIG. 9a. Data shown in Figure 5 with random noise added.

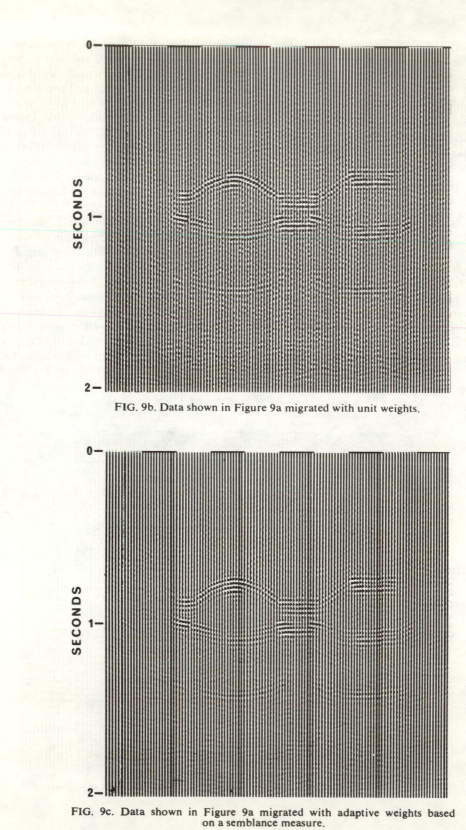

FIG. 9b. Data shown in Figure 9a migrated with unit weights.

FIG. 9c. Data shown in Figure 9a migrated with adaptive weights based on a semblance measure.

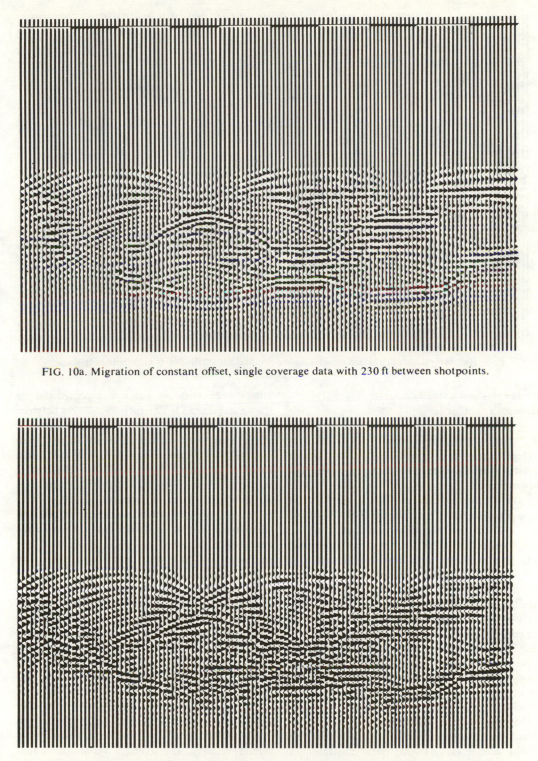

FIG. 10a. Migration of constant offset, single coverage data with 230 ft between shotpoints.

FIG. 10b. Migration of constant offset, single coverage data with 345 ft between shotpoints.

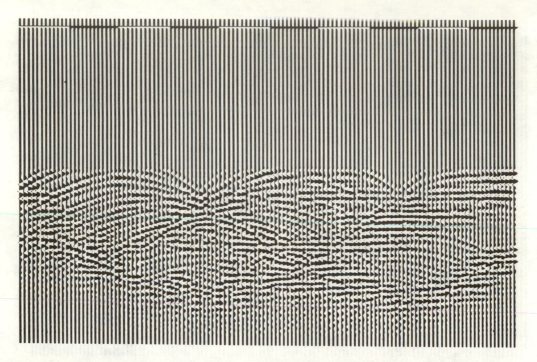

FIG. 10c. Migration of constant offset, single coverage data with 460 ft between shotpoints.

and 460 ft, show a steady decrease in quality with increase in spacing.

VELOCITY ANALYSIS

Migration of a constant-offset seismic profile requires that a correspondence be postulated between point P and traveltime curve S. Once any traveltime curve is postulated, some sort of correspondence between the curve and the point can be achieved through migration, but no definitive test exists to prove that the attempted correspondence results in the correct image.

However, additional data enables one to propose a definitive test. For example, if two profiles have been recorded with different constant offsets, the possible correspondences between point P and curve S are severely restricted by the requirement that the migrated images be as nearly alike as possible. In practice, it appears that a unique image can often be obtained. However, multiple reflections or a ringing pulse may complicate the problem in some cases.

This method of determining velocity and focusing on the correct image can be illustrated for the model experimental data illustrated in Figure 5. Two profiles with different constant

offsets, 1125 ft and 6405 ft, were recorded and are shown in Figure 11. The correspondences between points P and curves S were calculated for each offset; a constant velocity of sound above P was assumed. Each profile was then migrated for a range of velocities above and below the correct velocity. In an attempt to convey the results in a simple visual form, the locus of the first positive peaks of each migrated image was traced and these are presented as dashed line curves for the near offset migrated profiles and as solid line curves for the large offset migrated profiles. The comparison in shapes and alignments of near versus far offset migrated profiles for low velocity (Figure 12a), correct velocity (Figure 12b), and high velocity (Figure 12c) shows that a definitive test has indeed been demonstrated. In particular, the correct velocity produces both the most accurate geologic cross-sectional view and the most coincident vertical alignment between near and far offset migrated images.

The main difference between the migrated images for the large and small offset is a vertical displacement along the time axis. For any assumed velocity, the average vertical misalignment can be measured. Figure 13 shows this mis-

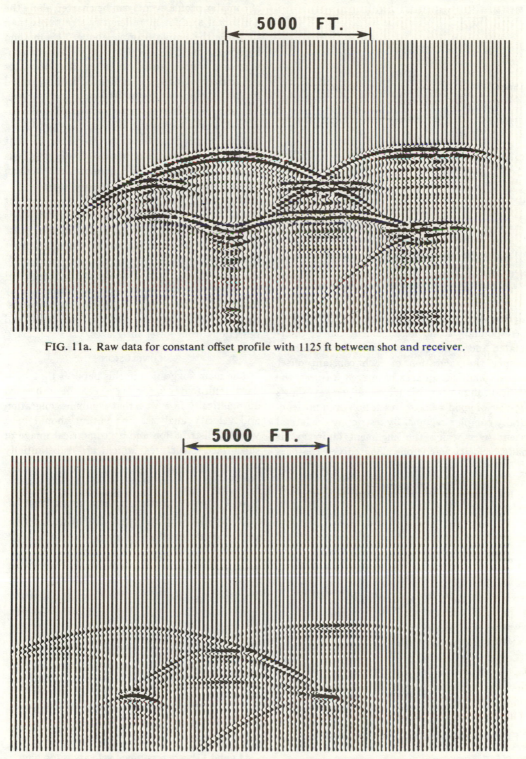

FIG. 11a. Raw data for constant offset profile with 1125 ft between shot and receiver.

FIG. 11b. Raw data for constant offset profile with 6405 ft between shot and receiver.

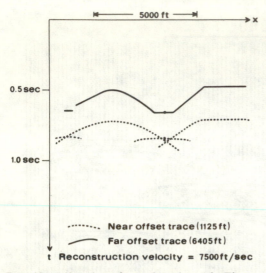

FIG. 12a. Migration of raw data shown in Figures 11a and 11b when assumed velocity is too low.

of complex profile events can be charted along the horizontal axis for more lengthy structures than shown in the simple example above. This method becomes almost identical to the method of conventional velocity analysis when the interfaces are horizontal and the comparison of migrated images is made on a trace-to-trace comparison. The similarity is even more apparent if one considers 24 or 48 constant offsets, each producing a migrated profile. The main differences are that the present method is applicable to dipping beds or complex structures and that the focusing can be done over a window in space and time.

It should be noted that the frequency content of the signature of the migrated pulses depends on the offset distance. The far offset imaged pulse is stretched in time relative to the near offset imaged pulse (see Appendix B). This is similar to the effect described by Dunkin and Levin (1973) for stacking CDP data. The phenomenon may be turned to advantage if we can use it to distinguish between later peaks in the pulse and reflections from lower interfaces.

alignment plotted against velocity. The velocity corresponding to no misalignment can be interpolated and is the correct velocity for migration.

Since the migration of each constant offset profile comprises an area of imaged features, the vertical alignment of migrated images shows promise of good velocity resolution even in structurally complex regions. By restricting the lateral extent over which the alignment of images is made, velocity variations along curved contours

CONCLUSION

The main suggestion in this paper is that standard multifold seismic data can be processed automatically in a way that combines migration and velocity analysis. The output should be a velocity distribution and a geometrical image of subsurface structures which are substantially

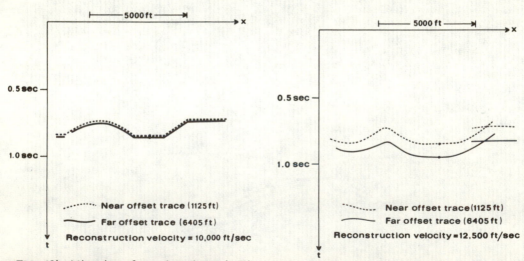

FIG. 12b. Migration of raw data shown in Figures 11a and 11b when assumed velocity is almost correct.

FIG. 12c. Migration of raw data shown in Figures 11a and 11b when assumed velocity is too high.

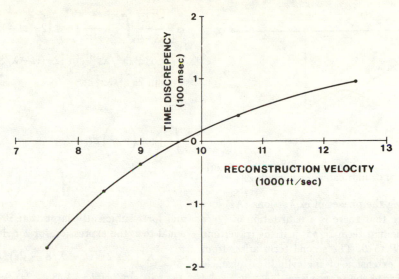

FIG. 13. For any assumed reconstruction velocity the migrated images of the near and far offset data will be out of alignment by a time almost proportional to the error in the assumed velocity. Interpolation to a zero time discrepancy gives the correct velocity for the event.

correct even in complex regions. Extension of the method to three-dimensional migration should result in improved accuracy and resolution.

REFERENCES

Dix, C. H., 1952, Seismic prospecting for oil: New York, Harper & Brothers, p. 124 ff.
Dunkin, J. W., and Levin, F. K., 1973, Effect of normal moveout on a seismic pulse: Geophysics, v. 38, p. 635–642.
French, W. S., 1974, Two-dimensional and three-dimensional migration of model experiment reflection profiles: Geophysics, v. 39, p. 265–277
Gardner, L. W., 1949, Seismograph determination of salt-dome boundary using well detector deep on dome flank: Geophysics, v. 14, p. 29–38.
Green, C. H., 1938, Velocity determinations by means of reflection profiles: Geophysics, v. 3, p. 295–305.
Rayleigh, 1926, The theory of sound: London, Macmillan, Ltd., v. 2, p. 310.
Rieber, F., 1937, Complex reflection patterns and their geologic source: Geophysics, v. 2, p. 132–160
Schneider, W. P., 1971, Developments in seismic data processing and analysis (1968–1970): Geophysics, v. 36, p. 1043–1073.
Taner, M. T., and Koehler, F., 1969, Velocity spectra—Digital computer derivation and applications of velocity functions: Geophysics, v. 32, p. 859–881.

APPENDIX A

The scattered field of a cylindrical obstacle in a plane wave field

We consider first the case of a uniform medium in which there is a single cylindrical obstacle P whose radius R is small compared with the wavelength (Figure 14). The incident pulse is a plane wave traveling in the direction of r. Taking one Fourier component of the pulse at the datum level $r = 0$ as

$$\cos 2\pi vt/\lambda, \qquad (A1)$$

we can write the scattered signal received at B, the origin of the coordinates, approximately as (Rayleigh, 1926)

$$K \frac{R^2}{r^{1/2}\lambda^{3/2}} \cos \left[2\pi(vt - 2r - \lambda/8)/\lambda\right], \quad (A2)$$

where K depends on the mechanical parameters of the cylinder and the surrounding medium.

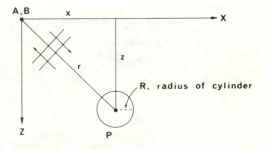

FIG. 14. Schematic diagram showing pulse source at A, receiver at B, and horizontal cylindrical wire at P.

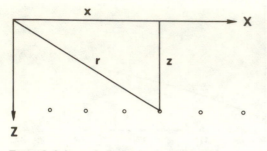

FIG. 15. Schematic diagram showing a line of horizontal cylindrical wires at depth Z.

Comparing the phases of expressions (A1) and (A2), we see that there is a retardation of $(2r/v + \lambda/8v)$. The first term $2r/v$ is just the traveltime from A to P to B. The second term arises from the infinite extension of the cylindrical obstacle perpendicular to the (x, z) plane.

The scattered field of a line of cylindrical obstacles

Next we consider a line of cylindrical obstacles as shown in Figure 15. Adding the scattered fields of the cylindrical obstacles, each given appropriately by equation (A2), we obtain the limit for closely spaced cylinders in the form

$$\phi = \frac{KR^2}{\lambda^{3/2}} \int_{-\infty}^{\infty} \frac{1}{r^{1/2}} \cos 2\pi(vt - \lambda/8 - 2\sqrt{z^2 + x^2})/\lambda \, dx. \quad \text{(A3)}$$

Using the approximation

$$2\sqrt{z^2 + x^2} \doteq 2z + x^2/z \quad \text{(A4)}$$

and the identity, $\cos (A - B) = \cos A \cos B + \sin A \sin B$, we find

$$\phi = \frac{KR^2}{\lambda^{3/2}} \int_{-\infty}^{\infty} \frac{1}{r^{1/2}} \left\{ \cos \left[2\pi(vt - \lambda/8 - 2z)/\lambda \right] \right.$$
$$\cdot \cos 2\pi x^2/\lambda z + \sin \left[2\pi(vt - \lambda/8 - 2z)/\lambda \right]$$
$$\left. \cdot \sin 2\pi x^2/\lambda z \right\} dx. \quad \text{(A5)}$$

Since

$$\int_0^{\infty} \cos \left(\frac{\pi}{2} t^2 \right) dt$$

$$= \int_0^{\infty} \sin \left(\frac{\pi}{2} t^2 \right) dt = 1/2,$$

and for r sufficiently large that it can be taken equal to z, the expression for ϕ reduces to

$$\phi = K' \{ \cos 2\pi(vt - \lambda/8 - 2z)/\lambda$$
$$+ \sin 2\pi(vt - \lambda/8 - 2z)/\lambda \} \quad \text{(A6)}$$
$$= \sqrt{2} \, K' \{ \sin \pi/4 \cos 2\pi(vt - \lambda/8 - 2z)/\lambda$$
$$+ \cos \pi/4 \sin 2\pi(vt - \lambda/8 - 2z)/\lambda \}$$
$$= \sqrt{2} \, K' \{ \sin 2\pi(vt - 2z)/\lambda \}.$$

This last expression coincides in phase with the reflection from a *sheet* of material at a depth z below the datum line. In other words, the scattered field from an array of equally spaced parallel cylinders approaches the field reflected by a plane sheet in the limit that the distance between the cylinders shrinks to zero.

APPENDIX B

Effect of offset distance on the shape of a migrated pulse

The curve S corresponding to point P is given by

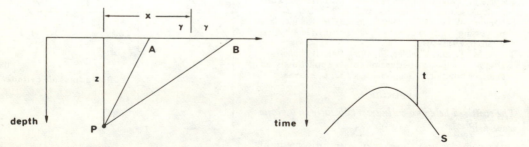

FIG. 16. Schematic diagram showing source A, receiver B, scattering point P, and the traveltime curve S for the path APB.

$$t = \frac{1}{v}\sqrt{z^2 + (x-l)^2}$$
$$+ \frac{1}{v}\sqrt{z^2 + (x+l)^2}, \tag{B1}$$

where v is the velocity and $2l$ the offset distance AB in Figure 16. Assuming $(x \pm l)$ is small compared with z, we see that S can be approximated by

$$t = 2z/v + (x^2 + l^2)/zv. \tag{B2}$$

If we take the wave field to be $\cos 2\pi vt/\lambda$, where λ is the wavelength, then the migrated field is, approximately,

$$2\int_0^\infty \cos 2\pi [2z + (x^2 + l^2)/z]/\lambda \, dx$$

$$= 2\int_0^\infty \{\cos [2\pi(2z + l^2/z)/\lambda] \cos[2\pi x^2/z\lambda]$$
$$\tag{B3}$$

$$- \sin [2\pi(2z + l^2/z)/\lambda] \sin [2\pi x^2/z\lambda]\} \, dx$$

$$= (1/2)\sqrt{z\lambda}\{\cos 2\pi(2z + l^2/z)/\lambda$$
$$- \sin 2\pi(2z + l^2/z)/\lambda\}$$

$$= (1/2)\sqrt{z\lambda} \cos 2\pi(2z + l^2/z + \lambda/8)/\lambda.$$

The wavelength of the migrated field Δz is given by the change in z which changes the argument of the cosine terms by 2π. Hence Δz is given by

$$2\Delta z - l^2\Delta z/z^2 = \lambda. \tag{B4}$$

Hence, solving for Δz,

$$\Delta z = \tfrac{1}{2}\lambda(1 + l^2/2z^2). \tag{B5}$$

We see then that any one frequency component of a wave field is migrated to a wavelength that is multiplied by the factor $1 + l^2/2z^2$, where $2l$ is the offset distance AB. Hence, in the first approximation, any pulse is stretched by this factor during migration.

Reprinted from GEOPHYSICS, **40**, 961-980.

COMPUTER MIGRATION OF OBLIQUE SEISMIC REFLECTION PROFILES

WILLIAM S. FRENCH*

A reflecting interface with irregular shape is overlain by a material of constant velocity V_T. Multifold reflection data are collected on a plane above the reflector and the reflector is imaged by first stacking then migrating the reflection data. There are three velocity functions encountered in this process: the measured stacking velocity V_{NMO}; the true overburden velocity V_T; and a profile migration velocity V_M, which is required by present point-imaging migration programs. Methods of determining V_{NMO} and, subsequently, V_T are well-known. The determination of V_M from V_T, on the other hand, has not been previously discussed. By considering a line-imaging migration process we find that V_M depends not only on the true section velocity but also on certain geometrical factors which relate the profile direction to the structure. The relation between V_M and V_T is similar to, but should not be confused with, the known relation between V_{NMO} and V_T. The correct profile migration velocity is always equal to or greater than the true overburden velocity but may be less than, equal to, or greater than the best stacking velocity. When a profile is taken at an angle of $(90-\theta)$ degrees to the trend of a two-dimensional structure, then the appropriate migration velocity is $V_T/\cos\theta$ and is independent of the magnitude of any dips present. If, in addition, the two-dimensional structure plunges along the trend at an angle γ, then the correct migration velocity is given by $V_T/(1-\sin^2\theta\cos^2\gamma)^{1/2}$. The time axis of the migrated profile for the plunging two-dimensional case must be rescaled by a factor of $[(1-\sin^2\theta\cos^2\gamma)/\cos^2\theta\cos^2\gamma]^{1/2}$, and structures on the rescaled profile must be projected to the surface along diagonal lines to find their true positions. When three-dimensional data are collected and automatic three-dimensional migration is performed, the geometrical factors are inherently incorporated. In that case, the migration velocity is always equal to the true velocity regardless of whether the structure is two-dimensional, plunging two-dimensional, or three-dimensional. Processed model data support these conclusions.

The equations given above are intended for use in conventional migration-after-stack. Recently developed schemes combining migration-before-stack with velocity analysis give V_M directly. In that case, the above equations provide a method of determining V_T from V_M.

INTRODUCTION

This paper considers the migration of seismic reflections from three classes of structures:

(1) *True two-dimensional (2-D) structures,* which display thickness variations in one horizontal direction only. A reflecting surface in a true two-dimensional structure can be considered as a collection of parallel horizontal lines, e.g., the surface of the ridge and fault model shown in Figure 1.

(2) *Plunging two-dimensional structures,* which can be swept out by parallel displacements of a straight nonhorizontal generating

Paper presented at the 44th Annual International SEG Meeting, November 14, 1974 in Dallas. Manuscript received by the Editor December 27, 1974; revised manuscript received May 15, 1975.

* Amoco Production Research Co., Tulsa, Okla. 74102; formerly Oregon State University, Corvallis, Ore.

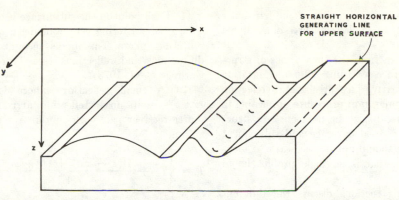

FIG. 1.   Ridge and fault model. The model is a true 2-D structure when the generating line is horizontal (parallel to the y-axis). This is a plunging 2-D structure if the generating line makes an angle γ with the y-axis while remaining perpendicular to the x-axis.

line. For example, if the structure shown in Figure 1 were made to dip in the y-direction it would be a plunging two-dimensional structure.

(3) *Three-dimensional (3-D) structures,* which may display variations in any horizontal direction and which must be described point by point. The upper surface of the model shown in Figure 2 is an example. Portions of three-dimensional surfaces may, however, represent true or plunging two-dimensional structures. The fault plane in the model shown in Figure 2

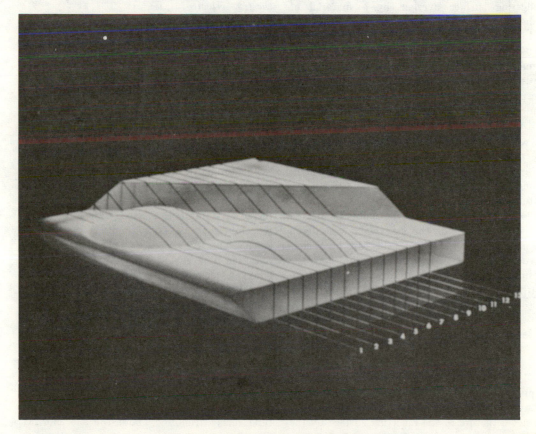

FIG. 2.   Photograph of 3-D model.

is, in effect, a true two-dimensional structure which is part of a larger three-dimensional surface.

Reflection data from the three types of structures can be processed using different migration techniques. Clearly, any areal data collection and computer migration procedure applicable to 3-D structures can also be used on 2-D structures. However, areal coverage and subsequent 3-D processing should not be necessary to focus 2-D structures, that is, a single profile should suffice.

The following sections discuss the different methods of migrating data from the three classes of structures. Three-dimensional wave fields (point sources and point receivers) are considered in all cases, even though the reflecting structure may be 2-D or 3-D.

A MODEL OF THE SUBSURFACE

The migration of exploration data can be studied with a model consisting of a single reflecting surface overlain by a material of constant velocity V_T (referred to as the true velocity). The conclusions to be drawn from consideration of this simple model are quite general, qualitatively if not quantitatively, because the following traveltime calculations could be performed using a depth dependent velocity function for the overburden.

In discussing migration techniques, it is easiest to treat the general case first. A brief review of certain aspects of 3-D migration follows. This process has been described in detail elsewhere (French, 1974; Sattlegger, 1975). The intuitive description of migration used below follows from certain analytic solutions of the wave equation. For the case being considered (constant V_T) the wave field can be reconstructed by the Kirchhoff integral, and the resultant reflector images are meaningful to within the validity of Claerbout's Principle (see Claerbout, 1971; Claerbout and Doherty, 1972). Appendix B is devoted to a further discussion of this point.

THREE-DIMENSIONAL MIGRATION

Zero offset single coverage reflection data (or equivalent stacked data) collected over an area above the reflecting surface are used to construct a picture of the reflecting surface by the following process of migration:

(1) Each point in the subsurface is considered a possible scattering center modeled as a small reflecting sphere. This model accounts for both diffractions and reflections (see Trorey, 1970; Gardner et al, 1973).

(2) A small reflecting sphere located at (x,y,z) results in a reflection arrival for the source-receiver pair with coordinates $(x_0, y_0, 0)$ at time

$$T = \left\{ T_0^2 + \frac{4[(x - x_0)^2 + (y - y_0)^2]}{V_T^2} \right\}^{1/2}, \quad (1)$$

where

$$T_0 = \frac{2z}{V_T}.$$

Each subsurface point thus has a corresponding reflection-time surface in the data. (This same equation with the y-terms omitted is the basis of present 2-D point-imaging migration programs.) *The moveout in equation (1) is governed by the true velocity.*

(3) A sum of the data values which lie on the reflection-time surface for a particular subsurface point provides a measure of the probability that scattered seismic energy (diffracted or reflected) emanated from that point. (That is, if a scattering center existed at a particular point then the reflection data on the reflection-time surface corresponding to that point will stack coherently.)

(4) Repeating step (3) for all subsurface points results in an image of the reflector (see French, 1974, for further details and examples).

The above description of migration implies that the geologic interface acts as a diffuse (or rough) reflector—that is, each point is assumed to scatter finite energy all along its corresponding reflection-time surface. In seismic exploration, however, significant geologic interfaces are smooth over the dimensions of a wavelength, and the phenomenon is actually one of specular (mirror-like) rather than diffuse reflection. For specular reflection, finite energy is reflected only in the vicinity of that ray (the specular ray) for which the angle of incidence on the surface is equal to the angle of reflection from the surface. Thus, coherent reflection arrivals exist along only that portion of the reflection-time surface centered about the specular ray. *The data collection area must contain the spec-*

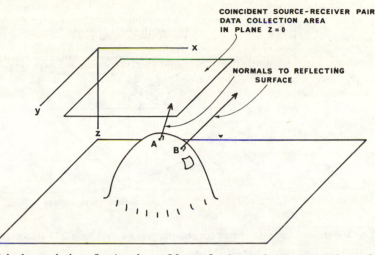

Fig. 3. Model of specularly reflecting dome. Most reflecting surfaces are specular rather than diffuse reflectors. Point A will be seen in the migrated results, but point B will not be seen since no significant energy is reflected back to the indicated data collection area for coincident source-receiver pairs.

ular raypath from a subsurface reflection point if that point is to be seen in the results of migration. For a zero offset source-receiver pair, the specular raypath is the normal to the reflecting surface. Therefore, if the normal to a reflecting surface at a point on that surface does not pass through the data collection area, then that point of the reflecting surface cannot be reconstructed by migration. This effect is illustrated in Figure 3 for reflections from a domal surface. This caveat holds also for the 2-D cases.

TRUE TWO-DIMENSIONAL STRUCTURES

The objective is to reconstruct the points on the reflecting surface which lie directly under the profile line. Figure 4 shows a case in which the profile line is perpendicular to the trend

of a true 2-D structure; in this case specular reflections from the subsurface points under the line all pass through the data collection line (if it is long enough). Thus, the migration method described above can be used to image the desired reflector cross-section. On the other hand, Figure 5 indicates that when the profile makes an angle other than 90 degrees with the trend of a true 2-D structure, specular reflections from points directly under the profile line can miss the profile line altogether. Migration techniques which try to image these points fail because data concerning these points are not recorded. What is recorded in these instances is reflected energy from offside points such as point P′ in Figure 5. Thus, either 3-D recording and processing must be used (wherein reflected energy from offside points is properly treated)

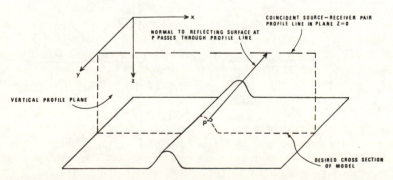

Fig. 4. Model of a true 2-D structure. Specular reflections from all points beneath the profile line are recorded when the profile is long enough and is perpendicular to the trend of true 2-D structures.

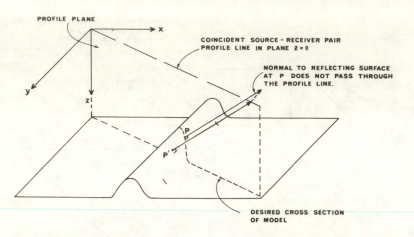

FIG. 5. Origin of offside energy for oblique profiles over true 2-D structures.

or the concept of point imaging of 2-D structures must be altered.

True 2-D surfaces can be considered as a collection of parallel horizontal lines. Hence, it seems reasonable to test for the existence of scattering lines rather than scattering points. Scattering points are modeled as small reflecting spheres, and scattering lines are modeled as reflecting cylinders of small radius. Figure 6a depicts the reflection-time surface for a particular scattering line and for zero offset (or stacked) seismic reflection data covering an area of the xy-plane above the scattering line. This time surface in 3-D data space is given by

$$ T = \frac{2[(x - x_0)^2 + z_0^2]^{1/2}}{V_T}, $$

where the scattering line is parallel to the y-axis and intersects the plane $y = 0$ at the point (x_0, z_0). The coherence of the data along this surface is a measure of the probability that the particular scattering line exists.

When reflection data are collected along a line rather than over an area, the data are examined for coherence along the curve shown in Figure 6b rather than the surface shown in Figure 6a. If η represents distance along the profile, then

$$ x = \eta \cos \theta, $$

where the angle θ is shown in the figure. The summation curve in the plane of the profile is given by

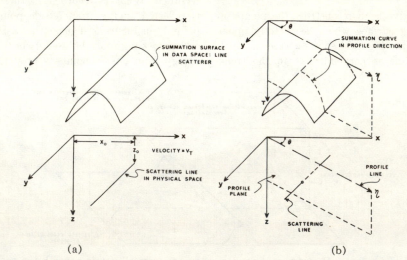

(a)   (b)

FIG. 6. (a) Summation surface for a line scatterer. (b) Summation surface along the profile direction.

$$T = \frac{2[(\eta - \eta_0)^2 \cos^2 \theta + z_0^2]^{1/2}}{V_T},$$

which can be written

$$T = \left[T_0^2 + \frac{4(\eta - \eta_0)^2}{(V_T/\cos \theta)^2} \right]^{1/2}, \qquad (2)$$

where $T_0 = 2z/V_T$ is the two-way vertical traveltime.

Equation (2) has exactly the same form as that derived from the point-imaging concept [equation (1)] except for a change in the velocity value. Thus, *existing point-imaging migration programs can be used to reconstruct profile-line cross-sections of true 2-D structures. The velocity to be used in these programs, however, may differ from the true velocity. The correct migration velocity is the true velocity divided by $\cos\theta$, where (90-θ) degrees is the angle between the profile line and the trend of the 2-D structure.*

Note: the profile direction is related to structural trend for consistency. For a true 2-D structure, θ is the angle between the profile line and a dip line. However, for a plunging 2-D structure, the dip line azimuth varies with the dip magnitude. The structural trend direction is constant for both cases.

The above theory has been verified by model experiments similar to those described elsewhere (French, 1974). Reflection profiles were taken over the structure shown in Figure 7 for $\theta = 0$, 30, 45, and 60 degrees. The raw data profiles with equivalent field dimensions are shown in Figures 8a, 9a, 10a, and 11a, respectively. Corresponding profiles migrated using the true section velocity are shown in Figures 8b, 9b, 10b, and 11b. Migrated profiles using the true velocity divided by $\cos\theta$ are shown in Figures 9c, 10c, and 11c. A comparison with Figure 7 shows that the migrations using $V_T/\cos\theta$ give the correct structural cross-sections under the appropriate profile lines.

The structural picture is correct in Figure 11c, but the quality of the migration is reduced from that of Figure 8b. The reason is that the migration aperture was held constant (48 trace scan) for Figures 8b, 9c, 10c, and 11c. Consideration of Figure 6b, however, shows that both the velocity and the aperture should be increased by $1/\cos\theta$. Thus, the aperture for Figure 11c should have been $48/\cos 60° = 96$ traces in order to produce a migration result equivalent in quality to Figure 8b.

The model represents a low-velocity-layer structure. The acoustic velocity of the model is 6400 ft/sec, while that of the overburden is 9860 ft/sec. The flat bottom of the model is

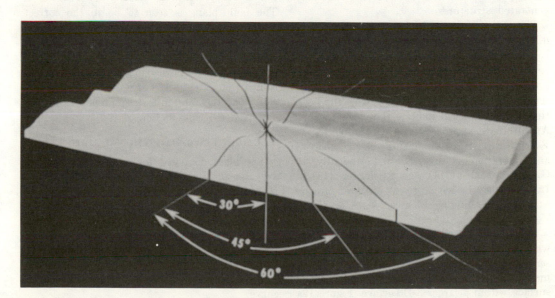

Fig. 7. Photograph of 2-D model showing the angle θ between the experimental profile lines and a line perpendicular to the structural trend.

A: RAW DATA   B: MIGRATION WITH
                               TRUE VELOCITY = 9860 FT/SEC

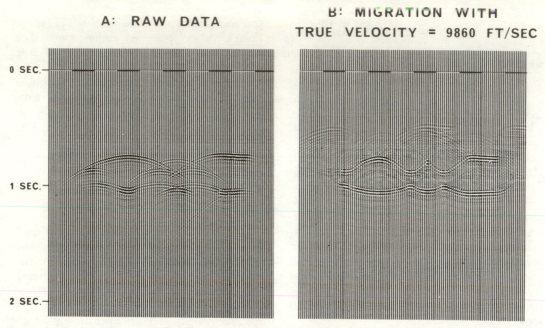

FIG. 8.   Profiles for $\theta = 0$ degrees.

distorted in the time profile and displays apparent structure generated by the velocity anomaly. Migration, however, focuses these false structures to produce an approximate vertical time section as shown in Figure 8b. Figures 9c, 10c, and 11c indicate that the corrected velocity, $V_T/\cos\theta$, is the proper migration velocity for both real and velocity anomaly generated structures.

We wish to emphasize that all attempts to migrate single profile data must take this geometrical factor (velocity change) into account. Geologic or geophysical knowledge of the area may provide an a priori estimate of θ, or several values of θ may be tried with the hope that the results will indicate the correct value. An alternate approach is to determine $V_M = V_T/\cos\theta$ directly by combining velocity analysis and migration as suggested independently by Gardner et al (1974) and Sattlegger (1975). In that case, the ideas presented above provide a method for finding V_T given V_M (i.e., $V_T = V_M \cos\theta$).

Application of the above theory is effective when only portions of a complex surface represent true 2-D structures. The fault scarp in the model shown in Figure 2 is effectively a 2-D surface within a larger 3-D surface. Figure 12a

is the reflection profile along line no. 11 of Figure 2; this line crosses the trend (strike) of the fault plane at 45 degrees. Figure 12b is a migrated profile using the true velocity V_T. The fault information is incorrectly placed down dip and is partly obscured by other sideswipes. Figure 12c is the migrated profile resulting from the correct migration velocity ($V_T/\cos 45°$). The fault plane is seen to be in the correct position in Figure 12c. Clearly, this 2-D process cannot remove the other 3-D sideswipe events which clutter up the figure. In this example, a priori knowledge of θ was used. For comparison, Figure 12d is a 3-D migration result along the same line (see French, 1974, for details of the construction of 12d).

PLUNGING TWO-DIMENSIONAL STRUCTURES

The ideas discussed for true 2-D structures are also involved in the proper migration of reflecting data over plunging 2-D structures. As shown in Appendix A, the differences lie in an additional modification to the shape of the summation curve in the direction of the profile (additional correction to migration velocity) and in the fact that the summed data values are not placed at the apex of the summation curve.

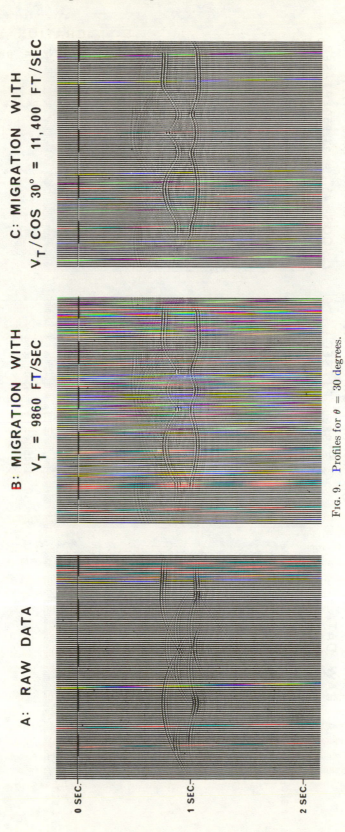

A: RAW DATA

B: MIGRATION WITH V_T = 9860 FT/SEC

C: MIGRATION WITH V_T/COS 30° = 11,400 FT/SEC

0 SEC.—   1 SEC.—   2 SEC.—

FIG. 9. Profiles for θ = 30 degrees.

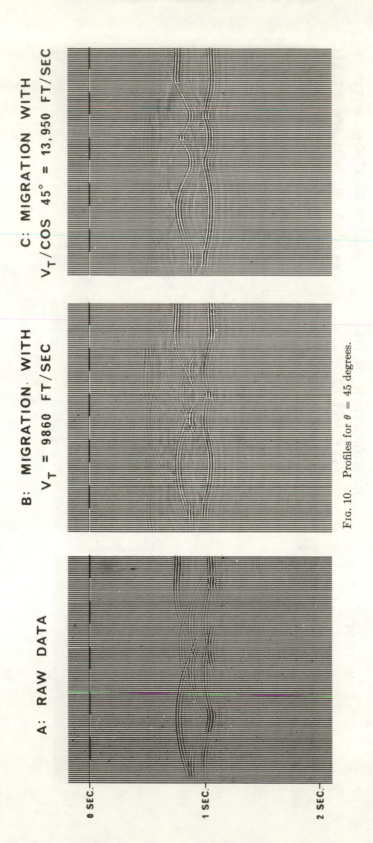

Fig. 10. Profiles for $\theta = 45$ degrees.

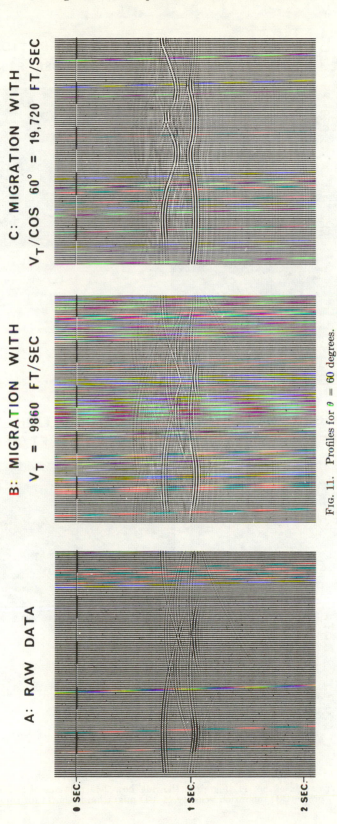

Fig. 11. Profiles for $\theta = 60$ degrees.

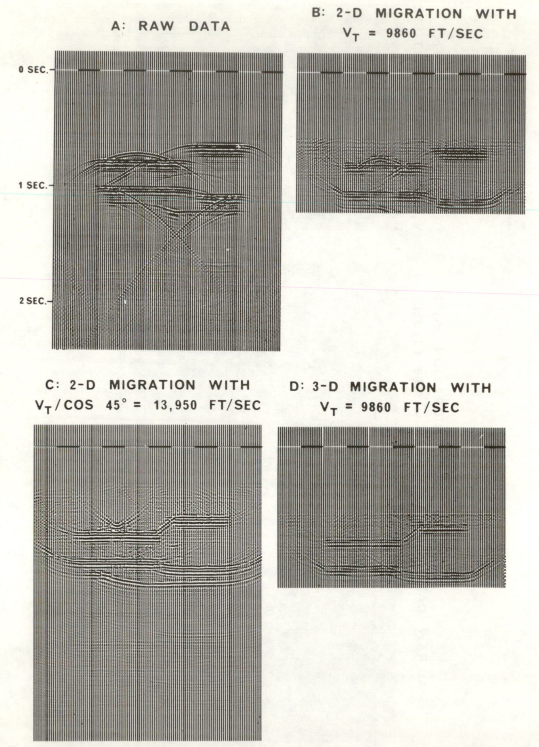

A: RAW DATA

B: 2-D MIGRATION WITH
$V_T = 9860$ FT/SEC

0 SEC.

1 SEC.

2 SEC.

C: 2-D MIGRATION WITH
$V_T / COS\ 45° = 13,950$ FT/SEC

D: 3-D MIGRATION WITH
$V_T = 9860$ FT/SEC

FIG. 12.   Profiles of cross-section no. 11, Figure 2, showing how 2-D portions of 3-D structures can be focussed with proper 2-D migration velocity. A 3-D migration is shown for comparison.

Use of the theory presents two possibilities. Prior knowledge is required of both the angle which the profile makes with the trend direction and the angle at which the structure is plunging along trend, or processing can be repeated for several values of the angles with the hope that the results will indicate the correct values. It is unlikely that the second method will be feasible due to the large number of possibilities for combinations of the two unknown parameters. Without prior knowledge of the angle values, it would seem best to turn directly to 3-D techniques which do not require these values.

Appendix A outlines the theory and processing techniques for plunging 2-D structures.

CONCLUSIONS

There are essentially three velocities with which a seismic interpreter must deal. These are the CDP velocity V_{NMO}; the true velocity V_T; and the migration velocity V_M, which is to be used in existing point-imaging migration programs. The values of these three velocities are, in general, all different.

The CDP velocity is the best stacking velocity as determined from a CDP spread. Levin (1971) has shown that the true section velocity can be determined from V_{NMO}. In this paper we have shown how to use the true velocity to determine the migration velocity V_M for three different cases. The delineation of 3-D structures requires an areal coverage of reflection data, and the correct migration velocity is equal to the true velocity. 2-D structures can be delineated from a single reflection profile provided a priori knowledge of trend direction and plunge angle are available. Otherwise, one must reprocess with a suite of angles and hope that the results will indicate the correct values. The 2-D migration velocity is always equal to or greater than the true velocity and may be either higher or lower than the stacking velocity, depending upon the specific geometry involved. The above statements can be directly extended to the case where the overburden velocity is a function of depth. In that case, V_T in equation (1) must be replaced by the vertical root-mean-square velocity function V_{VRMS}, which defines the moveout on a diffraction curve above a hypothetical scatterer. The migration velocity is then given by $V_M = V_{\text{VRMS}}/\cos\theta$. Also, θ may vary with depth.

The above theory explains, in part, why migration does not always produce improved images of structure—we simply haven't been using the correct migration velocities in many cases. It also explains why "unreasonably high" velocities have on occasion been required to collapse diffractions. The profile simply crossed the diffracting edge at an angle other than 90 degrees!

ACKNOWLEDGMENTS

I express my appreciation to the management of Gulf Research and Development Co. for permission to publish this part of the research on 3-D seismic exploration conducted at GR&DC. In particular, I thank Dr. G. H. F. Gardner and P. G. Mathieu for helpful discussions.

Special thanks are due Dr. M. O. Marcoux, with whom I worked on related experimental and theoretical projects. Dr. Marcoux had addressed the fundamental principles of reflector imaging prior to their independent appearance in the literature.

Support funds for preparation and publication of the manuscript were provided by the Oregon State University Oceanography Industrial Cooperation Program. I thank Dr. R. W. Couch for critically reading the manuscript.

Note: It has recently come to the attention of the author that the effect of oblique traverse over a horizontal line diffractor was mentioned in an article by Lindsey and Herman (1970).

REFERENCES

Baker, B. B., and Copson, E. T., 1950, The mathematical theory of Huygens Principle: London, Oxford University Press.
Claerbout, J. F., 1970, Numerical holography, *in* Acoustical holography, 3: edited by A. F. Metherell, New York, Plenum Press.
—— 1971, Toward a unified theory of reflector mapping: Geophysics, v. 36, p. 467–481.
Claerbout, J. F., and Doherty, S. M., 1972, Downward continuation of moveout corrected seismograms: Geophysics, v. 37, p. 741–768.
Farr, J. B., 1968, Earth holography, a new seismic method: Paper presented at the 38th Annual International SEG Meeting, Denver.
Fontanel, A., and Grau, G., 1969, Application of impulse seismic holography: Paper presented at the 39th Annual International SEG Meeting, Calgary.
French, W. S., 1974, Two-dimensional and three-dimensional migration of model-experiment reflection profiles: Geophysics, v. 39, p. 265–277.
Gardner, G. H. F., French, W. S., and Matzuk, T., 1974, Elements of migration and velocity analysis: Geophysics, v. 39, p. 811–825.

Levin, F. K., 1971, Apparent velocity from dipping interface reflections: Geophysics, v. 36, p. 510–516.

Lindsey, J. P., and Herman, A., 1970, Digital migration: Oil and Gas Journal, v. 38, p. 112–115.

Maginness, M. G., 1972, The reconstruction of elastic wavefields from measurements over a transducer array: J. of Sound and Vibration, v. 20, p. 219–240.

Peterson, R. A., 1969, Seismography 1970, the writing of the earth waves: preprint from SEG sympos., Los Angeles.

Sattlegger, J. W., 1975, Migration velocity determination: part I. Philosophy: Geophysics, v. 40, p. 1–5.

Schneider, W. A., 1971, Developments in seismic data processing and analysis (1968–1970): Geophysics, v. 36, p. 1043–1073.

Sondhi, M. M., 1969, Reconstruction of objects from their sound-diffraction patterns: J. Acoust. Soc. Am., v. 46:5, part 2, p. 1158–1164.

Trorey, A. W., 1970, A simple theory for seismic diffractions: Geophysics, v. 35, p. 762–784.

APPENDIX A

THEORY OF MIGRATION FOR PLUNGING TWO-DIMENSIONAL STRUCTURES

As mentioned in the text, the generating line for a plunging two-dimensional structure is a straight line which is perpendicular to the x-axis and dips at an angle γ with respect to the y-axis. We wish to find the shape of the reflection-time curve that such a line would produce on a profile taken at an angle θ with the x-axis. The geometry of the problem is shown in Figure 13. As shown, η is the distance measured along the profile direction. The scattering line intersects the profile plane below η_0 at a depth h. This is the point on the profile where we want to

plot the number which measures the probability of existence of the scattering line. We calculate the reflection-time curve by finding the minimum (perpendicular) two-way traveltime from any point on the profile line to the scattering line.

The scattering line is given in parametric form by

$$x = \eta_0 \cos \theta,$$
$$y = \eta_0 \sin \theta + \zeta \cos \gamma,$$
$$z = h + \zeta \sin \gamma,$$

where ζ is the distance along the scattering line measured from the point of intersection with the profile plane. The positive direction of the ζ-axis is on the same side of the profile plane as the positive direction of the y-axis.

The distance from any point η on the profile line to the point with parameter ζ on the scattering line is given by

$$D = [(\eta - \eta_0)^2 + \zeta^2 + h^2$$
$$- 2(\eta - n_0)\zeta \sin \theta \cos \gamma + 2h\zeta \sin \gamma]^{1/2}. \tag{A-1}$$

The perpendicular (shortest) distance from a point on the profile line to the scattering line can be found by minimizing D with respect to ζ. The value of ζ at minimum D is

$$\zeta_{\min} = (\eta - \eta_0) \sin \theta \cos \gamma - h \sin \gamma. \tag{A-2}$$

Substitution of equation (A-2) into (A-1) gives (after algebraic simplification)

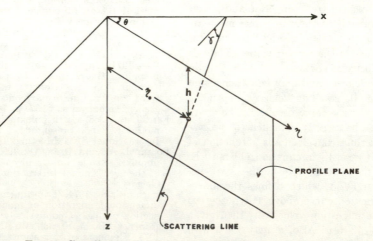

Fig. 13. Coordinate system for a plunging line scatterer.

$$D = \{(\eta - \eta_0)^2 + h^2$$
$$- [(\eta - \eta_0) \sin \theta \cos \gamma - h \sin \gamma]^2\}^{1/2}. \tag{A-3}$$

The two-way perpendicular traveltime is thus

$$T = \frac{2D}{V_T} = 2\{(\eta - \eta_0)^2 + h^2$$
$$- [(\eta - \eta_0) \sin \theta \cos \gamma - h \sin \gamma]^2\}^{1/2}/V_T. \tag{A-4}$$

Equation (A-4) is the desired reflection-time curve along the profile.

The first thing we notice about this curve is that its apex is not at η_0. In fact, we find the location of the apex as follows:

Let $\rho = \eta - \eta_0$. Then,

$$\left.\frac{dT}{d\rho}\right|_{\rho=\rho_0} = 0 = \frac{2\rho_0 - 2[\rho_0 \sin \theta \cos \gamma - h \sin \gamma] \sin \theta \cos \gamma}{V_T\{\rho_0^2 + h^2 - [\rho_0 \sin \theta \cos \gamma - h \sin \gamma]^2\}^{1/2}},$$

or

$$\rho_0 = \frac{-h \sin \theta \cos \gamma \sin \gamma}{1 - \sin^2 \theta \cos^2 \gamma}.$$

Thus,

$$\eta_{\text{apex}} = \eta_0 - \frac{h \sin \theta \cos \gamma \sin \gamma}{1 - \sin^2 \theta \cos^2 \gamma}. \tag{A-5}$$

We now wish to rewrite equation (A-4) in terms of δ, the distance from the apex. From (A-5) we have

$$\eta - \eta_0 = \eta - \eta_{\text{apex}} - \frac{h \sin \theta \cos \gamma \sin \gamma}{1 - \sin^2 \theta \cos^2 \gamma}$$
$$= \delta - \frac{h \sin \theta \cos \gamma \sin \gamma}{1 - \sin^2 \theta \cos^2 \gamma}. \tag{A-6}$$

Upon substituting (A-6) into (A-4) and simplifying, we obtain

$$T = \left\{T_0^2 \frac{\cos^2 \gamma \cos^2 \theta}{1 - \sin^2 \theta \cos^2 \gamma}\right.$$
$$+ \left.\left[\frac{2\delta}{V_T/(1 - \sin^2 \theta \cos^2 \gamma)^{1/2}}\right]^2\right\}^{1/2}$$
$$= \left\{\bar{T}_0^2 + \left[\frac{2\delta}{V_T/(1 - \sin^2 \theta \cos^2 \gamma)^{1/2}}\right]^2\right\}^{1/2}, \tag{A-7}$$

where

$$T_0 = \frac{2h}{V_T}$$

and

$$\bar{T}_0 = T_0\left[\frac{\cos^2 \gamma \cos^2 \theta}{1 - \sin^2 \theta \cos^2 \gamma}\right]^{1/2}.$$

Equation (A-7) has the same form as the reflection-time curve for a point [see equation (1) of the text]. Therefore, we can use existing migration programs to perform the sums over the hyperbolic curves provided we use the migration velocity given by

$$V_M = V_T/(1 - \sin^2 \theta \cos^2 \gamma)^{1/2}.$$

Care must be taken in interpreting the resulting section, however, as the program will operate as if the data were plotted in $(x, \bar{T}_0)$ space rather than the desired (x, T_0) space. We must, therefore, rescale the time axis of the migrated output by the factor

$$[(1 - \sin^2 \theta \cos^2 \gamma)/\cos^2 \gamma \cos^2 \theta]^{1/2}.$$

The profile can then be converted to a depth section upon multiplication of the time axis by $V_T/2$.

There is an additional interpretational complication. Equation (A-5) shows that on this depth section, objects directly beneath each other in physical space will lie along lines inclined at an angle ϕ with the vertical.

$$\phi = -\tan^{-1}\left[\frac{\sin \theta \cos \gamma \sin \gamma}{\mu(1 - \sin^2 \theta \cos^2 \gamma)}\right]$$

and is measured positively from the vertical toward increasing η. μ is the vertical-to-horizontal exaggeration of the plotted migrated depth section. The interpreter must follow structures to the surface along these diagonal lines in order to find the true position of the structure under the profile line. Clearly, the situation is more complex when a variable velocity function is introduced.

When $\gamma = 0$, the above equations reduce to the description for true two-dimensional structures as given in the text.

APPENDIX B

The purpose of this appendix is to demonstrate that the intuitive concept of reflection profile migration as discussed in the text and elsewhere (Schneider, 1971; French, 1974; Gardner et al, 1974) is founded on Kirchhoff's solution of the wave equation. The discussion will be limited to the case of constant overburden velocity. This case allows an exact mathematical description of the migration process based upon currently available theoretical results and is directly applicable to results in this paper and other published model results (French, 1974; Gardner et al, 1974). No new theory is involved, and our purpose here is to relate the numerical manipulations actually performed when migrating reflection data to well-known mathematical results. The framework to be used in the following discussion has been published (Claerbout, 1970, 1971; Claerbout and Doherty, 1972). The only difference is our use of an analytic solution of the wave equation to propagate wave fields backwards in time as opposed to Claerbout's use of computer simulations of the wave equation for the same purpose. As discussed by Claerbout and Doherty (1972), the mathematical part of the problem consists of using reflection data recorded at the surface and the wave equation to calculate hypothetical reflection data which would simulate recordings by geophones buried at all locations in the earth. The physical argument for forming an image of the reflecting surfaces involves the evaluation of the signals for each buried geophone at a time equal to the propagation time from the original source to the buried geophone of interest. This, of course, is the time that a reflection would arrive at the buried geophone from a reflector (if one exists) just below the buried geophone (see Figure 14). Subsequent division by the amplitude of the direct wave at each subsurface point yields approximate reflection coefficients (see Claerbout, 1971). We refer to this imaging concept as Claerbout's principle. We turn now to the problem of reconstructing the buried geophone signals.

The assumptions to be used are stated explicitly below:

(1) The scalar wave equation adequately

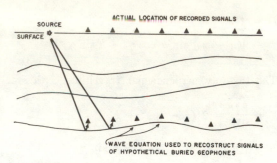

Fig. 14. Diagrammatic representation of Claerbout's principle of reflector imaging.

describes seismic wave propagation for the usual reflection recording. That is, shear waves can be ignored, and for convenience we treat the recorded signals as recordings of pressure fluctuations.

(2) The velocity can be considered constant.

(3) Reflecting interfaces within the constant velocity earth may have arbitrary configurations but are weak reflectors so that transmission effects at the interfaces can be ignored.

(4) The free surface effects on the signals recorded at the earth's surface can be corrected prior to our use of the data. Thus, the equations to follow utilize the pressure field incident upon the surface (i.e., Claerbout's upgoing waves) and not the total pressure, which includes the reflected waves and actually vanishes at the free surface. This means that conceptually the empty half-space above the earth's surface can be replaced by a homogeneous material equal to that in a reflector-free region of our earth model. Assumption (4) is unnecessary to the theory but simplifies the following discussion.

The geometry of our earth model is shown in Figure 15. The plane $z = 0$ represents the surface of the earth on which all actual sources are located and on which the incident pressure field is known. By assumption (4), the constant velocity material has been extended above this plane throughout the half-space $z < 0$. The data are the reflected pressure waves generated by an impulsive source and observed at each point on the plane $z = 0$; the direct arrival is eliminated by temporal gating of the receivers. We consider such a data set to be recorded separately for a number of source locations. Let

$\mathbf{s}^j = (x^j, y^j, 0)$ be the location of the jth shot point,

$\mathbf{r}^{ij} = (x^{ij}, y^{ij}, 0)$ be the location of the ith receiver for the jth shot (in practice these are discrete, but we treat the receiver location as a continuous variable below),

$\mathbf{r} = (x, y, z)$ be an arbitrary point, and

$u(\mathbf{r}^{ij}, t) =$ pressure field from the jth shot incident on the plane $z = 0$ at $\mathbf{r}^{ij}$, where t is a time variable which is reset to zero at the instant of each successive shot.

We state the following theorem (known as Kirchhoff's formula) from Baker and Copson (1950, p. 37).

Theorem: Let $u(x,y,z,t)$ be a solution of the equation

$$\nabla^2 u = \frac{1}{c^2}\frac{\partial^2 u}{\partial t^2}$$

Let the first and second order partial derivatives of u be continuous within and on a closed surface S, and let (x_1, y_1, z_1) be a point within S. Then,

$u(x_1, y_1, z_1, t)$

$$= \frac{1}{4\pi}\int\int_S \left\{ [u]\,\frac{\partial}{\partial n}\left(\frac{1}{r}\right) \right.$$
$$\left. -\frac{1}{cr}\frac{\partial r}{\partial n}\left[\frac{\partial u}{\partial t}\right] - \frac{1}{r}\left[\frac{\partial u}{\partial n}\right] \right\} dS, \qquad \text{(B-1)}$$

where r is the distance from (x_1, y_1, z_1) to a typical point of S, $\partial/\partial n$ denotes differentiation along the inward normal to S, and square brackets indicate retarded values. If, however, the point (x_1, y_1, z_1) lies outside S, the value of the integral is zero.

The retarded values mentioned in the theorem are defined as follows:

If ϕ is a function of the coordinates (x,y,z) of a variable point Q and of the time t, say

$$\phi = \phi(x, y, z, t),$$

and if r is the distance of Q from a fixed point P, we write

$$[\phi] = \phi\left(x, y, z, t - \frac{r}{c}\right)$$

and call $[\phi]$ the retarded value of ϕ. In the theorem, the point P is at (x_1, y_1, z_1) and the point Q varies over the surface S.

In using Kirchhoff's formula, we can completely ignore actual physical sources since (B-1) generates a wave field identical to that from the physical source. The only fact about our physical source which we use is that it is localized and of finite energy so that all wave fields to be considered fall off like $f(ct\text{-}R)/R$ for large distances R from the physical source region. We shall ignore the rare cases where the reflector geometry is such as to generate a collimated beam.

Let us apply equation (B-1) to a forward problem. Assume that we want to use the reflection data from the jth shot $u(\mathbf{r}^{ij}, t)$ to calculate the pressure disturbance which will pass the point $\mathbf{r}_k(z_k < 0)$. We assume that the wave field satisfies the conditions of the theorem, and we construct the surface S so that it consists of the plane $z = 0$ and the infinite hemispherical surface enclosing the point $\mathbf{r}_k$, as shown in Figure 15. Equation (B-1) is directly applicable. Since the integrand of (B-1) is evaluated at retarded times, the value of $u(x_k, y_k, z_k < 0, t_1)$ will depend upon the values of u on S for times less than t_1. Thus, since the hemispherical portion of S is infinitely distant from the transient source region, the retarded values of the integrand of (B-1) will be zero on the hemispherical surface for finite t_1. Equation (B-1) reduces to an integral over the data recorded in the plane $z = 0$. To evaluate the wave field at $\mathbf{r}_k$ for time t_1, we determine the corresponding retarded time for

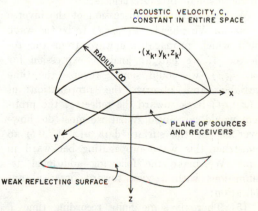

FIG. 15. Schematic of earth model and surface S for use with Kirchhoff's formula in the forward problem.

each receiver location, form the integrand of (B-1) for each receiver at the appropriate retarded time, and integrate these numbers across the receivers. Figure 16 shows several retarded time curves along which the data are combined according to equation (B-1) for forward propagation. In Figure 16, only a single cross-section of the data is shown. Notice that the retarded time curves are all identical except for different constant shifts in the t direction.

We have assumed that we know $u(\mathbf{r}, t)$ and, hence, $\partial u/\partial t$ on $z = 0$; but not $\partial u/\partial n$. However, this unknown quantity is not independent and can be easily calculated from the known values in the case that the data are collected on a plane (Baker and Copson, 1950). Specifically, if we apply the theorem to the image point of $\mathbf{r}_k$ through the plane $z = 0$, the integral vanishes showing that the third term of the resultant integral is the negative of the sum of the first two terms. Since the integrands for $\mathbf{r}_k$ and its image point differ only in the sign of the first two terms, we have immediately

$$u(x_k, y_k, z_k < 0, t)$$

$$= \frac{1}{2\pi} \iint_{z=0} \left\{ [u] \frac{\partial}{\partial n}\left(\frac{1}{r}\right) - \frac{1}{cr}\frac{\partial r}{\partial n}\left[\frac{\partial u}{\partial t}\right] \right\} dS$$

$$= -\frac{1}{2\pi} \iint_{z=0} \frac{1}{r}\left[\frac{\partial u}{\partial n}\right] dS. \qquad (B-2)$$

The first of the equations (B-2) is in terms of known quantities. The second of equations (B-2) requires knowledge of vertical gradients in the incident pressure field.

We now turn to a discussion of the inverse problem. We consider $u(x,y,z,t)$ to be the wave field which propagates upward from the reflector. If we had an analytic expression for $u(x,y,z,t)$, we could simply reverse the time variable and observe the propagation of $u(x,y,z,t)$ back toward the reflector; the problem would be solved. What we must do, however, is use the surface data $u(x^{ij}, y^{ij}, 0, t)$ to construct this wave propagating backward in time. We make the following additional assumptions with a view toward practical application:

(5) There is some finite recording time T after which there is no measurable reflected energy crossing the plane $z = 0$. (This actually

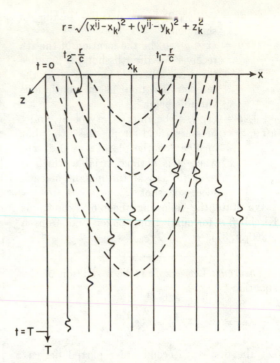

$$r = \sqrt{(x^{ij} - x_k)^2 + (y^{ij} - y_k)^2 + z_k^2}$$

FIG. 16. Summation curves in data space for forward propagation of wave fields according to Kirchhoff's formula. Only one profile of data space is shown.

follows from the previous assumptions that the source is transient and that all wave fields fall off as $f(ct-R)/R$ for large R, where R is the distance from the physical source region.)

(6) We assume that $u(x,y,z,t)$ when propagated backward in time is not singular on the reflector (where it will be evaluated by Claerbout's principle).

We now note that if $u(x,y,z,t)$ is a solution of a wave equation in t, then $g(x,y,z,\tau) = u(x,y,z,T-\tau)$ is a solution of a wave equation in τ. Thus, Kirchhoff's formula in τ provides a method of constructing $g(x,y,z,\tau)$ from measurements on the plane $z = 0$.

We know nothing about any singularities in $g(x,y,z,\tau)$. We do know from (6) that we are interested in $g(x,y,z,\tau)$ only at nonsingular points. Hence, since Claerbout's principle requires that we search the entire subsurface for the reflector, we will look for a solution to the wave equation in τ which satisfies the measured data on the plane $z = 0$ and whose partial derivatives of the first and second orders are continuous within and on the surface of the lower half-space of Figure 14: we call this function $g_0(x,y,z,\tau)$ and assume that g_0 differs negligibly from g for mapping purposes except near possible singularities of g.

If we now let $\mathbf{r}_k$ be a point in the lower half-space of Figure 15, then

$$g_0(x_k, y_k, z_k > 0, \tau)$$

$$= \frac{1}{4\pi} \iint_S \left\{ [g_0] \frac{\partial}{\partial n}\left(\frac{1}{r}\right) - \frac{1}{cr}\frac{\partial r}{\partial n}\left[\frac{\partial g_0}{\partial \tau}\right] \right.$$

$$\left. - \frac{1}{r}\left[\frac{\partial g_0}{\partial n}\right] \right\} dS, \tag{B-3}$$

where the closed surface S consists of the plane $z = 0$ and an infinite hemisphere enclosing the lower half-space. According to an earlier assumption, all wave fields diminish as $f(ct - R)/R$ so the contribution to (B-3) from the infinite hemisphere vanishes. By arguments similar to those leading to equation (B-2) we have, finally,

$$g_0(x_k, y_k, z_k > 0, \tau)$$

$$= \frac{1}{2\pi} \iint_{z=0} \left\{ [g_0] \frac{\partial}{\partial n}\left(\frac{1}{r}\right) - \frac{1}{cr}\frac{\partial r}{\partial n}\left[\frac{\partial g_0}{\partial \tau}\right] \right\} dS$$

$$= -\frac{1}{2\pi} \iint_{z=0} \frac{1}{r}\left[\frac{\partial g_0}{\partial n}\right] dS. \tag{B-4}$$

Here the brackets refer to retarded τ values and are evaluated at $(\tau - r/c)$. Figure 17 shows several retarded time curves along which the data are combined according to equation (B-4). We note that all the retarded time curves are identical except for a shift in the τ direction. In both the forward and reverse cases, the entire time signal can be reconstructed at a point by a *static shift* and stack over the data traces. We see that the wave field reconstruction process does not involve the *time distorting* features of NMO processes. This will come later as a natural consequence of imaging.

According to Claerbout's principle, we do not use the entire reconstructed time history $g_0(x_k, y_k, z_k, \tau)$. Rather, we use only that value at $\tau_k = T - t_k$, where t_k is the propagation time from the actual source at $\mathbf{s}^i$ to the point $\mathbf{r}_k$ (T is the total recording time). Thus, we use only the data combined along the particular dashed curve of Figure 17 corresponding to τ_k. If we consider a point $\mathbf{r}_m = (x_k, y_k, z_m)$ with $z_m > z_k$, then the times history $g_0(\mathbf{r}_m, \tau)$ will be recon-

structed by integrating along retarded time curves such as those in Figure 18. The retarded time curves are slightly flatter in Figure 18 than in Figure 17 owing to the slower change in r with x as z increases. We note that all the hyperbolas along which the data are combined to reconstruct the wave field at $\mathbf{r}_m$ are identical except for a shift in the τ direction. In other words, $g_0(\mathbf{r}_m, \tau)$ is also reconstructed by a static shift with a subsequent stack of the data, and no time-stretching is involved. We form the image at $\mathbf{r}_m$ by evaluating $g_0(\mathbf{r}_m, \tau)$ at the time given by Claerbout's principle. This image value corresponds to an integration of the data along the particular retarded time curve indicated by a heavy line in Figure 18. The retarded time surfaces of Figures 17 and 18 used for imaging the respective points are shown together in Figure 19; the data have been time reversed and plotted in terms of clock time t. In order to combine the data along these curves (and the corresponding curves for points at other depths) a nonlinear stretching of the time axis is indicated. This is a natural consequence, not of wave field reconstruction, but of the principle of

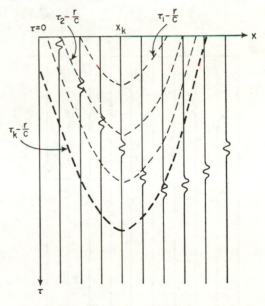

Fig. 17. Summation curves in data space for inverse propagation of wave fields according to Kirchhoff's formula. Notice that the data are the time reversed version of the actual recorded data shown in Figure 16. The heavy dashed summation curve corresponds to Claerbout's principle of reflector mapping.

reflector mapping, and attempts to eliminate the effect are not consistent with this imaging scheme.

Figure 19 suggests the migration process described by Schneider (1971) when applied to a single field record. In fact, equation (B-4) provides the relationship between the imaging methods discussed in this paper and elsewhere (Schneider, 1971; French, 1974; Gardner et al, 1974; and Sattlegger, 1975) and solutions of the wave equation. These imaging methods can be used for quantitative as well as qualitative results. In order to utilize the data from other source locations, we repeat the use of equation (B-4) and Claerbout's principle for each source and add the resulting images as Claerbout suggests (Claerbout, 1971).

Rarely, however, is the migration process employed before stack as justified above; it is usually done after stacking. The process of stacking is used to provide a good estimate (from a signal-to-noise point of view) of what a coincident source-receiver pair would record in the absence of multiples. Thus, migration of stacked profiles is explained as a partial imaging method and is justified by going through the

above process keeping track only of the operations performed on the zero-offset traces.

For practical application, the above ideas are extended to inhomogeneous media by the use of root-mean-square velocities in the calculation of retarded times. It should be reiterated, however, (see Gardner et al, 1974) that in actual use where organized noise (surface waves, shear waves) and random noise are present, the integrand of equation (B-4) may not represent the best use of the recorded data. Additional elements from information theory may be useful.

In closing, I would like to point out that the finite-difference equation approach and the Kirchhoff integral approach are not the only possible implementations of the wave equation and the basic imaging principles. A number of possibilities are shown in Figure 20. For the constant velocity case, these methods are mathematically identical as long as each indicated linear operation is properly carried out. From this point of view, the holography-like processes are a single temporal frequency component of our imaging method.

For a discussion of the top path through Figure 20 see Claerbout (1971). This Appendix discusses the second path from the top (Figure

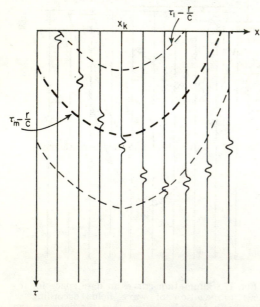

FIG. 18. Summation curves for wave field reconstruction at a point deeper in the earth than that of Figure 17. The heavy dashed curve corresponds to Claerbout's principle for this deeper point.

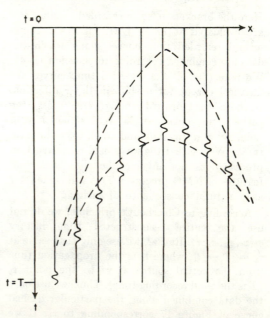

FIG. 19. Summation curves of Figures 17 and 18 corresponding to Claerbout's principle replotted in terms of real clock time.

20). The third path (uppermost holography path) is discussed by Claerbout (1970). For the center holographic path through Figure 20 see Sondhi (1969). For a discussion of the lower holographic path see Maginness (1972). The aplanat principle of reflector imaging is discussed by Peterson (1969) along with several other topics pertinent to Figure 20. From the point of view expressed in Figure 20, the method of Fontanel and Grau (1969) is not a holographic process. It is, rather, an approximate graphic scheme for going through the second path from the top of the figure in the case of a plane-layered earth. The methods discussed by Farr (1968) are holographic wave-field reconstruction schemes but do not include a reflector imaging principle.

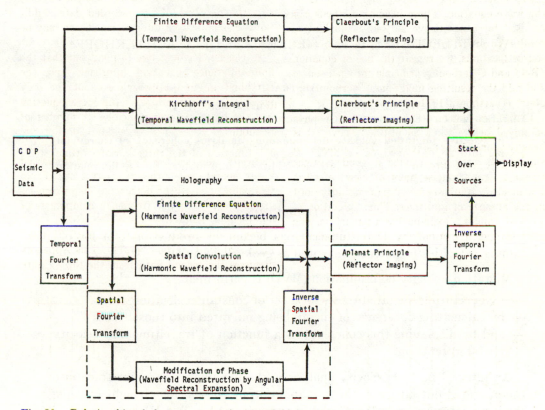

Fig. 20. Relationship of alternate acoustic wave field reconstruction and reflector imaging schemes.

Reprinted from *Geophysical Prospecting*, **24**, 650-659.

DIP SELECTIVE MIGRATION VELOCITY DETERMINATION *

BY

J. W. SATTLEGGER**, P. K. STILLER*** and J. A. ECHTERHOFF***

ABSTRACT

SATTLEGGER, J. W., STILLER, P. K., and ECHTERHOFF, J. A., 1976, Dip Selective Migration Velocity Determination, Geophysical Prospecting 24, 650-659.

Dip selective migration velocity determination calculates coherency of energy migrating into an output trace or 'MVD-axis' as a function of the three variables time, velocity, and dip. In doing so, it provides additional information, the dip information, relative to standard MVD.

INTRODUCTION

Migration Velocity Determination has become a production tool for obtaining velocities for migration as well as geological studies.

Migration velocity determination (MVD) is performed

— by generating migrated traces for a set of constant velocities
— by calculating coherency of energy being migrated into those traces
— and by displaying this coherency as a function of traveltime and velocity for interpretation.

An integral part of velocity analysis is the imaging of an input trace into the migrated output trace, the migrated output trace being thought of being the locus along which velocity analysis is performed, called MVD axis.

This imaging requires computation of the raypath as indicated in figure 1,

where

— S is the source location
— R is the receiver location
— A and a unit vector **a** define the locus of the MVD analysis or MVD axis
— T is the trace observed with S and R.

---

* Paper presented at the 37th Meeting of the EAEG, Bergen, June 1975.
** Geophysical Consultant, 4470 Meppen, Kleiststrasse 20, F.R. Germany.
*** Preussag Leibniz-Rechenzentrum GmbH, 3000 Hanover 1, P.O. Box 4827, F.R. Germany.

The number of the sample to be transferred from the input trace into, say, point E of the MVD axis is obtained by calculating the traveltime SER for the given velocity and deviding by the input data sampling increment.

Because of the raypath being available, the dip of a reflection element that may have generated this sample is also available; because of the law of reflection the perpendicular to the reflection element devides the angle between rays SE and RE into halves.

IMAGING OF AN INPUT TRACE INTO AN OUTPUT TRACE

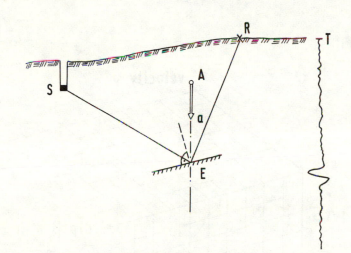

Fig. 1. Imaging of trace recorded with source S and receiver R into output trace defined by point A and direction **a**.

These ideas have led to the development of the dip selective MVD:

— why not making use of the dip information in order to separate energy migrating into the MVD axis, thus obtaining

- better event separation perhaps
- additional dip information for certain

— and, simply from intuition, does not a loss of information have to be expected when the high number of, typically 500 or 1000 contributing traces of an MVD analysis are imaged in only a two-dimensional (time, velocity)-space. Dip selective MVD images this information into a three-dimensional space of time, velocity and dip.

Dip selective MVD algorithm

The space into which input traces are imaged is depicted in figure 2 as already mentioned, it is a three-dimensional space, discretely sampled of course.

A description of how the contribution of an arbitrary input trace to a selected point E in this three-dimensional grid is calculated will make the entire process clear.

TIME VELOCITY DIP SPACE OF DIP SELECTIVE MVD

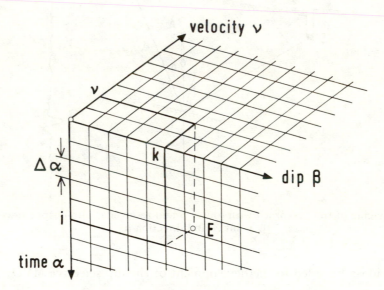

Fig. 2. (time, velocity, dip)-space of dip selective MVD.

Consider figure 3, where source, receiver MVD axis, and point E are depicted. Let E in the (time, velocity, dip) $= (\alpha, \nu, \beta)$-space be given by

— its sample number in time direction i
— its discrete velocity ν.

Let the sampling increment in time in the (α, ν, β)-space be $\Delta\alpha$.
Let the input data sampling increment be $\Delta\tau$.

The coordinates of point E below the surface may be calculated by the formula

$$E = A + a \, i \, \Delta\alpha \, v/2 \tag{1}$$

The length of the path $_{SER}$ a pulse has to travel is then given by

$$\lambda_1 + \lambda_2 = |\mathbf{E} - \mathbf{S}| + |\mathbf{E} - \mathbf{R}| \tag{2}$$

where $\mathbf{E}$, $\mathbf{S}$, $\mathbf{R}$ are location vectors, and the traveltime by

$$\tau = (\lambda_1 + \lambda_2)/v. \tag{3}$$

RAYPATH AND TRAVEL TIME τ FROM S TO E TO R AND REFLECTION DIP β

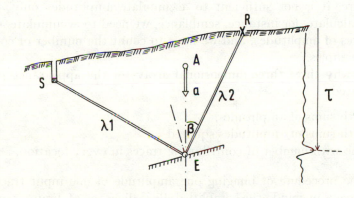

Fig. 3. Calculating raypath and traveltime from S to E to R.

Consequently, the pulse emanating from source S, reflected in point E and recorded at receiver R will be found at time τ on the input trace.

Provided this sample is truly reflected energy, the dip of the reflection element is also obtainable from the raypath. The two unit vectors from source and receiver respectively to the reflection point E are

$$\mathbf{s} = (\mathbf{E} - \mathbf{S}) / |\mathbf{E} - \mathbf{S}| \tag{4}$$

$$\mathbf{r} = (\mathbf{E} - \mathbf{R}) / |\mathbf{E} - \mathbf{R}|. \tag{5}$$

A perpendicular to the reflection element is then

$$\mathbf{e} = (\mathbf{s} + \mathbf{r}) / |\mathbf{s} + \mathbf{r}|. \tag{6}$$

In order to separate dips we will have to decide on a dip range $(-\beta, \beta)$ and a dip sampling increment $\Delta\beta$. The subscript of point E in the dip space of figure 2 is given by

$$k = \text{integer } ((\text{atan } \mathbf{e}_1/\mathbf{e}_3 + \beta)/\Delta\beta) + 1 \qquad (7)$$

for $-\beta \leq \text{atan } \mathbf{e}_1/\mathbf{e}_3 \leq \beta$.

Knowing the subcripts

— i in direction of time
— k in direction of dip

and the discrete velocity ν we know where to accumulate the input trace amplitude interpolated earlier.

At this point we must remember that our final objective is to calculate a measure of coherency as a function of time, velocity, and dip for analysis and interpretation.

Therefore, it is not sufficient to accumulate amplitudes only; rather, in order to calculate, for instance, semblance, we need to accumulate amplitudes and squares of amplitudes, and we have to count the number of contributing non-zero samples.

This is why three three-dimensional arrays of the appearance of figure 2 are generated:

— one with sums of amplitudes,
— one with sums of amplitudes squared,
— and one with number of contributing traces in every location.

With the procedure of imaging one amplitude of one input trace into the (α, ν, β)-space in mind, the simplified flow diagram of figure 4 will clarify the total procedure.

Three arrays A, B, and C have to be erased in order to accomodate three three-dimensional matrices of appearance 2. This is performed in box 4.1.

Processing is then organized in the form of three nested loops 4.2, 4.3, 4.4 running

— through all samples (4.4)
— of all input traces (4.3)
— for all velocities (4.2).

At this point

— sums of amplitudes, say $\Sigma\sigma$
— sums of amplitudes squared, $\Sigma\sigma^2$
— numbers of contributing samples μ

are available as three sampled functions of the independent variables time, velocity, and dip.

Hence, semblance as a coherency measure may now be calculated as a function of the three same independent variables with the formula

$$\Omega(i \; \nu, k) = \frac{\displaystyle\sum_{j=i-\omega}^{i+\omega} (\Sigma\sigma)^2 - \sum_{j=i-\omega}^{i+\omega} \sigma^2}{(\mu - 1) \displaystyle\sum_{j=i-\omega}^{i+\omega} \sigma^2} \qquad (8)$$

FLOW DIAGRAM

IMAGING INPUT TRACES
INTO THREE DIMENTIONAL
OUTPUT MATRICES

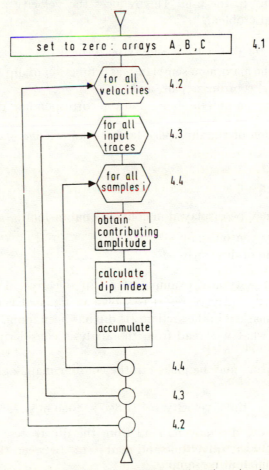

Fig. 4. Flow of first section of MVD analysis generation: imaging input traces into three dimensional output arrays.

In (8), ω is half the time gate, given in samples, for semblance calculation. The total time gate is then

$$gate = 2\omega\,\Delta\alpha\ s$$

or

$$igate = 2\omega + 1\ samples$$

where ω is an input parameter.

Having generated semblance as a function of time, velocity, and dip we are faced with the problem of displaying a function of three variables.

We solve this problem by having the algorithm perform interpretation, i.e. maximum semblance search in one direction, the direction of dip, and by leaving selection of maxima in the two other directions, i.e. time and velocity, or "interpretation" to the user. This reduces the velocity picking process to a two-dimensional problem.

Elimination of the dip space is simply performed by

— determining the maximum semblance as a function of dip for every possible (time, velocity) coordinate pair, and
— storing this maximum semblance and the corresponding dip.

Thus, elimination of the dip space leads to knowledge of two functions of two variables each:

— semblance and
— reflector dip.

These results may be displayed in conventional fashion:

— semblance as a contour map
— dip in the form of dip traces.

Figure 5 shows a typical example of a dip selective MVD display. The section to which this display refers is shown in fig. 6. Event A interpreted in the analysis is marked in the section. Its dip has been found to be $-16°$. This agrees well with what was read from the analysis considering the resolution chosen.

Time-velocity pairs may be picked in the contour map, e.g. that of event A with

$$(time, velocity) = (1.850\ s,\ 2900\ m/s).$$

For this event the dip may be read from the dip traces: project event A to the right at constant traveltime and read $-12°$ between the two dip traces of velocities 2800 and 3000 m/s.

DIP SELECTIVE MVD DISPLAY

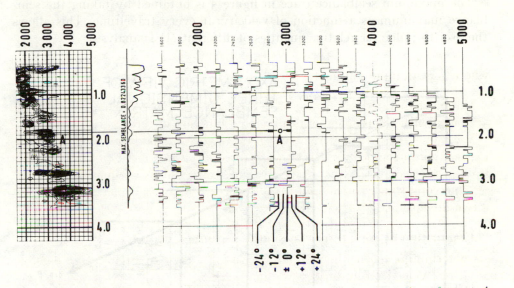

Fig. 5. Dip selective migration velocity analysis display with interpretation of event *A*.

SECTION OF NORTHERN GERMANY

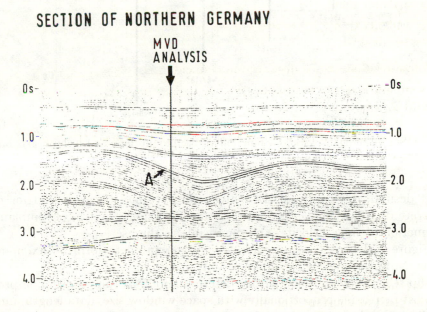

Fig. 6. Section from Northern Germany showing location of MVD analysis of figure 5.

Thus, a picked event is now defined by three parameters: time, velocity, and dip.

The maximum semblance trace in figure 5 is obtained by taking the semblance maximum as a function of velocity at every traveltime. This allows the user to judge the relative qualities of different MVD analyses.

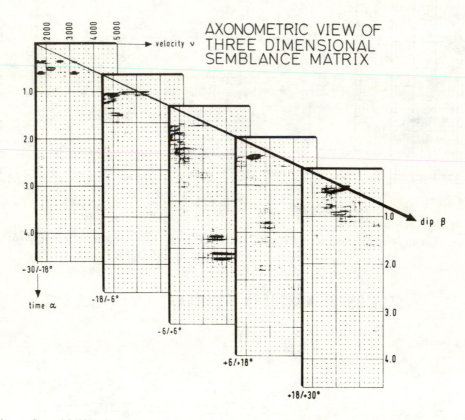

Fig. 7. Set of MVD displays as a function of discrete dip mounted to form axonometric view of three-dimensional semblance matrix.

If desired, a separate MVD analysis display may be generated for every discrete dip. This makes the entire three-dimensional matrix of semblance as a function of time, velocity, and dip visible to the user.

Figure 7 depicts a set of such displays mounted to form an axonometric view. Note the good separation of events according to dip.

Dip selective migration velocity determination is a time consuming process, its cost increasing proportionally with space window size, data length, number of discrete velocities, and angular resolution.

Conclusions

Dip selective MVD provides dip as additional information for many purposes, for instance to judge events in the contoured velocity display or for inverse modeling.

Higher semblance is obtained because of the separation of energy into separate reflector dip ranges.

Acknowledgement

The authors wish to thank Gewerkschaften Brigitta Elwerath Betriebs-führungs GmbH for the permission to make use of the sample section shown.

This project benefits from substantial support by the Bundesministerium für Forschung und Technologie of the Federal Republic of Germany.

References

NEIDELL, N. S., and TANER, M. T., 1971, Semblance and other coherency measures for multichannel data, Geophysics 36, 482-497.

SATTLEGGER, J. W., 1975, Migration velocity determination: Part I: Philisophy, Geophysics 40, 1-5.

DOHR, G. P., and STILLER, P. K., 1975, Migration velocity determination: Part II: Applications, Geophysics 40, 6-16.

Reprinted from *Geophysical Prospecting*, **25**, 738-745.

TIME MIGRATION—SOME RAY THEORETICAL ASPECTS *

BY

P. HUBRAL**

ABSTRACT

HUBRAL, P., 1977. Time Migration—Some Ray Theoretical Aspects, Geophysical Prospecting 25, 738-745.

Using an elementary theory of migration one can consider a reflecting horizon as a continuum of scattering centres for seismic waves. Reflections arising at interfaces can thus be looked upon as the sum of energy scattered by interface points. The energy from one point is distributed among signals upon its reflection time surface. This surface is usually well approximated by a hyperboloid in the vicinity of its apex. Migration aims at focusing the scattered energy of each depth point into an image point upon the reflection time surface. To ensure a complete migration the image must be vertical above the depth point. This is difficult to achieve for subsurface interfaces which fall below laterally inhomogeneous velocity media. Migration is hence frequently performed for these interfaces as well by the Kirchhoff summation method which systematically sums signals into the apex of the approximation hyperboloid even though the Kirchhoff integral is in this case not strictly valid. For a multilayered subsurface isovelocity layer model with interfaces of a generally curved nature this can only provide a complete migration for the uppermost interface. Still there are various advantages gained by having a process which sums signals consistently into the minimum of the reflection time surface. The position of the time surface minimum is the place where a ray from the depth point emerges vertically to the surface. The Kirchhoff migration, if applied to media with laterally inhomogeneous velocity, must necessarily be followed by a further time-to-depth migration if the true depth structure is to be recovered. Primary normal reflections and their respective migrated reflections have a complementary relationship to each other. Normal reflections relate to rays normal to the reflector and migrated reflections relate to rays normal to the free surface. Ray modeling is performed to indicate a new approach for simulating seismic reflections. Commonly occuring situations are investigated from which lessons can be learned which are of immediate value for those concerned with interpreting time migrated reflections. The concept of the 'image ray' is introduced.

INTRODUCTION

Migration is usually most useful in areas of complex geology if performed with 3-D migration schemes on 3-D recorded seismic data. The process aims at focusing unstacked or stacked seismic reflections into data more suitable for interpretation. These data are still presented as a function of two-way time.

---

* Received August 1976.

** Bundesanstalt für Geowissenschaften und Rohstoffe, 3 Hannover-Buchholz, Stilleweg 2.

Scattered energy is usually contracted and weak segmented reflections often appear in a continuous and geologically more reasonable form. Faults show up frequently more distinct. To account for all observed phenomena the theory of migration (Hagedoorn 1954, Claerbout and Doherty 1972, French 1974) must certainly involve more than can be reasoned with ray-theoretical considerations alone. Still, ray theory is entirely sufficient to investigate the relationship between travel times of unmigrated and migrated reflections as done in this work. The conclusions at which we have arrived are in fact quite simple. They have, to our knowledge, not been stated in this form previously elsewhere.

Exploration seismologists base their interpretational skill and judgement largely on the study of primary reflections for selected key horizons. These are normally available to them in either a stacked or time-migrated form. In the presence of structure both types of reflections may considerably differ from each other. Migrated reflections often provide a more realistic picture of the geology which may not that simply be inferred from stacked reflections. Migrated reflections need, however, by no means always provide a more truthful picture of subsurface reflectors. This results from extending the Kirchhoff summation process which sums the scattered signals into the apex to media with lateral inhomogeneous velocity. All migration schemes are related to the wave equation (Larner and Hatton 1976). As they transform one wave field into another they have been extensively studied from various wave theoretical points of view. Little attention has, however, been given to the relationship which the travel times of migrated primary reflections have to their actual reflectors at depth. Rays lack physical existence and seem to be most inappropriate for a theory concerned with diffraction phenomena. Still, they contribute in clarifying this particular aspect.

THEORY

If the geology is complex it is most unlikely that it can be approximated by a 2-D model. It is for this reason that our theoretical considerations are initially based on 3-D models though subsequent examples have for reason of simplicity a two-dimensional nature. Fig. 1 shows a 3-D isovelocity layer model with interfaces of a generally curved nature. Three interfaces are shown from an arbitrary number that may be permitted. All media are isotropic. It is this earth model for which the theory discussed here is valid. Wavefronts are thus always perpendicular to rays. Some source-receiver pairs are placed upon the free surface. The position of the jth pair is indicated by R_j. One scattering point D has been selected on the third interface. Rays from various source-receiver pairs have been drawn to D. Along each path a wave travels to D and part of the energy is scattered back to the receiver. Rays obey Snell's

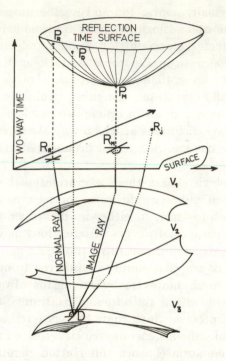

Fig. 1. 3-D subsurface isovelocity layer model featuring an interface scatterer D and its reflection time surface.

law at interfaces. Plotting the two-way time along a ray as a function of the source-receiver position provides the reflection time surface for D (fig. 1). Each subsurface scatterer usually results in a different reflection time surface. Three points upon it have special significance: P_R relates to a free surface point R_R where a ray normal to the interface in D emerges. The two-way time along this normal incidence ray is the primary reflection time to the interface in D recorded with a source-receiver pair in R_R. This particular time is usually well approximated by the stacked primary reflection time of a CDP gather with common mid point in R_R. P_D is a point within the time surface vertical above D. If the scattered energy of the coherent signals within a reflection time surface is summed into a signal in P_D then the migration is complete in the sense that a subsequent time-to-depth conversion is achieved by only correcting the migrated travel times. No further 'depth migration' is involved. A search for the image P_D within the reflection time surface requires, however, considerable decision making as its position with respect to the time surface minimum may differ for different depth points. It is thus for practical reasons that the 'image' is always chosen to coincide with a signal in P_M, the apex of the reflection time surface. The reflected signal recorded in R_R thus migrates to P_M rather than P_D. Practically, the image falls in fact into the apex of the approximation hyperboloid to the time surface. Both points are, however, very close to each other. It therefore appears reasonable to define *point image migration* as a process which establishes the depth point image *per definition* in the minimum of its reflection time surface even for models for which the Kirchhoff integral is not strictly valid. In this way one honours more the actual procedure commonly applied rather than the desired result which can not be obtained by this particular migration scheme. From this practical definition one can immediately conclude that a migration of the type considered is by no means complete for reflectors falling below interfaces which are not horizontal. Point image migration has hence to be followed by an additional time-to-depth migration if the true depth structure is to be recovered. This additional migration is necessary as P_M is usually not vertical above D (fig. 1). In spite of the apparent drawbacks this practical definition proposed here has also some interesting ray theoretical aspects. One can conclude that the position of P_M coincides with the position of the ray which emerges from D vertical to the free surface. This can be proven as follows:

The ensemble of rays of fig. 1 is necessarily identical with the one related to a wave originating in D. As wavefronts are perpendicular to rays the travel time of a wave from D to reach the surface is therefore half the time attributed to the reflection time surface. Both time surface minima have a horizontal tangent. Their position must therefore coincide with the position of the vertically emerging ray from D. *Migrated primary reflection times for a given depth model are consequently obtained by tracing rays vertically down from the free surface to the desired reflector at depth while plotting the two-way times at the respective surface positions of the rays*. Rays which emerge vertically at the free surface are subsequently referred to as *image rays*. They are naturally vertical only in a medium with constant—or only vertically inhomogeneous—velocity which may lie above (but not below) one with more complex velocity distribution. Below a curved interface they are refracted in a similar way as normal incidence rays. We have called them image rays as they connect a depth point with the surface position of its image. The signal positions and two-way times related to image rays are not affected by the migration process proposed here. Signals for all other rays (fig. 1) are migrated. *Unlike normal incidence rays image rays conform with each other irrespective of the interface to which they belong*. Both kinds of rays are complementary in the sense that normal rays are normal to the selected interface while image rays are normal to the free surface. The two-way primary reflection time for a shot-receiver pair in the free surface pertains to a ray normal to the selected interface. This time is transformed by the migration process into the time by which a wave reflected at the free surface is recorded by a shot-receiver pair within the normal incidence point of the normal ray. Based upon these simple considerations some commonly occurring 2-D models have been studied to demonstrate the importance of image rays as a means of interpreting and simulating migrated reflections. Image rays relate to migrated reflections in a similar way as normal incidence rays relate to stacked primary reflections (Taner, Cook, and Neidell 1970). They contribute to understanding the theory and limitations of the time migration process from a ray theoretical point of view. When one computes interval velocities from migration velocities they play the same fundamental role which normal incidence rays play in connection with computing interval velocities from stacking velocities. The complementary relationship between normal rays and image rays holds whenever the velocity distribution is isotropic and inhomogeneous.

EXAMPLES

Migration in its usually accepted sense poses no problem for the chosen earth model if an arbitrarily curved subsurface reflector falls below a system of plane horizontal isovelocity layers. Image rays are then strictly vertical down to the uppermost curved interface and a performed migration is complete. Quite often, however, curved interfaces fall below other curved velocity interfaces and image rays deviate from strictly vertical rays. The more refracted they are the less complete is in fact the migration. Though usually most effective if the structure is complex, migration may also be most incomplete in such cases. Errors introduced by predicting the true depths by only scaling the time-migrated reflections to depth can then be quite severe. As the following examples show it is quite likely that for migrated reflections puzzling situations may arise which are similar to those of stacked primary reflections of nonlinear subsurface reflectors (Taner et al. 1970). There is, however, one positive aspect which reduces the diversity of migrated primary travel time functions and deserves particular mentioning: It is that they remain invariant with the depth of the structure. This does not apply to stacked primary reflections which may be quite sensitive to the depth of a nonlinear reflector. It is well known that a trough near the surface may look very different on a stacked section from a trough at some depth. After section migration it will, however, look the same. Fig. 2 shows some plane horizontal interfaces below typical geological structures. Fig. 2b shows the normal incidence rays to the plane interfaces and fig. 2a shows the respective image rays. The two-way travel times along both types of rays are plotted in fig. 2c. The time-migrated primary reflections are dotted. Normal rays to the first interface are not shown. The structure of the uppermost interface changes the linearity of all other reflectors in a way which gets worse with the depth of the reflectors. The amount of travel time distortion for the migrated reflections becomes in

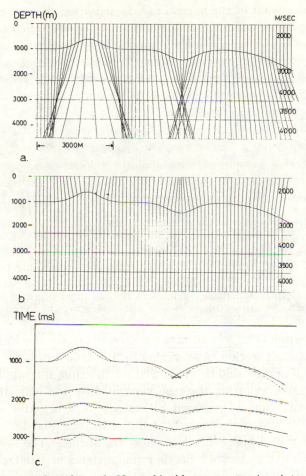

Fig. 2. a. Image rays to interfaces; b. Normal incidence rays to plane interfaces; c. Normal and time-migrated (dotted) reflections.

fact larger than that for stacked reflections. It is only the primary reflections for the first interface which are truely migrated in the conventional sense. All other depth reflectors are 'time-migrated' if one accepts the definition of 'migration' proposed in this work. To recover the linearity of all interfaces deeper than the uppermost one a further time-to-depth migration has to be applied in order to correct the non vertical image rays.

Fig. 3 shows a North German salt diapir where normal rays and image rays have been traced down to the Permian formation. The migrated primary reflections relate to *almost vertical* image rays. They thus provide a better approximation to the true depth model. As can be judged from the non vertical image rays there will, however, be some lateral distortion left in the structure

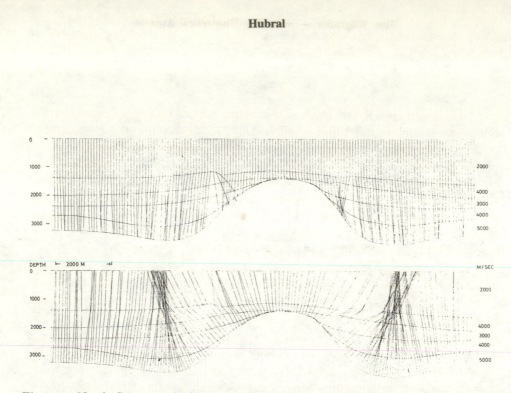

Fig. 3. a. North German salt dome model with normal incidence rays to Permian formation; b. North German salt dome model with image rays to Permian formation.

after time migration. The normal rays exaggerate largely the width of the salt dome on the stacked section. As revealed by the image rays the actual width of the dome will appear slightly decreased on the time migrated section. The deviation of image rays from the true vertical direction will often affect the interpretation of primary migrated reflections in a negative way. Caution must hence be exercised particularly if the reflectors used for interpretation fall below curved or dipping interfaces near the surface. The systematic summing of signals into the apex of reflection time surfaces may in fact occasionally create migration problems which are more severe than those existent on the original stacked section. An image ray describes the subsurface locations from which information is gathered that is displayed in a point image migrated trace. As image rays may cross each other one depth point may in fact be imaged into two different points on the section. This should disappoint only those readers who believe that the Kirchhoff summation can be applied to vertically and laterally inhomogeneous velocity media without having to pay any penalties. It cannot disappoint readers who accept the definition of point image migration given in this work—a definition valid for subsurface models more general than those upon which the Kirchhoff integral is based.

CONCLUSIONS

If signals scattered at subsurface points are summed into the minimum of their respective reflection time surfaces, the migration in its usual sense is only complete for depth points falling below plane horizontal isovelocity layers. Scattered signals for depth points below curved interfaces provide then no longer a complete migration when summed into the reflection time surface minima. A new ray theoretical definition is hence proposed for point image migration which extends the Kirchhoff summation method to all reflectors of the 3-D layered inhomogeneous earth model used in this work. As shown in this paper this process can be considered as pushing source-receiver pairs into the subsurface along image rays. It thus differs from the finite difference wave equation method which is generally viewed as a process which aims at pushing source receiver pairs vertically down into the subsurface. This aim is to our knowledge not as yet achieved by any time migration scheme. Point image migration as defined above is hence incomplete in its conventional sense. It offers, however, a strict complementary logic to the process of stacking. It provides consistent surface measurements which relate to subsurface reflectors in a unique way and which can be used for a further time-to-depth migration if the true depth-structure is to be recovered. The introduction of image rays contributes in our opinion much to clarifying the theory of migration from a ray theoretical point of view. Image rays offer themselves for interpretive seismic modeling. By accepting their existence one immediately realizes that time migrated sections—after having been scaled to depth—must not necessarily overlay depth sections which have been obtained by conventional time-to-depth migration methods which consider Snell's law and use known interval velocities and travel times of stacked primary reflections. Though a point image migration is not complete it resolves generally much better the reflecting horizons due to the contraction of disturbing diffraction patterns. Using primary time migrated reflection times and available interval velocities and performing a time-to-depth migration with image rays should result in a clearer picture of the true reflecting horizons as can generally be obtained from conventional time to depth migration methods that use stacked primary reflections and normal rays.

References

CLAERBOUT, J. F. and DOHERTY, S. M., 1972, Downward continuation of moveout corrected seismograms, Geophysics 37, 741-768.

FRENCH, W. S., 1974, Two-dimensional and three-dimensional migration of model-experiment reflection profiles, Geophysics 39, 256-277.

HAGEDOORN, J. G., 1954, A process of seismic reflection interpretation, Geophysical Prosp. 2, 85-127.

LARNER, K. and HATTON, L., 1976, Wave equation migration: Two approaches, Paper presented at the 8th Annual Offshore Technology Conference, Houston, Texas, 1976.

TANER, M. T., COOK, E. E. and NEIDELL, N. S., 1970, Limitations of the reflection seismic method; lessons from computer simulation, Geophysics 35, 551-573.

Reprinted from *Geophysical Prospecting*, **24**, 380-399.

THE WAVE EQUATION APPLIED TO MIGRATION *

BY

D. LOEWENTHAL, L. LU, R. ROBERSON, and J. SHERWOOD **

ABSTRACT

LOEWENTHAL, D., LU, L., ROBERSON, R., and SHERWOOD, J., 1976, The Wave Equation Applied to Migration, Geophysical Prospecting 24, 380-399.

Claerbout's method has been implemented for the migration of stacked seismic data. A simplified description of the method is given together with an account of some of the practical programming problems and the types of inaccuracy encountered. Routine production results are considered to be comparable or superior to the results derived from alternative migration techniques. Particular advantages are 1) the possibility of using a detailed velocity model for the migration and 2) the preservation of the amplitude and character of the seismic events on the migrated time section.

INTRODUCTION

In recent years Jon Claerbout and his students at Stanford University have published many articles on numerical studies of approximations to the wave equation. Although their work has been strongly oriented towards applications in exploration seismology it seems that the geophysical industry has not been overly enthusiastic to experiment with or implement these techniques in routine data processing. With the conviction that maximizing the extraction of geophysical information from seismic data is ultimately dependent on the use of the wave equation, we felt that it was desirable to initiate an appropriate development program.

As an initial step we have implemented Claerbout's method for the migration of stacked seismic data. A simplified summary of the method will be given here, together with examples of typical results.

ASSUMPTIONS AND OBJECTIVE

The data consist of a CDP stack section which we will define by its amplitude $A(x, t)$ as a function of the two-way travel time t and the position x along a straight surface line. Provided that reasonable stacking velocities and amplitude control have been used, the primary reflections on this section approximate

---

* Paper read at the 36th meeting of the European Association of Exploration Geophysicists, Madrid, Spain, June 1974.

** Digicon, Inc., 3701 Kirby Drive, Houston, Texas 77098.

what would be generated by a point source—receiver combination yielding independently recorded seimic traces at some constant increment along the x axis. For simplicity we will not discuss the effect of the migration on multiple reflections or on reflections from locations outside the plane of the section. Finally, an inherent assumption is that the basic seismic pulse is a simple impulse and that its amplitude is invariant as it is transmitted through the earth layers (see Foster 1975, for a pertinent discussion of this subject). One should note that this latter assumption and the neglect of multiples are compatible with an earth which is only weakly inhomogeneous.

Having made all these assumptions, our objective now is to modify the CDP section into a form more representative of a cross-section through the earth. This parallels the effort of research seismologists to estimate the detailed acoustic velocity and density variations in the subsurface. This is a difficult problem in statistical estimation and is beyond the scope of the present discussion. We will lower our objectives here and state that we wish to utilize our gross velocity estimates in order to derive from the CDP stack section $A(x, t)$ a function $B(x, z)$ where B is related to the variation of the earth's acoustic impedance with horizontal position x and depth z. In fact, because of the relative inaccuracy of our velocity information, we will content ourselves with deriving a function $B(x, \tau)$, where the depth variable z has been replaced by a less sensitive parameter τ representing the *vertical* two-way travel time.

In the following sections we will show that, for an earth with the assumed properties, a relatively simple transformation of $A(x, t)$ will yield a reasonable approximation to the desired migrated section $B(x, \tau)$.

METHOD

The CDP stack section may be considered as a wave field measured at the surface of the earth. Given the approximate velocity variations within the earth we are going to downward continue this wavefield into the subsurface and elucidate the source of the reflected and diffracted seismic events. The detailed procedure for this draws heavily on Claerbout's elegant and economic method for propagating a wave field using finite difference approximations to the wave equation (Claerbout and Johnson 1971, Claerbout and Doherty 1972). To achieve these favorable economics it is necessary for the significant components of the wavefield to be traveling within some small angular distribution around a given direction, which in our case is vertically upwards. At first sight, this requirement implies that Claerbout's technique is inapplicable to a CDP stack section, which is produced by the upward reflection and diffraction of originally downward traveling waves. Fortunately, this is not the case, since simplistic considerations permit us to visualize the CDP stack section as a superposition of upward traveling waves.

Let us quantize the subsurface so that it consists of small diffracting points, each with an appropriate scattering strength. Normally, of course, the strongest scatterers, or diffractors, will be distributed along strong reflecting interfaces.

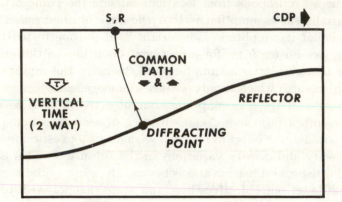

Fig. 1. The earth cross section.

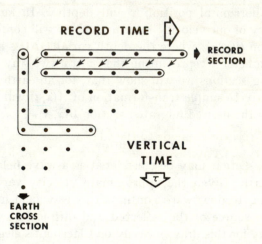

Fig. 2. Schematic diagram of the wave equation migration process.

Now consider a single point source and receiver combination, the impulsive source being initiated at time zero (see figure 1). The downward ray path and travel time $t/2$ to a diffracting point is identical with the return upward path and travel time $t/2$ to the receiver. The amplitude of the return will be proportional to the strength of the diffracting element. It is clear that an equivalent recording will result from a source initiated at time zero at the diffractor, with the strength of the diffractor, provided that the velocity $c(x, \tau)$ in the medium is halved. The CDP stack section can then be considered as due to upcoming waves from the totality of quantized diffractor sources, each initiated at time zero with an appropriate strength, and with the instantaneous velocity throughout the medium being $c(x, \tau)/2$.

A discussion of this equivalent mechanism for generating a CDP stack section will clarify the steps in the inverse procedure of migration. In figure 2, the horizontal axis is the observed two-way travel time t, the vertical axis is the *vertical* two-way travel time τ, and the CDP location x is perpendicular to the plane of the figure. The top left hand point is the origin $t = \tau = 0$. Thus, the vertical left hand column represents an earth cross-section composed of the strengths of the hypothetical diffractor sources initiated at time $t = 0$.

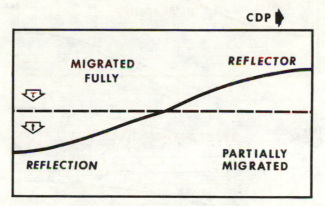

Fig. 3. An intermediate stage.

The top horizontal row corresponds to the CDP record section observed at the earth's surface, $\tau = 0$. In mathematical notation we can define the complete wave field as $A(x, t, \tau)$. Given that the wave-field is basically upward traveling, and given a particular initial source distribution $A(x, 0, \tau)$, it is possible to utilize Claerbout's algorithm to determine the complete wave field $A(x, t, \tau)$ using incremental steps Δt in the observational time direction. The surface observations $A(x, t, 0)$ correspond to a conventional seismic cross-section. Alternatively, if this seismic cross-section $A(x, t, 0)$ is given, it is possible to extrapolate the wave field downwards using incremental steps $\Delta \tau$. The subset of derived values $A(x, 0, \tau)$ corresponds to the diffractor source distribution and provides the desired migrated section result.

The actual order in which we choose to execute the computations is depicted by the L shaped outline in figure 2. This illustrates the data stored in the computer system at an intermediate stage of the processing. It consists of the completely migrated results down to a "depth" τ, followed by the remnant partially migrated data which still requires propagation through the deeper layers of the subsurface. This intermediate stage is exhibited with more clarity in figure 3. The upper part of the data field has been fully migrated, original reflection data having been transformed into the equivalent reflector positions. The lower part of the data is only partially migrated and is the seismic wave field that would be observed if the point source and detector combinations were positioned along the dashed line rather that at the surface.

MODEL STUDY RESULTS

We will first examine the results of the wave equation migration process using simple synthetic data. Figure 4 shows an earth model, consisting of

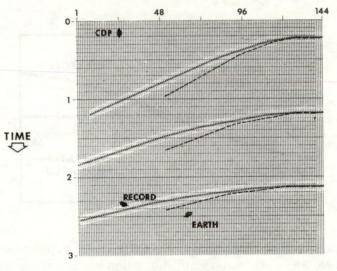

Fig. 4. Earth model.

three continuous reflectors $A(x, 0, \tau)$, superimposed on which is the corresponding seismic record section. The latter was derived from the earth cross-section by means of hand calculations using ray theory. The computer inputs were formed by inserting impulses at appropriate points (x, τ) and (x, t) and then convolving these impulses with a simple zero phase filter wavelet (8 Hz low cut, 30 Hz high cut) in order to avoid any drastic aliasing problems during the migration procedure. The model parameters were a constant instantaneous velocity of 3048 m/s, a horizontal or trace interval of $\Delta x = 24.4$ m, and a time dimension sampling rate of 4 ms. The maximum "dip" in the earth model is 12.5 ms/trace corresponding to a dip of about 39°. This leads to a reflection event on the record section having a dip of 10.0 ms/trace.

For the sake of economy, it is desirable to speed up the migration procedure by stepping down through earth layers which have individual two way travel times $\Delta\tau$ considerably greater than 4 ms. Figure 5 shows the result of migrating the synthetic record using a layer thickness $\Delta\tau$ of 200 ms after first resampling the data to an 8 ms increment. The comparison of the migrated result with the original earth model illustrates the inadequacy of these parameters. Distortion is particularly apparent for the higher frequencies at large dips. In this region the finite difference approximation to the wave equation propagates the wave disturbance at too slow a velocity, resulting in under-migration of the data. The discontinuities on the migrated record section are caused

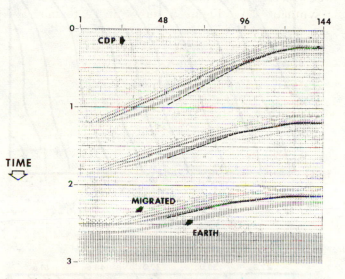

Fig. 5. Migration of the synthetic record section using $\Delta t = 8$ ms and $\Delta\tau = 200$ ms.

by use of the overly large layer thickness of 200 ms. It seems reasonable to believe that a finer sampling would produce more acceptable results. A quantitative investigation of the migration errors as a function of these sampling increments is appropriate at this point. For the convenience of the reader we will delegate detailed mathematics to the appendix and will discuss only the concept underlying the error analysis in the main text.

MIGRATION ERROR ANALYSIS

For simplicity we consider a uniform medium in which a plane wave with frequency f is propagating upwards with acoustic wave velocity c at an angle θ to the vertical. This plane wave corresponds to an exact solution of the wave equation for our idealized homogeneous, nondispersive medium. Now let us take this wave field as observed at the surface $z = 0$ and use Claerbout's finite difference equation as it would appear at the bottom of a layer of thickness

$\Delta z = c\Delta\tau/2$. The analysis in the appendix shows that this extrapolated wave field is consistent with a plane wave whose velocity and angle deviate from c and θ. The migration error for the frequency f can then be expressed as the relative decrease in velocity ($\Delta c/c$) and the decrease in angle $\Delta\theta$. This velocity

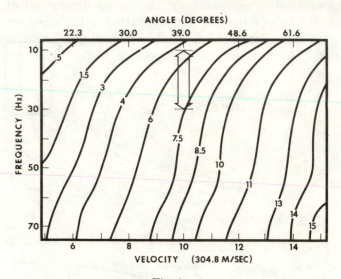

Fig. 6a

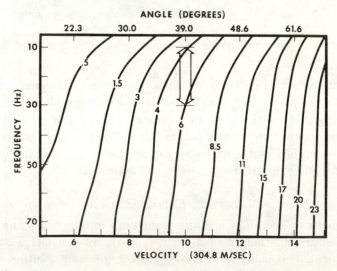

Fig. 6b

Fig. 6. Migration error estimates at dip = 10 ms/trace using $\Delta t = 8$ ms and $\Delta\tau = 200$ ms. (a) velocity error (%), (b) angular error (degrees).

dispersion $\Delta c/c$ is plotted in figure 6a for the highest synthetic record section dip of 10 ms/trace, and the parameters $\Delta x = 24.4$ m, $\Delta t = 8$ ms, and $\Delta \tau = 200$ ms. The vertical axis is frequency in Hz and the horizontal axis is acoustic velocity in units of 304.8 m/s. Since the dip is constant for this display and

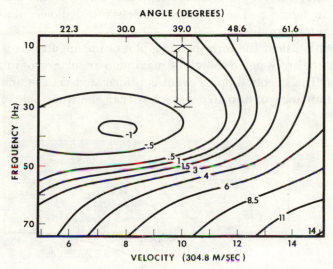

Fig. 7a

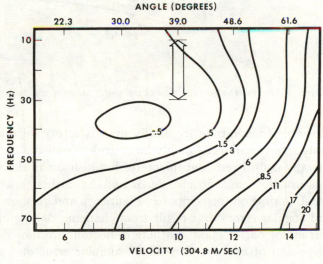

Fig. 7b

Fig. 7. Migration error estimates at dip = 10 ms/trace using $\Delta t = 4$ ms and $\Delta \tau = 20$ ms. (a) velocity error (%), (b) angular error (degrees).

is expressible as $(2\Delta x \sin \theta/c)$, we can also calibrate the horizontal axis linearly in $\sin \theta$. It is clear that the velocity dispersion increases with frequency and with acoustic velocity. The region indicated by the arrow shows the maximum velocity errors pertaining to the synthetic migration results in figure 5. It is emphasized that the 7-13% velocity dispersion refers to the steepest dip on the section in figure 4. The dispersion is less for shallower dips, becoming precisely zero for zero dip.

In figure 6b we show the corresponding plot of the angular error. The region indicated by the arrow again shows the maximum angular error in the previous model study. The steeply dipping event is propagated via the finite difference equations at an angle of 6 to 10 degrees less than the true angle of 39 degrees.

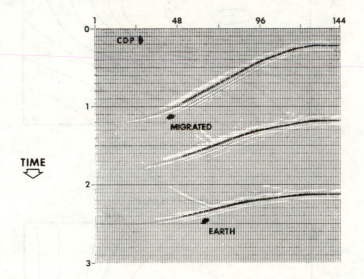

Fig. 8. Migration of the synthetic record section using $\Delta t = 4$ ms and $\Delta \tau = 20$ ms.

Errors of the above magnitude are totally unsatisfactory for quality processing and mainly result from using overly large grid increments Δx, Δt, and $\Delta \tau$. It is not generally considered convenient to diminish Δx in data processing. However, it is usually simple to change Δt and $\Delta \tau$, although at the cost of increased execution time on the computer. Figures 7a and 7b show the velocity dispersion and angular errors that result from changing Δt to 4 ms and $\Delta \tau$ to 20 ms. The errors for our steepest synthetic reflection are now diminished to a velocity dispersion of about 1%, with an angular error of 1 degree. It is believed that such errors are quite satisfactory for normal seismic processing. The result of using these improved values for the parameters Δt and $\Delta \tau$ are shown in figure 8. It is clear that most of the data have been satisfactorily migrated. For shallow dip the wavelet character is essentially unchanged by this wave equation method, a fact which is of considerable convenience for an interpreter wishing to correlate between events on the CDP record section and the migration display. For steep dips, especially at later times, some change in wavelet shape is apparent. Part of this event stretch is anticipated and correct. For example: in a uniform medium simple geometrical considerations show that a reflection event becomes stretched by a factor of $(1/\cos \theta)$ when migrated, θ being the true dip of the layering. Thus part of the character change for the steepest reflector in figure 8 is valid. The remainder is due to errors arising partly from Claerbout's approximations to the wave equation and partly from the finite sampling increments Δt, Δx, and $\Delta \tau$.

PROGRAM CODING CONSIDERATIONS

The migration algorithm originally disclosed by Claerbout and Johnson (1971) and Claerbout and Doherty (1972) requires that the CDP stack data be multiplexed in time. In our original computer implementation it was convenient to manipulate only 140 seismic traces in this manner, mainly because of the limited random access storage available. Hence, we chose to partition the original seismic data into sections of this size and migrate each section independently, allowing the data to migrate outwards about 72 traces on each side of the section. Each resultant section would then be demultiplexed independently, prior to merging the now overlapping sections back together. At this point the migrated seismic data was in conventional trace sequential form and was ready for display or any additional processing.

This particular implementation of the process was satisfactory for obtaining initial experience and yielded good quality results for probably 75 per cent of our production workload. The limitation in the remaining cases was the inability of the data to migrate more than 72 traces away from the edge of a section.

For small values of Δx, say 25 m or less, the deeper high dip events which should have migrated far outside a section would instead reflect back into the section (a minor example of this is the weak spurious event with opposite dip which occurs around 2.2 seconds in the central part of figure 8). Decimation of the data in the x direction can obviously diminish this problem but in practice this proved to result in an inferior quality for the final migrated section.

In view of these initial results we transferred the migration algorithm to a computer with substantially more random access storage available. This enabled us to segment the data into sections of about 1520 traces, and the migrated results are allowed to expand 256 traces on each side. With this implementation we have not observed any significant problems. During production usage of it we have observed legitimate seismic reflections, at times of 4.0 to 5.0 seconds, migrate over horizontal distances of up to 7500 meters. For the common horizontal sampling increment of 25 meters, this corresponds to a migration over 300 seismic traces.

MIGRATION EXAMPLES

Example 1

Figure 9a is a 36-fold conventional stack of our offshore Louisiana data recorded using a 3600 meter streamer. The data have been deconvolved before stack and filtered and equalized after stack. The sampling interval is 4 ms, the trace interval is 50 m. The horizontal scale is about 32 traces to a mile and the vertical exaggeration is roughly 3 to 1. The rms velocity in this area is generally 1830 m/s at 1 second, 2740 m/s at 4 seconds, and 3050 m/s at 6 seconds. In this section, there are two major depositional faults, with a shale sheath under the left fault.

Figure 9b is the migrated time section. It seems to clarify the original stack. First of all, the faults are better defined and the diffractions are clearly migrated back to the fault interface. A more simplified picture is obtained. For example, in the region indicated by a box on both sections, we notice that considerable energy has been moved up-dip. We believe that this fault plane is now closer

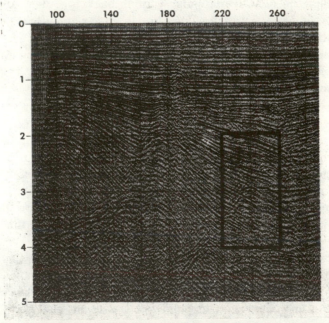

Fig. 9a. 36 fold stack.

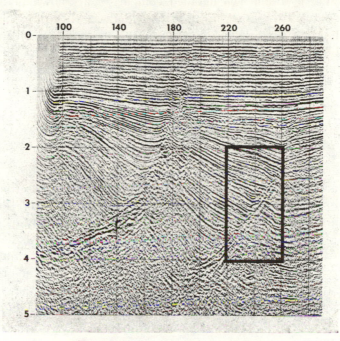
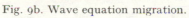

Fig. 9b. Wave equation migration.

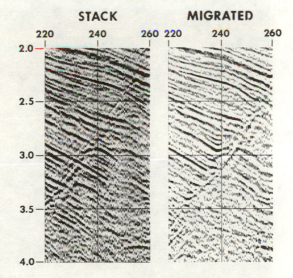

Fig. 9c. Delineation of fault.

to its actual location. The detailed comparison is shown in figure 9c. The
strong energy in the taper region caused high amplitude artificial parabolas.
The boundary effects always exist on both sides and at the bottom of a section
migrated by the wave equation migration process. The migration errors in this
example are less than 2% in velocity and 2° in angle for dips less than or equal
to 10 ms/trace. A 4 ms sample rate and 100 ms layer thickness were used for
the migration.

Example 2

In figure 10a and 10b, there are two interesting examples (indicated in
boxes A and B) in addition to the delineation of faults as described in example
1. The first example (enlarged in figure 10c) shows focusing diffractions by
migration. The energy in the diffractor region is apparently increased due to
focusing. The second example (enlarged in figure 10d) shows delineation of a
syncline by migration. The original section indicated very complicated events
crossing one another. It is not obvious that this is a syncline or buried focus.
The migrated section clearly shows the structure. Some of the steeply dipping
events (20 ms/trace) in the original section might be dispersed during migration.

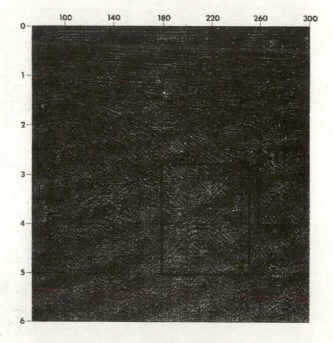

Fig. 10a. Stack.

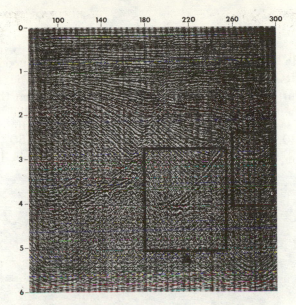

Fig. 10b. Wave equation migration.

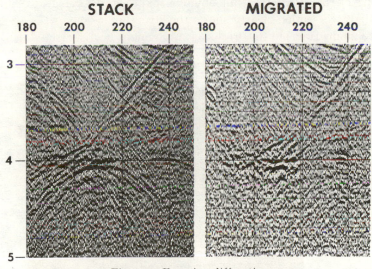

Fig. 10c. Focusing diffraction.

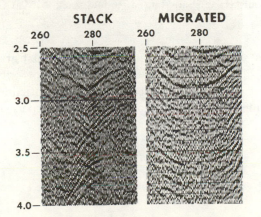

Fig. 10d. Delineation of syncline.

Example 3

Figure 11a shows a typical Gulf coast section with good delineation of a growth fault on the migrated section (figure 11b). The strong diffraction pattern occurring at shot point 70 around 4.2 seconds on the stack section is probably due to out-of-profile returns from the fault plane.

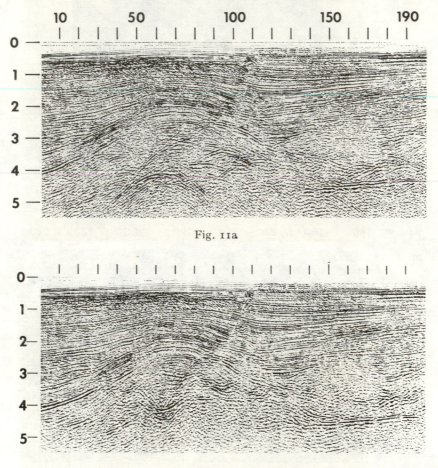

Fig. 11a

Fig. 11b

Fig. 11. Louisiana offshore data. a. 36 fold stack, b. Wave equation migration.

Example 4

Figure 12 is an interesting example of a Gulf coast graben, bounded by two growth faults. The migration better delineates not only faults but also the structure. First of all, it is easy to see that the anticline is more localized by curvature changes after migration, showing sharper structural relief. Apex

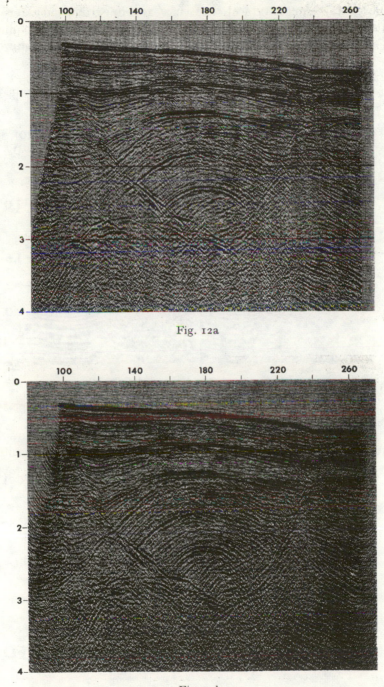

Fig. 12a

Fig. 12b

Fig. 12a, b. Comprised of stacked (a) and migrated data (b).

amplitudes on the migrated section are increased due to the curvature focusing. Some minor faults also become more obvious by focusing the diffractions. The boundary effects are also noticeable on both sides of this secti

Example 5

Figure 13 demonstrates the improved delineation of a salt dome flank and the preservation of wavelet character of the migrated events.

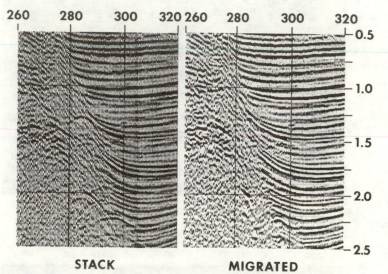

STACK   MIGRATED

Fig. 13. Preservation of character.

Example 6

In figure 14, the migrated bright spot shows detailed event character which is nearly identical with that of the unmigrated data. This is the result which is

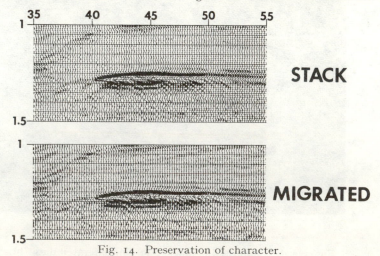

Fig. 14. Preservation of character.

expected because the migration process should not affect any flat event in a weakly inhomogeneous medium. Focusing effect due to weak diffractions is almost unnoticable in this example. Along with the amplitude preservation, we notice that the frequency content has been preserved as well. Due to the character preservation, it is now possible to model with migrated true amplitude sections.

CONCLUSION

Results have been shown of applying the wave equation form of processing to conventional exploration data. Resultant data quality has been very good and gives one an incentive for extending the use of this type of algorithm in production data processing. One obvious route to pursue is the application to the migration of unstacked seismic data, which could well lead to more accurate velocity analysis in addition to an improved migrated section (Sattlegger 1973). We believe that wave equation methods will play a significant role in the continual development of high quality, high resolution seismic surveys.

APPENDIX

In this appendix we briefly derive the migration error analysis, results of which were quoted in the main text.

Claerbout and Johnson (1971), in their equation (23), give the transfer characteristic across a layer of thickness Δz as

$$\exp(i\,\Phi) = \{1 - [aT(1 + Z)/(1 - Z)]\}/\{1 + [aT(1 + Z)/(1 - Z)]\}, \quad (1)$$

where
$$a = c\Delta t\Delta z/(8\Delta x^2). \quad (2)$$

This transfer characteristic applies to their simple finite difference approximation to the wave equation. Since there is no energy loss the transmission involves only a phase shift Φ at a frequency f. $(T/\Delta x^2)$ represents some finite difference approximation to the differential operator $(-\partial^2/\partial x^2)$. In our particular computer algorithm we have used a recursive operator for T, namely,

$$T = D/(1 - D/6), \quad (3)$$

where D is the double difference operator corresponding to the three point convolver $(-1, 2, -1)$.

The parameter Z denotes the conventional Z transform time delaying operator and is equivalent to

$$Z = \exp(-2\pi\,i\,f\Delta t). \quad (4)$$

We convert from the thickness of the layer Δz to the two way travel time across the layer using

$$\Delta z = c\Delta\tau/2. \quad (5)$$

We recall from the section on METHOD that the migration algorithm requires the replacement of the true velocity c in the medium by $(c/2)$. With these changes

$$a = c^2 \Delta t \Delta \tau / (32 \Delta x^2). \qquad (6)$$

Now let us consider a plane wave traveling upward at some angle θ to the vertical. In the continuous medium this wave is expressible as:

$$A = \exp\left[2\pi i \, f\left(t - \frac{2x}{c}\sin\theta + \frac{2z}{c}\cos\theta\right)\right]. \qquad (7))$$

The phase shift due to crossing a layer with thickness Δz is with (5)

$$\Gamma = 2\pi f \Delta \tau \cos\theta. \qquad (8)$$

Claerbout and Johnson (1971) derive their finite difference equations using a time frame referenced to a vertically traveling wave ($\theta = 0$). Changing to a fixed time frame will cause the phase shift Φ, from equation (1), to be modified to

$$\psi = \Phi + 2\pi f \Delta \tau. \qquad (9)$$

The phase shifts Γ and ψ in (8) and (9) will not normally be equal and their difference provides the error we wish to estimate. However, we prefer to investigate this error numerically from a somewhat different viewpoint.

In propagating the wave field on the finite difference grid the (x, t) dependence remains constant. This implies that the disturbance can still be visualized as a plane wave, but that its velocity and angle of propagation are distorted due to the error in the phase change across the layer. Let the apparent velocity be s and the angle be γ. Then the apparent plane wave becomes

$$B = \exp\left[2\pi i \, f\left(t - \frac{2x}{s}\sin\gamma + \frac{2z}{s}\cos\gamma\right)\right], \qquad (10)$$

where, because of the invariance of the horizontal wave number, we have

$$\sin\theta/c = \sin\gamma/s. \qquad (11)$$

The phase shift across the layer is $(4\pi f \Delta z \cos\gamma/s)$ which, when substituted into (9), yields

$$4\pi f \Delta z \cos\gamma/s = \Phi + 2\pi f \Delta \tau. \qquad (12)$$

Substituting in equation (5) and rearranging gives

$$\frac{c}{s}\cos\gamma = 1 + \frac{\Phi}{2\pi f \Delta \tau}. \qquad (13)$$

Equations (11) and (13) can be solved for the ratio of apparent velocity to true velocity:

$$(s/c) = \left[\sin^2 \gamma + \left(1 + \frac{\Phi}{2\pi f \Delta \tau} \right)^2 \right]^{-\frac{1}{2}} \tag{14}$$

Alternatively, (11) and (13) can be solved for γ, the apparent angle of propagation:

$$\tan \gamma = \sin \theta \left/ \left(1 + \frac{\Phi}{2\pi f \Delta \tau} \right) \right. . \tag{15}$$

Equations (14) and (15) were used in deriving figures 6a, 6b, 7a, and 7b. The phase is evaluated from equations (1), (3), (4), and (6), the quantity D in equation (3) being evaluated as

$$D = 2 \left[1 - \cos \left(2\pi \, \frac{2f}{c} \, \sin \theta \, \Delta x \right) \right]. \tag{16}$$

ACKNOWLEDGEMENT

We wish to thank Digicon Inc. for permission to publish this paper.

REFERENCES

CLAERBOUT, J. F., 1970, Coarse grid calculations of waves in inhomogeneous media with application to delineation of complicated seismic structure, Geophysics 35, 407-418.

CLAERBOUT, J. F., 1971, Numerical holography, in Acoustical Holography, v.3, New York Plenum Press, p. 273-283.

Claerbout, J. F., 1971a, Toward a unified theory of reflector mapping, Geophysics 36, 467-481.

CLAERBOUT, J. F. and DOHERTY S. M., 1972, Downward continuation of moveout corrected seismographs, Geophysics 37, 741-768.

CLAERBOUT, J. F. and JOHNSON, A. G., 1971, Extrapolation of time dependent waveforms along their path of propagation, Geophys. J. R. Astr. Soc. 26, 285-293.

FOSTER, M., 1975, Transmission effects in the continuous one-dimensional seismic model, Geophys. J. R. Astr. Soc. 42, 519-527.

SATTLEGGER, J. W., 1973, Migration velocity determination in two and three dimensions, Presentation at Annual SEG Meeting in Mexico City.

Chapter 5

Wave Equation Developments

The mathematical simplicity of representing a geologic model and the corresponding stacked section as initial and boundary values for a differential equation is exploited in the papers in this chapter. The velocity of propagation is assumed known; the location of the reflectors and the magnitude of the reflection coefficients are the unknowns.

For constant velocity, Stolt's $f - k$ migration algorithm is clearly the most efficient, and it makes migration of a stacked 3-D data set possible even with a small computer system (Gardner and Kotcher, 1978; Jacubowicz and Levin, 1983; Gibson et al., 1983). For a general velocity distribution $c(x,y)$, Gazdag's phase-shift method may prove popular in the future, but there are rival recursive and differential methods (Berkhout and de Jong, 1981). It is also possible that a combination of wavefront and ray tracing methods may be practical and economic, especially when velocity analysis and migration are linked (Sattlegger et al., 1980; Sattlegger, 1982; Ameely et al., 1983). However, in the following papers, the underlying mathematical concept is the exploding reflector model and numerical methods to simulate wave propagation.

Stolt's (1978) paper introduced the use of fast Fourier transforms (FFTs) for migrating stacked seismic profiles, assuming a constant velocity of propagation. FFTs of the measured data $B(x,t)$ with respect to x and t give Fourier coefficients from which a wave equation solution $p(x,y,t)$ is formed with the property $p(x,0,t) = B(x,t)$. The Fourier transform of the initial value function $p(x,y,0)$ is evaluated by interpolating and scaling these Fourier coefficients. The geologic model is then given by an inverse Fourier transform. This method is fast because it transforms the data into the answer in a single step. The speed is lost when local variations in velocity must be incorporated, because many small steps are then required. The *Chun and Jacewitz (1981)* paper describes this method with as little recourse to mathematical equations as possible.

Schneider (1978) demonstrated that the Kirchhoff integral formulation of the diffraction-stack migration for stacked seismic data is identical with wave equation migration when the velocity is constant. Again, the exploding reflector model is the underlying concept. The measured boundary values $B(x,t)$ are integrated along diffraction curves to construct a wave equation potential $\psi(x,y,t)$ with the property that $B(x,y)$ is equal to the derivative of ψ with respect to y when $y = 0$; the derivative of ψ with respect to y gives $p(x,y,t)$. The geologic model is interpreted from a plot of $p(x,y,0)$. Moderate changes in velocity, both vertically and laterally, can be incorporated in Kirchhoff migration (Carger and Frazer, 1984) and in wave equation migration (Castle, 1982).

Gazdag (1978 and 1980) and *Gazdag and Squazzero (1984)* developed the use of Fourier transforms for migrating stacked sections even when there were large changes in velocity. The wave field $p(x,y,t)$ was constructed by moving each frequency component $F(x,\omega)$ of the measured data $B(x,t)$ from one y level to a neighboring y level. This extrapolation, done by multiplying the Fourier coefficients of $F(x,\omega)$ with respect to x by phase-shift factors (complex numbers of unit magnitude), is exact if the velocity is constant, no matter how large the change in the y level. When velocity varies laterally an interpolation to the true velocity from the solutions for several constant reference velocities is an effective depth migration scheme. One-way wave propagation is modeled in these methods.

Berryhill (1979) showed how data collected on a rough topography could be extrapolated to a horizontal plane. *Junson et al. (1980), Schulz and Sherwood (1980), Larner et al. (1981),* and *Hatton et al. (1981)* described schemes for depth migration both before and after stack. *Berkhout (1981)* gave a review of previous migration procedures.

Kosloff and Baysal (1983) described an accurate depth migration in which each frequency component of the boundary measurements is integrated with respect to depth using a Runge-Kutta scheme and in which space derivatives are obtained using Fourier transforms. The waves undergo multiple internal reflections in this method because the full (two-way) acoustic wave equation is used and not a one-way approximation. At time zero, the upward traveling component of the wave field is taken as the migrated image. Since this method uses the full acoustic wave equation, both velocity and density can be varied with x and y. In order to reduce internal reflections, the density can be assigned so the acoustic impedance (density times velocity) is constant throughout the model (Baysal et al., 1984). An advantage, compared with one-way equations, is that the wavefronts can bend through more than 90 degrees as may be required when large-velocity gradients are present in the model.

McMechan (1983) and *Baysal and Kosloff (1983)* described a method based on integration with respect to time instead of depth. The measurements $B(x,t)$ are used as boundary conditions on $p(x,y,t)$ i.e., $p(x,0,t) = B(x,t)$, as p is stepped backward in time until $t = 0$. The values $p(x,y,0)$ are interpreted as the reflectivity distribution. Any velocity distribution can be used at each time step, and the time step can be made using finite differences. An advantage in using reverse time migration is that the propagating equations are the same as for forward modeling. Finite-difference equations (Kelly et al., 1976) and Fourier transform methods (Gazdag, 1981; Kosloff and Baysal, 1983) can be applied for migration.

The central theme of the final chapter is migration of stacked data. Perhaps the next great advance will be efficient schemes for migration before stack combined with velocity analysis, or some stacking scheme that does not require a velocity analysis (Hubral, 1983; Ottolini and Claerbout, 1984).

References

Ameely, L., Krey, Th., Marhtadie, F., Rau, H-F., and Rist, H., 1983, Migration in the presence of a rugged interface with high velocity contrast: Geophys. Prosp., **31**, 561-573.

Baysal, E., Kosloff, D. D., and Sherwood, J. W. C., 1984, A two-way nonreflecting wave equation: Geophysics, **49**, 132-141.

Berkhout, A. J., and de Jong, B. A., 1981, Recursive migration in three dimensions: Geophys. Prosp., **29**, 758-781.

Carter, J. A., and Frazer, L. N., 1984, Accommodating lateral velocity changes in Kirchhoff migration by means of Fermat's principle: Geophysics, **49**, 46-53.

Castle, R. J., 1982, Wave equation migration in the presence of lateral velocity variations: Geophysics, **47**, 1001-1011.

Gardner, G. H. F., and Kotcher, J., 1978, An innovative 3-D marine seismic survey: Presented at the Ann. Mtg., Europ. Ass'n. Explor. Geophys., Dublin.

Gazdag, J., 1981, Modeling of the acoustic wave equation with transform methods: Geophysics, **46**, 854-859.

Gibson, B., Larner, K., and Levin, S., 1983, Efficient 3-D migration in two steps: Geophys. Prosp., **31**, 1-33.

Hubral, P., 1984, Simulating true amplitude reflections by stacking shot records: Geophysics, **49**, 303-306.

Jakubowicz, H., and Levin, S., 1983, A simple exact method of 3-D migration-theory: Geophys. Prosp., **31**, 34-56.

Kelly K. R., Ward, R. W., Treitel, S., and Alford, R. M., 1976, Synthetic Seismograms — A finite-difference approach: Geophysics, **41**, 2-27.

Kosloff, D. D., and Baysal, E., 1982, forward modeling by a Fourier method: Geophysics, **47**, 1402-1412.

Ottolini, R., and J. F. Claerbout, 1984, The migration of common midpoint slant stacks: Geophysics, **49**, 237-249.

Sattlegger, J. W., Stiller, P. R., Echterhoff, J. A., and Hentschke, M. K., 1980, Common offset plane migration (COPMIG): Geophs. Prosp., **28**, 859-871.

Sattlegger, J., 1982, Migration of seismic interfaces: Geophys. Prosp., **30**, 71-85.

Reprinted from GEOPHYSICS, **43**, 23-48.

MIGRATION BY FOURIER TRANSFORM

R . H . S T O L T *

Wave equation migration is known to be simpler in principle when the horizontal coordinate or coordinates are replaced by their Fourier conjugates. Two practical migration schemes utilizing this concept are developed in this paper. One scheme extends the Claerbout finite difference method, greatly reducing dispersion problems usually associated with this method at higher dips and frequencies. The second scheme effects a Fourier transform in both space and time; by using the full scalar wave equation in the conjugate space, the method eliminates (up to the aliasing frequency) dispersion altogether. The second method in particular appears adaptable to three-dimensional migration and migration before stack.

INTRODUCTION

The migration of seismic data has been improved in recent years by application of the theory of scalar waves. Both the difference equation techniques pioneered by Jon Claerbout (1971, 1972, 1976) and integral equation techniques such as those developed by William French (1974, 1975) have been successful as applied.

Described below are two new schemes for the migration of seismic data. Both operate in momentum (i.e., wavenumber or spatial frequency) space in the horizontal (basement) direction. The first scheme is a high-accuracy, high-frequency, steep dip extension of the Claerbout finite difference algorithm. By formulating this algorithm in momentum space, we are able to (a) eliminate a matrix inversion without loss of accuracy, (b) migrate separately each momentum component using an algorithm tailor-made for each, so as to (c) reduce dispersion (within sampling limitations) to negligible proportions.

The second scheme is also based on the scalar wave equation but does not employ finite differences; rather, the exact wave equation is used to predict a transformation in frequency-momentum space. Subject to the sampling limitations of the data, dips of any angle can be migrated correctly and without dispersion.

Emphasis will be placed on digital migration of stacked seismic cross-sections. In addition, the second scheme will be shown to be adaptable to migration before stack and three-dimensional migration.

THEORETICAL FRAMEWORK

General

In what follows, we consider the earth to be a two-dimensional half-space. We assume sound to travel as a scalar field with velocity at point (x, z) of $c(x, z)$. Every point in the earth has the ability to transform downgoing sound waves into upgoing sound waves. This property is characterized by a reflection strength $R(x, z)$ whose angular dependence we ignore. Multiple reflections are also ignored.

Measurements are taken at the earth's surface by placing a source at point (x_s, z_s) and a receiver at (x_0, z_0). The reflected sound wave[1] observed at (x_0, z_0) is represented by $\psi(x_s, z_s, x_0, z_0, t)$, where

---

[1] ψ may represent a pressure, a displacement or velocity potential, or some other suitably defined parameter. Spatial derivatives of compressibility and density will be largely ignored in what follows, so the distinctions between the various fields will not be of concern here. ψ may be thought of as an impulse response or Green's function.

Manuscript received by the Editor September 8, 1976; revised manuscript received September 30, 1977.
*Continental Oil Co., Ponca City, OK 74601.

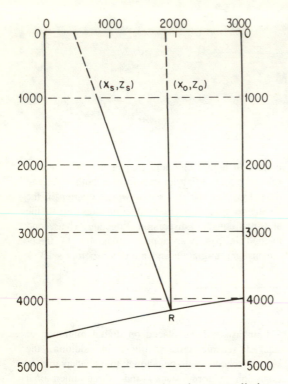

FIG. 1. Migration may be viewed as a prediction of changes in the seismic field as sources and receivers are moved into the earth.

t is the two-way traveltime from source to receiver. For a flat earth, z_s and z_0 are zero during the measurement.

By migration, one attempts to determine the reflection strength $R(x, z)$ from $\psi(x_s, 0, x_0, 0, t)$ at the earth's surface. This is done by predicting what ψ would be for sources and receivers inside the earth. Then (Claerbout, 1971),

$$R(x, z) \sim \psi(x, z, x, z, 0). \qquad (1)$$

That is, (x_s, z_s) and (x_0, z_0) are extrapolated to the common point (x, z) as shown in Figure 1. As they approach each other, the traveltime between source and receiver approaches zero, and ψ, subject to limitations in source bandwidth, becomes proportional to the reflection strength at that point.

The changes in ψ as source and receiver migrate into the earth can be predicted by the scalar wave equation. We require (using subscripts to denote partial derivatives),

$$\psi_{x_s x_s} + \psi_{z_s z_s} - \psi_{tt}/c(x_s, z_s)^2 = 0, \qquad (2)$$

and

$$\psi_{x_0 x_0} + \psi_{z_0 z_0} - \psi_{tt}/c(x_0, z_0)^2 = 0, \qquad (3)$$

at all points in space-time. That is, the scalar wave equation governs ψ with respect to small changes in receiver *or* source location.

Migration of stacked sections

These two equations can be simplified considerably for the migration of stacked sections, if we pretend that "stacked" sections are equivalent to normal incidence sections, where $(x_s, z_s) = (x_0, z_0)$. We define midpoint coordinates,

$$X = (x_s + x_0)/2 \text{ and } Z = (z_s + z_0)/2, \qquad (4)$$

and relative, or offset, coordinates,

$$x = (x_0 - x_s)/2 \text{ and } z = (z_0 - z_s)/2. \qquad (5)$$

Setting $\psi(X, x, Z, z, t) = \psi(x_s, z_s, x_0, z_0, t)$, the stacked section in the new coordinate system corresponds to $\psi(X, 0, 0, 0, t)$ and the migrated section to $\psi(X, 0, Z, 0, 0)$. Equations (2) and (3) become

$$\psi_{XX} + \psi_{ZZ} + \psi_{xx} + \psi_{zz} - 2\psi_{xX}$$
$$- 2\psi_{zZ} - 4/c(X - x, Z - z)^2 \psi_{tt} = 0, \qquad (6)$$

$$\psi_{XX} + \psi_{ZZ} + \psi_{xx} + \psi_{zz} + 2\psi_{xX}$$
$$+ 2\psi_{zZ} - 4/c(X + x, Z + z)^2 \psi_{tt} = 0.$$

If we ignore derivatives with respect to x and z,[2] we are left with the single equation

$$\psi_{XX} + \psi_{ZZ} - 4/c(X, Z)^2 \psi_{tt} = 0. \qquad (8)$$

Equation (8) differs from (2) and (3) by the factor of 4 in the ψ_{tt} term. This difference is due to the fact that in (8), both source and receiver coordinates are required to migrate synchronously, whereas in (2) and (3), one set of coordinates is kept fixed. The form of (8) can be made identical to (2) and (3) by redefining t in (8) to be one-way traveltime.

The Claerbout coordinate transformation

In the Claerbout approach, the wave equation (8) is converted to a difference equation which can be

[2]Strictly speaking, this is hard to justify, though we can argue that as $x, z \to 0$, first derivatives with respect to x and z should vanish, and second derivatives are moveout generators which produce mainly gradual changes of amplitude with time when x and z are fixed at zero.

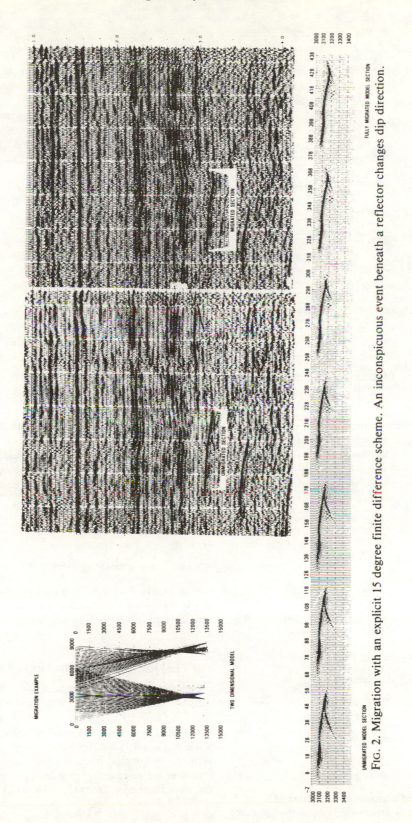

FIG. 2. Migration with an explicit 15 degree finite difference scheme. An inconspicuous event beneath a reflector changes dip direction.

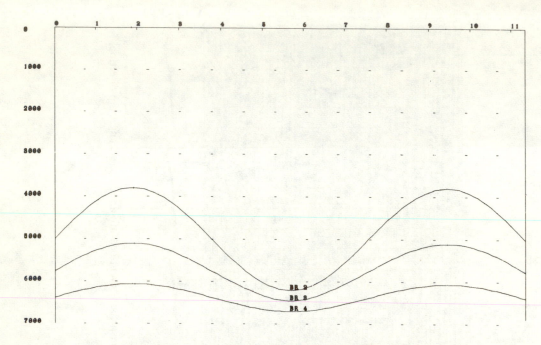

FIG. 3. A two-dimensional earth model. Velocity is 9600 fps in all layers.

used to gradually sink the source-receiver midpoint coordinates (X, Z) into the earth. To make such a scheme practical, Claerbout defines a new coordinate system in which ψ varies less rapidly with depth. If c were constant, one such transformation would be

$$D = ct/2 + Z; \, d = Z. \tag{9}$$

In this coordinate system, D is the depth of some reflection point, while d is the depth of the source-receiver midpoint.

Setting

$$\phi(x, d, D) = \psi(X, 0, Z, 0, t), \tag{10}$$

the wave equation (8) becomes (Claerbout, 1976, p. 211)

$$\phi_{XX} + \phi_{dd} + 2\phi_{dD} = 0. \tag{11}$$

The stacked section corresponds to

$$\begin{aligned} \phi(X, 0, D) &= \phi(X, 0, ct/2) \\ &= \psi(X, 0, 0, 0, t), \end{aligned} \tag{12}$$

and the migrated section to

$$\begin{aligned} \phi(X, D, D) &= \phi(X, Z, Z) \\ &= \psi(X, 0, Z, 0, 0). \end{aligned} \tag{13}$$

Of course, c will normally be a function of X and Z.

A coordinate system which casts the migration problem into a velocity independent form similar to that in equations (11)–(13) will be discussed later.

To help understand equation (11), consider a plane wave of angular frequency $\omega = 2\pi f$ traveling upward at angle θ to the vertical. According to equation (8), such a wave will take the form

$$\psi = e^{i2\omega\{X \sin\theta - Z \cos\theta - ct/2\}/c}, \tag{14}$$

or, in the new coordinate system,

$$\phi = e^{i2\omega\{X \sin\theta + d(1 - \cos\theta) - D\}/c}. \tag{15}$$

From (15) it follows that for upward traveling waves ϕ_{XX} is always greater than ϕ_{dd}. For waves traveling near the vertical, ϕ_{dd} will be negligible compared to ϕ_{XX}.

THE CLAERBOUT FINITE DIFFERENCE METHOD

Following Claerbout's approach, we now convert the wave equation (11) into a difference equation. Since $\phi_d(X, 0, D)$ is not known a priori, the equation should not involve second differences in d. Since ϕ_{dd} is very small for waves traveling near the vertical, the simplest thing to do is ignore it. This results in the so-called 15 degree approximation to the wave equation (Claerbout, 1976, p. 211)

$$\phi_{XX} + 2\phi_{dD} = 0. \tag{16}$$

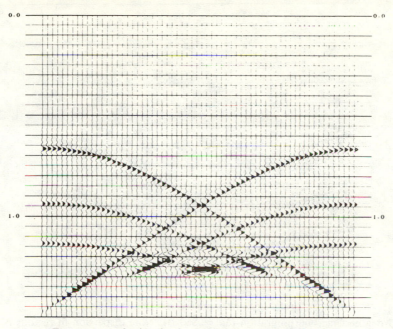

FIG. 4. A synthetic seismic section. Trace spacing is 120 ft.

We now define the discrete variables j and k by the relations

$$D_j = j\Delta D \text{ and } d_k = k\Delta d, \qquad (17)$$

where ΔD and Δd are the increments in reflector depth (transformed traveltime) and source-receiver depth, respectively. We also adapt the shorthand notation

$$\phi(X)_j^k \equiv \phi(x, d_k, D_j). \qquad (18)$$

The conversion of Equation (16) to a difference equation is not unique. Two possible lowest-order forms are

$$(1 - T)\,\phi_j^{k+1} = -(1 - T)\,\phi_{j+1}^k$$

$$+ (1 + T)(\phi_{j+1}^{k+1} + \phi_j^k), \qquad (19a)$$

and

$$\phi_j^{k+1} = -\phi_{j+1}^k + (1 + 2T)(\phi_{j+1}^{k+1} + \phi_j^k), \qquad (19b)$$

where

$$T = \frac{\Delta D \Delta d}{8}\frac{d^2}{dX^2}, \qquad (20)$$

is an operator in X. In practice, T may be a second (or higher) difference operator in X. The form (19a)

is referred to as an implicit solution for ϕ_j^{k+1}, since in order to solve for that quantity, it is necessary to invert the operator $1 - T$. Since no inversion is re-required in (19b), we call it an explicit solution for ϕ_j^{k+1}.

The implicit form (19a) is a more expensive algorithm than (19b) but has the capacity for greater accuracy at steep dips.

Under many circumstances, the cheap explicit form (19b) will adequately migrate a stacked section. Figure 2 is an example of such a migration in which an apparent downturn of a surface at a fault is converted into an upturn. Migration of a simple model shows the upturn to have developed from an inconspicuous event beneath the reflector. The event appears less prominent on the actual section. This is partly attributable to interference from another event beneath it and partly to losses of diffractive energy during stack. The migration depth increment Δd used to migrate the model was the equivalent of about 500 msec (that is, six steps were required to migrate an event at 3 sec). The actual section was migrated using a smaller increment.

Figures 3 to 5 provide an example of the limitations of the explicit 15 degree algorithm. Figure 3 shows a two-dimensional model consisting of three reflecting surfaces whose depths vary sinusoidally. The maximum dip of the bottom reflector is 15 degrees; that of the middle, 30 degrees; and that of the top, 45

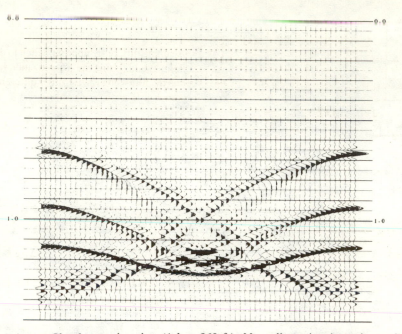

FIG. 5. A 15 degree Claerbout migration (Δd = 860 ft). Note dispersion in regions of steeper dip.

degrees. Velocity is constant at 9600 fps in all layers. Figure 4 is a synthetic normal incidence section constructed from this model, filtered 4–40 Hz. Trace spacing is 120 ft. The 15 degree Claerbout migration of this section is shown in Figure 5. As one might expect, the bottom reflection is migrated properly. The middle reflection, however, shows strong evidence of dispersion (that is, different frequencies are migrated to different places), and the top reflection is less migrated than mangled. Migration step size used was about 860 ft or 180 msec. Use of a smaller Δd step would not, in this case, improve the migration.

Higher order approximations to equation (16) are possible. Little is gained, however, unless the ϕ_{dd} term neglected in (16) is included.

MOMENTUM OR K-SPACE MIGRATION

For waves traveling at large angles to the vertical, ϕ_{dd} (though still smaller than ϕ_{XX}) is not negligible. Since it is not desirable to include ϕ_{dd} explicitly, an approximation must be found.

It is convenient to take a Fourier transform of ϕ with respect to X at this point, defining $p = 2\pi K$ to be the Fourier conjugate of X. Though not necessary for the development of a higher order approximation, there are several advantages to this step. First, the computer time disadvantage of an implicit solution

to the wave equation disappears, since the operator inversion becomes a simple division in the wave-number domain. Second, a simpler algorithm is allowed because the wave equation does not mix traces with different p values. Third, each p value can be migrated separately, using an algorithm individually tailored to it. Finally, the second derivative ϕ_{xx} is well approximated clear up to the spatial Nyquist frequency.

In the wavenumber domain, equation (11) takes the form

$$p^2 \phi = \phi_{dd} + 2\phi_{dD}. \qquad (21)$$

We can approximate the effect of ϕ_{dd} by differentiating equation (21) with respect to D,

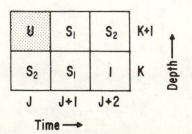

FIG. 6. The two-coeffiecint K-space migration operator.

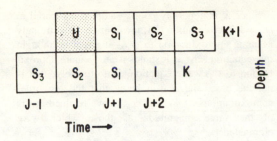

FIG. 7. The three-coefficient K-space migration operator.

$$p^2 \phi_D = 2\phi_{dDD} + \phi_{ddD}, \qquad (22)$$

and with respect to d:

$$p^2 \phi_d = 2\phi_{ddD} + \phi_{ddd}. \qquad (23)$$

Neglecting ϕ_{ddd} in (23) allows us to write the single equation,

$$2p^2 \phi_D - p^2 \phi_d = 4\phi_{dDD}. \qquad (24)$$

Equation (24) represents what is commonly called a 45 degree approximation to the scalar wave equation (22).

A simple difference approximation to equation (24) is easily made. Using the discretization formula (18), define the following lowest-order difference operators centered at $(j + 1, k + 1/2)$,

$$\frac{1}{4\Delta D} \left(\phi_{j+2}^{k/1} + \phi_{j+2}^{k} - \phi_{j}^{k+1} - \phi_{j}^{k} \right) \sim \phi_D, \qquad (25)$$

$$\frac{1}{\Delta d} (\phi_{j+1}^{k+1} - \phi_{j+1}^{k}) \sim \phi_d, \qquad (26)$$

$$\frac{2}{\Delta d (\Delta D)^2} (\phi_{j+2}^{k+1} - \phi_{j+1}^{k+1} + \phi_{j+1}^{k}$$

$$- \phi_j^k) \sim \phi_{dDD} + \frac{4}{\Delta d \Delta D} \phi_D. \qquad (27)$$

Substitution of (25), (26), and (27) into (24) yields the difference equation

$$\phi_j^{k+1} - \phi_{j+2}^{k} + S_1 (\phi_{j+1}^{k+1} - \phi_{j+1}^{k})$$

$$+ S_2 (\phi_{j+2}^{k+1} - \phi_j^k) = 0, \qquad (28)$$

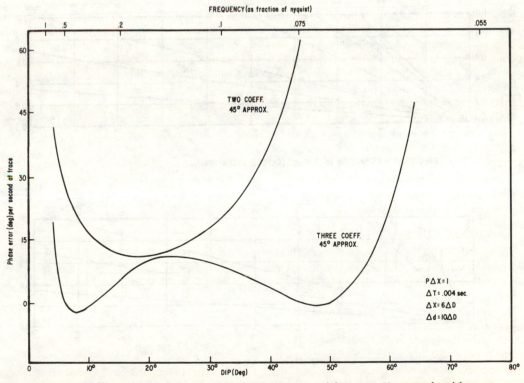

FIG. 8. An illustration of the phase error to be expected from the K-space algorithms.

$$S_1 = -2(8 - p^2 \Delta D^2)/(8 + p^2 \Delta D \Delta d),$$
$$(29)$$

$$S_2 = (8 - p^2 \Delta D \Delta d)/(8 + p^2 \Delta D \Delta d).$$
$$(30)$$

Equation (28) allows solution for an unknown ϕ_j^{k+1} in terms of the five known ϕ values illustrated in Figure 6. Stability is assured provided the polynomial,

$$1 + S_1 Z + S_2 Z^2,$$

has no roots inside the unit circle; i.e., provided its Levinson reflection coefficients (Claerbout, 1976, p. 55–57) are less than one in magnitude. This imposes the constraints

$$|S_2| < 1,$$
$$(31)$$

and

$$|S_1| < |1 + S_2|,$$
$$(32)$$

which are automatically met for any migration step size Δd larger than ΔD.

Since X is generally poorly sampled compared to D, and since equation (28) incorporates an extremely accurate approximation to ϕ_{xx}, it might be thought that equation (28) as it stands is an accurate approximation to the wave equation. Unfortunately, that is not the case, the reason being that the low-order approximations to ϕ_D and ϕ_{dDD} have retained errors of the same magnitude as those which were eliminated.

Conversion of (28) to a high accuracy equation is accomplished in two steps. First, we bring in more points along the D-axis. We write,

$$\phi_j^{k+1} - \phi_{j+2}^k + S_1(\phi_{j+1}^{k+1} - \phi_{j+1}^k)$$
$$+ S_2(\phi_{j+2}^{k+1} - \phi_j^k) + S_3(\phi_{j+3}^{k+1} - \phi_{j-1}^k) = 0.$$
$$(33)$$

That is, the unknown point ϕ_j^{k+1} is determined from the seven points illustrated in Figure 7. Stability is assured provided the polynomial,

$$1 + S_1 Z + S_2 Z^2 + S_3 Z^3,$$

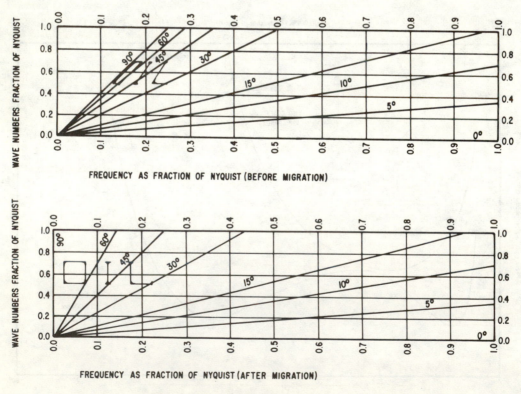

FIG. 9. Migration in F-K space.

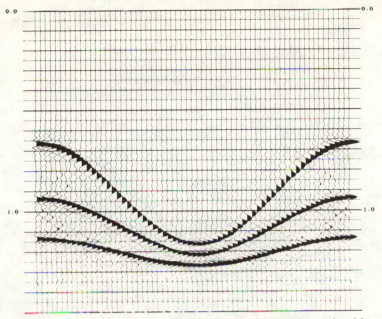

Fɪɢ. 10. Migration of a synthetic section using the 45 degree Claerbout algorithm. Maximum phase error is less than 8 degrees for frequencies below the aliasing frequency.

has no roots inside the unit circle. Constraining the Levinson reflection coefficients to be less than one in magnitude, we find

$$|S_3| < 1, \tag{34}$$

$$|S_2 - S_1 S_3| < |1 - S_3^2|, \tag{35}$$

and

$$|S_1 - S_2 S_3| < |1 - S_3^2 + S_2 - S_1 S_3|. \tag{36}$$

The appearance of the anti-causal term ϕ_{j-1}^k in equation (33) may be somewhat disquieting, but is really no cause for alarm. It merely reflects the fact that when dealing with bandlimited data, higher order approximations to D-derivatives at the point $j + 1$ will use more points on both sides of $j + 1$.

The coefficients S_1, S_2, and S_3 may be determined by choosing higher order analogs to the difference operators (25)–(27). However, greater accuracy may be achieved by choosing S_1, S_2, and S_3 so that the difference equation (33) best approximates the exact wave equation (21) rather than the 45 degree approximation (24).

To do this, we look at individual plane wave solutions to equation (21) setting

$$\phi = e^{i(qd - 2\omega D/c)} \tag{37}$$

Equation (21) then gives the dispersion relation (for upcoming waves)

$$q = \frac{2\omega}{c} - \sqrt{\frac{4\omega^2}{c^2} - p^2}. \tag{38}$$

The difference equation (33), on the other hand, gives the relation

$$\sin\left(\frac{\bar{q}\Delta d}{2} + \frac{2\omega\Delta D}{c}\right) + S_1 \sin\frac{\bar{q}\Delta d}{2} + S_2 \sin\left(\frac{\bar{q}\Delta d}{2} - \frac{2\omega\Delta D}{c}\right) + S_3 \sin\left(\frac{\bar{q}\Delta D}{2} - \frac{4\omega\Delta D}{c}\right) = 0; \tag{39}$$

or, solving for $\bar{q}$,

$$\bar{q} = \frac{2}{\Delta d} \arctan \frac{(S_2 - 1)\sin 2\omega\Delta D/c + S_3 \sin 4\omega\Delta D/c}{S_1 + (S_2 + 1)\cos 2\omega\Delta D/c + S_3 \cos 4\omega\Delta D/c}. \tag{40}$$

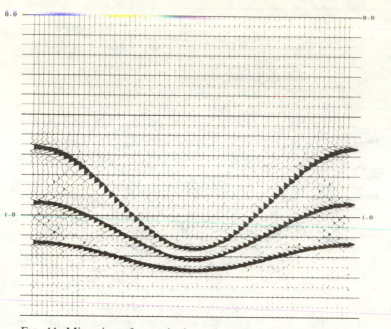

FIG. 11. Migration of a synthetic section using the *F-K* algorithm.

Since $\bar{q}$ will in general be different from the correct value q, a plane wave of frequency ω and wavenumber p [dip $\theta = \arcsin (pc/2\omega)$] will develop an error in phase ϵ proportional to the difference $\bar{q} - q$ and to the distance traveled:

$$\epsilon = (\bar{q} - q)d. \qquad (41)$$

Values for $S_1, S_2,$ and S_3 may be chosen so as to (in some sense) minimize ϵ. ϵ may be forced to zero at any three frequencies by substituting equation (38) into equation (39) and solving the resultant linear set of equations for $S_1, S_2,$ and S_3. Substitution of q at more than three frequencies results in an overdetermined system which can be solved by least squares. If θ_m is the maximum dip present in the data and ω_m the maximum frequency, the frequencies chosen should lie in the range

$$pc/2 \sin \theta_m < \omega < \omega_m. \qquad (42)$$

Figure 8 illustrates the phase errors to be expected from the two-coefficient difference equation (28) and the three-coefficient equation (33). In this example, $p\Delta X = 1$ radian, $\Delta X = 6\Delta D$, and $\Delta d = 10 \Delta D$. For the two-coefficient equation, $S_1 = -1.926174$, $S_2 = .932886$. For the three-coefficient equation, $S_1 = -1.913213$, $S_2 = .914654$, and $S_3 = .010563$. These values were chosen to give zero-phase error at $\omega_1 = $ Nyquist/2, $\omega_3 = pc/2\sin 45$

degrees, and $\omega_2 = (\omega_1 + \omega_3)/2$. Phase error is plotted in degrees per second of trace, assuming a .004 sec sample interval. Two things are apparent in this illustration: (1) equation (33) is more accurate than equation (28) over the entire range of dips and frequencies; and (2) by forcing three zero crossings for equation (33), we have actually gotten four (the fourth zero crossing occurs at about 50 degree dip), significantly extending its region of accuracy. This suggests that (a) modification of the coefficients of equation (28) could not produce accuracy comparable to that of equation (33); and (b) adding a fourth coefficient to equation (33) is not likely to significantly improve accuracy.

In practice, equation (33) is found to be accurate and stable for dips up to 45–55 degrees. Beyond this, accuracy may require an extremely small Δd, and stability problems may be encountered. Note that in general the coefficients $S_1, S_2,$ and S_3 and also the step size Δd will be different for every spatial frequency p.

The phase error defined in equation (41) does not include error introduced by a finite sample interval ΔX in X. In principle, these errors are zero, provided the maximum dip angle $\theta_{\max}$ obeys

$$\theta_{\max} < \text{arc sin} \, (\Delta D/\Delta X). \qquad (43)$$

If the maximum frequency in the data is $f_{\max} = \nu$

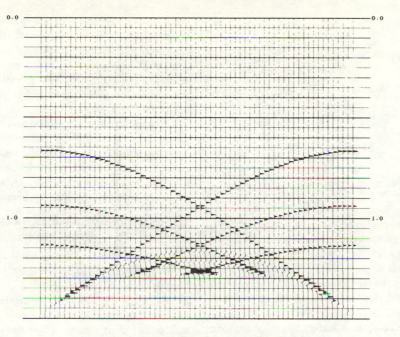

FIG. 12. A synthetic seismic section.

fnyq, where *fnyq* is the Nyquist frequency (125 Hz for $\Delta t = 4$ msec), then the above relation becomes less restrictive:

$$\theta_{\max} < \text{arc} \sin \left(\Delta D / v \Delta X \right). \quad (44)$$

Dips θ greater than $\theta_{\max}$ are migrated correctly as long as frequency obeys

$$f < fnyq \cdot \Delta D / DX \sin \theta \equiv f_a. \quad (45)$$

f_a will be referred to as the aliasing frequency.

At larger frequencies, the dip will be interpreted as smaller than $\theta_{\max}$ and will be migrated incorrectly.

Examples of migration using the equation (33) algorithm will be deferred until after a discussion of the $F - K$ migration scheme.

MIGRATION IN F–K SPACE

Suppose we take a two-dimensional Fourier transform of the surface field $\phi(X, 0, D)$:

$$A(p, \omega) = \frac{1}{2\pi} \int dX \int dD \, e^{i(pX - 2D/c)} \phi(X, 0, D), \quad (46)$$

so that

$$\phi(X, 0, D) = \frac{1}{2\pi} \int dp \int d\omega e^{-i(pX - 2\omega D/c)} A(p, \omega). \quad (47)$$

For upcoming waves, equation (47) generalizes at positive depth to

$$\phi(X, d, D)$$
$$= \frac{1}{2\pi} \int dp \int d\omega e^{-i(pX + qd - 2\omega D/c)} A(p, \omega), \quad (48)$$

where, to satisfy the wave equation (11),

$$q = \frac{2\omega}{c} - \sqrt{\frac{4\omega^2}{c^2} - p^2}. \quad (49)$$

The migrated section $\phi(X, D, D)$ then has the form

$$\phi(X, D, D)$$
$$= \frac{1}{2\pi} \int dp \int d\omega A(p, \omega) e^{-i\left\{pX - \sqrt{\left(\frac{4\omega^2}{c^2} - p^2\right)}D\right\}}. \quad (50)$$

The substance of equations (46) and (50) is that migration may be accomplished by a double Fourier transform of the original data from (X, D) space into (p, ω) space (46), followed by the more complicated transformation (50). If equation (50) could be converted into a double Fourier transform, a practical migration scheme could result.

Fortunately, a simple change of variable from ω to

$$k = \sqrt{\frac{4\omega^2}{c^2} - p^2}$$

does the trick:

$$\phi(X, D, D) = \frac{1}{2\pi} \int dp \int dk\, B(p, k)\, e^{-i(pX - kD)},$$
$$(51)$$

where

$$B(p, k) = \frac{1}{\sqrt{1 + p^2/k^2}}\, A\left(p, \frac{kc}{2}\sqrt{1 + p^2/k^2}\right).$$
$$(52)$$

The transformation (52) represents, for a fixed p, a shift of data from frequency ω to a lower frequency $\omega' = \sqrt{\omega^2 - p^2 c^2/4}$ (in fact, a "moveout correction" where ω takes the place of time, and p of offset), plus a change of scale $k/\sqrt{k^2 + p^2} = \omega'/\omega$. The frequency shift, depicted in Figure 9, effects only what migrators have always known, namely that an apparent dip of θ_A before migration translates into a dip,

$$\theta_M = \text{arc sin tan } \theta_A, \qquad (53)$$

after migration.

The operations (46), (51), and (52) could easily be done in an analog system. On a digital computer, the Fourier transforms (46) and (51) will be carried out as FFTs. The transformation (52) then involves a dangerous interpolation of the data in the frequency domain. To avoid ghost events appearing on the section, it is usually necessary at least to double the trace length by adding zeros before performing the initial time FFT.

No phase error or dispersion should be seen in double Fourier transform migration, since the exact wave equation is used. The aliasing problem discussed in the last section will still be present, though it is now possible to predict exactly where aliasing may exist and conceivably even unravel it.

EXAMPLES OF FOURIER TRANSFORM MIGRATION

Migrations of the synthetic section of Figure 4 are shown in Figures 10 and 11. Figure 10 results from k-space finite difference algorithm (33). Maximum-phase error at the bottom of the section was held to less than 8 degrees. All three reflectors have assumed their proper shapes. Little dispersion is evident, except for a loss of high frequencies in the region of 45 degree dip. This would be expected, since the aliasing frequency at that dip is 28.3 Hz. Figure 11 shows a migration using the double Fourier transform $(F\text{-}K)$ algorithm. It appears very similar to the K-space finite difference migration (Figure 10).

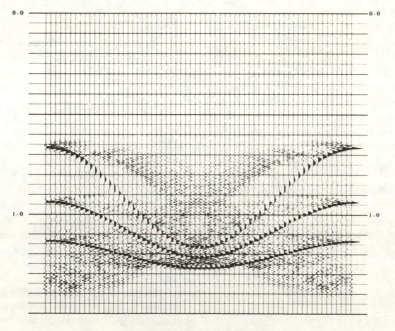

FIG. 13. A K-space migration.

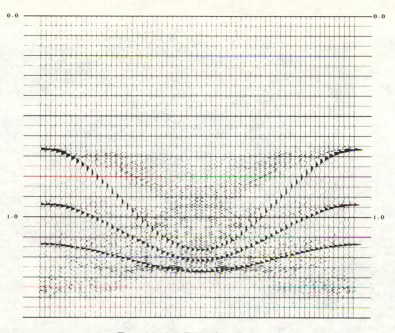

FIG. 14. An *F-K* migration.

Figures 12 to 14 show more clearly what happens to frequencies above the aliasing frequency. Figure 13 shows a *K*-space finite difference migration of the same model as before with a frequency range expanded to 0–90 Hz. Figure 14 is an *F-K* migration in the range 0–125 Hz. The similarity between the two migrations is still apparent. Frequencies below the aliasing frequency are in both cases migrated properly, while frequencies above it are undermigrated, mostly remaining close to their original position.

The remaining examples involve seismic field data. Maximum dip on Figure 15 is of the order of 35 degrees. Trace spacing is 220 ft, with rms velocities ranging from 7000 to 13,000 fps. Figure 16 is the *K*-space finite difference migration (maximum phase error <3 degrees); Figure 17 the *F-K* migration. Though some differences are apparent in the two migrations (due in part to slightly different parameterizations), they give substantively the same reasonable subsurface picture.

Figure 18 is a somewhat more complex section. The 15 degree Claerbout migration, shown in Figure 19, leaves several crossing events and some dispersion. The *K*-space and *F-K* migrations in Figures 20 and 21 give a more satisfying picture with a substantively different interpretation. Residual crossing events indicate a three-dimensional structure.

In Figure 22, the structure is actually three-dimensional, involving dips in excess of 45 degrees. To complicate matters, crucial data from the steep north flank of the structure were not recorded. Trace spacing is 220 ft, with rms velocities in the 7000–14,000 fps range. The *K*-space and *F-K* migrations in Figures 23 and 24 are again very similar. In this case, the results are predictably less than perfect but do provide some basis for interpretation.

The concluding section, Figure 25 is a regional seismic line. Its *F-K* migration, Figure 26, is a dramatic illustration of steep dip migration.

MIGRATION BEFORE STACK

The *F-K* migration scheme in principle can be used to effect moveout correction, stack, and migration of data in one process. In this case, we use a three-dimensional Fourier transform in the coordinate system specified in equations (4) and (5):

$$A(P, p, \omega) = (2\pi)^{-3/2} \int dt \int dX \cdot$$

$$\cdot \int dx\, e^{i(PX + px - \omega t)}\, \psi(X, x, 0, 0, t). \tag{54}$$

We assume the field to be governed by the two wave equations (2) and (3), where, for simplicity, *c* is assumed constant. Then,

FIG. 15. Five-fold CDP section maximum dip 35 degrees.

$$\psi(X, x, (z_0 + z_s)/2, (z_0 - z_s)/2, t)$$

$$= (2\pi)^{-3/2} \int d\omega \int dP \int dp \cdot$$

$$\cdot e^{-i(PX + px - q_s z_s - q_0 z_0 - \omega t)} A(P, p, \omega), \quad (55)$$

where, from equations (2) and (3)

$$q_s = \omega/c \sqrt{1 - (P - p)^2 c^2 / 4\omega^2};$$

$$q_0 = \omega/c \sqrt{1 - (P + p)^2 c^2 / 4\omega^2}. \quad (56)$$

The migrated section is

$$\psi(X, 0, ct/2, 0, 0)$$

$$= (2\pi)^{-3/2} \int d\omega \int dP \cdot$$

$$\cdot \int dp \, e^{-i(PX - (q_s + q_0)ct/2)} A(P, p, \omega). \quad (57)$$

To put (57) in the form of a Fourier transform, a coordinate transformation is required. We define two new variables which will replace p and ω,

$$u = q_s + q_0; \quad v = q_s - q_0. \quad (58)$$

From equations (56) and (58) follow the relations,

$$p = uv/P, \quad (59)$$

$$\omega = sgn(\omega)(c/2P)\sqrt{P^4 + P^2(u^2 + v^2) + u^2 v^2}, \tag{60}$$

and

$$\psi(X, 0, ct/2, 0, 0)$$

$$= (2\pi)^{-3/2} \int dP \int du \int dv\, e^{-i(PX - uct/2)} \cdot$$

$$\cdot A(P, p, \omega)c^2(u^2 - v^2)/4\omega P. \tag{61}$$

Again, a simple transformation accomplishes the migration process. Note that the Fourier p-integral becomes a simple integration over v, since only zero offset is relevant after migration.

MIGRATION IN THREE DIMENSIONS

The three-dimensional analog of equation (8) is

$$\psi_{XX} + \psi_{YY} + \psi_{ZZ} - 4/c^2\, \psi_{tt} = 0. \tag{62}$$

When c is constant, we can write

$$\psi(X, Y, Z, t) = (2\pi)^{-3/2} \int dP\, e^{iPX} \int dQ\, e^{iQY} \cdot$$

$$\cdot \int d\omega\, e^{-i\omega t} A(P, Q, \omega, Z), \tag{63}$$

where

$$A(P, Q, \omega, 0) = (2\pi)^{-3/2} \int dX \int dY \cdot$$

$$\cdot \int dt\, \psi(X, Y, 0, t)\, e^{-i(PX + QY - \omega t)}, \tag{64}$$

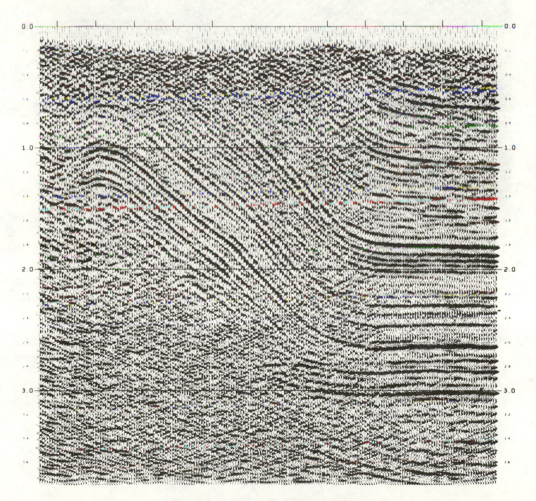

FIG. 16. A K-space migration of Figure 11.

FIG. 17. An *F-K* migration of Figure 11.

is the triple Fourier transform of the unmigrated three-dimensional data. Now, $A(P, Q, \omega, Z)$ satisfies the transformed wave equation

$$A_{ZZ} = (P^2 + Q^2 - 4\omega^2/c^2)A, \tag{65}$$

which has upcoming solutions,

$$A(P, Q, \omega, Z) = A(P, Q, \omega, 0)e^{-iZ\sqrt{4\omega^2/c^2 - P^2 - Q^2}}. \tag{66}$$

EXTENSION TO A VARIABLE VELOCITY

The K-space migration scheme described above relies on a velocity which is X-independent. The derivation of the F-K algorithm was even more restrictive, requiring velocity to be constant. In order to use these schemes in the presence of a variable velocity, it is necessary to transform to a coordinate

Hence, we can write the migrated field as,

$$\psi(X, Y, Z, 0) = (2\pi)^{-3/2} \int dP \int dQ \int d\omega\, B(P, Q, \omega)e^{i(PX + QY - 2\omega Z/c)}, \tag{67a}$$

where

$$B(P, Q, \omega) = A\{P, Q, \omega\sqrt{1 + (P^2 + Q^2)c^2/4\omega^2}\}/\sqrt{1 + (P^2 + Q^2)c^2/4\omega^2}. \tag{67b}$$

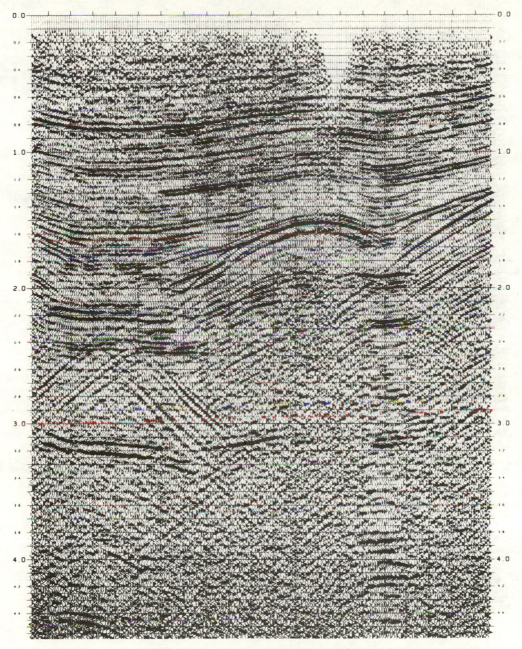

FIG. 18. A complex 10-fold CDP section.

FIG. 19. A 15 degree Claerbout migration of Figure 14.

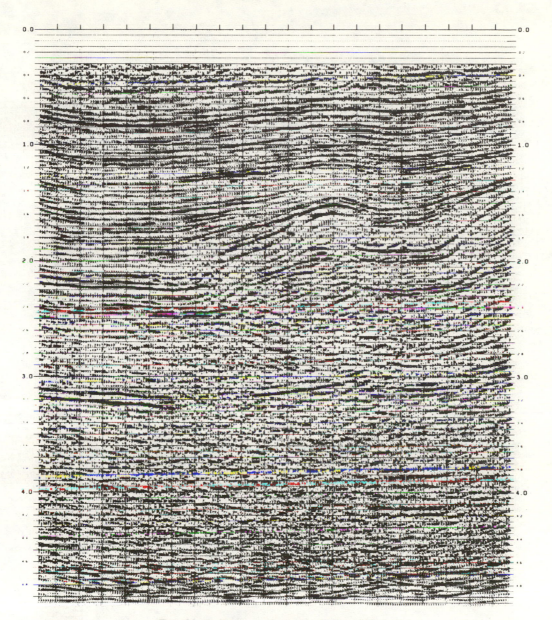

FIG. 20. A *K*-space migration of Figure 14.

system in which both the wave equation and the boundary conditions are velocity independent.

Beginning with the coordinate system X, Z, t of equation (8), we first define new time and depth coordinates

$$t' = \frac{t}{2} + \int_0^z \frac{dZ}{c}, \qquad (68)$$

$$Z' = \frac{1}{c_0} \int_0^z dZc. \qquad (69)$$

Since Z' represents the apparent depth to a reflector at Z in a layered medium, we may expect this to be a useful coordinate system for migration. Neglecting velocity derivatives, the wave equation becomes

$$\psi'_{xx} + c^2/c_0^2 \psi_{z'z'} + 2/c_0 \psi'_{z't'} = 0. \qquad (70)$$

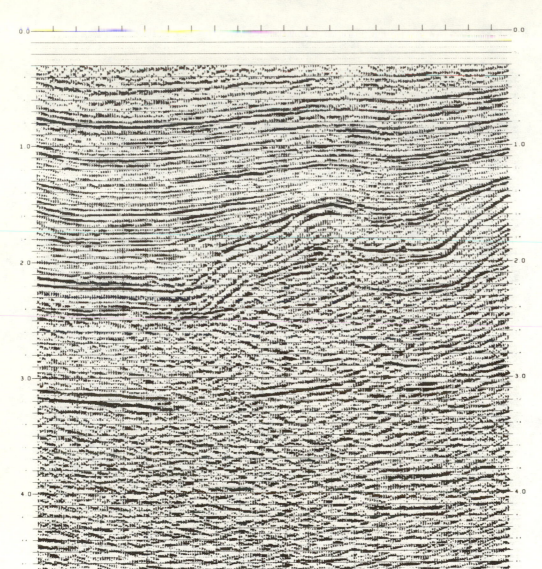

FIG. 21. An *F-K* migration of Figure 14.

Note that the coefficient of the dominant ψ_{Zt} term is now constant. To define the migration limits in this coordinate system, we define two new variables $\zeta(t')$ and $\eta(t')$

$$t' = \int_0^\zeta \frac{dZ}{c},$$ (71)

and

$$\eta = \int_0^\zeta c\, dZ.$$ (72)

Then before migration the limits are

$$Z' = 0,\ t' > 0,$$ (73)

and the limits upon migration are

$$Z' = \eta(t')/c_0,\ t' > 0.$$ (74)

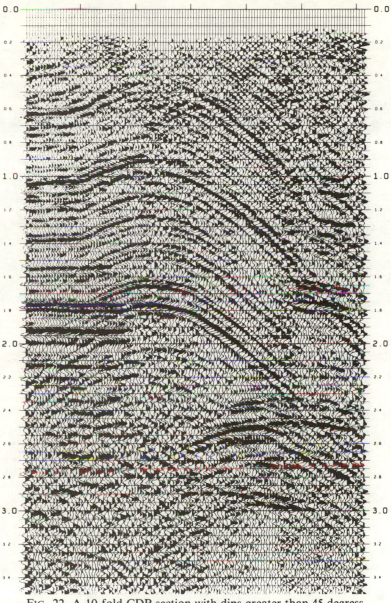

FIG. 22. A 10-fold CDP section with dips greater than 45 degress.

Because Equation (74) is time dependent, this coordinate system is not in general useable for K-space or F-K migration. However, one more change of variables will do the trick. Define

$$D = \sqrt{2 \int_0^{t'} \eta(t') \, dt'} \qquad (75)$$

and

$$d = Z' c_0 D / \eta. \qquad (76)$$

Setting $\phi(X, d, D) = \psi(X, Z, t)$, the wave equation takes the form

$$\phi_{XX} + W(X, d, D) \phi_{dd} + 2 \phi_{dD} = 0, \qquad (77)$$

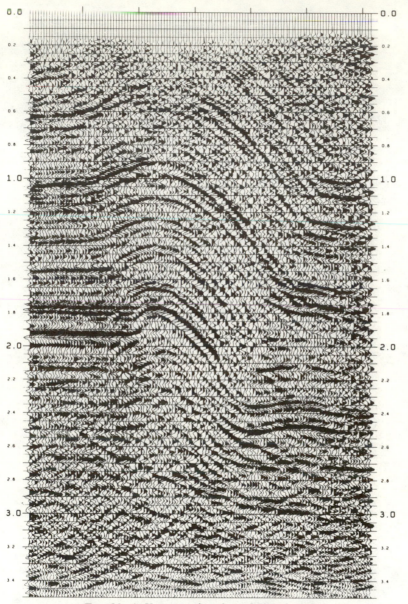

FIG. 23. A K-space migration of Figure 18.

where

$$W = \frac{c^2(X, Z)D^2}{\eta^2} + \frac{2d}{D}\left(1 - \frac{c^2(X'z)D^2}{\eta^2}\right). \quad (78)$$

Migration proceeds from the half-plane,

$$d = 0, D > 0, \quad (79)$$

to the half-plane,

$$d = D, D > 0. \quad (80)$$

All explicit dependence on X and Z now resides in the coefficient W of ϕ_{dd}. $W \neq 1$ reflects the fact that diffractions are not pure hyperbolas in a layered medium. Since ϕ_{dd} is ordinarily small, it is usually justifiable to replace W with an average constant value (usually a number between .5 and 1) for a given section. Then (77) has almost the form of equation

FIG. 24. An *F-K* migration of Figure 18.

(11), and the derivation of the K-space and F-K migration algorithms proceeds as outlined above, except for a slight modification of the dispersion relation (38) or (49).

Use of K-space or F-K migration in practice, then, involves a time to depth conversion (75). Even though simple to effect, this gives rise to practical difficulties in that (a) the frequency content of the data is altered; and (b) incorrect lateral velocity variations will distort the reflecting surfaces and cause improper migration. These problems can be lived with, however, and at present the K-space and F-K migration schemes appear to be practical and useful.

ACKNOWLEDGMENTS

I wish to thank the management of Continental Oil Co., for permission to publish this paper. Jerry Ware, Pierre Goupillaud, and Bill Heath have through

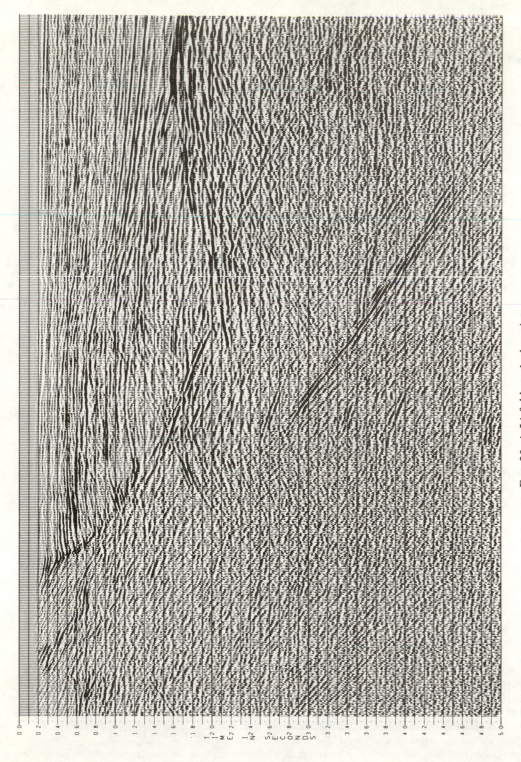

FIG. 25. A 24-fold stacked section.

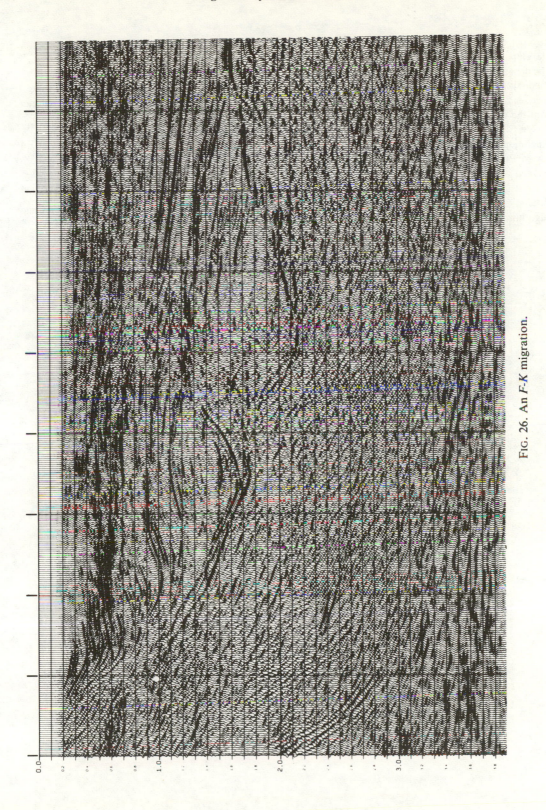

FIG. 26. An *F-K* migration.

several discussions contributed to the theoretical development of this work. Several staff members of the data processing division of Conoco's Exploration-Geophysics Department have, through discussion, criticism, and programming assistance, helped convert the theory to practice.

The development of the high-accuracy K-space migration scheme owes a great deal to the work of Bjorn Enquist at the Stanford Exploration Project. Francis Muir of Chevron has provided several helpful comments and suggestions in addition to the stability criteria and a more efficient form for the K-space algorithm. Of course, this paper would not have been possible but for the original inspiration of Jon Claerbout.

REFERENCES

Claerbout, J. F., 1971, Toward a unified theory of reflector mapping: Geophysics, v. 36, p. 467–481.
———— 1976, Fundamentals of geophysical data processing: New York, McGraw-Hill Book Co., Inc.
Claerbout, J. F., and Doherty, S. M., 1972, Downward continuation of moveout corrected seismograms: Geophysics, v. 37, p. 741–768.
French, W. S., 1974, Two-dimensional and three-dimensional migration of model-experiment reflection profiles: Geophysics, v. 39, p. 265–287.
———— 1975, Computer migration of oblique seismic reflection profiles: Geophysics, v. 40, p. 961–980.

Reprinted from GEOPHYSICS, **46**, 717-733.

Fundamentals of frequency domain migration

Joong H. Chun* and Chester A. Jacewitz*

ABSTRACT

Frequency domain migration is founded upon the wave equation, and so includes diffractions and other effects. This paper seeks to motivate and illuminate frequency domain migration using straightforward geometric techniques and simple frequency domain observations.

INTRODUCTION

Migration of seismic data is a process of mapping one time section onto a second time section in which events are repositioned under the appropriate surface location and at the correct time. Thus, a migration output should be a time section of the geologic depth section. No current migration technique handles all the difficulties of noise, rapidly varying velocities, steep dips, and other problems perfectly. Techniques vary greatly in performance relative to these problems.

Three of the major techniques of migration are diffraction, finite-difference, and frequency domain migration. Diffraction migration is also known as Kirchhoff integral migration when high-order approximations are used. The finite-difference approach is commonly known as time domain or wave equation migration. Frequency domain migration may also be referred to as frequency-wavenumber migration or Fourier transform migration.

The diffraction type migration process is often described as a statistical approach. A particular datum might have originated from many possible subsurface locations. All such possible origins are treated as equally likely in this approach. The major advantage of diffraction migration is good performance with steep dip. One disadvantage is poor performance under low signal-to-noise (S/N) ratio conditions.

Finite-difference migration is a deterministic approach (Claerbout, 1976). The migration procedure is modeled by the wave equation. This equation is then approximated by a simpler type of equation appropriate for migration. This last equation is then approximated by a finite-difference algorithm. An advantage of the finite-difference method is its good performance with a low S/N ratio. Disadvantages of this method include a relatively long computing time and difficulty in handling steep dip data.

Frequency domain migration is also based upon a deterministic approach via the wave equation (Stolt, 1978). Instead of utilizing finite-difference approximations, the two-dimensional (2-D)

Fourier transform is the fundamental technique of this method. The advantages of this method include fast computing time, good performance under low S/N ratio conditions, and excellent performance for steep dip. Disadvantages include difficulties with widely varying velocity functions. Others discussing aspects of frequency domain migration include Bolondi et al (1978), Hood (1978), and Whittlesey and Quay (1977).

This article is meant as an overview for those working in the geophysical industry who wish to know more about migration. In particular, the authors hope to provide new insights into the fundamental aspects of migration with emphasis upon frequency domain migration. There will be much reliance upon the geometry behind the usual physical and mathematical treatments of migration. The intimate geometric relations between the migration of a dipping event in time and the counterpart migration in the frequency domain will be explained. Certain general effects of migration upon seismic data will be considered. Some specific parameters are considered to determine how they affect migration.

Migration in the frequency domain has been known for some time. Stolt's insights have made it a practical method. We intend to illuminate his results from the geometric viewpoint.

This article covers the fundamentals of migration. Any implementation has problems of its own. Frequency domain migration has the usual problems of finitely sampled data. Working in the frequency domain has special problems. These can be handled effectively, but it is not our purpose to cover such material.

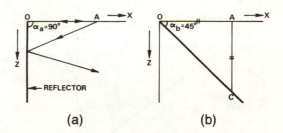

FIG. 1. A 90-degree reflector model. (a) Earth section, (b) record section.

Presented at the 31st Annual Midwest SEG Meeting, March 8, 1978, in Midland. Manuscript received by the Editor May 19, 1978; revised manuscript received September 12, 1980.
*Seismograph Service Corp., P.O. Box 1590, Tulsa, OK 74102.

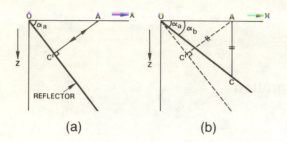

(a)   (b)

FIG. 2. A dipping reflector model. (a) Earth section, (b) record section.

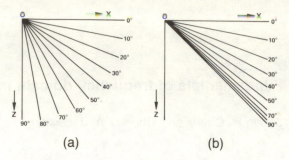

(a)   (b)

FIG. 3. The fan earth model. (a) Some representative earth dips—open fan. (b) Corresponding record section—folded fan.

MIGRATION IN THE DEPTH DOMAIN

We will begin the study of migration fundamentals with special cases, then work toward a complete understanding of the geometry of migration. A step-by-step procedure for performing migration of a seismic event will be discussed. This procedure is similar to mapping used in the frequency domain to accomplish migration.

Consider the vertical earth model ($\alpha_a = 90$-degree reflector) of Figure 1a. Suppose a seismic source is at A with the signal recorded at the same point. Then the only energy that can be recorded at A is from the horizontal path in the ray theory approach. Any nonhorizontally traveling wave will be reflected downward and not reach A. Map the travel distance of the horizontally traveling ray on the depth plane (x-z) in the z-direction as in Figure 1b. Since $OA = AC$, the dip angle α_b of the reflection is equal to 45 degrees. Thus, for a 90-degree reflector the reflection takes place only at one point at the surface and the recorded reflections are mapped along a 45-degree line on the depth plane as A moves along the surface. Migration maps the 45-degree reflection of Figure 1b onto the 90-degree reflector of Figure 1a.

Next, consider the dipping earth model in Figure 2a. Again, assume that source and receiver are both at A. The wave from A will be reflected at C' and will be recorded at A. The travel dis-

tance AC' is mapped vertically on the x-z plane as in Figure 2b to the segment AC. The travel distance is thus $AC' = AC$. The earth model in Figure 2a is superimposed on Figure 2b as a dotted line. From Figure 2b, we easily see the following relationship:

$$\sin \alpha_a = \frac{AC'}{OA} = \frac{AC}{OA} = \tan \alpha_b. \tag{1}$$

The equation above is a familiar description of the relationship between the migrated dip (α_a) and recorded dip angle (α_b). Since C maps to C' under migration, this process moves data updip.

The use of reflections explains many of the observed phenomena. The concept of diffraction is required for deeper understanding of migration. Diffractions are usually associated with discontinuities. Reflections may be considered as a superposition of diffractions. We are using geometric ideas to motivate and explain the exact solution. For this purpose, the term reflection will generally be adequate. Thus, given a reflector, the reflection process gives rise to a reflection event on the record section. Since sections will be considered before and after migration, the general process of mapping a reflector or diffractor to the record section will be described as the diffraction process. Migration proceeds from a record section to the earth model. Diffraction proceeds in the di-

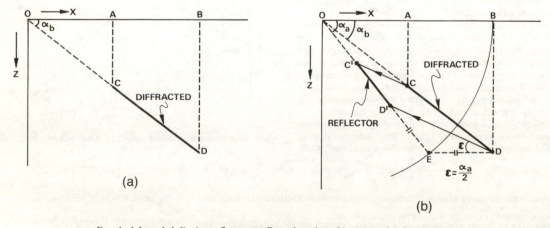

(a)   (b)

FIG. 4. A bounded dipping reflector. (a) Record section, (b) construction for migration.

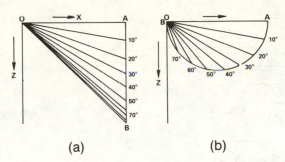

(a)   (b)

FIG. 5. A correspondence between the record section and the depth section. (a) Wedge of events before migration, (b) half-disc of events after migration.

rection opposite to migration. For example, given a point diffractor, the diffraction process gives rise to a diffraction on the record section. Just as reflection may refer to the reflection process or reflection data, so too with diffraction. The context should make the distinction clear.

Consider an earth model with varying dip angles. This "fan model" appears in Figure 3a. In this model, the velocity is constant for each layer between the reflectors or ribs of the fan. Each layer has a different density. A reflection will be received from each boundary because of the density contrast. If the source and receiver are placed at the same point along the surface, the recorded seismic section will be a diffracted seismogram of Figure 3b. For this fan earth model, one can view the migration process as an unfolding (extending) of a folded fan (diffraction model). The diffraction process can be viewed as a folding of a fully extended fan (the earth model). Note that the intensity of folding is such that the steeper the dip of the unfolded fan, the more densely it is folded on the diffraction model and vice versa.

A simple mechanical method of migration is illustrated in Figures 4a and 4b which uses equation (1). An understanding of this method is helpful for understanding frequency domain migration. A reflection segment to be migrated appears in Figure 4a. To migrate the event, one may follow this procedure:

(1) Extend line segment CD toward the surface to intersection point 0.
(2) From point D draw a vertical line to the surface. The point of intersection is B in this example.

(3) With point 0 as origin, draw a circle of radius $0B$.
(4) From point D draw a horizontal line toward the circle and find the intersecting point between the circle and the line (point E).
(5) Connect origin 0 and point E, to find the migrated dip angle α_a.
(6) Construct point D' by making $ED = ED'$. D' is the migrated point of D. The angle of line DD' is $\varepsilon = \alpha_a/2$. By projecting line CD with this angle to line $0E$, the migration process is completed.

Validation of this migration process is given in Appendix A.

Thus, migration requires a two-step procedure. The first is to find the correct line which gives the migrated dip angle, and the second is to map reflection points to their appropriate migrated points linearly. This linear mapping property is important for understanding frequency domain migration.

The seismic section is a bounded diffraction model (the lengths in x and z are limited). Consider the bounded fan diffraction model of Figure 5a. By migrating this section, we obtain the section of Figure 5b. For this model, an event inside the triangle ABO is migrated within a semicircle whose diameter is equal to OA.

To understand the correspondence between Figures 5a and 5b, the reader should review Figure 2b. In this figure, the raypath AC' is perpendicular to the reflector OC'. Return to Figure 5. The triangles of Figure 5a correspond to a traveltime and diffracted line pair. In Figure 5b the distance from A to the endpoint of a segment corresponds to the raypath as in Figure 2a. Thus, each angle in Figure 5b (A to endpoint to O) must be a right angle. Recall the geometric fact that the collection of all points representing an apex of a right triangle formed with a fixed hypotenuse OA must necessarily form a semicircle. This should help explain the relationship between Figures 5a and 5b.

The basic concept of migrating a fan model in the depth domain is useful for understanding the relationship between the migration procedure in the time domain and a similar procedure in the frequency domain.

MIGRATION IN THE FREQUENCY DOMAIN

We have discussed a special type of seismic section, which was described as a fan model. This meant that the unmigrated section closely resembled a partially opened oriental fan. A real seismic section does not resemble this, but it can be imagined as being a set of scattered diffraction fans. A curved event may be considered as consisting of many very small straight-line segments. A most important aspect of the 2-D Fourier transform is that, how-

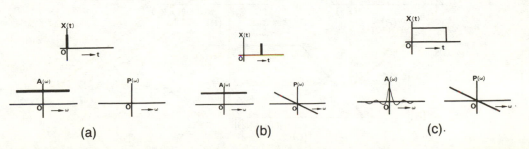

(a)   (b)   (c)

FIG. 6. Review of Fourier transforms—time domain, frequency domain amplitude, and phase. (a) Spike at time 0, (b) spike shifted in time, and (c) boxcar.

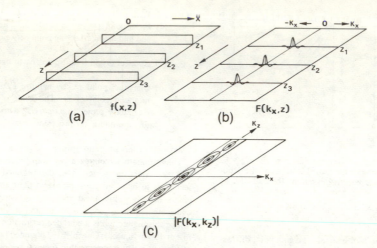

FIG. 7. A 3-D sketch of three bounded horizontal reflectors. (a) Line spike model in the depth domain. (b) Fourier transform in the x-direction. (c) Typical amplitude contours in the 2-D frequency domain.

ever scattered events are in the depth domain, the Fourier trans-formed data will be well organized and become a nicely gathered fan. Thus, by applying the migration process to the gathered fan while in the frequency domain, all the scattered dipping events in the depth domain are handled easily. We will review a few funda-mental properties of the Fourier transform in both the 1- and 2-D cases, then consider how the folded fan earth model can be un-folded in the frequency domain.

It is well known that an impulse (spike) at $t = 0$ in the time domain transforms to a constant for all frequencies in the frequency domain as in Figure 6a. When the spike is located at $t \neq 0$, the amplitude spectrum remains the same but the phase spectrum changes. Thus, information regarding the time of the spike is pre-served in the phase spectrum as in Figure 6b. It is also known that

the Fourier transform of a rectangular pulse (boxcar) is a sinc func-tion in the frequency domain as in Figure 6c.

Consider the seismic section $f(x, z)$ of Figure 7. Assume that there are three horizontal reflections (line spike series) located at Z_1, Z_2, and Z_3. Figure 7a is a three-dimensional (3-D) sketch of this section. The Fourier transform of this 2-D section with respect to the variable x appears in Figure 7b. This is a bounded seismic section. In this figure, as the range of x tends to ∞, the sinc func-tions located at Z_1, Z_2, Z_3 will be reduced to spikes. If we per-form the Fourier transform once again in the z-direction, we would obtain the 2-D transform of $f(x, z)$ of Figure 7c. Since $F(k_x, k_z)$ is a complex function, we have only presented the contours of the amplitude spectrum $|F(k_x, k_z)|$ in Figure 7c. Thus, we see that a horizontal event in the x-z plane is mapped along the k_z-axis,

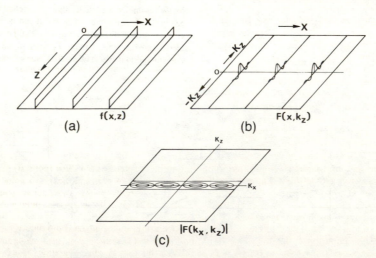

FIG. 8. A 3-D sketch of three bounded vertical reflectors. (a) Line spike model in the depth domain, (b) Fourier transform in the depth direction, and (c) typical amplitude contours in the 2-D frequency domain.

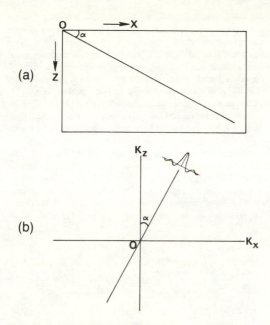

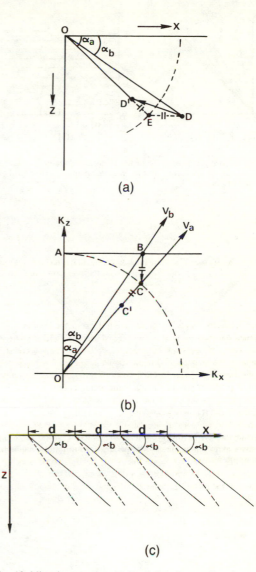

FIG. 9. Bounded dipping reflection segment. (a) Reflection in the depth domain, (b) frequency domain representation.

FIG. 10. Migration mapping. (a) Depth domain mapping, (b) construction of frequency domain mapping, and (c) preservation of spatial frequency.

and the width of the mapped event is inversely proportional to the length of the horizontal event in the x-z plane. For the width of an event, one might use the size of a band containing 90 percent of the energy. By applying the same logic, if there are three vertical events on the $f(x, z)$ section of Figure 8a, they will be mapped along the k_x-axis as in Figures 8b and 8c.

These examples illustrate that the 2-D transform maps *normal* to the dip direction of the events in the x-z plane. In other words, horizontal events are mapped along the k_z-axis and vertical events are mapped along the k_x-axis. The width of a mapped event is inversely proportional to its extent in the x-z plane.

Consider the dipping reflection in the x-z plane of Figure 9a. It will be mapped on the $k_x - k_z$ plane along the normal direction to the dip angle in Figure 9b. The width will be inversely proportional to the extension of the event in the x-z plane.

Thus, any event in the x-z plane is mapped normal to its dip angle in the $k_x - k_z$ plane. Therefore, the dip angle of both diffracted and migrated events in the x-z plane must be preserved in the $k_x - k_z$ plane. Also, all events of the same dip are gathered together in the frequency domain along a single line.

When events on the x-z plane are not line spikes but have some waveform, the mapping of such events is more complex.

Figure 10b shows the Fourier transform of the diffracted and migrated events of Figure 10a. As mentioned before, the diffracted event (OD) and migrated event (OD') of Figure 10a will be mapped along the V_b and V_a normal vectors of Figure 10b. In the frequency domain, if V_b is given, the V_a vector may be found by following a procedure similar to that of the depth domain. This procedure is:

(1) Draw a circle whose radius is equal to OA (A is an arbitrary point on the k_z-axis). The center is at O.
(2) Draw a horizontal line from A and find the intersecting

point on the V_b line (point B).
(3) Draw a vertical line from point B and find the intersecting point with the circle (point C).
(4) Draw a line that passes through the origin and point C.

This will give us a migrated dip line (V_a).

Note that frequency components along line V_b must be mapped along the line V_a after migration. However, we do not know exactly where the point B is mapped on the line V_a. If we follow the same procedure used for the depth domain migration, it

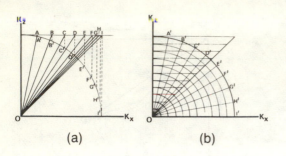

(a)   (b)

FIG. 11. Migration mapping in the frequency domain. (a) A line of constant K_z frequency and its migrated mapping. (b) a grid of curves of constant K_z.

would appear that point B on the line V_b maps at point C' on the line V_a. That procedure was illustrated by Figures 4a and 4b earlier. However, in the frequency domain, point B is directly mapped to C. In other words, this is a *vertical* mapping (projection) while it was a slanted projection with the angle of $\alpha_a/2$ in the depth domain (see Appendix B). The basic reason for this vertical mapping is illustrated in Figure 10c. The dark lines represent lines of fixed dip before migration. The results of migration are indicated by the dashed lines. The spacing between these lines is unchanged by migration. Thus, for such a seismic section with a single dip, the frequency in the x-direction is the same before and after migration. In terms of k_x and k_z as in Figure 10b, this means the component of k_x cannot change under the migration mapping. This mapping must be vertical in the $k_x - k_z$ plane.

In Figure 11a, the horizontal line intersects all the dipping lines. If we project frequency components on this line to the circle, all the points on the line are properly migrated. Therefore, if we

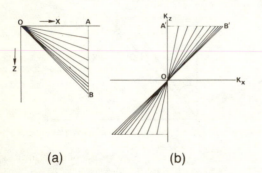

(a)   (b)

FIG. 12. Wedge of dips prior to migration. (a) A wedge of events in the depth domain, (b) the frequency domain representation of the wedge.

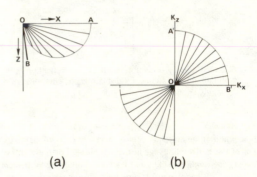

(a)   (b)

FIG. 13. Wedge of dips after migration. (a) A semi-disc in the depth domain, (b) the frequency domain equivalent.

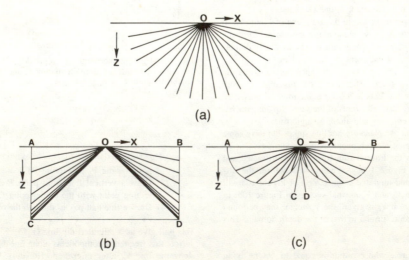

(a)

(b)   (c)

FIG. 14. The full fan earth model. (a) The full fan, (b) the bounded fan of the record section, and (c) the migrated correspondence of (b).

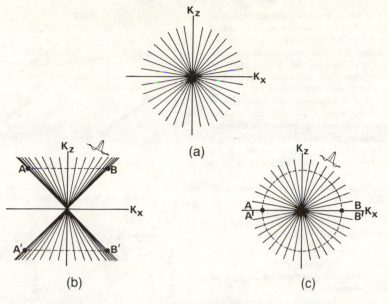

(a)

(b)

(c)

FIG. 15. The frequency domain representation of Figure 14. (a) Fourier transform of the full fan, (b) Fourier transform of the bounded fan with representative pulse spreading, and (c) migration of (b). Note the dotted lines in (b) and (c) correspond under the migration mapping, as do points A, A', B, and B'.

project frequency components on a series of horizontal lines to their corresponding circles vertically as in Figure 11b, the migration process in the frequency domain is completed. A mathematical expression of this process may be written as

$$\overline{F}(k_x, k_z) = \frac{k_z}{\sqrt{k_x^2 + k_z^2}} \, F\left(k_x, \sqrt{k_x^2 + k_z^2}\right). \quad (2)$$

Function F is the 2-D Fourier transform of the original section. The transform of the migrated section is given by $\overline{F}$. Thus the migrated section may be computed directly in terms of the original seismic section. The reason for the scaling factor $k_z/\sqrt{k_x^2 + k_z^2}$ in the frequency domain migration is explained in Appendix B.

The Fourier transform pair of the diffracted fan earth model of Figure 5a is presented in Figure 12. Since the model is bounded in both x- and z-directions, the real transform should be a broad-

ened line as explained before. For simplicity, it is presented as a series of dipping lines (Figure 12b).

Figure 13 shows a Fourier transform pair of the migrated fan model obtained from the diffracted fan model of Figure 12. Note that the frequency domain migration may be viewed as unfolding (Figure 13b) of the folded fan (Figure 12b). Also, in the depth domain the triangle ABO of Figure 12a is mapped within a semicircle of Figure 13a whose diameter is equal to OA. In the frequency domain, the triangle $A'B'O$ of Figure 12b is mapped within an area of quarter of a circle of Figure 13b.

We thus conclude the following:

(1) The dip angles of reflectors, both diffracted and migrated, in the depth domain are preserved in the frequency domain. However, they are normal to the depth domain dip angle.

(2) The point-to-point mapping from the diffracted event to a migrated event is a slanted projection with the angle of $\alpha_a/2$ in the depth domain. It is a vertical projection in the frequency domain.

(3) For the case of a complete 180-degree fan earth model case of Figure 14a, the diffracted fan model is a split 45-degree folded fan model as in Figure 14b. If we impose boundaries $ABCD$ as in Figure 14b and migrate back, then the fans are confined within two semicircles as in Figure 14c. The Fourier transform of the models in the depth domain are presented in Figure 15. Figure 15a is the Fourier transform of Figure 14a, and Figure 15b is that of Figure 14b. Imposing the boundaries shown in Figure 14b results in broadening of frequency components of each dipping event as explained before. Figure 15c is the Fourier transform of Figure 14c.

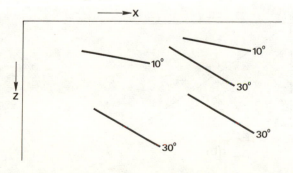

FIG. 16. Synthetic data: record section.

An example of scattered fans is depicted in Figure 16. It is as if the fan model of Figure 14b were scattered across the section. Migrating the scattered fan in the depth domain, by extending each reflection to the surface and using the geometric migration construction, is cumbersome. Because the Fourier transform gathers together events of the same dip, the transform of Figure 16 resembles that of the gathered fan model of Figure 15b.

The most important aspect of frequency domain migration is that the scattered fan in the depth domain becomes a nicely gathered fan in the frequency domain. The information regarding the spatial position of a dipping reflection in the depth domain is preserved in phase relations within the complex numbers in the frequency domain. By applying the simple migration process to the model of Figure 16 in the frequency domain, then performing an inverse Fourier transform, we obtain the migrated scattered fan model of Figure 17.

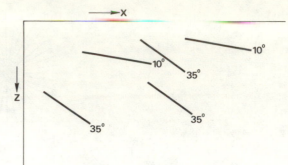

FIG. 17. Migrated section from Figure 16.

MIGRATION PARAMETERS

For real data, there are many unknowns. We must consider the effects each of these parameters has upon migration. Then, there will be a coherent picture available of the interaction among these variables and their relative importance.

To consider various migration parameters, it is necessary to distinguish the dip angles on the time and depth sections. Our notation will be θ_t for the dip angle on the time section, θ_z for the corresponding dip angle in depth, $\bar{\theta}_z$ for the migrated dip angle in depth, and $\bar{\theta}_t$ for the migrated dip angle in time (Figure 18).

Converting from time to depth via velocity yields $\tan \theta_z = (V \tan \theta_t)/2$. The time is the two-way time, while the depth is the one-way depth, hence the division by 2. Migration in the depth domain yields the usual formula $\sin \bar{\theta}_z = \tan \theta_z$. Reconversion to time is performed via $\tan \bar{\theta}_t = (2 \tan \bar{\theta}_z)/V$. These operations combine to produce:

$$\tan \bar{\theta}_t = (2 \tan\{\sin^{-1}[V(\tan \theta_t)/2]\})/V, \qquad (3)$$

so that

$$\bar{\theta}_t = \tan^{-1}\left(2 \tan\{\sin^{-1}[V(\tan \theta_t)/2]\}/V\right). \qquad (4)$$

It is important to realize that the resultant migrated dip angle $\bar{\theta}_t$ depends upon the velocity of conversion V, not merely the original dip angle θ_t. The velocity V does *not* cancel in this equation. An alternate formula from Appendix C is

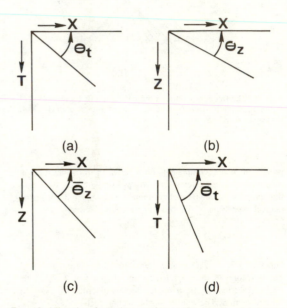

FIG. 18. Migration from time section to time section. (a) Input record section, (b) conversion to depth domain, (c) migration in depth domain, and (d) reconversion to the time domain.

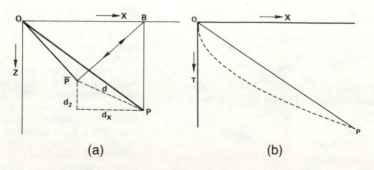

FIG. 19. Migration path as a function of velocity. (a) Components of migration in the depth domain, (b) migration of a point under different velocities.

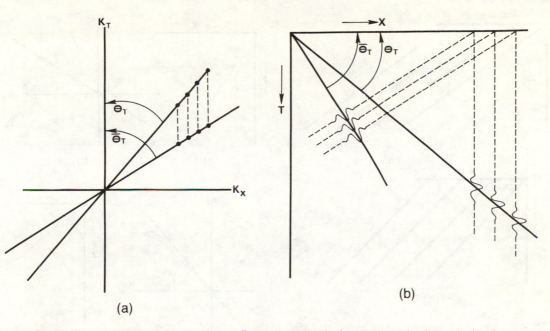

(a)   (b)

FIG. 20. Frequency changes under migration. (a) Frequency domain, (b) time domain—migrating a seismic pulse.

$$\tan \bar{\theta}_t = \frac{\tan \theta_t}{\sqrt{1 - \left(\dfrac{V}{2}\right)^2 \tan^2 \theta_t}} . \qquad (5)$$

From this formula, for any fixed θ_t, $\bar{\theta}_t$ will approach θ_t for small velocities. That is, small velocities produce little noticeable change in dip angle. As the velocity is increased, $\bar{\theta}_t$ will become 90 degrees. Beyond this range, there is *no* real angle for migration.

Next, consider the migration position as a function of the angle on an apparent depth section θ_z and the apparent depth of a point z. The following fundamental formulas from Appendix D are illustrated in Figure 19a. If $\mathbf{d} = (d_x, d_z)$ represents the components of the displacement of a point P by migration, then

$$d = |\mathbf{d}| = 2 z \sin (\bar{\theta}_z/2) = \sqrt{d_x^2 + d_z^2} , \qquad (6)$$

$$d_x = z \sin \bar{\theta}_z, \qquad (7)$$

and

$$d_z = z (1 - \cos \bar{\theta}_z). \qquad (8)$$

These relations are only of preliminary interest as we are interested in changes from a time section to a migrated output time section. However, note that in a depth section, a point on a dipping reflection is displaced by migration linearly and proportional to its depth before migration. We can also leave the velocity and dip angle constant, but vary the depth. For a constant velocity, a uniform change in time is equivalent to a change of depth (with no change in θ_z). Thus, in the time section, the corresponding displacements d, d_x, and d_t are linearly proportional to the time shift.

We may return to the time domain again to see how a point migrates as a function of velocity. Using the same fundamental equations, and the fact that the depth of a reflection segment on

the time section is a function of velocity, one can derive (see Appendix E)

$$d_x = V^2 T \frac{\tan \theta_t}{4}, \qquad (9)$$

and

$$d_t = T \left[1 - \sqrt{1 - \left(\frac{V}{2} \tan \theta_t\right)^2} \right]. \qquad (10)$$

The first equation indicates that a reflection segment on a seismic time section migrates a horizontal distance proportional to the square of the migration velocity. The second equation is more difficult to visualize. Instead, we may eliminate the variable $\tan \theta_t$ to study the migration path of a point (see Appendix F).

Let us describe the migration path of a point P on a dipping reflection segment of a time section as velocity changes. The migrated point $\bar{P}$ is on the parabolic path connecting P and O and symmetric about the x-axis (Figure 19b). For small velocities, $\bar{P}$ will be found near P, while for large velocities, $\bar{P}$ will be found near O.

The change of frequency as a function of velocity can be explained more easily from the frequency domain viewpoint. The 2-D frequency domain mapping given for migration in equation (2) leaves the k_x component unchanged. Velocity changes also leave the k_x component unchanged. Therefore, in migration of a dipping reflection on a seismic time section, only the k_t component will be affected by velocity variation. Using a notation of F_x, F_t for input frequencies in k_x and k_t, one can apply the conversions and mapping to determine

$$\bar{F}_t = \sqrt{F_t^2 - \frac{V^2}{4} F_x^2} \qquad (11)$$

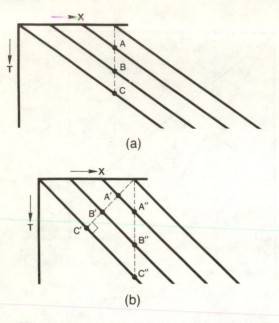

FIG. 21. Reflections with a single dip. (a) Before migration, (b) after migration.

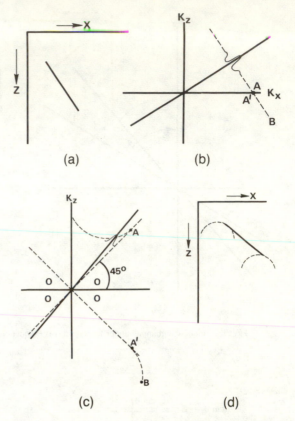

FIG. 22. Diffraction of a bounded reflector. (a) Depth domain earth model, (b) frequency domain, (c) diffracted (inverse process of migration), and (d) back to depth domain (record section).

for the output time frequency $\bar{F}_t$. As is expected for a small velocity, $\bar{F}_t \approx F_t$. As discussed, this reflects the fact that the seismic time section is only slightly altered under migration by a low velocity. For V large enough, $\bar{F}_t \approx 0$. The dip of the reflection is now 45 degrees in the depth section, so that it migrates to a vertical event with $\bar{F}_t \approx 0$. For velocities above the velocity which would make the dip of a reflection 45 degrees in the depth section, the reflection will be eliminated from the resulting section. This phenomenon will be described in detail later. It is not merely conversion to depth that alters F_t. Migration itself will alter this frequency even if one works only in depth domain. In this case the mapping is

$$\bar{F}_z = \sqrt{F_z^2 - F_x^2}. \qquad (12)$$

Figure 20a illustrates the two key points of the frequency domain mapping. First, the mapping takes lines of one apparent dip into lines of the true dip. Second, the mapping is vertical. That is, the frequency in x is unchanged by the migration mapping.

Let us return to the time domain to understand these two points. Figure 20b shows an event with a single dip. The associated waveform is complex. Nonetheless, after migration, the event still has a single dip. The preservation of spatial frequencies is illustrated in Figures 21a and 21b. The horizontal spacing of the dipping reflections is unchanged by the migration process. However, spacings A, B, and C before migration must be the same as A', B', and C' after migration, respectively. Thus, the spacings A'', B'', and C'' must be greater. This shows that the frequency in the time direction must be reduced by migration. This may be confirmed by reconsidering the frequency domain mapping directly.

We may follow the processing through a complete cycle to investigate the loss of input data under the frequency domain

mapping. Start with a geologic section, then consider a record section. Figure 22a is a representative geologic section. This event is a bounded dipping reflector. In the frequency domain, as seen earlier, the event will be modulated by a sinc function. An example is drawn with two points A and B labeled in Figure 22b.

Consider a record section generated by the diffraction mapping which is the inverse of migration mapping (take $V = 2$ for simplicity). Note in Figure 22c that the entire area occupied by dips steeper than 45 degrees (*supersteep*) has been zeroed. This is because line OA, which has a 90-degree dip in Figure 22b, is mapped to a 45-degree line on Figure 22c. No dip greater than 45 degrees can exist on a record section. Also, note that no information has been lost via the diffraction mapping. The representative sinc function of Figure 22b has merely been split in Figure 22c. No part of it has been lost. Figure 22d shows the section which results from such a typical geologic model. These figures illustrate that the diffraction patterns of Figure 22d must be related to the distortion of the sinc function in Figure 22c.

If one performs migration via the frequency domain approach with the input section of Figure 22d, one merely retraces the steps successively to Figure 22c, Figure 22b, and, finally, Figure 22a, the correct geologic model. Migration appears to entail no loss of information, but there is a more subtle consideration.

Proceed in a different direction. Start with the bounded dipping

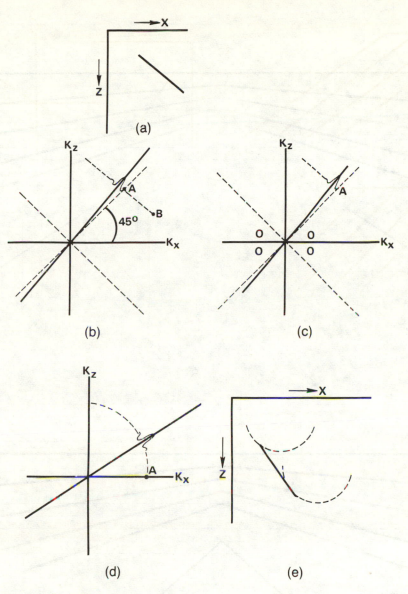

FIG. 23. Bounded dipping reflection model without edge diffraction. (a) Depth domain model (record section), (b) frequency domain, (c) truncation, (d) migration, and (e) back to depth domain after migration.

reflector on the section of Figure 23a. The frequency domain illustration in Figure 23b is again modulated by a representative sinc function. The migration mapping will not use any data in the supersteep region indicated by O in Figure 23c. These data are lost. Figure 23e illustrates the result when viewed in the geologic plane after the migration. The dashed curves of Figure 23e represent low-amplitude data caused by the zeroing and stretching of the sinc function of Figure 23b. If we start with the geologic model of Figure 23e and retrace the steps, we never recover the original section of Figure 23a. This is as it should be. The record section of Figure 23a cannot be generated from any geologic

model because of the supersteep dips shown in Figure 23b.

The cause of the supersteep data which is not of geologic origin is worth investigating. One cause of supersteep events is clearly noise which can occur at any dip, even supersteep. A second cause of supersteep events is a velocity which is too high. Any event will become supersteep if the velocity used for processing is taken high enough. Thus, a poor choice of velocity may cause geologically valid data to be interpreted as supersteep, and hence it will be discarded.

We have seen how migrated seismic sections change as one varies certain parameters. Some results which might appear

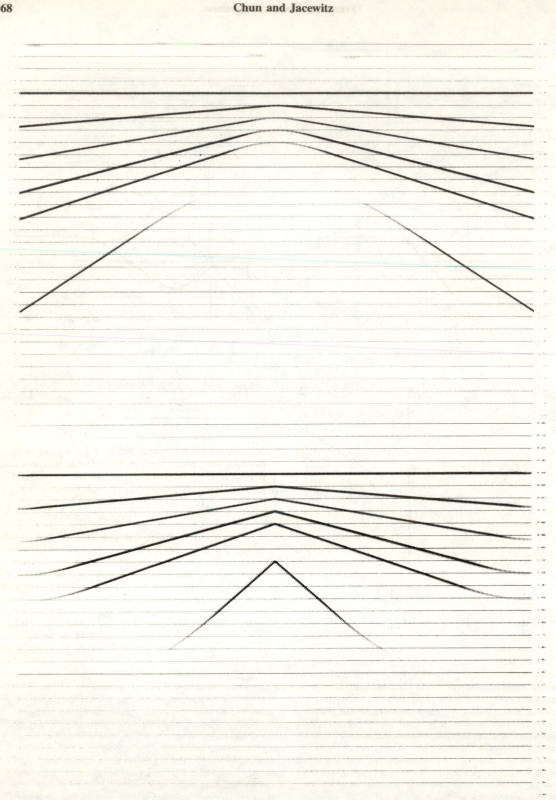

FIG. 24. Synthetic sections. (a) Top, before migration. (b) bottom, after frequency domain migration.

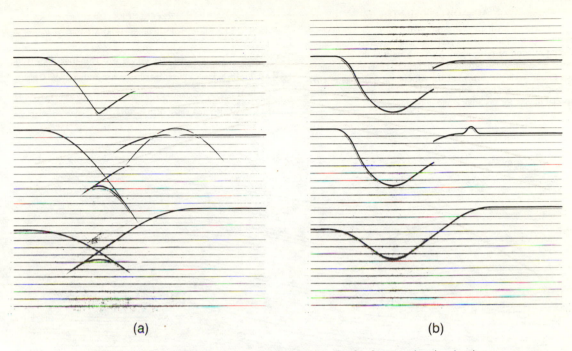

(a)   (b)

FIG. 25. Complex synthetic section. (a) Before migration, (b) after frequency domain migration.

negative have been seen to be natural consequences of the migration process.

APPLICATIONS TO REAL AND SYNTHETIC DATA

A few frequency domain migration examples are presented to show that the scattered fans were migrated correctly.

The section in Figure 24 contains bounded dipping reflections. Notice the separation between the terminators of corresponding events in Figure 24a. After migration, the events have been connected making the reflectors continuous. Furthermore, the events have moved updip and the dips themselves have increased. This section is relatively easy to visualize as a scattered fan. It has been gathered in the frequency domain and migrated in the simple manner we have discussed. Note that the reflections do not terminate sharply. Sharp terminations on the geologic section generate diffractions on the record section. But these synthetic reflections were sharply terminated, and the necessary diffractions were absent. Note also the lowering of frequency along the time axis after migration. This effect follows from the frequency domain migration mapping being downward toward lower k_z. This effect increases with increasing dip.

The section of Figure 25 is more complex. Before migration, Figure 25a is a bewildering collection of crossing arcs. After migration in Figure 25b, the reflectors are seen to be smooth curves. This section contains few straight-line segments. Instead, the curved lines may be considered to be short line segments of rapidly changing dip. Of course, the principle of superposition allows an easy treatment of even this complicated section.

A real seismic section of Figure 26 is the most complicated section in our illustrations. Nonetheless, migration via the fre-

quency domain easily handles events of any dip and location. Migration with a nonconstant velocity is a delicate and complicated problem in itself. Still, even the simple approximations using smoothed stacking velocities did a reasonably adequate job.

CONCLUSIONS

Whenever one is dealing with problems of a mathematical nature, it is always prudent to select the "natural" coordinate system if possible. In this coordinate system, the statement of the problem and methods of solution should become exceptionally clear. We pursue the idea that the coordinates of the Fourier transform domain are the natural coordinates for dealing with migration. In particular, no matter where they are located in space, all events of the same dip appear along lines normal to that dip when viewed in the frequency domain. Events of the same dip are migrated together. Thus, dip lines through the origin in the Fourier transform domain are part of a coordinate system natural to migration. In the diffracted transform domain, straight lines of constant frequency in depth are quite natural. These transform into circles, which are curves of constant frequency along the normal.

The use of events of a single dip greatly simplifies our understanding of migration. Using the Fourier transform, every event on a seismic section may be considered as a sum of events of a single dip and the section a sum of all of such events. Although it may be more difficult to visualize what is occurring in this more complicated case, one knows that the principle of superposition applies. Thus, the entire migration process may be considered as the simple procedure just described applied to a vast sum of simple events.

ACKNOWLEDGMENTS

The authors wish to thank the management of Seismograph Service Corp. for their cooperation and encouragement during the preparation of this paper. Also, we wish to thank the many other colleagues who provided both help and data as valuable input to our work. In particular, the editors and reviewers of GEOPHYSICS have been most helpful. Dr. Enders A. Robinson deserves special thanks for his advice and encouragement.

REFERENCES

Bolondi, G., Rocca, F., and Savelli, S., 1978, A frequency domain approach to two-dimensional migration: Geophys. Prosp., v. 26, p. 750–772.
Claerbout, J. F., 1976, Fundamentals of data processing: New York, McGraw-Hill Book Co., Inc.
Hood, P., 1978, Finite difference and wave number migration: Geophys. Prosp., v. 26, p. 773–789.
Stolt, R. H., 1978, Migration by Fourier transform: Geophysics, v. 43, p. 23–48.
Whittlesey, J. R., and Quay, R. G., 1977, Wave equation migration:

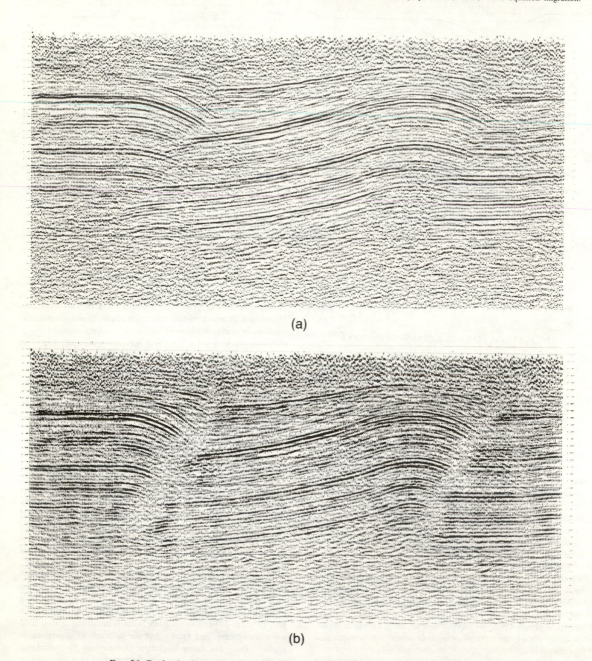

FIG. 26. Real seismic section. (a) Before migration, (b) after frequency domain migration.

Presented at the 47th Annual International SEG Meeting, September 20, in Calgary.

APPENDIX A
MIGRATION MAPPING IN THE DEPTH DOMAIN

Appendix A serves to prove the validity of the construction illustrated in Figure 4b. Steps (1)–(4) are straightforward geometric procedures. Because of construction of a circle in step (3), $OB = OE$. Let d be the distance from E to the x-axis. By step (4), $d = BD$. Thus, combining these facts,

$$\tan \alpha_b = \frac{BD}{OB}$$
$$= \frac{d}{OE}$$
$$= \sin \alpha_a.$$

This justifies step (5).

Now one must show that $\varepsilon = \alpha_a/2$ and that D' is the migrated counterpart of D. It is somewhat easier to demonstrate these results in reverse. That is, suppose D' is the migrated point, making $BD' = BD$. What are the angle ε and the distance ED'? Note that BD' is perpendicular to OE. Thus, $BD'E$ and BDE are both right triangles. Since they have the same hypotenuse BE, and have one other side equal in length, the third sides must be equal. Thus $ED = ED'$ as required. Computing ε is then easy. Since ED is horizontal, it is parallel to OB. Thus the exterior angle at E is α_a. The interior angle is then 180 degrees $- \alpha_a$. Since the triangle is isosceles, the remaining angles are both ε. The angles of a triangle sum to 180 degrees. Thus, 180 degrees $- \alpha_a + \varepsilon + \varepsilon = 180$ degrees or $\varepsilon = \alpha_a/2$.

APPENDIX B
MIGRATION MAPPING IN THE FREQUENCY DOMAIN

Appendix B has two purposes. The first is to show the correctness of the geometric procedure described for migration in the frequency domain. The second is to indicate the validity of the algebraic equivalent of equation (2).

To validate the procedure, the relation $\tan \alpha_b = \sin \alpha_a$ must hold in the construction. Referring to Figure 10b, by step (1), $OA = OC$. Let d be the perpendicular distance from C to the k_z-axis. By steps (2) and (3), $d = AB$. Thus

$$\tan \alpha_b = \frac{AB}{OA}$$
$$= \frac{d}{OC}$$
$$= \sin \alpha_a.$$

This shows the construction produces the correct migrated angle α_a when starting with α_b. Thus, when considered as a mapping between lines through the origin, the construction takes lines of dip α_b to lines of dip α_a. One must still show the construction is correct as a mapping between points in the frequency domain.

There are many ways to indicate the validity of migration mapping. Stolt's important paper (1978) first established this via the wave equation and a change of variables in the Fourier transform. Since the formula holds, a more descriptive indication of why it should hold based upon a raypath approach will be presented.

First, the mapping must map sections of dip α_b to sections of dip α_a. Certainly, each line of dip α_b is mapped to one of dip α_a

by the geometric procedure. Now apply the principle of superposition. Since each section of dip α_b is a combination of lines of this dip, such a section must migrate to a section of dip α_a.

One need consider only sections with a single dip. Under migration, each line is moved about its intersection with the surface. This point of intersection remains fixed. In particular, this means that the frequency in x, the spatial frequency, is unchanged by migration. Thus, if the mapping which implements migration takes the form

$$(\bar{K}_x, \bar{K}_z) \rightarrow (K_x, K_z),$$

one must have $K_x = \bar{K}_x$. The mapping leaves the spatial frequency invariant.

A point (K_x, K_z) in the frequency domain corresponds to an event of dip θ with $\tan \theta = -K_x/K_z$. This is because the Fourier transform maps events to their normals. Thus the relation

$$\tan \alpha_b = \sin \alpha_a$$

may be applied to the case of one point $(\bar{K}_x, \bar{K}_z)$ to give

$$\bar{K}_x/\bar{K}_z = K_x/\sqrt{K_x^2 + K_z^2}.$$

Since $\bar{K}_x = K_x$, this may be rewritten

$$\bar{K}_x^2 + K_z^2 = \bar{K}_z^2$$

or

$$K_z = \sqrt{\bar{K}_z^2 - \bar{K}_x^2}.$$

This validates mapping points vertically in Figure 10b. In addition, the equivalence between equation (2) and the geometric construction of Figure 10b has been demonstrated.

We still must discuss the weighting factor in equation (2). This is not included in the geometric construction, but only in the algebraic formulation. Recall that each section with a single dip, no matter how complex, migrates to another section with a single dip. Under the mapping of Figure 4b, the points of triangle OBD migrate to those of triangle OBD'. But

$$\text{Area } (OBD') = (1/2)(BD')(OD')$$
$$= (1/2)(BD)(OB \cos \alpha_a)$$
$$= \cos \alpha_a \text{ Area } (OBD).$$

This means, roughly, that before migration the event occupies more area so that it is overemphasized. Thus, after migration, the event must be multiplied by the factor of $\cos \alpha_a$ to restore the correct balance. In the frequency domain, the event maps to its normal so that

$$\cos \alpha_a = K_z/\sqrt{K_x^2 + K_z^2}.$$

But this is precisely the factor which occurs in equation (2).

APPENDIX C
THE TIME SECTION ANGLE BEFORE AND AFTER MIGRATION

Appendix C derives a formula relating θ_t (the premigration dip angle on the time section) with $\bar{\theta}_t$ (the postmigration angle). The notation follows migration parameters. Begin with the basic migration relation for dip angles on the depth section:

$$\tan \theta_z = \sin \bar{\theta}_z$$
$$= \cos \bar{\theta}_z \frac{\sin \bar{\theta}_z}{\cos \bar{\theta}_z}$$
$$= \frac{1}{\sqrt{1 + \tan^2 \bar{\theta}_z}} \tan \bar{\theta}_z.$$

Here we have used the relations

$$\tan \theta = \frac{\sin \theta}{\cos \theta}$$

and

$$\cos \theta = 1/\sqrt{1 + \tan^2 \theta}.$$

The first is the definition of the tangent. The second is easily checked by using

$$\sin^2 \theta + \cos^2 \theta = 1.$$

One has, after converting from depth to time,

$$\frac{V}{2} \tan \theta_t = \frac{V}{2} \tan \bar{\theta}_t \Big/ \sqrt{1 + \left(\frac{V}{2}\right)^2 \tan^2 \bar{\theta}_t},$$

or

$$\tan \theta_t = \tan \bar{\theta}_t \Big/ \sqrt{1 + \frac{V^2}{4} \tan^2 \bar{\theta}_t}.$$

Then solve for $\tan \bar{\theta}_t$. The result is

$$\tan \bar{\theta}_t = \tan \theta_t \Big/ \sqrt{1 - \frac{V^2}{4} \tan^2 \theta_t}.$$

APPENDIX D
DISPLACEMENT OF A POINT UNDER MIGRATION

Appendix D refers to Figure 19a. Here we derive the formulas for d_x and d_z. Let h be the length from $\bar{P}$ to the x-axis, and let $z = BP$. Migration means in particular that $B\bar{P} = BP$. The angle $OB\bar{P}$ is 90 degrees $- \bar{\theta}_z$.

Thus,

$$h = B\bar{P} \sin(90 \text{ degrees} - \bar{\theta}_z),$$

and

$$h = z \cos \bar{\theta}_z.$$

Therefore,

$$d_z = z - h,$$
$$d_z = z - z \cos \bar{\theta}_z,$$

and

$$d_z = z(1 - \cos \bar{\theta}_z).$$

The relation for d_x also follows from basic trigonometry:

$$d_x = B\bar{P} \cos(90 \text{ degrees} - \bar{\theta}_z)$$

so that

$$d_x = z \sin \bar{\theta}_z.$$

Finally, the triangle $\bar{P}BP$ is isosceles with vertex angle $\bar{\theta}_z$. Drop a perpendicular from B to $\bar{P}P$. Then one sees that

$$\frac{d}{2} = z \sin(\bar{\theta}_z/2),$$

or

$$d = 2z \sin(\bar{\theta}_z/2).$$

APPENDIX E
MIGRATION OF A TIME SECTION AS A FUNCTION OF VELOCITY

Appendix E uses the notation of Figure 19a. In addition, d_t is the displacement in the time section. Equations (9) and (10) will be proved here. Consider a fixed point at time T on a reflector of dip θ_t on the time section. Note that d_x is the same whether measured on the time section or on the depth section. After converting to depth, the point is located at depth $z = (V/2)T$ on a reflector dip θ_z. From Appendix D,

$$d_x = z \sin \bar{\theta}_z,$$
$$= \left(\frac{VT}{2}\right) \tan \theta_z,$$
$$= \left(\frac{VT}{2}\right)\left(\frac{V}{2} \tan \theta_t\right),$$

or

$$d_x = (V^2 T \tan \theta_t)/4.$$

Also, the displacement in z is given from Appendix D by

$$d_z = z(1 - \cos \bar{\theta}_z).$$

Using the relation between cosine and sine,

$$d_z = \frac{VT}{2}[1 - (1 - \sin^2 \bar{\theta}_z)^{1/2}].$$

By the migration relation,

$$d_z = \frac{VT}{2}[1 - (1 - \tan^2 \theta_z)^{1/2}].$$

Reconverting to time,

$$d_z = \frac{VT}{2}\{1 - [1 - (V^2 \tan^2 \theta_t)/4]^{1/2}\}.$$

Thus, since

$$d_t = \frac{2}{V} d_z,$$
$$d_t = T\{1 - [1 - (V^2 \tan^2 \theta_t)/4]^{1/2}\}.$$

APPENDIX F
MIGRATION OF A POINT AS A FUNCTION OF VELOCITY

Appendix F refers to Figure 19b and the proof that a point migrates along a parabola as the migration velocity changes. Continuing from Appendix E, one has

$$d_x = V^2 T \frac{\tan \theta_t}{4}$$

for the x displacement, and

$$d_t = T\left[1 - \left(1 - \frac{V^2}{4} \tan^2 \theta_t\right)^{1/2}\right]$$

for the t displacement. The last equation may be rewritten as

$$T\left(1 - \frac{V^2}{4} \tan^2 \theta_t\right)^{1/2} = T - d_t.$$

So, squaring the last expression,

$$T^2\left(1 - \frac{V^2}{4} \tan^2 \theta_t\right) = T^2 - 2Td_t + d_t^2.$$

Thus,

$$\frac{V^2}{4} T^2 \tan^2 \theta_t = 2Td_t - d_t^2.$$

From the expression for d_x, one has

$$\frac{V^2}{4} T^2 \tan^2 \theta_t = T \tan \theta_t d_x.$$

Combining these last two expressions,

$$T \tan \theta_t d_x = 2Td_t - d_t^2.$$

Rearranging this equation,

$$d_t^2 - 2Td_t + T \tan \theta_t d_x = 0.$$

This is a parabolic relation between d_x and d_t. Clearly for $V = 0$, $d_x = 0$ and $d_t = 0$. One may rewrite the relation as

$$(d_t - T)^2 = T^2 - T \tan \theta_t d_x.$$

For the choice of $d_t = T$,

$$0 = T^2 - T \tan \theta_t d_x.$$

But this may be rewritten as

$$\tan \theta_t = T/d_x.$$

Hence, the point (X, T) has migrated to $(0, 0)$.

The form of the relation shows that d_x has a maximum when $d_t = T$. Thus, the parabolic path is uniquely defined by the following two properties: (1) The parabola passes through (X, T) and $(0, 0)$, and (2) the parabola is tangent to the t-axis at $(0, 0)$.

Reprinted from Geophysics, **43**, 49-76.

INTEGRAL FORMULATION FOR MIGRATION IN TWO AND THREE DIMENSIONS

W I L L I A M   A .   S C H N E I D E R *

Computer migration of seismic data emerged in the late 1960s as a natural outgrowth of manual migration techniques based on wavefront charts and diffraction curves. Summation (integration) along a diffraction hyperbola was recognized as a way to automate the familiar point-to-point coordinate transformation performed by interpreters in mapping reflections from the x, t (traveltime) domain into the x, z (depth domain).

We will discuss the mathematical formulation of migration as a solution to the scalar wave equation in which surface seismic observations are the known boundary values. Solution of this boundary value problem follows standard techniques, and the migrated image is expressed as a surface integral over the known seismic observations when areal or 3-D coverage exists. If only 2-D seismic coverage is available, wave equation migration is still possible by assuming the subsurface and hence surface recorded data do not vary perpendicular to the seismic profile. With this assumption, the surface integral reduces to a line integral over the seismic section, suitably modified to account for the implicit broadside integral. Neither the 2-D or 3-D integral migration algorithms require any approximation to the scalar wave equation. The only limitations imposed are those of space and time sampling, and accurate knowledge of the velocity field.

Migration can also be viewed as a downward continuation operation which transforms surface recorded data to a deeper hypothetical recording surface. This transformation is convolutional in nature and the transfer functions in both two and three dimensions are developed and discussed in terms of their characteristic properties. Simple analytic and computer model data are migrated to illustrate the basic properties of migration and the fidelity of the integral method. Finally, applications of these algorithms to field data in both two and three dimensions are presented and discussed in terms of their impact on the seismic image.

INTRODUCTION

Migration of seismic data has been a basic tool of interpreters since at least the 1940s. The classic work of Hagedoorn (1954) provided firm theoretical basis for the migration of time sections in two or three dimensions based upon the use of wavefront charts and diffraction curves. In the late 1960s, numerous computer implementations of Hagedoorn's migration principle became available for commercial use in seismic data processing. In the main, these programs accomplished migration by summation of stacked trace amplitudes along hyperbolic trajectories governed by the rms velocity distribution.

A recent revival in migration theory and practice stems principally from the work of Jon Claerbout (1970, 1972) and his colleagues at Stanford University, who first formulated a finite-difference algorithm for migration based upon the scalar wave equation. Commercial programs are now available in industry to implement finite-difference migration of seismic data based on Claerbout's original work and extensions thereof. These techniques are variously called "wave equation" migrations.

This paper develops an alternate view of wave equation migration in which the problem is posed as a boundary value problem, which leads naturally to

Presented at the 46th Annual International SEG Meeting, October 24, 1976 in Houston. Manuscript received by the Editor January 7, 1977; revised manuscript received June 26, 1977.
*Formerly Texas Instruments, Dallas, TX.; currently Colorado School of Mines, Golden, CO 80401.

an integral or summation algorithm for migration in either two or three dimensions. As will be seen, the integral solution has strong historic ties to the "conventional" diffraction summation approach of the late 1960s. The differences are subtle but significant in terms of amplitude and waveform reconstruction, faithful to the scalar wave equation.

THEORY

For completeness, the integral migration algorithm will be derived from first principles starting with the scalar wave equation,

$$\nabla^2 U - \frac{1}{C^2} U_{tt} = -4\pi q(r, t).$$

The complete solution to the inhomogeneous wave equation in an arbitrary volume V_0 is given by a surface integral on S_0 enclosing V_0 involving the boundary values, and a volume integral over V_0 involving both source terms and initial values. This result is well known in the mathematical physics literature (see, for example, Morse and Feshback, 1953) and derives from an application of Green's theorem. For our purposes, the volume integral may be ignored since the initial values are assumed to be zero before the shot instant, and there are no real sources in the subsurface image space, just reflectors and scatterers. Thus, we are left with the homogeneous wave equation and inhomogeneous boundary conditions of the Dirichlet type. The remaining surface integral is given by,

$$U(r, t) = \frac{1}{4\pi} \int dt_0 \int dS_0 \left[G \frac{\partial}{\partial n} U(r_0, t_0) \right.$$

$$\left. - U(r_0, t_0) \frac{\partial}{\partial n} G \right]. \tag{2}$$

The specific geometry of interest is shown in Figure 1 with **n** the outward normal vector to the surface S_0. It includes the recording surface $Z = 0$ place, and a hemisphere extending to infinity in the subsurface. Contributions from the distant hemisphere are ignored, and the boundary value representation reduces to an integral over the surface involving the wave field on S_0 and a suitable Green's function G. Since $U(r_0, t_0)$ in equation (2) is equated to the observed seismic data, we require that $G = 0$ on S_0 in order to eliminate the gradient of U, which may not also be independently specified, nor is it measured in current seismic practice. A Green's function having the desired properties at the free surface consists of a point source at r_0 and its negative image at r'_0, or

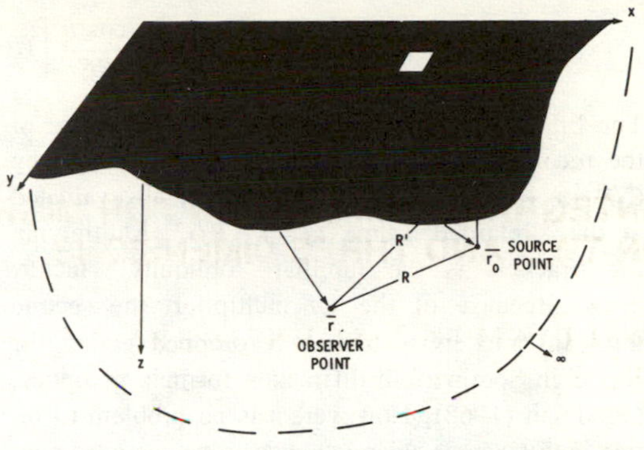

FIG. 1. Geometry for boundary value solution.

$$G(r, t|r_0, t_0) = \frac{\delta\left(t - t_0 - \dfrac{R}{C}\right)}{R} - \frac{\delta\left(t - t_0 - \dfrac{R'}{C}\right)}{R'}, \tag{3}$$

where

$$R = \sqrt{(z - z_0)^2 + (x - x_0)^2 + (y - y_0)^2},$$

and

$$R' = \sqrt{(z + z_0)^2 + (x - x_0)^2 + (y - y_0)^2}.$$

Other choices of G are possible for image reconstruction purposes as discussed by Kuhn and Alhilali (1976). Substitution of equation (3) into equation (2) and simplification yields the following integral representation for the wave field $U(r, t)$ at any point in image space in terms of observations of the wave field $U(r_0, t_0)$ on the surface,

$$\cdot U(r, t) = \frac{1}{2\pi} \int dt_0 \int dA_0$$

$$\cdot U(r_0, t_0) \frac{\partial}{\partial z_0} \left[\frac{\delta\left(t - t_0 - \dfrac{R}{C}\right)}{R} \right]. \tag{4}$$

This is a rigorous statement of Huygen's principle and is commonly called the Kirchhoff integral. From the form of the kernel of equation (4), we recognize the transformation as a three-dimensional convolution of the observed wave field with a space-time operator related to the point source solution to the wave equation. We will return to this point subsequently. Before doing so, however, it is instructive to re-express equation (4) by performing the indicated Z_0 differentiation and t_0 integration resulting in an equivalent expression.

$$U(r, t) = \frac{1}{2\pi} \int dA_0 \frac{\cos \theta}{RC} \left[U'(r_0, t_0) + \frac{C}{R} U(r_0, t_0) \right]_{t_0 = t - (R/C)} . \tag{5}$$

The bracketed term contains the time derivative of the recorded data plus the recorded data scaled by C/R or $1/t$ the reciprocal traveltime, all evaluated at the "retarded" time $t_0 = t - R/C$. Multiplying the brackets is the familiar "obliquity" factor, $\cos \theta$. Because of the $1/t$ multiplier, the second term in brackets is frequently dropped giving the Rayleigh-Sommerfeld diffraction formula of optics, Goodman (1968). However, it is no problem to retain both terms in seismic applications; we need only to differentiate the seismic section and add to it the same section scaled by $1/t$ in order to implement equation (5) exactly. Trorey (1970), Hilterman (1970, 1975), and Berryhill (1976) make extensive use of the Kirchhoff integral [equation (5)] in forward modeling studies of diffraction and other propagation complexities in two and three dimensions. For a lucid discussion of the historic role of equation (5) and its many variants in optical, acoustic, and seismic imagery, the author recommends the excellent treatment given by Walter and Peterson (1976).

Still another representation of equation (4) is possible by interchanging the Z_0 derivative with a Z derivative which may then be taken outside the integral, giving

$$U(r, t) = -\frac{1}{\pi} \frac{\partial}{\partial z} \int dA_0 \frac{U\left(r_0, t - \frac{R}{C}\right)}{R} . \tag{6}$$

This is the most compact form and clearly demonstrates that the integral transformation is a solution to the 3-D wave equation by virtue of the form of the kernel $f(t - R/C)/R$. Now let us return to the convolutional aspects of this transformation. As noted previously, the integral transformation equation (4) may be written symbolically as a three-dimensional convolution,

$$U(x, y, z, t) = U(x, y, z_0, t)$$
$$* \frac{1}{2\pi} \frac{\partial}{\partial z_0} \left[\frac{\delta\left(t \pm \frac{r}{C}\right)}{r} \right], \tag{7}$$

where

$$r^2 = \Delta z^2 + x^2 + y^2,$$

which translates the observed wave field from one

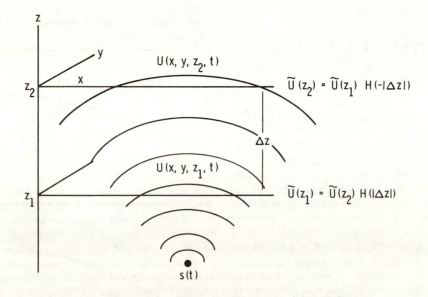

MIGRATION USES $H(+|\Delta z|)$ TO EXTRAPOLATE CONVERGING WAVES

FIG. 2. Extrapolation of converging and diverging waves.

Z-plane to another. If we Fourier transform expression (7) over x, y, and t (Appendix A), the operation becomes complex multiplication in the frequency wavenumber domain, giving

$$\bar{U}(k_x, k_y, z, \omega) = \bar{U}(k_x, k_y, z_0, \omega)$$
$$H(k_x, k_y, \Delta z, \omega), \qquad (8)$$

where

$$H = e^{\pm i |\Delta z|} \sqrt{\left(\frac{\omega}{C}\right)^2 - k_x^2 - k_y^2}. \qquad (9)$$

The transfer function H is seen to be a pure phase operator embodying the exact dispersion relation for the scalar wave equation. The operator H, expressed either in the space-time domain or frequency wavenumber domain, enables us to extrapolate waves in space, which we will see shortly is basic to seismic image reconstruction. The choice of sign in equation (7) and (8) is important insofar as it determines the direction of extrapolation. To clarify the choice, consider Figure 2 which depicts a spherical wave radiating from S and two observation surfaces at Z_1 and Z_2. The wave field at Z_2 can be obtained from the field at Z_1, which is closer to the source, by using expression (8) with the negative sign in operator H

to reflect the phase delay in propagation across the slab of thickness ΔZ. Conversely, we can make the clock run backward and compute the field at Z_1 from the field at Z_2 by use of the positive sign in H to reflect the phase advance in moving a distance ΔZ closer to the source. In the operation of migration we must use the positive sign in operator H to extrapolate *converging waves* back toward their origins. With these basic mathematical tools to move data around, let us now review the principle of migration based on these integral transformations.

First, it is important to recognize that the wave field extrapolation equations developed thus far are not suitable for application to field records. While it is not difficult to pose the problem to accommodate shot-to-detector offset [see, for example, Timoshin (1970) and French (1975)], the mathematics are somewhat messier. For simplicity, we limit this discussion to the familiar CDP stack representation which approximates coincident source/receiver geometry as illustrated in Figure 3. Furthermore, the equations are cast in one-way traveltime so we can either divide our stacked section time scales by 2 or use a velocity in migration equal to 1/2 the true velocity. With these two assumptions, it becomes clear that the "physical" experiment we are approximating with stacked data is one in which the re-

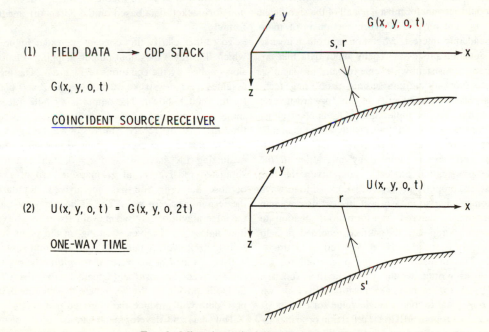

(1) FIELD DATA ⟶ CDP STACK

G (x, y, o, t)

COINCIDENT SOURCE/RECEIVER

(2) U (x, y, o, t) = G (x, y, o, 2t)

ONE-WAY TIME

FIG. 3. Migration principle: steps 1 and 2.

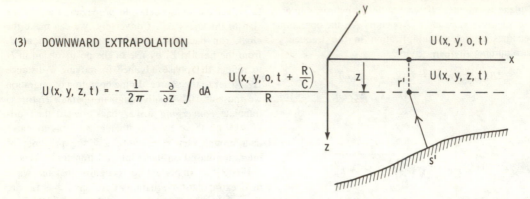

(3)   DOWNWARD EXTRAPOLATION

$$U(x, y, z, t) = -\frac{1}{2\pi} \frac{\partial}{\partial z} \int dA \; \frac{U\left(x, y, o, t + \frac{R}{C}\right)}{R}$$

(4)   IMAGING PRINCIPLE — EXTRAPOLATE RECEIVERS FOR ALL $Z > 0$ AT $t = 0$

$$U(x, y, z, o) = -\frac{1}{2\pi} \frac{\partial}{\partial z} \iint dx \, dy \; \frac{U\left(x, y, o, \frac{R}{C}\right)}{R} \quad = \underline{\underline{\text{3D MIGRATION}}}$$

Fig. 4. Migration principle: steps 3 and 4.

ceivers are located on the surface, the sources are positioned along the reflecting interfaces, their strengths are proportional to the reflection coefficients, and they are fired in unison. That such a physical (though not necessarily practical) experiment could account for most if not all of the significant events present on a CDP stack section is of more than academic interest. Migration and other inverse wave equation processes require input data that are reasonably consistent with some forward propagation process. Not all current seismic processing techniques preserve the integrity of this forward-inverse relationship. For example, fast AGC applied either before or after stack can dramatically alter the amplitude of complex wave interferences which, if undisturbed, can be unscrambled by migration in the inverse propagation process. Thus, given that the CDP stack, properly processed, is amenable to wave equation manipulation, we next insert this data into our previously derived transformation equation to downward extrapolate the surface recorded data to successively deeper levels, as depicted in Figure 4, step 3. This in itself is not yet migration, for the equation as written would give us a time function $U(x, y, z, t)$ for each x, y, z position. Instead, we really only want to map a single value for each position, a value proportional to the reflection or scattering strength at that subsurface location, or in the context

of our "physical" experiment, we wish to map the equivalent source strength at all subsurface positions at the shot instant $t = 0$. Therefore, we must fix $t = 0$ and evaluate the integral for all x, y, z of interest as indicated in step 4 of Figure 4. This is *3-D migration* for stacked data based on the Kirchhoff integral formula.

To further clarify this concept, consider Figure 5 which illustrates the input-output mapping relationship implied by the equations of Figure 4. The input assumes we have stacked data over the $Z = 0$ plane for the model shown. The output is a single trace at some x, y location plotted versus Z and vertical time Z/C. As the receiver moves down through successive positions, a point is mapped at each step by evaluating the integral with $t = 0$. For example, simulated receiver r_1 at z_1 maps a zero at $t = 0$ because the reflection has not arrived. Similarly, the response is zero at z_2, and as becomes obvious, the integral will be zero when evaluated at zero traveltime unless the receiver is sitting on top of or very near the reflector. When it is, the reflection wavelet will be mapped at the vertical traveltime below the surface receiver position (actually below the CDP midpoint position). We recognize that this mapping procedure will produce the migrated picture.

Thus far, the development has assumed seismic observations are available over an area of sufficient

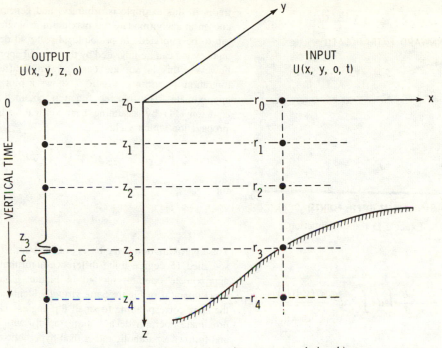

OUTPUT
$U(x, y, z, o)$

INPUT
$U(x, y, o, t)$

FIG. 5. Migration principle: input-output relationship.

extent to perform the indicated surface integrals or their discretized equivalents. This requires a 3-D seismic survey in which data are acquired over an entire prospect with space sampling of the order of one-half the shortest wavelengths of interest. While surveys of this kind have been and are now being conducted on a limited scale, the bulk of current seismic information is still 2-D, having been collected along widely spaced lines or traverses. In order to use the foregoing theory to migrate data from a single line and retain the benefits of a wave equation algorithm, we must make some assumption about the nature of the surface data $U(x, y, 0, t)$ where we did not measure it. The most common practice is to assume the wave field at the surface is only 2-D; that is, if the line was shot in the x-direction, then

$$U(x, y, 0, t) = U(x, 0, t),$$

independent of y. For this to be true, two conditions must be met: (1) the subsurface geology must be independent of y, and (2) the source must either be a line source in the y-direction or the source and receiver must be colocated as is approximately the case in CDP stack. If these conditions are met, the appropriate 2-D transfer functions can be obtained

from equations (7) and (8), by either integrating out the y dependence in the 3-D space-time operator or setting $k_y = 0$ in the frequency-wave number operator. The corresponding 2-D transfer functions are given below:

$$h(x, \Delta z, t) = \frac{1}{\pi} \frac{\partial}{\partial z_0} \frac{H\left(t \pm \dfrac{r}{c}\right)}{\sqrt{t^2 - \left(\dfrac{r}{c}\right)^2}}, \quad (10)$$

where

$$H = \text{unit step function},$$
$$r = \sqrt{(z - z_0)^2 + x^2},$$

and

$$H(k_x, \Delta z, \omega) = e^{\pm i|\Delta z| \sqrt{\left(\frac{\omega}{c}\right)^2 - k_x^2}}. \quad (11)$$

The resulting expressions (10) and (11) are, of course, Fourier transform pairs and bear the same relationship to the 2-D wave equation solution as the 3-D transfer functions equations (7) and (8) bear to the 3-D wave equation. The 2-D migration algorithm obtained by convolving equation (10) with $U(x, 0, t)$ and setting $t = 0$ as required by the mapping principle gives:

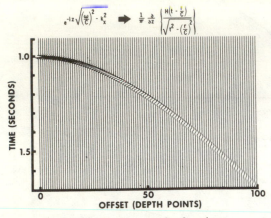

FIG. 6. Exact 2-D transfer function.

$$U(x, z, 0) = -\frac{1}{\pi} \frac{\partial}{\partial z} \int dx \int_{r/c} d\tau \frac{U(x, 0, \tau)}{\sqrt{\tau^2 - \frac{r^2}{c^2}}}. \quad (12)$$

This expression is somewhat more complicated looking than its 3-D counterpart of Figure 4, because the y integral has been replaced by a time integral along the trace. In order to more fully appreciate the relationship between the 2-D and 3-D migration expressions, consider the following hypothetical experiment. First, consider migrating a seismic section using the 2-D expression given in equation (12). Next, imagine replicating that same input section many times to simulate shooting parallel lines in the dip direction of a 2-D subsurface model. Then 3-D migrate these parallel lines using the expression in Figure 4. The results of the two migrations will be identical; that is, expression (12) actually accomplishes 3-D wave equation migration under the special circumstances that the surface recorded data are independent of one surface variable. When the above is not true, then equation (12) is not a valid migration, and as every interpreter should know, user beware!

Before leaving the mathematics of migration, it is instructive to examine the behavior of the 2-D transfer function, equation (10), in more detail. Figure 6 shows a plot one-half the exact space-time operator as it would appear when being convolved with the section for an output value at 1 sec vertical traveltime. The operator shown is band-limited in both space and time appropriate to the sampling. The hyperbolic trajectory is predictable from simple ray theory considerations; however, the amplitude and phase behavior are not. The aperture width, ±100

traces in this example is arbitrary, and generally is chosen to accommodate the maximum geologic time dip to be migrated. In principle, dips to 90 degrees and beyond can be migrated by the integral approach; however, this is not the case for finite-difference migration algorithms. Figure 7 shows a plot of an approximate 2-D transfer function obtained from equation (11) by assuming near vertical incidence propagation which yields

$$e^{-iz\sqrt{\left(\frac{\omega}{c}\right)^2 - k_x^2}} \Rightarrow e^{-i\frac{z\omega}{c}\left(1 - \frac{1}{2}\left(\frac{ck_x}{\omega}\right)^2\right)} \quad (13)$$

for

$$\frac{\omega}{c} \gg k_x.$$

This approximation is the basis of Claerbout's (1972) so-called 15 degree finite difference algorithm. The approximate operator plotted in Figure 7 and the exact operator in Figure 6 are virtually identical near the apex corresponding to small dip angles. The approximate operator decays more rapidly with offset and follows a parabolic rather than hyperbolic trajectory. Both these factors, plus frequency dispersion associated with finite differencing schemes, limit the accuracy and fidelity of finite difference migration in steeply dipping situations. While it is true that higher order approximations are possible and have been discussed by Claerbout (1976) and others, in the limit they can only approach the exact transfer function which the integral method achieves with ease. Next, let us examine the application of these migration algorithms to both model and field data.

MODEL RESULTS

First, consider the analytical migration of a plane

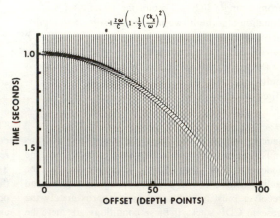

FIG. 7. Approximate 2-D transfer function.

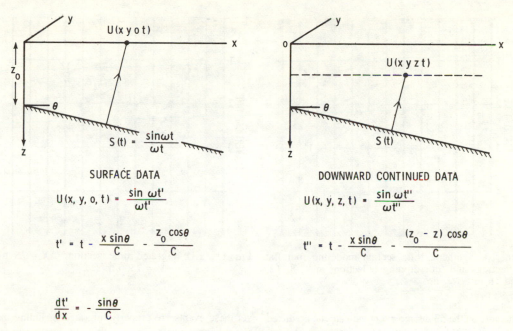

SURFACE DATA

$$U(x, y, o, t) = \frac{\sin \omega t'}{\omega t'}$$

$$t' = t - \frac{x \sin\theta}{C} - \frac{z_0 \cos\theta}{C}$$

$$\frac{dt'}{dx} = -\frac{\sin\theta}{C}$$

DOWNWARD CONTINUED DATA

$$U(x, y, z, t) = \frac{\sin \omega t''}{\omega t''}$$

$$t'' = t - \frac{x \sin\theta}{C} - \frac{(z_0 - z) \cos\theta}{C}$$

FIG. 8. Plane dipping reflector example.

dipping reflection depicted in Figure 8. On the left of the figure, we postulate a band-limited signal $S(t)$ emanating from a bed dipping at angle θ in the x, z plane. The surface recorded data $U(x, y, 0, t)$ is a delayed version of this signal, and the observed time dip dt'/dx is the familiar quantity $\sin\theta/c$. For this analytical signal we can actually analytically downward continue our receiver to a depth z using either equation (7) or (8). The result, $U(x, y, z, t)$, is not unexpected and could have been arrived at by inspection. Since the receiver is a distance z closer to the reflector, the traveltime delay is reduced by $z \cos\theta/c$. Now to obtain the migrated time picture we must invoke the mapping principle by setting $t = 0$, and change variable from depth z to vertical time $\tau = z/c$ as shown by $U(x, y, \tau, 0)$ in Figure 9. In migrated time space, the time dip $d\tau'/dx$ after migration becomes $\tan\theta/c$, and the bandwidth of the signal is reduced by $\cos\theta$. Since migration is a loss-less process, the latter is purely a geometrical effect due to rotation of the reflection. Put another way, migration increases the time dip of a reflector by $\cos\theta$ and decreases the apparent signal frequency by the same factor so as to preserve horizontal wavenumber.

Now let us examine the computer migration of several simple synthetic sections. Figure 10 models four flat reflections and four dipping reflections with

time dips of 4, 8, 12, and 16 msec/trace, respectively. The reflection wavelet is a zero phase, 0–80 Hz bandwidth pulse. A 2-D migrated picture is shown in Figure 11. A trace spacing of 25 m and velocity of 2500 m/sec were used resulting in structural dips ranging from 12 to 53 degrees. The steepest event has migrated some 200 traces, or about 5 km.

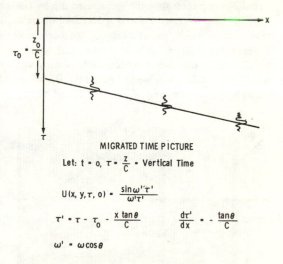

MIGRATED TIME PICTURE

Let: $t = o$, $\tau = \frac{z}{C} =$ Vertical Time

$$U(x, y, \tau, o) = \frac{\sin \omega' \tau'}{\omega' \tau'}$$

$$\tau' = \tau - \tau_0 - \frac{x \tan\theta}{C} \qquad \frac{d\tau'}{dx} = -\frac{\tan\theta}{C}$$

$$\omega' = \omega \cos\theta$$

FIG. 9. Analytical migration of dipping reflector.

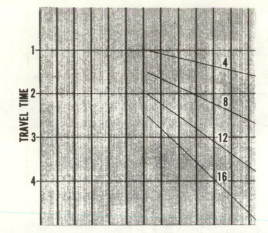

FIG. 10. Synthetic time section modeling four flat reflections and four dipping reflections at 4, 8, 12, and 16 msec/tr dip.

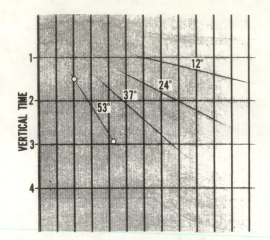

FIG. 11. 2-D migrated time section, $\Delta X = 25$ m, $V = 2500$ m/sec.

The dots on the 53 degree event indicate the predicted migrated end points for the 16 msec/trace reflection in Figure 10, the correspondence is excellent. The slight tails on each of the migrated events result from not including diffractions in the input model. The details of the result are more evident in Figures 12 and 13 which show enlarged portions of the input section and migrated section, respectively. The low level jitter on the input are sidelobes associated with the sharp cutoffs in the wavelet design. The background noise in the output is a combination of the above and migration noise. The expected results are also evident from these figures; namely, (1) the migrated dip is greater than the unmigrated dip by cos θ, and (2) the migrated pulse is reduced in apparent bandwidth by cos θ, thereby keeping the horizontal wavenumber invariant. It is also interesting to note the 12 and 16 msec/trace reflections have a significant portion of their bandwidth beyond the 1/2 wavelength Nyquist space sampling limit of 42 and 32 Hz,

yet these events are properly migrated including the aliased components. The question of migrating under sampled data is more complex than one might expect, nor is it independent of the algorithm. With the integral approach it is possible to correctly migrate aliased data by not spatially bandlimiting the migration operator shown previously in Figure 6. The risk in doing this is to generate migration background noise which in the previous example is sufficiently low level to be unobjectionable. However, the effect is very sensitive to the space sampling ΔX. Figure 14 shows another migration of the input model in Figure 10 in which the trace interval was doubled to 50 m; that is, the migration program was told the ΔX was 50 m instead of 25 m, but the identical section was migrated. Of course, the implied structural picture is different, and, as before, the events are correctly migrated, including aliased frequencies. The most notable difference, however, is the increase in migration noise from both flat and

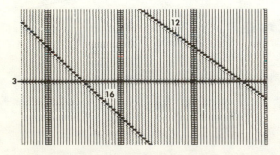

FIG. 12. Detail of input section in Figure 10.

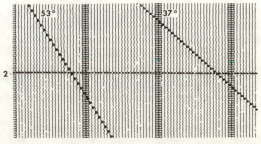

FIG. 13. Detail of migrated section in Figure 11.

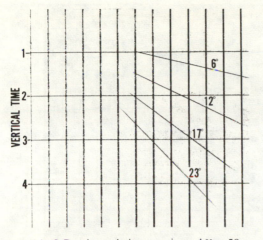

FIG. 14. 2-D migrated time section, $\Delta X = 50$ m, $V = 2500$ m/sec.

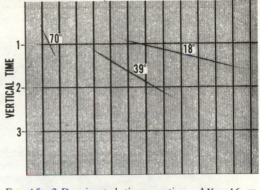

FIG. 15. 2-D migrated time section, $\Delta X = 16$ m, $V = 2500$ m/sec.

dipping events. This is basically a leakage problem caused by approximating the integral [equation (12)] by a discrete summation. The coarser the ΔX, the poorer the summation approximates the integral and the greater the leakage. While ΔX is a critical parameter, the leakage also depends on frequency, velocity, migration aperture, and traveltime. In general, the leakage worsens with increasing ΔX, increasing frequency, increasing aperture, decreasing velocity, and decreasing traveltime. The problem also scales as the ratio of $V / \Delta X$. In other words, the 50 m model with a 5000 m/sec velocity would have the same low noise level as the 25 m, 2500 m/sec migration in Figure 11. To guard against this problem on coarsely sampled data (whether or not it is aliased), the migration operator must be spatially bandlimited as in Figure 6 or, equivalently, a more sophisticated numerical integration must be used in lieu of discrete summation. Before leaving this sample plane dipping model, however, we will migrate it one more time using a ΔX of 16 m giving the result as shown in Figure 15. The third dipping event now appears as a greatly compressed 70 degree segment which has migrated some 300 traces horizontally and 1.0 to 1.5 sec in time to its correct subsurface position. The missing fourth reflection does not represent a possible reflection in this model because its 16 msec/trace time dip exceeds the maximum of 13 msec/trace for a 90 degree reflector; hence it is not imaged.

As is apparent from these examples, the integral method has no algorithmic limitations on dip. Reflections can be migrated to 90 degrees and beyond in the presence of vertical velocity gradients. The issues

of velocity and cost ultimately become the limiting factors, but before addressing the questions of velocity inhomogeneity, another slightly more realistic model is of interest.

Figure 16 shows a synthetic zero offset time section for three reflecting horizons of moderate complexity computed using a forward wave theory approach described by Trorey (1970). A constant 8000 ft/sec velocity was used, the trace interval is 50 ft, and the wavelet bandwidth is approximately 0–60 Hz. Many of the classic diffraction phenomena so often seen on stacked sections are present. The 2-D integral migration shown in Figure 17 is virtually a perfect reconstruction of the subsurface acoustic impedance with accurate representation of the amplitude and waveform, structural attitude, curvature, and bed terminations. Were the world so simple, seismic processing would be a closed book. Unfortunately, real seismograms are infinitely more complex than the constant velocity model depicted here, and much progress remains to be made in seismic processing techniques before we can accurately image in heterogeneous media.

Some of the more practical aspects of migrating seismic data are knowing what velocity to use, how to estimate it from the data, and how accurate it must be. None of these are trivial issues, nor shall I attempt to provide comprehensive answers. Certainly the issues of estimating seismic velocity for stacking and more recently for migration have received ample attention in the literature and in professional society meetings. I will not attempt to summarize the current art in this mature activity except to point out there is a trend away from CDP stack based velocity analysis toward migration based techniques [Sattlegger (1975), Dohr (1975)]. The trend will undoubtedly accelerate as migration of unstacked

FIG. 16. Wave theory, zero offset time section modeled at 8000 ft/sec and 50 ft trace spacing.

data gradually replaces the CDP stack as the standard seismic image.

The question of (post stack) migration velocity sensitivity is somewhat more easily addressed, and I use the model in Figure 16 to illustrate the effects of slightly under and over migration. The section was remigrated with a velocity of 7600 ft/sec (5 percent low) and 8400 ft/sec (5 percent high). The results are shown in Figures 18 and 19, respectively. In a gross sense the pictures are very similar to Figure 17 migrated with the correct velocity. In detail they differ; for example, the fault terminations are blurred, the flanks of the synclines are in error by several hundred feet, and the small bump on the second reflector is severely distorted. Other distortions in amplitude and waveform are not readily

discernible, and, as expected, the flat reflections are totally insensitive to velocity. While not a comprehensive answer to the velocity sensitivity question, we may readily conclude that: the more complex the subsurface, the more diffraction-like is the time section, and the more accurate must the velocity be. Even with faulted simple geology, relatively small errors in migration velocity will improperly collapse the diffraction tails and blunt fault resolution.

VELOCITY INHOMOGENEITY

In the foregoing development, starting with the scalar wave equation and resulting in integral migration algorithms based thereon, it has been tacitly assumed the techniques could be successfully applied to waves propagating in a variable velocity medium

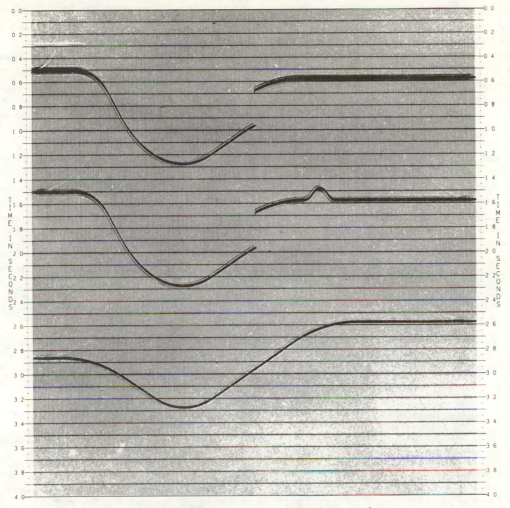

FIG. 17. 2-D migrated synthetic wave theory section.

even though the derivation assumed constant velocity. This is certainly not intuitively obvious, but as will be shown, the method is remarkably robust with regard to vertical velocity variations.

First, consider the 2-D migration of a single sinusoidal trace with constant velocity. The result in Figure 20 shows the familiar windshield wiper pattern. The dark bands trace out the wavefronts at 2, 3, and 4 sec of traveltime. The trace amplitude decays with time to compensate for decreasing wavefront curvature, and the decay with offset along the wavefront reflects the obliquity factor $\cos \theta$. Now if we vary the velocity $C = C(z)$ for each depth or vertical time step in the migration algorithm [equation (12)], the wavefront pattern takes on a very different shape as shown in Figure 21. For this linear increasing

velocity $V = 1800 + 600t$ m/sec, the wavefronts are flatter for small dip angles and swoop up more rapidly for steep dips. When the wavefront approaches vertical, the structural dip is 90 degrees. This occurs at the surface for constant velocity, but well below the surface with an increasing velocity function. The question of accuracy of the variable velocity result can best be answered from simple model studies.

The first model is described by the instantaneous, average and rms velocities shown in Figure 22. Figure 23 compares the true curved ray wavefront with those generated by the integral migration algorithm using both the average and rms velocity distributions of Figure 22. The exact curve was computed by integrating the traveltime equations as described by Musgrave (1961) For traveltimes of 1, 2, 3, and 4 sec, the wavefronts are virtually identical for dips less than 20 degrees. The rms velocity curve continues to track the exact wavefront to about 40 degrees and then departs gradually as the dip angle increases. Even out to approximately 60 degrees the offset error is only about 1–2 percent, which implies the velocity is too slow by the same amount.

A second model, Figure 24, presents a more complicated velocity distribution consisting of a deep water layer over a high-velocity subsurface. The wavefronts shown in Figure 25 tell a similar story; the errors are slightly greater due to the large discontinuity at the water bottom, yet migration using

the rms velocity appears quite satisfactory to about 60 degrees, considering the expected accuracy of seismic velocity estimation.

A final model, Figures 26 and 27, shows a first order velocity discontinuity at depth between two linearly increasing functions. The errors are of the same magnitude as in the previous two models and suggests that for this class of linear increasing velocity functions (with or without discontinuities), the strategy of using the vertical rms velocity in the integral migration algorithm [equation (12)] will produce quite accurate migrations to the order of 60 degrees structural dip. The errrors increase with angle and total distance traveled. At early times, less than 2 sec for these models, accurate migrations

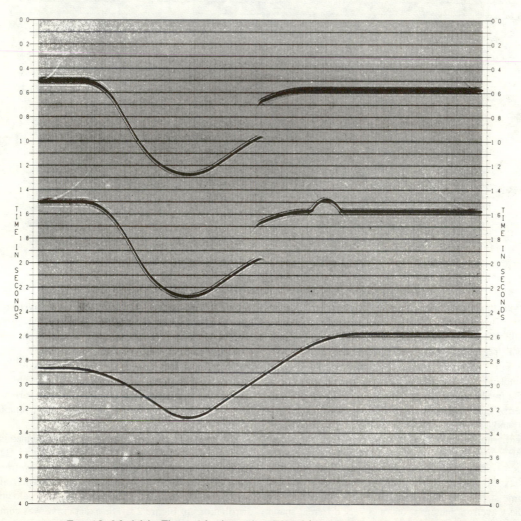

FIG. 18. Model in Figure 16 migrated at 7600 ft/sec, 5 percent low velocity.

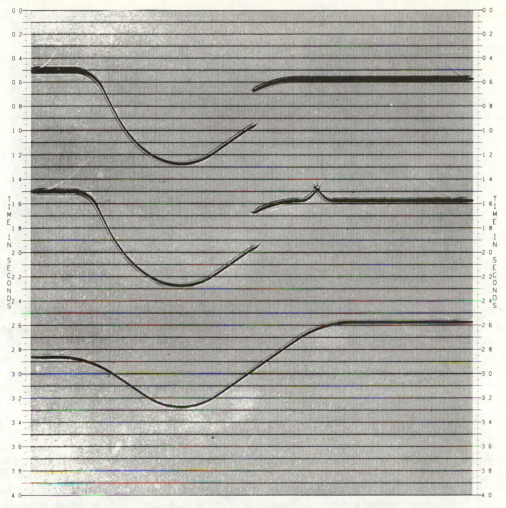

FIG. 19. Model in Figure 16 migrated at 8400 ft/sec, 5 percent high velocity.

can be obtained well beyond 60 degrees. Refinements are possible to improve accuracy at steeper dips by using ray-tracing strategies or modifying the rms velocity as a function of angle to approximate the rms velocity along the ray, which would give the exact result. At present, however, we do not believe these refinements are warranted until further advances are made in estimating velocity with accuracy of the order of 1 percent in structurally complex geologic settings.

Horizontal velocity gradients present an additional complication to both migration and velocity estimation techniques. They can be "handled" from a mechanical point of view in much the same way as are the vertical gradients, by allowing the rms velocity term in the 2-D and 3-D migration algo-

rithms to vary with x, y, and z. The errors in migration caused by lateral velocity gradients are not as well understood as those caused by vertical gradients, and the matter is still being actively researched.

Let us now leave the theory and models and examine several field examples of 2-D and 3-D migration.

FIELD EXAMPLES

The first 2-D migration example comes from the Gulf of Mexico. Figures 28 and 29 show, respectively, the stack section and migrated section. The trace interval is 50 m, typical Gulf velocities were used, and the display bandwidth is about 40 Hz shallow and 20 Hz at depth. Technically the result is very clean with virtually no migration noise or

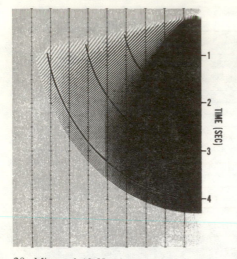

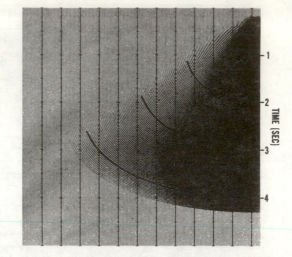

FIG. 20. Migrated 40 Hz sinewave at constant 2500 m/sec velocity.

FIG. 21. Migrated 40 Hz sinewave with varying velocity $V = 1800 + 600\,t$ m/sec.

artifacts, and waveform character has been well preserved. Geophysically, the main value of 2-D migration in this example is to enhance fault resolution in the relatively simple sand/shale section. At depth, while some simplification occurs in the structural picture, it would be remarkable if the assumptions for valid 2-D migration were met; that is, sections parallel to this one would look exactly alike.

A second 2-D example comes from the North Sea. The CDP stack and 2-D migrated sections are shown in Figures 29 and 30, respectively. Trace spacing in this case is 25 m and the display bandwidth is higher than in the previous Gulf Coast example. The stacked section exhibits a simple Tertiary-Cretaceous section down to about 1 sec. Below the Cretaceous-Jurassic unconformity the data are complex, discontinuous, and exhibit numerous diffraction events. The 2-D migrated picture reveals a much more interpretable Jurassic section between 1 and 2 sec, indicating major block faulting and tectonic activity probably related to salt movement. In particular, several small fault blocks on the left of the section and also just right of center are virtually obscured by diffractions on the stacked section. After migration, they stand out with remarkable clarity, as does the uncomformity at the base of the Cretaceous. Overall, the waveform and character of the input section are faithfully preserved in the migrated picture, due principally to the wave equation formulation of the algorithm. This is perhaps the most significant difference between this "Kirchhoff" based migration and the earlier diffraction-summation techniques.

Whether the structural picture portrayed here by migration is correct cannot be answered from this result alone; additional seismic control is necessary. In general, from our experience we know complex geology is seldom sufficiently two-dimensional to satisfy the assumptions required for 2-D migration. While French (1975) extends the range of applicability of 2-D migration to both oblique profiles and plunging 2-D structures, there is no substitute for 3-D data and 3-D migration to correctly image seismic returns from complex geologic targets.

Until recently 3-D seismic data acquisition and processing were more of a research curiosity than a practical exploration tool. However, continuing advances in computer hardware and software coupled with innovations in seismic data acquisition over the

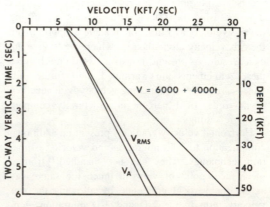

FIG. 22. Velocity function—model 1.

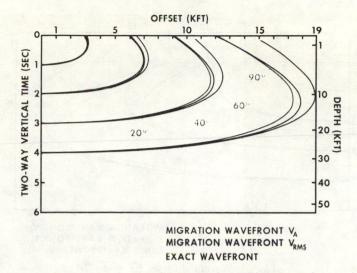

MIGRATION WAVEFRONT V_A
MIGRATION WAVEFRONT V_{RMS}
EXACT WAVEFRONT

FIG. 23. Wavefront curves for model 1 showing the exact and approximate wavefronts using the average and rms velocities in the Kirchhoff migration integral.

past several years now make it economically feasible to conduct 3-D surveys on both land and marine prospects. Tegland (1976) describes some of these significant advances which include streamer tracking systems to locate accurately the streamer relative to the boat for each shot, and efficient land 3-D collection techniques based on crooked traverses and perimeter shooting. An example of the latter applied to a California land prospect is shown in Figure 32. The technique, called Seisloop/Seisquare[1], obtains 3-D coverage by shooting around regular or irregular loops into geophones emplaced completely around the perimeter of the same loop. In this prospect, 14 rectangular loops covering 10 sq mi were shot using Vibroseis® source patterns at each of the large dots, into geophone arrays at each small dot. Each source-receiver midpoint location is calculated and the corresponding trace assigned to a unique "bin" 330 ft square, resulting in the regular CDP map of Figure 33. Traces common to a bin are stacked together after static and NMO corrections have been applied, yielding a set of stacked traces on a uniform 330 ft x, y grid with an average CDP fold of six. These data may be arranged for display in numerous ways. Figure 34 shows a north-south gather of CDP lines 19, 20, and 21 which are 330 ft apart and about 3 miles long. Data quality is fair to good in the shallow section. The geology is complex showing strong north dip, and well control indicates that the shallow

gas production is controlled by numerous small fault blocks. The prospect is an old field with 25 existing wells, generally thought to be drilled out. 3-D seismic was tried in an attempt to uncover additional secondary fault traps in a mature field development situation. As a result of the 3-D survey, five new drilling locations were identified, and as of this writing, two have been drilled and tested commercial gas with recoverable economic value about ten times the cost of the 3-D survey and drilling.

Of major significance in this project was the application of 3-D migration based on the formula of Figure 4 appropriately discretized for sampled data input. Figure 35 shows three output sections from the

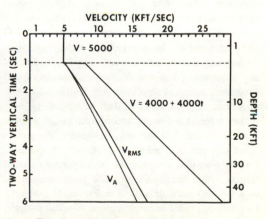

FIG. 24. Velocity function—model 2.

[1]Service Mark of GSI. U.S. Patent No. 3,867,713.
®Continental Oil Co.

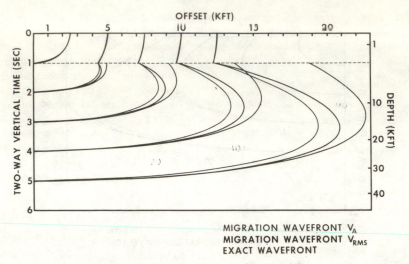

MIGRATION WAVEFRONT V_A
MIGRATION WAVEFRONT V_{RMS}
EXACT WAVEFRONT

FIG. 25. Wavefront curves for model 2.

3-D migration process at the same locations as the CDP stack lines in Figure 34. Each trace in the migrated picture results from a weighted sum over a square aperture of 10×10 input CDP stacked traces. The aperture size was determined by the maximum time dip rates as seen on the stacked sections in the north-south direction and between lines in the east-west direction.

For comparison, Figure 36 shows the same three lines migrated with the 2-D algorithm. The clarity and definition of the 3-D migration are superior to both the 2-D migration and CDP stack at all levels. In fairness to the latter, however, it should be noted that 6-fold conventional shooting is not a very heavy field effort by current standards. Better conventional 2-D results could have been obtained with higher fold shooting. That more conventional seismic would have contributed to finding additional reserves in this field is questionable since several generations of 2-D work had already been exploited to their fullest in discovering the known reserves. The key to finding any remaining pools was dense spatial sampling and accurate location of small untested fault blocks, both of which the 3-D migrated data addressed. Of secondary interest is a major unconformity seen at about 1.5 sec on the 3-D sections, but rather obscure on the CDP and 2-D migrated sections. This unconformity may play a significant role in deeper untested targets in the field. Its expression is more apparent on east-west lines illustrated in Figure 37, showing two 3-D migrated lines gathered across the prospect 660 ft apart. Also evident are numerous north-south

trending faults which control the shallow gas production.

Of considerable importance to the overall success of this project was the confidence placed in the 3-D migrated result by the geologist-interpreter because it tied subsurface well control, whereas the stacked sections did not. Finally, in retrospect, the 330 ft bin size was marginally adequate for the exploration objective in terms of structural dip and resolution implications. The space sampling intervals Δx, Δy are critical parameters in 3-D surveys and must be selected to meet the geologic and resolution objectives within the economic constraints imposed on the program. In general, 3-D seismic surveys will,

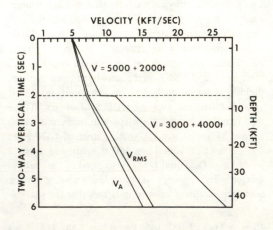

FIG. 26. Velocity function—model 3.

because of their custom problem-solving nature, require a much greater degree of preplanning and client-contractor interplay than conventional seismic surveys.

SUMMARY AND CONCLUSIONS

Our understanding of migration has come a long way from the era of wavefront charts and curves of maximum convexity. We now view the operation as a rigorous inverse wave propagation process subject only to the limitations of the scalar wave equation assumption, and our ability to estimate propagation velocity. Both these areas will undoubtedly be the focal points for further improvements in migration practice in the years ahead.

This discussion has centered on the integral formulation for migration. The finite-difference school also has its advocates and supporters and no attempt was made here to plead their case. Loewenthal (1974), Koehler (1976), Larner (1976), and others have discussed the latter in considerable detail. To claim one approach is vastly superior to the other is to ignore the fact that both integral and finite-difference migrations are based on the scalar wave equation. In the limit of no approximations in implementation they would yield the same results.

In the author's opinion, the integral method offers the following advantages:

1) The 2-D and 3-D algorithms can be implemented without approximating the scalar wave equation.

2) Data can be migrated to 90 degrees and beyond, velocity accuracy and cost being the only real limitations.

3) In 3-D applications, departure from a regular x, y grid can be easily accommodated by the integral method. This occurs frequently in both land and marine applications because of the difficulty in collecting seismic data exactly where you want it.

4) Finally, the integral method lends itself more readily to ad hoc weighting schemes which are meant to combat seismic noise not comprehended by any of the current wave equation formulations for migration.

ACKNOWLEDGMENTS

The author wishes to thank his colleagues, Cam Wason and Bruce Secrest, for their many helpful, theoretical discussions relevant to the concepts presented herein. My thanks also go to Frank Linville, Chyi Lu, and Bob Hester for their assistance in preparing the model results. And, finally, the author is grateful to Texas Instruments for permission to publish this work.

REFERENCES

Berryhill, J. R., 1976, Diffraction response for nonzero separation of source and receiver: Geophysics, v. 42, p. 1158–1176.

Claerbout, Jon F., 1970, Coarse grid calculations of waves in inhomogeneous media with application to delineation of complicated seismic structure: Geophysics, v. 35, p. 407–418.

(Text continued on p. 76)

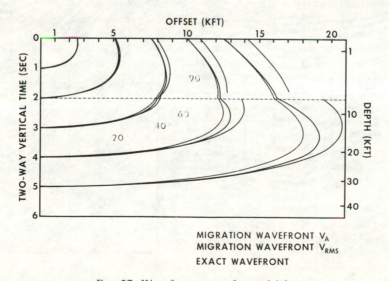

Fig. 27. Wavefront curves for model 3.

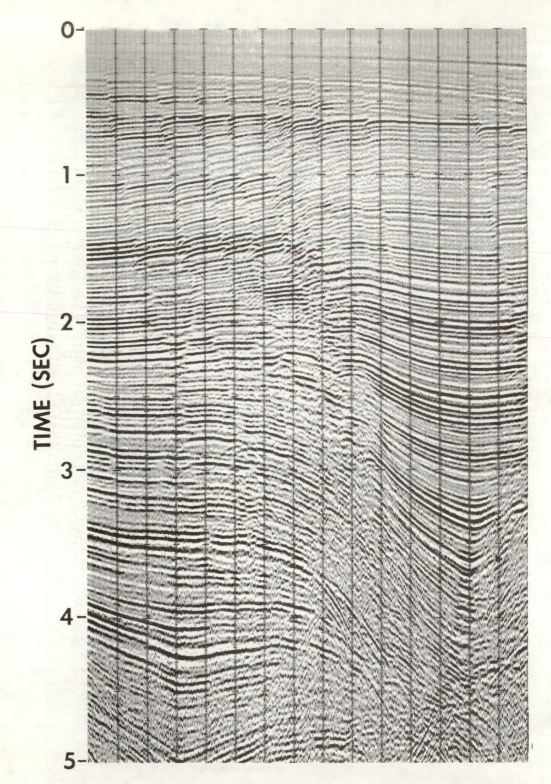

Fɪɢ. 28. Gulf of Mexico CDP stacked section.

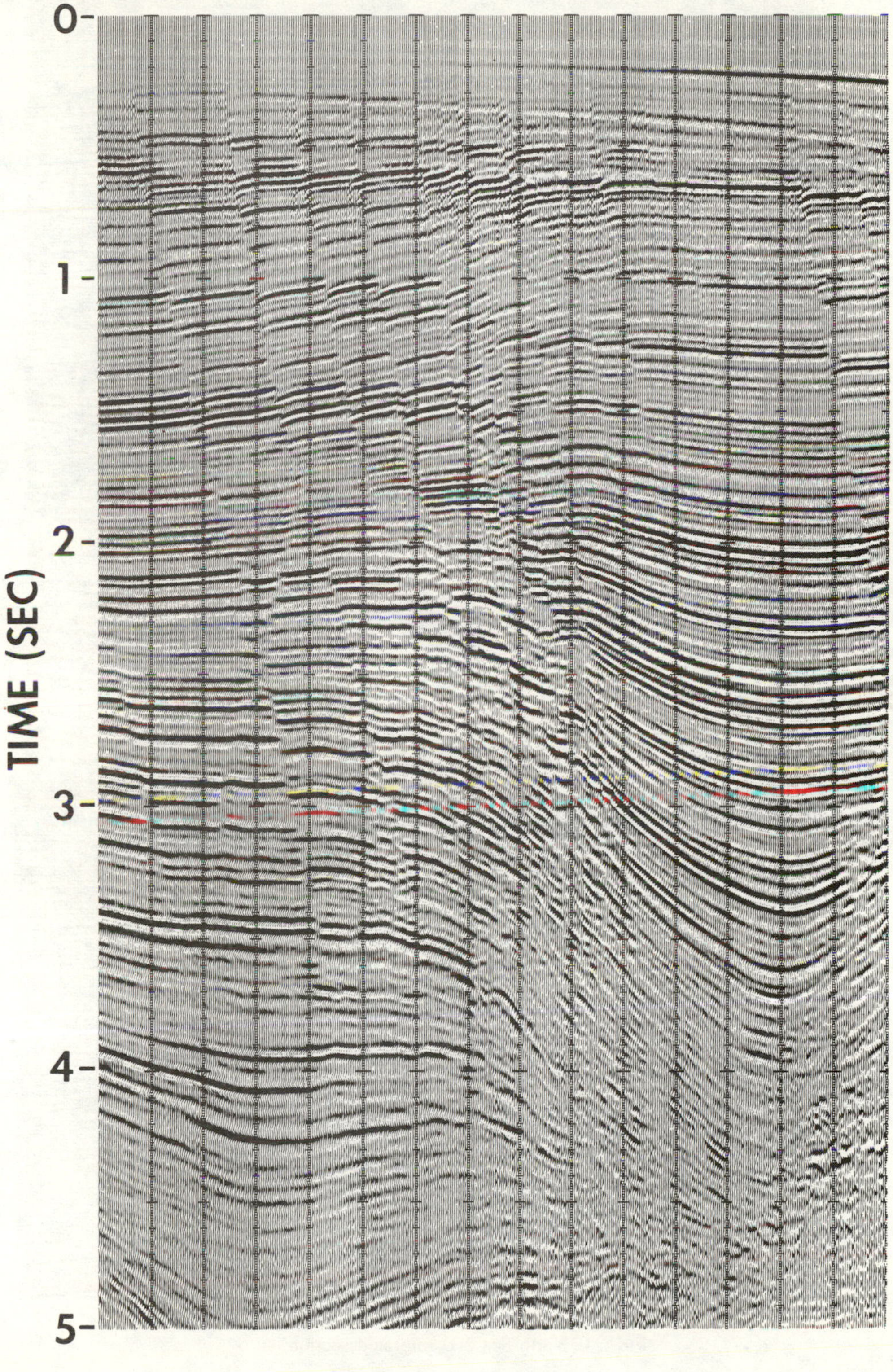
0
1
2
3
4
5
TIME (SEC)

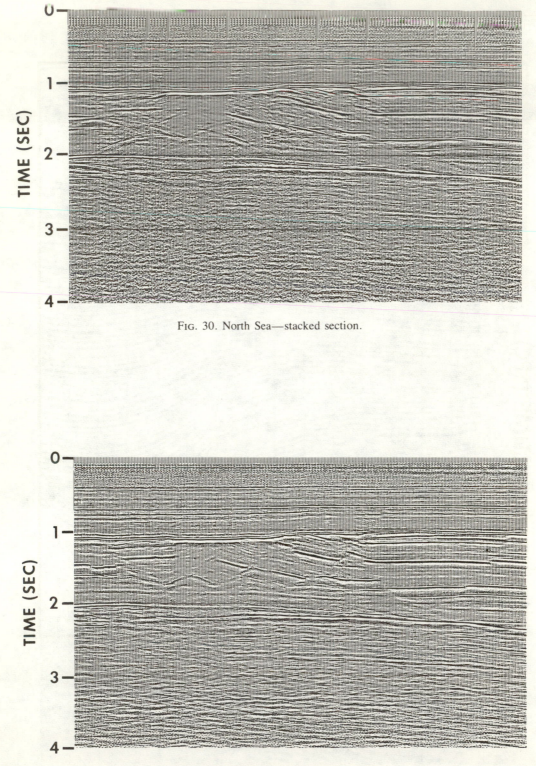

FIG. 30. North Sea—stacked section.

FIG. 31. North Sea 2-D migrated section.

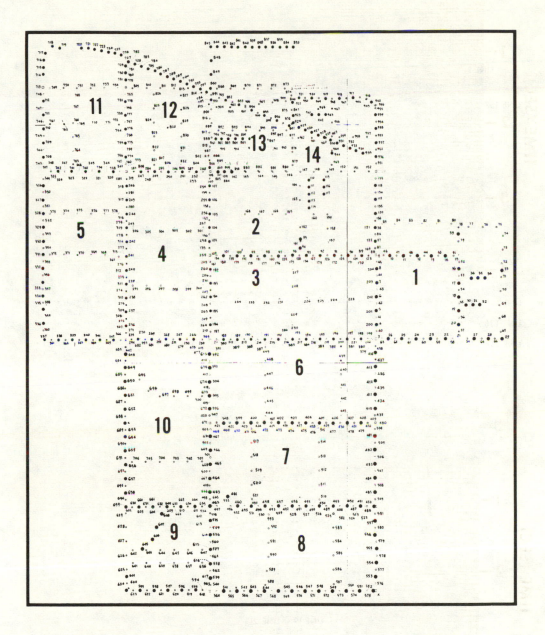

FIG. 32. California land 3-D prospect map showing location of source and receiver positions.

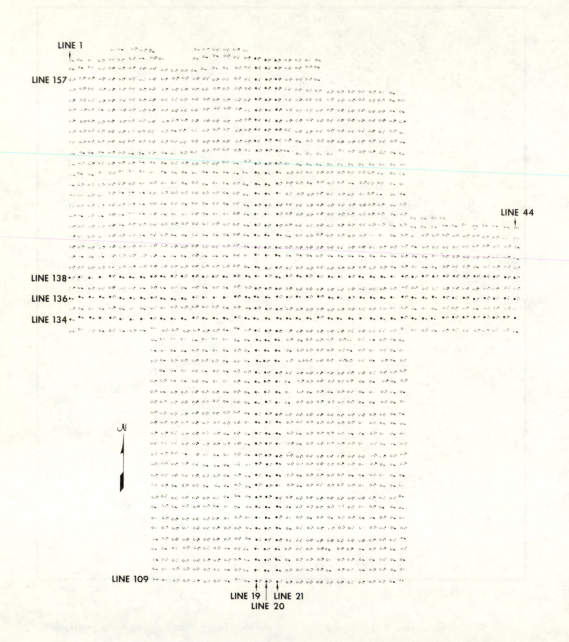

FIG. 33. California land 3-D prospect map showing grid of CDP bin center locations.

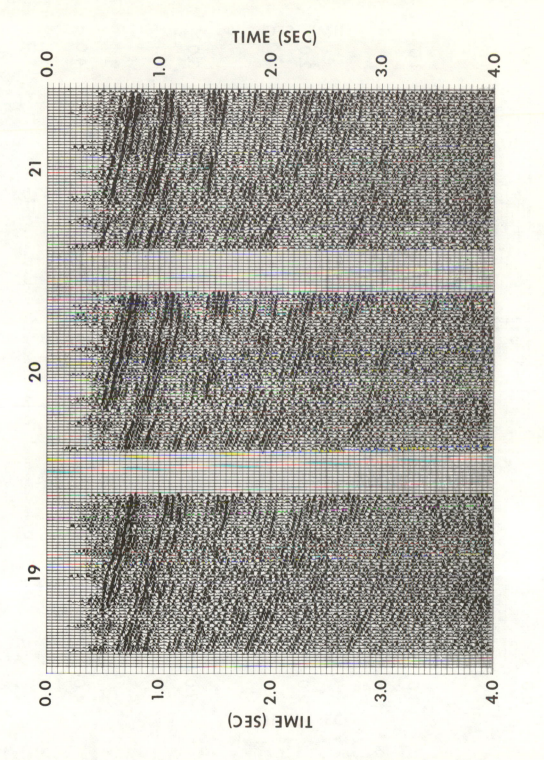

FIG. 34. California land 3-D prospect showing three north-south CDP stack sections 330 ft apart.

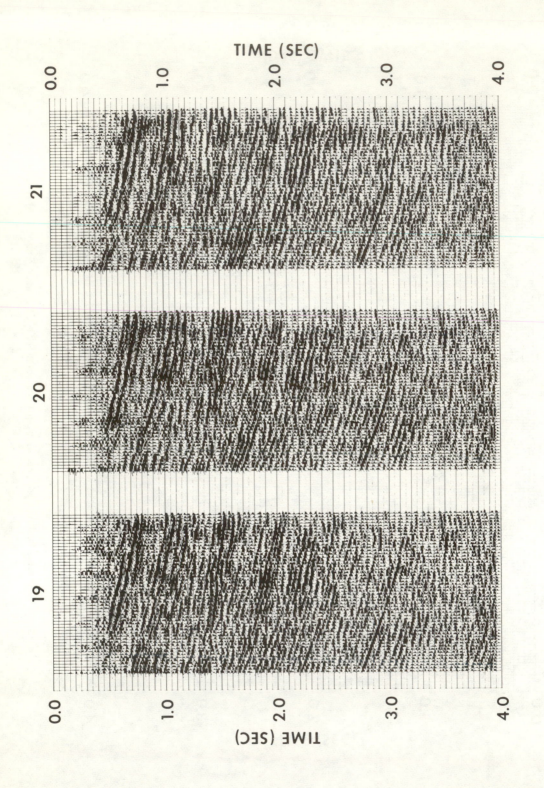

FIG. 35. 3-D migrated sections corresponding to the CDP stack sections in Figure 34.

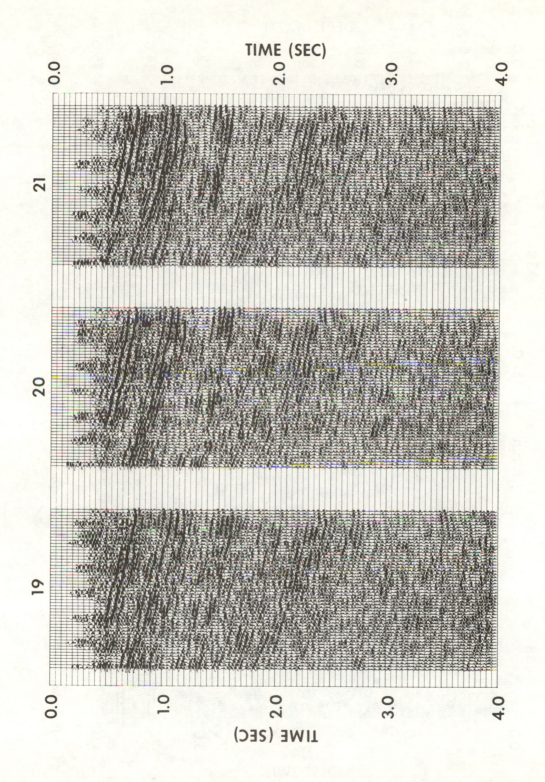

FIG. 36. 2-D migrated sections corresponding to the CDP stack in Figure 34.

DEPTH (KFT)

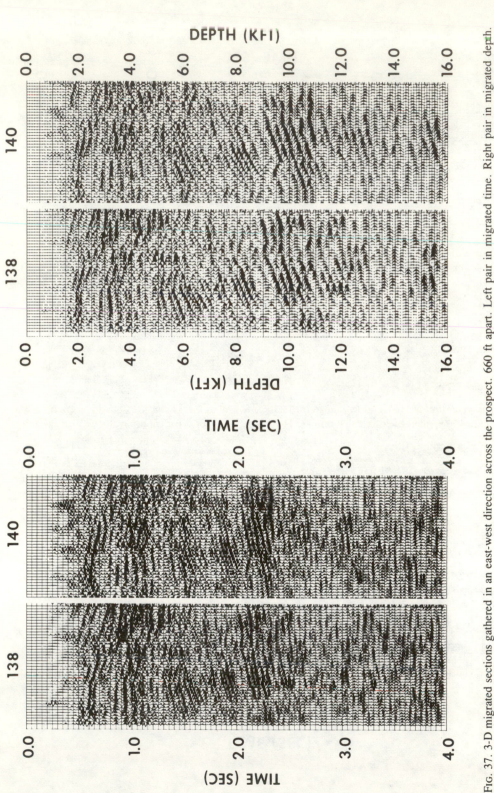

DEPTH (KFT)

TIME (SEC)

TIME (SEC)

Fig. 37. 3-D migrated sections gathered in an east-west direction across the prospect. 660 ft apart. Left pair in migrated time. Right pair in migrated depth.

—— 1976, Fundamentals of geophysical data processing: New York, McGraw-Hill Book Co., Inc.

Claerbout, Jon F., and Doherty, S. M., 1972, Downward continuation of moveout corrected seismograms: Geophysics, v. 37, p. 741–768.

Dohr, G. P., and Stiller, P. K., 1975, Migration velocity determination: Part II. Applications: Geophysics, v. 40, p. 6–16.

French, W. S., 1975, Computer migration of oblique seismic reflection profiles: Geophysics, v. 40, p. 961–980.

Goodman, J. W., 1968, Introduction to Fourier optics: New York, McGraw-Hill Book Co., Inc.

Hagedoorn, J. G., 1954, A process of seismic reflection interpretation: Geophys. Prosp., v. 2, p. 85–127.

Hilterman, F. J., 1970, Three-dimensional seismic modeling: Geophysics, v. 35, p. 1020–1037.

—— 1975, Amplitudes of seismic waves—a quick look: Geophysics, v. 40, p. 745–762.

Koehler, F., and Reilly, M.D., 1976, Interpretational benefits of wave equation migration: Presented at 46th International SEG meeting, October 27 in Houston.

Kuhn, M. J., and Alhilali, K. A., 1976, Weighting factors in the construction and reconstruction of acoustical wavefields: Geophysics, v. 42, p. 1183–1198.

Larner, K., and Hatton, L., 1976, Wave equation migration: two approaches: Presented at 46th International SEG meeting, October 24 in Houston.

Loewenthal, D., Lee, L., Robinson, R., and Sherwood, J., 1974, The wave equation applied to migration and water bottom multiples: Presented at 44th International SEG meeting, November 12 in Dallas.

Magnus, W., and Oberhettinger, F., 1954, Formulas and theorems for the functions of mathematical physics: New York, Chelsea Publishing Co.

Morse, P. M., and Feshback, H., 1953, Methods of theoretical physics, Parts I and II: New York, McGraw-Hill Book Co., Inc.

Musgrave, A. W., 1961, Wavefront charts and three-dimensional migration: Geophysics, v. 26, p. 738–753.

Sattlegger, J. W., 1975, Migration velocity determination: Part I. Philosophy: Geophysics, v. 40, p. 1–5.

Tegland, E. R., Bone, M. R., and Giles, B. F., 1976, 3-D high resolution data collection, processing and display: Presented at 46th International SEG meeting, October 27 in Houston.

Timoshin, Y. V., 1970, New possibilities for imagery: Soviet Physics—Acoustics, v. 15, p. 360–367.

Trorey, A. W., 1970, A simple theory for seismic diffractions: Geophysics, v. 35, p. 762–784.

Walter, W. C., and Peterson, R. A., 1976, Seismic imaging atlas 1976: United Geophysical Corp. publication.

APPENDIX

The system response function $H(k_x, k_y, \Delta z, \omega)$, which translates the scalar wave field across a slab of thickness $\Delta z = z - z_0$ in a constant velocity medium, is given by the 3-D Fourier transform of the convolutional operator given in equation (7). Therefore, we have

$$H(k_x, k_y, \Delta z, \omega) = -\frac{1}{2\pi} \frac{\partial}{\partial z} \iiint dx\,dy\,dt$$
$$\frac{\delta(t - r/C)}{r} e^{-i(\omega t + k_x x + k_y y)},$$

and integrating over t gives

$$H = -\frac{\partial}{\partial z} \int dy \left[\frac{1}{2\pi} \int dx \right.$$
$$\left. \frac{e^{-i\frac{\omega}{c}\sqrt{(z-z_0)^2 + x^2 + y^2}}}{\sqrt{(z-z_0)^2 + x^2 + y^2}} e^{-ik_x x} \right] e^{-ik_y y}.$$

The inner integral is a Hankel function of the second kind (Magnus and Oberhettinger, 1954) thus we are left with the following integral over y:

$$H = -\frac{\partial}{\partial z} \int dy - i\pi H_0^{(2)}\left(\sqrt{\left(\frac{\omega}{c}\right)^2 - k_x^2}\right.$$
$$\left. \sqrt{(z-z_0)^2 + y^2}\right) e^{-ik_y y}.$$

Now using the relation between cylindrical functions

$$H_0^{(2)} = J_0 - iN_0$$

where J_0 and N_0 are Bessel and Neumann functions, respectively, we substitute for $H_0^{(2)}$ in the y integral and evaluate the final transform as (Magnus and Oberhettinger, 1954)

$$H = \frac{\partial}{\partial z}\left[\frac{\sin(z - z_0)\sqrt{\left(\frac{\omega}{c}\right)^2 - k_x^2 - k_y^2}}{\sqrt{\left(\frac{\omega}{c}\right)^2 - k_x^2 - k_y^2}} \right.$$
$$\left. + i\frac{\cos(z - z_0)\sqrt{\left(\frac{\omega}{c}\right)^2 - k_x^2 - k_y^2}}{\sqrt{\left(\frac{\omega}{c}\right)^2 - k_x^2 - k_y^2}} \right]$$
$$= i\frac{\partial}{\partial z} \frac{e^{-i(z - z_0)\sqrt{\left(\frac{\omega}{c}\right)^2 - k_x^2 - k_y^2}}}{\sqrt{\left(\frac{\omega}{c}\right)^2 - k_x^2 - k_y^2}}$$
$$= e^{-i(z - z_0)\sqrt{\left(\frac{\omega}{c}\right)^2 - k_x^2 - k_y^2}}.$$

The two-dimensional transfer function [equation (11)] can be derived in a similar manner starting with the convolutional operator [equation (10)].

Reprinted from GEOPHYSICS, **43**, 1342-1351.

Wave equation migration
with the phase-shift method

JENÖ GAZDAG*

Accurate methods for the solution of the migration of zero-offset seismic records have been developed. The numerical operations are defined in the frequency domain. The source and recorder positions are lowered by means of a phase shift, or a rotation of the phase angle of the Fourier coefficients. For applications with laterally invariant velocities, the equations governing the migration process are solved very accurately by the phase-shift method. The partial differential equations considered include the 15 degree equation, as well as higher order approximations to the exact migration process. The most accurate migration is accomplished by using the asymptotic equation, whose dispersion relation is the same as that of the full wave equation for downward propagating waves. These equations, however, do not account for the reflection and transmission effects, multiples, or evanescent waves. For comparable accuracy, the present approach to migration is expected to be computationally more efficient than finite-difference methods in general.

INTRODUCTION

In recent years, migration methods based on the numerical solution of the wave equation have gained considerable acceptance. These migration techniques have their origin in the pioneering work of Claerbout. By defining the problem in a downward-moving coordinate system, Claerbout (1970, 1976) derived a simplified equation which lends itself to numerical treatment more conveniently than the full wave equation. This partial differential equation, which is often referred to as the 15 degree equation, has been solved so far by finite-difference methods.

More recently, Stolt (1978) used Fourier transform techniques for migration. Migration with Fourier transforms was also studied by Claerbout (1977) and Lynn (1977). In these studies, finite Fourier transforms are employed for obtaining a direct solution of the wave equation. Such direct solutions are known to exist for a large class of linear partial differential equations with constant coefficients. Thus, migration with these methods is limited to homogeneous media with constant-velocity function. In order to overcome this limitation, Stolt (1978) suggests coordinate transformations to cast the wave equation in velocity invariant form.

In this paper we develop solution methods for migration of seismic records in inhomogeneous media. This calls for the numerical solution of partial differential equations with variable coefficients. The numerical operations are defined in the frequency domain rather than in configuration space. The aim is to obtain the solution by operating on the Fourier coefficients of the seismic section. This permits us to formulate numerical procedures which are practically free from truncation errors (Gazdag, 1976). We find that the numerical algorithm formulated in the frequency domain is considerably simpler than any finite-difference method in configuration space. The reason for this is that the step-by-step process of lowering the source and the recorder positions is accomplished by implementing a phase change in the Fourier coefficients. Numerically, this is equivalent to a multiplication by a complex number of unit modulus. There is no stability condition associated with this operation. This means that the source and recorder positions can be lowered by any amount within one computational step.

In this paper we shall consider only migration of zero offset seismic data. In the following section, we set out the details of the phase shift method as applied

Manuscript received by the Editor July 25, 1977; revised manuscript received April 19, 1978.
* IBM Scientific Center, 1530 Page Mill Rd., Palo Alto, CA 94304.

to the 15 degree equation. The derivations apply to laterally homogeneous media. In the subsequent section, we extend this method to higher order equations. We obtain an asymptotic equation whose form is the same as the relativistic Schroedinger equation expressed in a downward-moving coordinate system. Next we attempt to generalize the phase shift method to laterally inhomogeneous media and discuss certain simplifications in the migration algorithm. The computations of the wave extrapolation and those of the inverse Fourier transformation with respect to the depth variable are combined into a single computational step. Finally, numerical results of migration examples are discussed.

THE PHASE SHIFT METHOD; SECOND-ORDER EQUATION

The zero offset seismic section $p(x, t, \tau)$ may be considered as a wave field measured at some specified depth from the surface of the earth. The variables x, t, and τ are the horizontal position, the two-way traveltime, and the two-way vertical traveltime, respectively. Computationally, the migration process can be regarded as a numerical approximation of the changes in the wave field as the sources and the recorders are moved downward into the earth. The seismic section $p(x, t, \tau = 0)$ recorded at the surface serves as the initial condition for the solution $p(x, t, \tau)$, the seismic section which would have been observed, had the sources and the recorders been positioned at depth τ. The subset $p(x, t = \tau, \tau)$ of all the computed seismic sections corresponds to the diffractor source distribution and provides the desired migrated section. This concept is illustrated by Claerbout (1976) in Figure (11-2-6), where the results are located along the diagonal of the (z, t) grid or, according to the present notation, the (τ, t) grid, as suggested by Loewenthal et al (1976).

We shall demonstrate the phase-shift method with the equation

$$p_{t\tau} = -\frac{v^2}{8} p_{xx}. \qquad (1)$$

This is a second-order approximation to the two-dimensional scalar wave equation written in a downward-moving coordinate system (Claerbout, 1970, 1976). It is also known as the 15 degree equation.

In order to keep the details of the derivation simple and tractable, we shall assume no lateral velocity variations, i.e., $v = v(\tau)$. Let the finite Fourier transform of p be defined as

$$P(k_x, \omega, \tau) = \frac{1}{4K\Omega} \sum\sum p(x, t, \tau) \cdot$$
$$\cdot \exp[-i(k_x x + \omega t)], \qquad (2)$$

in which

$$K = \pi/\Delta x \quad \text{and} \quad \Omega = \pi/\Delta t, \qquad (3)$$

where Δx and Δt are the grid spacings. The summation in (2) is carried out for all frequencies $|k_x| \leq K$ and $|\omega| \leq \Omega$.

In view of definition (2), the partial differential equation (1) expressed in the frequency domain becomes

$$P_\tau = \frac{-i \, v^2 k_x^2}{8\omega} P. \qquad (4)$$

The solution to (4) can be written in the following form

$$P(\tau + \Delta\tau) = P(\tau) \exp(-i\,\phi\,\Delta\tau), \qquad (5)$$

in which

$$\phi = \frac{k_x^2}{8\,\omega\,\Delta\tau} \int_\tau^{\tau+\Delta\tau} v^2 \, d\tau. \qquad (6)$$

If we define v_{rms} as the root-mean-square value of the velocity averaged between the interval τ and $\tau + \Delta\tau$, i.e.,

$$v_{\text{rms}}^2 = \frac{1}{\Delta\tau} \int_\tau^{\tau+\Delta\tau} v^2 \, d\tau, \qquad (7)$$

and let

$$m = 2\omega/v_{\text{rms}} \qquad (8)$$

for this interval of integration, then we can write ϕ in the simple form

$$\phi = \frac{\omega \, k_x^2}{2m^2}. \qquad (9)$$

Since solution (5) depends on the rms value of the velocity within any interval of integration, the velocity v in (4) can be replaced by its rms value v_{rms}. With the help of (8), we can write (4) in a somewhat more convenient form,

$$P_\tau = \frac{-i \, \omega \, k_x^2}{2m^2} P. \qquad (10)$$

The desired migrated section is given by the subset $p(x, t = \tau, \tau)$ of all the computed seismic sections. Therefore, after each $\Delta\tau$ step, we compute

$$P(x, t = \tau, \tau) = \sum\sum P(k_x, \omega, \tau) \cdot$$
$$\cdot \exp[i(k_x x + \omega\tau)]. \qquad (11)$$

It is emphasized that the phase shift $\phi\Delta\tau$ required for the extrapolation of a given wave component from τ to $\tau + \Delta\tau$ depends on the rms velocity across this layer of $\Delta\tau$ thickness.

Let NX and NT be the number of grid points along the x- and t-axes, respectively, over which $p(x, t, \tau)$ is defined. Then $P(k_x, \omega, \tau)$ has $NT \cdot NX/2$ complex data points. Approximately one-half of these Fourier coefficients can be set to zero, since those waves with

$$m^2 \le k_x^2 \qquad (12)$$

correspond to nonpropagating waves. One-half of the Fourier coefficients are nonphysical if $\Delta x = v_{\mathrm{rms}}\Delta\tau/2$. Normally, Δx is greater than that, so the number of the deleted Fourier coefficients represent a smaller fraction than one-half of the total. However, if the maximum dip in the data is less than 90 degrees, the number of coefficients which need not be considered increases. Therefore, the assumption that only one-half of all the Fourier coefficients need to be included in the computation is a reasonable one. Thus, the computations in (5) require $NX \cdot NT/4$ complex multiplications. The computations in (11) for one t value require approximately $NX \cdot NT/2$ additions as well as $NX \cdot NT/2$ multiplications. If $\Delta\tau \ne \Delta t$, for example, $\Delta\tau = r\Delta t$, then at each τ level (11) must be computed for r different t values. The total operation count for advancing one $\Delta\tau$ step [equation (5)] and reconverting the data into the (x, t) domain [equation (11)] requires

$$(r + 2)NX \cdot NT/2 \quad \text{multiplications} \quad (13)$$

and

$$(r + 1)NX \cdot NT/2 \quad \text{additions.} \qquad (14)$$

These figures do not include the computation of the complex multipliers and trigonometric functions, etc.

The remarkable advantage of solving (1) in the frequency domain is that the numerical integration of the Fourier transform (10) is reduced to a multiplication of P by a complex number of unit modulus. This requires less computing than in finite-difference methods. This is seen immediately when we consider the $\Delta\tau = \Delta t$ case, i.e., when $r = 1$ in (13) and (14). This means that one $\Delta\tau$ step requires approximately 1.5 multiplications and only 1 addition per data point. On the other hand, a simple (explicit) finite-difference scheme for (1) would have to involve no less than six neighboring grid points. Such a finite-difference expression would require about 2 multiplications and 5 additions per grid point. If $\Delta\tau$ is much

greater than Δt, i.e., $r \gg 1$, the operation count takes a turn in favor of the finite-difference method. In this case, however, the accuracy of migration with the finite-difference scheme cannot be expected to compare favorably with the accuracy of the phase-shift method. Moreover, the phase-shift method is equally suitable for solving higher-order equations to be discussed in the following sections, which could not be accomplished satisfactorily with finite-difference methods defined in terms of the variables x and t. Another important property of the phase-shift method is accuracy. The numerical procedure is free of truncation errors. Moreover, there is no stability condition imposed on the magnitude of $\Delta\tau$.

Another type of error which deserves attention is related to truncation in the frequency-wavenumber domain. It is known that the highest frequency and wavenumber which can be represented on a grid of spacings Δt and Δx are given by $\pi/\Delta t$ and $\pi/\Delta x$, respectively. This upper limit is commonly referred to as the Nyquist frequency or the "folding" frequency. When a continuous signal, say $\exp(i\omega' t)$ with $\omega' > \pi/\Delta t$, is sampled at a rate Δt, then the digitized data cannot be distinguished from the digitized version of one of its aliases, e.g., $\exp[i(\omega' - 2\pi/\Delta t)t]$. We are concerned here about aliasing or folding errors resulting only from the numerical solution method of (1). Therefore, we shall assume that the grid spacings Δt and Δx are sufficiently small for representing the unmigrated section in all important detail. In other words, the section is specified completely by a finite number of Fourier coefficients associated with the computational grid. The question to be answered is this: Are there any Fourier modes produced by the migration process whose frequencies exceed the folding frequencies $\pi/\Delta t$ and $\pi/\Delta x$? From (11), we obtain migrated results for some depth τ from a set of Fourier coefficients, which were subjected to phase changes whose magnitude is proportional to the same variable τ. This results in a frequency change in the migrated section corresponding to the dispersion relation of (1) (Claerbout, 1976),

$$k_\tau = \omega(1 - v^2 k_x^2/8\omega^2), \qquad (15)$$

where k_τ is the frequency (wavenumber) associated with the τ variable. From (15) and (12) we see that $k_\tau \le \omega$. This implies that the migrated results contain no higher frequencies than the unmigrated section. Thus, if both sections (before and after migration) are represented on the same grid, then there are no apparent aliasing or "folding" errors associated with the migration process.

FOURTH-ORDER AND ASYMPTOTIC EQUATIONS

Equation (1), whose solution we have considered so far, is characterized by the dispersion relation expressed in (15). This is a second-order approximation to the dispersion relation of the full wave equation, which is given by

$$k_\tau = \omega \left(1 - \frac{v^2 k_x^2}{4\omega^2} \right)^{1/2}. \qquad (16)$$

The fourth-order approximation to (16) is

$$k_\tau = \omega \left(1 - \frac{v^2 k_x^2}{8\omega^2} - \frac{v^4 k_x^4}{128\omega^4} \right). \qquad (17)$$

The partial differential equation with this dispersion relation is

$$P_{tt\tau} = -\frac{v^2}{8} P_{ttxx} - \frac{v^4}{128} P_{xxxx}. \qquad (18)$$

This equation corresponds to equation (10-3-17) of Claerbout (1976). We note that t represents the two-way traveltime, and that τ is the two-way vertical traveltime. The Fourier transform of (18) can be written as

$$P_\tau = \left[-i\omega \left(\frac{v^2 k_x^2}{8\omega^2} + \frac{v^4 k_x^4}{128\omega^4} \right) \right] P, \qquad (19)$$

which corresponds to equation (10-3-17) of Claerbout (1976). The numerical solution of (19) is as described for the second-order case, except for the amount of phase shift. If the velocity v can be regarded as some constant over the interval $(\tau, \tau + \Delta\tau)$, then ϕ is calculated for this layer from

$$\phi = \omega \left[\frac{v^2 k_x^2}{8\omega^2} + \frac{v^4 k_x^4}{128\omega^4} \right]. \qquad (20)$$

However, if there are significant velocity variations within the interval under consideration, then the correct expression for ϕ is

$$\phi = \frac{\omega}{\Delta\tau} \int_\tau^{\tau+\Delta\tau} \left[\frac{v^2 k_x^2}{8\omega^2} + \frac{v^4 k_x^4}{128\omega^4} \right] d\tau. \qquad (21)$$

The solution to (19) is expressed by (5) with ϕ given by (20) or (21).

If additional higher-order terms are used in computing ϕ, the solution (5) becomes a higher-order approximation to that of the full wave equation. With the help of the Taylor series approximation to the square root, which is

$$(1 - z^2)^{1/2} = 1 - \frac{z^2}{2} - \frac{z^4}{8} - \frac{z^6}{16} - \cdots, \qquad (22)$$

we can write the exact expression for ϕ in the following form

$$\phi = \omega \left[1 - \left(1 - \frac{v^2 k_x^2}{4\omega^2} \right)^{1/2} \right], \qquad (23)$$

providing that v is constant within the interval $(\tau, \tau + \Delta\tau)$ under consideration. If this is not the case, ϕ must be determined from the expression

$$\phi = \frac{\omega}{\Delta\tau} \int_\tau^{\tau+\Delta\tau} \left[1 - \left(1 - \frac{v^2 k_x^2}{4\omega^2} \right)^{1/2} \right] d\tau. \qquad (24)$$

When (24) is used to compute the rate of phase change ϕ, then (5) yields the numerical solution of the equation

$$P_\tau = i\omega \left[\left(1 - \frac{v^2 k_x^2}{4\omega^2} \right)^{1/2} - 1 \right] P, \qquad (25)$$

which is expressed in a downward-moving coordinate system. Equation (25), expressed in a stationary coordinate system, is

$$P_\tau = i\omega \left[1 - \frac{v^2 k_x^2}{4\omega^2} \right]^{1/2} P, \qquad (26)$$

which is known as the relativistic Schroedinger equation (Claerbout, 1976, p. 202).

The solution to the asymptotic equation (25) is obtained from (5) when the correct phase (24) is used. These results are characterized by a dispersion relation which is very close to that of the full wave equation. Naturally, the exact representation of a section is limited in that only a finite number of waves can be represented over any computational grid. This limitation is true for any computational method. An advantage of the present approach to migration (over finite-difference methods) is that all the wave components which are represented on the grid are extrapolated correctly, without truncation errors.

The frequency domain representation of a seismic section implies periodic boundary conditions for the partial differential equations under consideration. Consequently, there is a possibility for some objects to migrate across the boundaries of the computational domain. Such phenomena produce incorrect results in the neighborhood of these boundaries. It is important to note, however, that migrated results near the boundaries are always unreliable, and most often incorrect, with no regard to the boundary conditions and the numerical method being used. This can be seen by considering that accurate migration results near the boundary would require information from both sides of that boundary, which is not available. The effects of periodicity on the migrated section can be eliminated in practice by choosing the computational domain somewhat larger than the actual seismic section and padding the extra space with zeros.

LATERALLY INHOMOGENEOUS MEDIA

We have considered so far laterally homogeneous media only. In such cases the velocity is independent of the horizontal variable x. The numerical approaches described in this paper are particualrly suitable for this class of migration problems for two reasons. First, (25) and (26) represent a significantly better description of the migration process than lower-order equations do. The second reason is that the phase-shift method gives very accurate results within the limitations related to representing the seismic section as a finite set of double Fourier series, whose effects we have discussed above. When the medium is laterally inhomogeneous, we lose both of these important advantages. Therefore, a great deal of caution is required when dealing with lateral velocity variations. Notwithstanding these limitations, we can still expect to obtain practical results through simple modification of the phase-shift method based on judicious application of some physical principles.

Perhaps one of the most important examples of strong horizontal velocity variation is encountered at the ocean bottom which is inclined at some angle from the horizontal. In order to traverse such an interface, the source and recorder positions (τ) must be advanced downward at different rates depending on their horizontal positions. In order to incorporate this x-dependence into the solution, $P(k_x, \omega, \tau)$ must be Fourier transformed with respect to k_x in each $\Delta\tau$ step. Naturally, this requires additional computation. Alternatively, a different method (say finite-difference) could be used for traversing such critical regions. Even though the treatment of the ocean bottom requires special consideration, this extra cost is most likely offset by the very effective processing of data in reaching the bottom. It requires only one $\Delta\tau$ step for lowering the source and recorder positions down to any depth in the water itself.

One way to account for weak horizontal velocity variations is by undermigrating or overmigrating selectively, using some average velocity. Let us Fourier transform both sides of (1) with respect to the variable t only, which gives

$$p_\tau = i\,\frac{v^2}{8\omega}\,p_{xx}, \qquad (27)$$

whose solution can be expressed as

$$p(x, \omega, \tau + \Delta\tau) = p(x, \omega, \tau)$$
$$+ \int_\tau^{\tau+\Delta\tau} \frac{i\,v^2(x, \tau)}{8\omega}\,p_{xx}(x,\omega,\tau)\,d\tau. \qquad (28)$$

The first-order approximation of this integral can be written as

$$p(x, \omega, \tau + \Delta\tau) \cong p(x, \omega, \tau)$$
$$+ \frac{i\,v^2(x, \tau)}{8\omega}\,p_{xx}(x, \omega, \tau)\,\Delta\tau. \qquad (29)$$

Our goal is to remove the x-dependence from the velocity variable and to account for it in the variable representing the limit of the integration. We do this with the help of the lateral rms velocity $\bar{v}$, which is defined as

$$\bar{v}^2(\tau) = \frac{1}{X}\int_0^X v^2(x, \tau)\,dx, \qquad (30)$$

in which X is the horizontal length of the seismic section under consideration. Since we wish to use $\bar{v}$ as the migration velocity, rather than v, we define a new variable

$$\xi = \int_0^\tau (v^2/\bar{v}^2)\,d\tau, \qquad (31)$$

which will also set the upper limit of the definite integral in conjunction with $\bar{v}$. Substituting the derivative of (31), $d\xi/d\tau = v^2/\bar{v}^2$, into (29) we obtain

$$p(x, \omega, \tau + \Delta\tau) \cong p(x, \omega, \tau)$$
$$+ \frac{i\,\bar{v}^2}{8\omega}\,p_{xx}(x, \omega, \tau)\,\Delta\xi. \qquad (32)$$

In (32) the $v^2/\bar{v}^2$ ratio is absorbed in the increment $\Delta\xi = (d\xi/d\tau)\,\Delta\tau$, or equivalently, $\Delta\xi = (v^2/\bar{v}^2)\,\Delta\tau$. When (32) is written in the form of a definite integral, i.e., as

$$p(x, \omega, \tau + \Delta\tau) \cong p(x, \omega, \tau)$$
$$+ \int_\tau^{\tau+\Delta\xi} \frac{i\,\bar{v}^2}{8\omega}\,p_{xx}(x, \omega, \tau)\,d\tau, \qquad (33)$$

the increment $\Delta\xi$ appears in the upper limit of the integral rather than $\Delta\tau$ as in the case of (28). If we take the lower limit of the integral as zero, i.e., $\tau = 0$, and the depth of interest expressed by the two-way traveltime as τ', i.e., $\tau' = \tau + \Delta\tau$ in (33), we obtain the expression

$$p(x, \omega, \tau') \cong p(x, \omega, 0)$$
$$+ \int_0^{\xi'} \frac{i\,\bar{v}^2}{8\omega}\,p_{xx}(x, \omega, \tau)\,d\tau, \qquad (34)$$

in which the correspondence between ξ' and τ' is expressed in (31). The physical significance of (34)

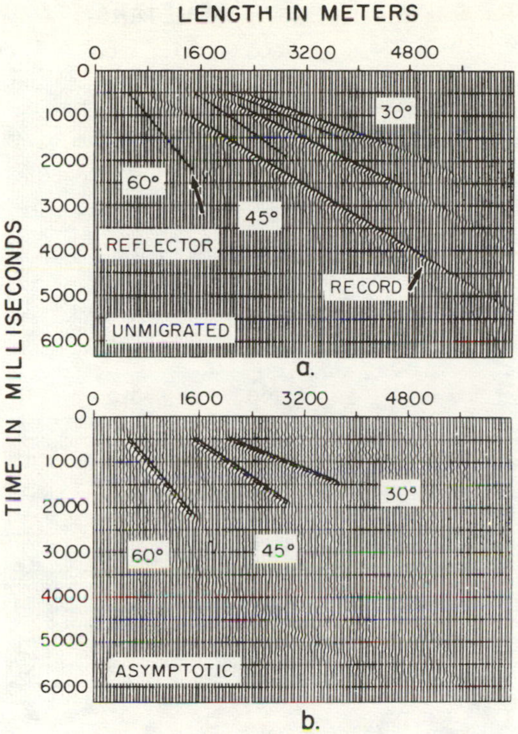
LENGTH IN METERS
0
1600
3200
4800
TIME IN MILLISECONDS
0
1000
2000
3000
4000
5000
6000
30°
60°
45°
REFLECTOR
RECORD
UNMIGRATED
a.
30°
60°
45°
ASYMPTOTIC
b.

FIG. 1. (a) Synthetic zero-offset record section of three reflectors directed at 30, 45, and 60 degrees. (b) Migration of (a) using the asymptotic equation.

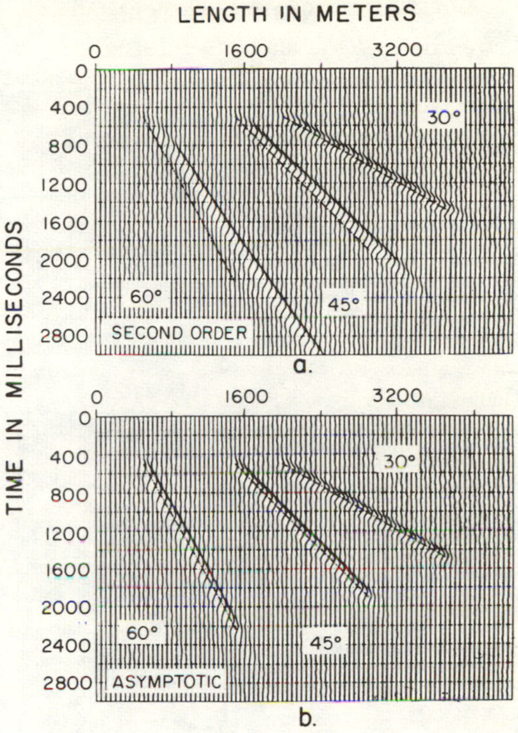
LENGTH IN METERS
0
1600
3200
TIME IN MILLISECONDS
0
400
800
1200
1600
2000
2400
2800
30°
60°
45°
SECOND ORDER
a.
30°
60°
45°
ASYMPTOTIC
b.

FIG. 2. Magnified views of the section shown in Figure 1a after migration using (a) the second-order equation, and (b) the asymptotic equation.

is that when v^2 is replaced by its lateral mean value $\bar{v}^2$, the migrated data at depth τ' is given by $p(x,\ t = \tau',\ \tau = \xi')$ rather than by $p(x,\ t = \tau',\ \tau = \tau')$ as before. The difference between ξ' and τ' represents the amount of overmigration ($\xi' > \tau'$) or undermigration ($\xi' < \tau'$) required in order to account for the difference between v^2 and its lateral mean $\bar{v}^2$.

PROGRAMMING CONSIDERATIONS FOR AN EFFICIENT ALGORITHM

As discussed above, the present approach to migration consists of (1) the extrapolation of the wave downward by operating on the Fourier coefficients, followed by (2) the inverse Fourier transformation of the correctly migrated data. These two operations expressed by (5) and (11) can be combined into a single operation. The solution to (4) can be expressed with reference to P at $\tau = 0$, i.e.,

$$P(\tau) = P(0)\,\exp(-i\psi\tau), \tag{35}$$

in which

$$\psi = \frac{k_x^2 v_{\text{rms}}^2}{8\omega}, \tag{36}$$

where

$$v_{\text{rms}}^2 = \frac{1}{\tau}\int_0^\tau v^2\,d\tau. \tag{37}$$

By letting

$$m = 2\omega/v_{\text{rms}}, \tag{38}$$

we obtain

$$\psi = \frac{\omega k_x^2}{2m^2}. \tag{39}$$

Notice that (39) is the same expression as (9). The difference is in the definition of the rms velocity, (7) and (37). By substituting (35) into (11), we obtain

$$p(x,\ t = \tau,\ \tau) = \sum\sum P(k_x,\ \omega,\ 0)\cdot \cdot \exp\{i[k_x x + (\omega - \psi)\tau]\}. \tag{40}$$

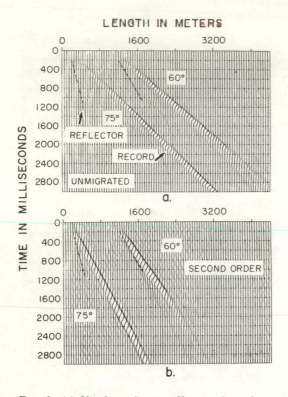

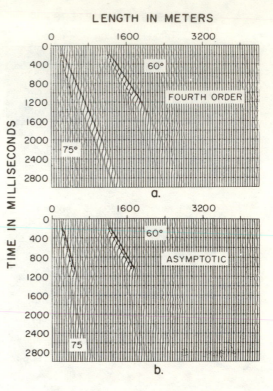

FIG. 3. (a) Unmigrated zero offset section of two reflectors inclined at 60 and 75 degrees. (b) Migration of (a) using the second-order equation.

FIG. 4. Migration of the synthetic record section shown in Figure 3a using (a) the fourth-order equation, and (b) the asymptotic equation.

By making use of (39), we can write (40) as

$$p(x, t = \tau, \tau) = \sum \sum P(k_x, \omega, 0) \cdot$$
$$\cdot \exp \left\{ i \left[k_x x + \left(1 - \frac{k_x^2}{2m^2} \right) \omega \tau \right] \right\}. \quad (41)$$

Equations (40) and (41) give results which are solutions to the second-order approximations given by (1). Higher-order solutions are obtained if instead of (39), ψ is computed from a higher-order approximation. In the case of the asymptotic equation, ψ is given by

$$\psi = \frac{\omega}{\tau} \int_0^\tau \left[1 - \left(1 - \frac{v^2 k_x^2}{4\omega^2} \right)^{1/2} \right] d\tau, \quad (42)$$

whose substitution into (40) gives

$$p(x, t = \tau, \tau) = \sum \sum P(k_x, \omega, 0) \cdot$$
$$\cdot \exp\{ i [k_x x + \omega \cdot$$
$$\cdot \int_0^\tau (1 - v^2 k_x^2 / 4\omega^2)^{1/2} d\tau] \}. \quad (43)$$

For each τ value, (41) or (43) represent approximately $NX \cdot NT/2$ additions and $NX \cdot NT/2$ multiplications, in addition to the operations required to compute the exponent itself. The computational complexity of (43) is roughly that of (11). However, in this direct approach the extra complex multiply (5) has been eliminated. Therefore, this direct method requires less than one-half of the operations necessary for the two-step procedure outlined previously.

In the special case when v is constant, (43) becomes,

$$p(x, t = \tau, \tau) = \sum \sum P(k_x, \omega, 0) \cdot$$
$$\cdot \exp[i(k_x x + k_\tau \tau)], \quad (44)$$

where k_τ is the frequency variable with respect to the variable τ, given by

$$k_\tau = \omega \left[1 - \frac{v^2 k_x^2}{4\omega^2} \right]^{1/2}. \quad (45)$$

Equation (44) is essentially the same as equation (50) of Stolt (1978). We note again that $k_\tau \leq \omega$, which means that the problem can be solved, in principle at least, without errors due to aliasing.

RESULTS

We tested the phase-shift method on synthetic zero-offset record sections. The record sections were generated by implementing the theory of Trorey (1970) for seismic diffractions. All numerical results were obtained by computing on a 128×128 grid, using $\Delta x = 50$ m, $\Delta t = \Delta \tau = 50$ msec. The migration velocity was constant, $v = 2000$ m/sec. When working with this velocity value, the two-way vertical traveltime τ measured in msec is numerically equivalent to the depth expressed in meters.

The migration examples represent results obtained from three different equations, which are solved numerically by the phase-shift method. The second-order approximation corresponds to the well-known 15 degree equation (1). Migration by means of (18) or (19) is called the fourth-order approximation. As the number of higher order terms (22) is increased without limit, the resulting approximation approaches equation (25) asymptotically. Therefore,

migration by (25) is referred to as the asymptotic approximation. The asymptotic equation has a dispersion relation which is identical to the full wave equation for downward-traveling waves in constant-velocity media. Therefore, it is clearly superior to lower-order approximations. This is demonstrated by our results beyond any doubt, particularly for steeper dips.

For the correct interpretation of the figures, it is important to emphasize that the record sections were *not* subjected to any kind of smoothing before migration. A synthetic trace is a sequence of pulses of very short duration. The time at which these pulses occur is usually between two grid points Δt apart. The pulse is represented on the grid by sharing its value between the two nearest grid points. Thus, the synthetic record contains a great deal of high-frequency components limited only by the Nyquist frequency. We found that we did not need to work with some kind of wavelets, whose high-frequency content is considerably less than that of a pulse of

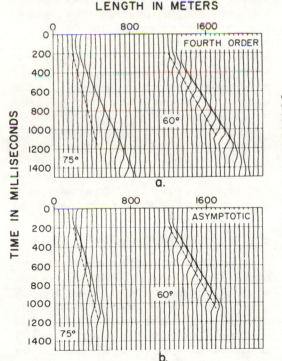

FIG. 5. Magnified views of the migrated sections shown in Figure 4.

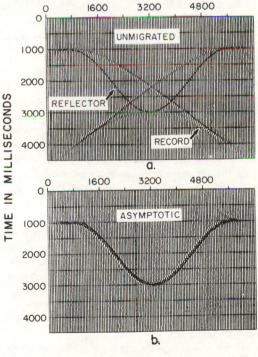

FIG. 6. (a) Synthetic zero-offset record section of a dipping reflector. (b) Migration of (a) using the asymptotic equation.

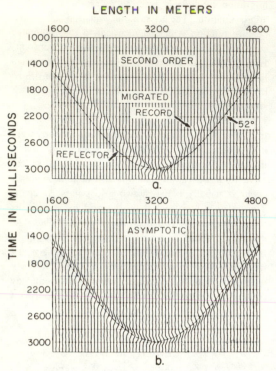

FIG. 7. Magnified views of the section shown in Figure 6 after migration using (a) the second-order equation, and (b) the asymptotic equation.

approximately Δt width. This is the consequence of migrating all waves accurately.

Figure 1 shows the synthetic zero-offset record section of three plane reflectors and the migrated section using equation (25). The migrated section using the second-order approximation is shown in Figure 2a. It is remarkable that the "15 degree equation" results in a rather good migration of the reflector directed at 30 degrees. This leads us to believe that this equation might have been called the "30 degree equation" if accurate numerical methods had been used initially for solving it. While the quality of migration with the 15 degree equation deteriorates for reflectors with higher dips, all reflectors are migrated properly by the asymptotic expression (25) as shown in Figure 2b (magnified view of Figure 1b).

We have also experimented with reflectors up to 75 degrees. The record section and the migrated sections processed by the three different equations are shown in Figures 3, 4, and 5. The reflectors had to be short in order to keep the record section within

the computational domain whose extent is 6400 × 6400 in units of m and msec. None of the lower-order approximations can migrate the 75 degree reflector with acceptable accuracy. With the asymptotic equation, however, we obtain good migration as shown in Figures 4b and 5b. This is quite remarkable if we consider that the 75 degree reflector spans only six traces.

We notice that the migrated records appear to spread out over several Δt lengths. This could suggest some shortcoming of the numerical methods. This is not the case, however. The reason for the widening of the migrated records with increasing angle is related to the resolution of the unmigrated data. The synthetic record section defined on a grid can only represent data whose spread (thickness) is at least the grid size Δt (or Δx). In practice, this thickness is usually wider than Δt due to the sharing of the pulse between two neighboring grid points. When these data are migrated, even with the best possible method, it will retain this "thickness." Thus, what we can expect to obtain from the migration is a reflector whose thickness is roughly $2\Delta t$. The cross-section of such reflector along the τ axis is $(1/\cos \alpha)$ times wider. For $\alpha = 60$ and 75 degrees, this corresponds to $4\Delta t$ and $7.7\Delta t$, respectively. This is approximately what we observe in Figure 5b. This broadening in τ can also be viewed as the corollary to the dispersion relation (16). Since $k_\tau \le \omega$, the frequency spectrum of the migrated section is narrower (in k_τ) than that of the unmigrated section (in the variable ω). The narrower frequency bandwidth implies broadening in τ, which is in complete agreement with our observation. Figures 6 and 7 show migration results of a dipping reflector consisting of 18 plane segments. The reflector is a piecewise linear approximation to a sinusoid with maximum dip of 52 degrees. For this synthetic section, Figure 7 provides a comparison of the migration accuracy of the 15 degree equation and that of the asymptotic expression.

CONCLUSIONS

We have described numerical methods for wave equation migration based on Fourier transform techniques. The algorithm is defined in the frequency domain, rather than in configuration space. The computations are simple, since the lowering of the source and recorder positions is implemented by incrementing the arguments of the complex Fourier coefficients. This amounts to a complex multiplication which is more economical than finite-difference methods. For laterally invariant migration velocities, the partial differential equations can be solved with high accuracy. The equations under consideration do not account for the reflection coefficients or evanescent waves. The most important result of this paper is that the phase-shift method can be used to solve the relativistic Schroedinger equation numerically, whose dispersion relation is identical to that of the wave equation for downward-traveling waves in constant velocity media. The phase shift method was tested on synthetic zero offset records, including reflectors with 75 degree dip, with excellent results.

ACKNOWLEDGMENTS

I wish to thank Prof. Jon F. Claerbout, Dr. Robert H. Stolt, and Dr. Ralph A. Wiggins for their valuable suggestions and comments.

REFERENCES

Claerbout, J. F., 1970, Coarse grid calculations of waves in inhomogeneous media with application to delineation of complicated seismic structure: Geophysics, v. 35, p. 407–418.
————— 1976, Fundamentals of geophysical data processing: New York, McGraw-Hill Book Co., Inc., Chaps. 10, 11.
————— 1977, Migration with Fourier transforms: Stanford Exploration Proj., Rep. no. 11, p. 3–5.
Gazdag, J., 1976, Time-differencing schemes and transform methods: J. Comput. Phys., v. 20, p. 196–207.
Loewenthal, D., Lu, L., Roberson, R., and Sherwood, J., 1976, The wave equation applied to migration: Geophys. Pros., v. 24, p. 380–399.
Lynn, W., 1977, Implementing F-K migration and diffraction: Stanford Exploration Proj., Rep. no. 11, p. 9–28.
Stolt, R. H., 1978, Migration by Fourier transform: Geophysics, v. 43, p. 23–48.
Trorey, A. W., 1970, A simple theory for seismic diffractions: Geophysics, v. 35, p. 762–784.

Reprinted from *Geophysical Prospecting*, **28**, 60-70.

WAVE EQUATION MIGRATION WITH THE ACCURATE SPACE DERIVATIVE METHOD*

J. GAZDAG**

ABSTRACT

GAZDAG J. 1980, Wave Equation Migration with the Accurate Space Derivative Method, Geophysical Prospecting 28, 60–70.

A stacked seismic section represents a wave-field recorded at regularly spaced points on the surface. The seismic migration process transforms this recorded data into a reflectivity display. In recent years, Jon F. Claerbout and his co-workers developed migration techniques based on the numerical approximation of the wave equation by finite difference methods. This paper describes an alternative method, termed ASD (for Accurate Space Derivative), and its application to the wave equation migration problem. In this approach to the numerical solution of partial differential equations, partial derivatives are computed by finite Fourier transform methods. This migration method can accommodate media with vertical as well as horizontal velocity variations.

INTRODUCTION

This paper presents a mathematical formulation of a migration scheme based on the ASD (for Accurate Space Derivative) method (Gazdag 1973, 1976). An important feature of this numerical approach is that instead of using some finite difference expressions, we compute the space derivative terms with very high accuracy by Fourier transform methods. The main source of error in the solution of initial value problems by finite difference methods is that the partial derivatives are approximated by expressions involving only few grid points. While most difference operators give good results for long waves, their accuracy is often inadequate for shorter wavelengths that are of importance to exploration seismologists. Finite difference operators defined over a

* Paper read at the 39th meeting of the European Association of Exploration Geophysicists., Zagreb, June 1977. Received October 1978.

** IBM Scientific Center, 1530 Page Mill Road, Palo Alto, California 94304.

small number of grid points cannot account accurately for the derivatives of all possible waves supported by the computational grid. The computation of the partial derivatives by finite Fourier transform methods permits, in principle at least, the accurate migration of all the waves which can be represented on the computational grid without ambiguity. Therefore, the truncation errors resulting from the ASD method, particularly for short waves, can be significantly smaller than in the case of finite difference methods. Another benefit that we derive from computing with transform methods is that it permits us to restrict ourselves to a given set of wave-numbers k_x depending on the magnitude of the temporal frequency ω. This way we can avoid processing data which correspond to non-physical waves.

Claerbout (1976) uses moving coordinate systems for deriving the migration equations. His goal is to separate downgoing waves from upcoming waves. Here we adopt a somewhat different approach. Upcoming and downgoing waves differ from each other in the sign of the first time derivative of the wave field, which in turn is related to the plus or minus sign of the dispersion relation. For homogeneous media, we obtain the migration equation through a simple derivation in a stationary coordinate system. For inhomogeneous media—including lateral velocity variations—we approximate the ideal solution by an equation resulting from series expansion. Since we always work with ω rather than t as one of the independent variables, we would not benefit at all from using moving reference frames. We find, furthermore, that the formulation of the problem in a stationary coordinate system helps to preserve clarity.

First we review the basic theory of migration for simple cases. Subsequently we generalize these ideas to accommodate lateral velocity variations. In the latter part of the paper we present the numerical procedure followed by the discussion of migration results.

Background

Consider the acoustic wave equation

$$\frac{\partial^2 p}{\partial t^2} = v^2 \left(\frac{\partial^2 p}{\partial x^2} + \frac{\partial^2 p}{\partial z^2} \right), \tag{1}$$

where p is the pressure, t is time, and x and z are the horizontal and vertical distances, respectively. For constant velocity v the solution to (1) can be expressed as

$$p(x, z, t) = \sum_{k_x} \sum_{k_z} P(k_x, k_z, t = 0) \exp\left[i(k_x x + k_z z + \omega t)\right], \tag{2}$$

where k_x and k_z are the wave-numbers—or spatial frequencies—whereas ω is the temporal frequency. For each wave vector (k_x, k_z) (2) has two solutions, corresponding to the two values of the dispersion relation

$$\omega = \pm v(k_x^2 + k_z^2)^{1/2}. \tag{3}$$

Waves with negative ω values travel in the direction of the wave vector (k_x, k_z). Wave components whose frequencies are positive travel in the opposite direction. If, for example, the solution is restricted to frequencies

$$\omega = vk_z[1 + (k_x/k_z)^2]^{1/2}, \tag{4}$$

then (2) represents waves whose direction of propagation is within plus and minus 90° about the negative z axis. In the coordinate system used in fig. 1,

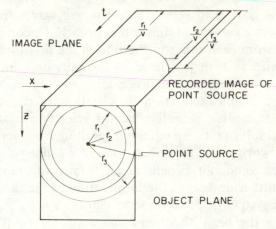

Fig. 1. Schematic of a coordinate system for migration.

(2) and (4) describe upward moving waves. In a homogeneous medium the solution satisfies (4) provided that we choose the second initial condition to (1) as

$$\frac{\partial P}{\partial t} = ivk_z[1 + (k_x/k_z)^2]^{1/2}P. \tag{5}$$

So long as we deal with homogeneous media, we may consider (5) as the necessary initial condition to the full wave equation (1), or equivalently, we may regard (5) as the equation describing upward propagating waves. Each point of view is perfectly valid.

By setting $z = 0$ in (2) we obtain the wave recorded at the surface as

$$p(x, z = 0, t) = \sum_{k_x} \sum_{k_z} P(k_x, k_z, t = 0) \exp[i(k_x x + \omega t)]. \tag{6}$$

These data, recorded at the image plane of fig. 1, represent the zero offset section.

In this paper, we assume that the variable t represents the one-way travel time, as opposed to the two-way travel time concept used with common depth point stacks. In view of this definition, the objective of migration is to reconstruct the wave field that existed in the medium at $t = 0$. This is equivalent to finding the initial wave field $p(x, z, t = 0)$ from the wave field $p(x, z = 0, t)$ observed at the surface. To obtain the desired expression, first we Fourier transform (6) with respect to the variables x and t. This results in

$$P(k_x, z = 0, \omega) = P(k_x, k_z, t = 0). \tag{7}$$

For $t = 0$, the substitution of (4) and (7) into (2) gives

$$p(x, z, t = 0) = \sum_{k_x} \sum_{\omega} P(k_x, z = 0, \omega) \exp \left\{ i \left[k_x x + \frac{\omega}{v} \left(1 - \left(\frac{k_x v}{\omega} \right)^2 \right)^{1/2} z \right] \right\}. \tag{8}$$

Equation (8) corresponds to equation (50) of Stolt (1978). In his method the Fourier coefficients are assigned modified frequencies before inverse transformation. The frequency change is proportional to the square root term in (8). Migration with this method is limited to homogeneous media with constant velocity. To overcome this limitation, Stolt (1978) suggests coordinate transformations to cast the wave equation in velocity invariant form.

One can readily verify that (8) is a solution to

$$\frac{\partial P(k_x, z, \omega)}{\partial z} = i \left(\frac{\omega}{v} \right) \left[1 - \left(\frac{k_x v}{\omega} \right)^2 \right]^{1/2} P(k_x, z, \omega) \tag{9}$$

at $t = 0$. Equation (8) expresses two distinct computational steps: 1) the solution to (9), which is

$$P(k_x, z, \omega) = P(k_x, z = 0, \omega) \exp \left\{ \frac{i\omega}{v} \left[1 - \left(\frac{k_x v}{\omega} \right)^2 \right]^{1/2} z \right\}, \tag{10}$$

followed by: 2) the inverse Fourier transformation

$$p(x, z, t = 0) = \sum_{k_x} \sum_{\omega} P(k_x, z, \omega) \exp (ik_x x). \tag{11}$$

Equation (10) represents the advancing of the wave field to greater depths by applying the proper amount of phase shift to the Fourier coefficients. The summation over the variable ω followed by the inverse Fourier transformation with respect to k_x as expressed in (11) gives the correct migrated result.

Let us consider now layered media, in which the velocity depends only on the depth variable, that is, $v = v(z)$. It is convenient to subdivide the z axis into N_z intervals

$$\zeta_j = \{z \mid z_j < z \cdot < z_{j+1}\}; j = 1, 2, \ldots, N_z. \tag{12}$$

Assuming that the velocity v_j is constant in each layer, we can express (10) as

$$P(k_x, z_{j+1}, \omega) = P(k_x, z_j, \omega) \exp\left\{ \frac{i\omega}{v} \left[1 - \left(\frac{k_x v_j}{\omega} \right)^2 \right]^{1/2} (z_{j+1} - z_j) \right\}. \tag{13}$$

We can apply (13) to successive layers of different interval velocities (Gazdag 1978, 1978a).

THE EQUATION FOR INHOMOGENEOUS MEDIA

In the absence of horizontal velocity dependence, a set of independent ordinary differential equations define the migration process in the (k_x, ω) domain. The existence of a simple analytic solution to (9), expressed in (10) and (13), forms the basis of the phase shift method (Gazdag 1978, 1978a). When the velocity $v(x, z)$ varies laterally, the phase shift method is no longer applicable. The square root expression in (9) rules out the possibility of accommodating horizontal velocity dependence. To circumvent this difficulty, we approximate (9) by series expansion, which gives

$$\frac{\partial P(k_x, z, \omega)}{\partial z} = \frac{i\omega}{v} \left[1 - \frac{v^2 k_x^2}{2\omega^2} - \frac{v^4 k_x^4}{8\omega^4} - \cdots \right] P(k_x, z, \omega). \tag{14}$$

In this paper we consider only the fourth order equation. This gives reliable migration results up to a dip of 50° (Gazdag 1978a). Dropping the fourth order term in (14) results in a second order migration equation whose reliability is limited approximately to a dip of 30°.

We can express (14) in the (x, ω) domain by Fourier transformation with respect to k_x. This equation becomes

$$\frac{\partial P(x, z, \omega)}{\partial z} = i \frac{\omega}{v} P + \frac{v}{2\omega} \frac{\partial^2 P}{\partial x^2} - \frac{v^3}{8\omega^3} \frac{\partial^4 P}{\partial x^4}. \tag{15}$$

In this equation the velocity dependence need not be restricted to the depth variable alone, but it may also include lateral variations, that is, $v = v(x, z)$.

In the next section we develop methods for the migration of zero offset sections using (15). We formulate the numerical algorithms in the (x, ω) plane. We could also express (15) using k_x rather than x. Consequently, we could formulate the algorithms in the (k_x, ω) domain rather than in the (x, ω) domain. At present, it is not clear which procedure is more desirable.

Although we plan to study the relative merits of these two approaches in the future, we restrict the scope of the present study to the (x, ω) domain.

The Numerical Method

We advance the field $P(x, z, \omega)$ from z to $z + \Delta z$ by a third order differencing scheme

$$P(z + \Delta z) = P(z) + \frac{\partial P(z)}{\partial z} \Delta z + \frac{\partial^2 P(z)}{\partial z^2} \frac{\Delta z^2}{2} + \frac{\partial^3 P(z)}{\partial z^3} \frac{\Delta z^3}{3!}. \tag{16}$$

We compute the first derivative in (16) from (15). We then differentiate (15) with respect to z to obtain the expressions for the second and third derivatives, which are

$$\frac{\partial^2 P}{\partial z^2} = i\left[\frac{\omega}{v} \frac{\partial P}{\partial z} + \frac{v}{2\omega} \frac{\partial^2}{\partial x^2}\left(\frac{\partial P}{\partial z}\right) - \frac{v^3}{8\omega^3} \frac{\partial^4}{\partial x^4}\left(\frac{\partial P}{\partial z}\right) \right] \tag{17}$$

and

$$\frac{\partial^3 P}{\partial z^3} = i\left[\frac{\omega}{v} \frac{\partial^2 P}{\partial z^2} + \frac{v}{2\omega} \frac{\partial^2}{\partial x^2}\left(\frac{\partial^2 P}{\partial z^2}\right) - \frac{v^3}{8\omega^3} \frac{\partial^4}{\partial x^4}\left(\frac{\partial^2 P}{\partial z^2}\right) \right], \tag{18}$$

respectively.

An important feature of this numerical method is that the derivatives with respect to x in (15), (17), and (18) are computed by using transforms. Let $P(k_x, z, \omega)$ be the double Fourier transform of the zero offset section defined at depth z. Then the second derivative is computed by implementing

$$\frac{\partial^2 P(x, z, \omega)}{\partial x^2} = \sum_{k_x} - k_x^2 P(k_x, z, \omega) \exp (ik_x x). \tag{19}$$

This method of evaluating the derivative terms assures very high accuracy. Moreover, it facilitates the selective exclusion of wave-numbers corresponding to non-physical wave components for any given temporal frequency under consideration.

Migration Examples

We tested the migration method on zero offset sections. In the first two examples there is a strong velocity change between the two media of uniform interval velocity. The reflectors are buried in the second medium as shown in figs 2 and 3. Their dip angle is 45 and $-45°$, respectively. Although the magnitude of the dip angles is the same in both cases, their angles with respect to the boundary line are 35 and 55°. In both examples the migrated

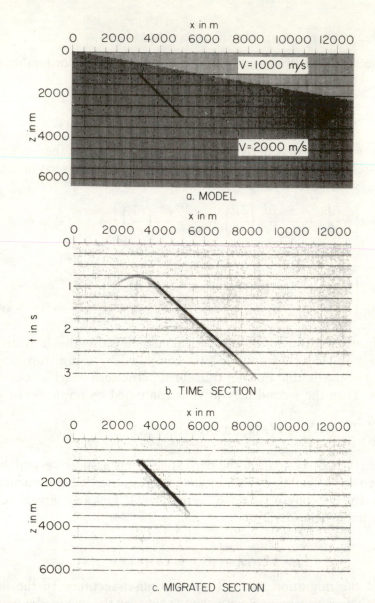

Fig. 2. (a) Model for reflector with dip of 45°. Shadings of domain signify velocity values as indicated. (b) Synthetic zero-offset section of (a). (c) Depth section after migration.

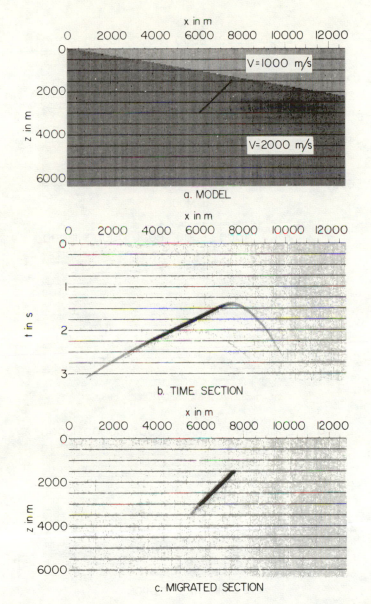

Fig. 3. (a) Model for reflector with a dip of −45°. (b) Synthetic zero-offset section of (a). (c) Depth section after migration.

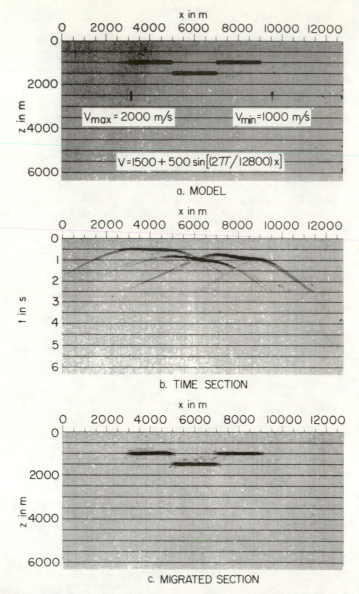

a. MODEL

b. TIME SECTION

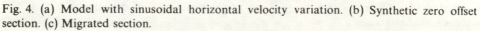

c. MIGRATED SECTION

Fig. 4. (a) Model with sinusoidal horizontal velocity variation. (b) Synthetic zero offset section. (c) Migrated section.

results reconstruct the reflector with reasonably good accuracy. The only apparent flaw in both cases is the short fading trail at the bottom of the reflector. The reason for this is that (15) has a dispersion relation which is only a fourth order approximation to (4). The diffraction from the lower edge of the reflector contributes strongly to wave components with propagation angles varying between 50 and 60°. For this range of angles the error in the dispersion relation of (15) becomes large enough to result in a visible trail suggesting undermigration.

In the example shown in fig. 4 we consider sinusoidal horizontal velocity variation

$$v(x, z)/v_0 = 1500 + 500 \sin [(2\pi/12800)x/x_0]. \tag{20}$$

For simplicity, there is no vertical velocity variation. The model consists of three horizontal reflectors shown in fig. 4-a. The zero offset section shown in fig. 4-b shows the effect of the velocity variation along x. We note that the maximum and minimum velocity positions are at $x = 3200$ m and $x = 9600$ m, respectively. The migrated results shown in fig. 4-c are in good agreement with the model. We observe that the migrated image of the left-most reflector has more noise than the others. This noise, which causes wiggles around the reflector, is due to undersampling of the time section in the high velocity region.

Conclusions

We have described a numerical scheme for wave equation migration based on Fourier transform methods. First we derived from simple physical considerations the exact migration equation whose application is limited to layered media. To accommodate horizontal velocity variations, expansion techniques were used to express the migration equation in the (x, ω) domain. The dispersion relation of the resulting migration equation was a fourth order approximation to that of the full wave equation. We tested the migration method on synthetic zero offset sections involving strong horizontal velocity variations. The migrated results were in good agreement with the models used. We found noticeable undermigration of wave components having propagation angles in the neighborhood of 60°. This was due to our fourth order approximation to the ideal dispersion relation.

References

Claerbout J.F. 1976, Fundamentals of Geophysical Data Processing, McGraw-Hill Book Co., New York.

Gazdag J. 1973, Numerical convective schemes based on the accurate computation of space derivatives, Journal of Computational Physics 13, 100–113.

GAZDAG J., 1976, Time-differencing schemes and transform methods, Journal of Computational Physics 20, 196–207.

GAZDAG J. 1978a, Wave equation migration with the phase shift method, Geophysics 43, 1342–1351.

GAZDAG J. 1978b, Extrapolation of seismic waveforms by Fourier methods, IBM Journal of Research Development 22, 481–486.

STOLT R.H. 1978, Migration by Fourier transform, Geophysics 43, 23–48.

Reprinted from GEOPHYSICS, **49**, 124-131.

Migration of seismic data by phase shift plus interpolation

Jeno Gazdag* and Piero Sguazzero‡

ABSTRACT

Under the horizontally layered velocity assumption, migration is defined by a set of independent ordinary differential equations in the wavenumber-frequency domain. The wave components are extrapolated downward by rotating their phases. This paper shows that one can generalize the concepts of the phase-shift method to media having lateral velocity variations. The wave extrapolation procedure consists of two steps. In the first step, the wave field is extrapolated by the phase-shift method using ℓ laterally uniform velocity fields. The intermediate result is ℓ reference wave fields. In the second step, the actual wave field is computed by interpolation from the reference wave fields. The phase shift plus interpolation (PSPI) method is unconditionally stable and lends itself conveniently to migration of three-dimensional data. The performance of the methods is demonstrated on synthetic examples. The PSPI migration results are then compared with those obtained from a finite-difference method.

INTRODUCTION

Migration is the process of constructing the reflector surface from the recorded seismic data. In a typical sequence of processing steps, the normal moveout (NMO) correction is applied first to a set of common midpoint (CMP) gathers. By summing these gathers over offset, one obtains a CMP stack. NMO correction and stacking is a process whereby the waves are shifted along the offset axis toward the zero offset. Therefore, a CMP stack is normally migrated as a zero-offset section.

Seismic migration consists of two steps: (1) wave extrapolation and (2) imaging. Downward extrapolation results in a wave field that is an approximation to the one that would have been recorded if both sources and recorders had been located at depth z. In the case of zero-offset data, imaging consists of mapping the data from zero time ($t = 0$) of the time section to the proper depth of the migrated section.

Since imaging is a trivial task, migration schemes differ in

their approach to the wave extrapolation problem, whose complexity depends largely on the migration velocity function. If the migration velocity has no horizontal variations, the extrapolation of the recorded seismic data can be expressed by an exact wave-extrapolation equation in the wavenumber-frequency domain. This equation has a simple analytic solution, whose implementation calls for a phase shift applied to the Fourier coefficients of the zero-offset section.

In the presence of lateral velocity variations, the exact wave-extrapolation equation is no longer valid. To circumvent this problem, the exact expression can be approximated by some series expansion (Hatton et al, 1981; Claerbout, 1980; Gazdag, 1980), which can accommodate horizontal velocity variations. These equations are then solved numerically, either in the space-time domain or in the space-frequency domain. Finite-difference migration methods formulated in the space-time domain or in the space-frequency domain are characterized by some of the following properties: (1) inaccurate dispersion relations for steep dips, (2) numerical errors resulting from the finite-difference approximation used, and (3) predisposition to numerical instability. The stability problem is kept under control by using implicit methods, which are also reasonably accurate and economical for two-dimensional (2-D) data. However, Claerbout (1980) finds that "in space dimensions higher than one the implicit method becomes prohibitively costly."

The aim of this paper is to develop a migration scheme whose cost and complexity do not depend significantly on the dimensionality of the data. In other words, the desired migration scheme is one that has no problems with higher dimensions. Another important goal is that the method be unconditionally stable. It may seem like a fortuitous coincidence that the wave-extrapolation algorithm that achieves the above objectives is also accurate. The wave extrapolation with a laterally varying velocity field consists of two steps. In the first step, the wave field is extrapolated by the phase-shift method (Gazdag, 1978), using ℓ laterally uniform velocity fields. The intermediate result is ℓ reference wave fields. In the second step, the actual wave field is computed by interpolation from the reference wave fields.

The organization of this paper is as follows. The first section is a review of the basic theory, and a description of a finite-

Manuscript received by the Editor April 19, 1983; revised manuscript received August 2, 1983.
*IBM Scientific Center, 1530 Page Mill Road, Palo Alto, CA 94304.
‡IBM Scientific Center, Via del Giorgione 129, Rome, Italy.

difference migration (FDM) method is given. Next the phase shift plus interpolation (PSPI) method is presented. This is followed by testing the PSPI method. Synthetic zero-offset sections are migrated by the FDM and the PSPI algorithms to contrast their performance. Finally, the migration results are discussed.

WAVE EXTRAPOLATION EQUATIONS

This section reviews the basic concepts of wave-extrapolation theory. The aim is to provide the reader with the background information to the PSPI method and the motivating reasons for its development. The simple case of depth-variable velocity is presented first. The treatment of laterally varying velocities is taken up next. This is followed by a description of a finite-difference migration method which serves as a means of comparison in the performance test of the PSPI method.

Depth-variable velocity

The theory of wave extrapolation is based on the assumption that the zero-offset pressure data, defined in the (x, t) domain, satisfy the scalar wave equation

$$\frac{\partial^2 p}{\partial z^2} = \frac{1}{v^2}\frac{\partial^2 p}{\partial t^2} - \frac{\partial^2 p}{\partial x^2} \tag{1}$$

with $p = p(x, z, t)$, where x is the midpoint variable, z is depth, t is two-way traveltime, and v is the half velocity. To obtain a better understanding of equation (1), it is helpful to express p as a double Fourier series

$$p(x, z, t) = \sum_{k_x}\sum_{\omega} P(k_x, z, \omega) \exp\left[i(k_x x + \omega t)\right], \tag{2}$$

where k_x is the midpoint wavenumber and ω is the temporal frequency. Substituting equation (2) into equation (1), one obtains

$$\frac{\partial^2 P}{\partial z^2} = -k_z^2 P, \tag{3}$$

which has an analytic solution

$$P(k_x, z + \Delta z, \omega) = P(k_x, z, \omega) \exp\left(i k_z \Delta z\right) \tag{4}$$

for each k_z, where k_z can be expressed as

$$k_z = \pm\frac{\omega}{v}\left[1 - \left(\frac{v k_x}{\omega}\right)^2\right]^{1/2} \tag{5}$$

in order to differentiate between two classes of solutions. In a downward extrapolation process, i.e., when Δz in equation (4) is positive, sign agreement between k_z and ω corresponds to waves that move in the negative t direction. On the other hand, when k_z and ω have opposite signs, equation (4) represents waves that move in the positive t direction. Since the downward extrapolation of recorded seismic data is an inverse process, one is interested only in

$$k_z = \frac{\omega}{v}\left[1 - \left(\frac{v k_x}{\omega}\right)^2\right]^{1/2}, \tag{6}$$

which corresponds to waves moving in the reverse-time direction.

It should be noted that the choice of sign in equation (5) does

not differentiate between upcoming and downgoing waves. This is decided by the sign of Δz in equation (4). Downward extrapolation requires that Δz be greater than zero. Once Δz is set, the two values of k_z in equation (5) correspond to forward-time and reverse-time wave propagation. Substituting equation (6) into equation (4), one obtains the desired expression for wave extrapolation

$$P(k_x, z + \Delta z, \omega) = P(k_x, z, \omega) \exp\left\{\frac{i\omega}{v}\left[1 - \left(\frac{k_x v}{\omega}\right)^2\right]^{1/2}\Delta z\right\}. \tag{7}$$

One can verify by inspection that equation (7) is the solution of

$$\frac{\partial P(k_x, z, \omega)}{\partial z} = i\left(\frac{\omega}{v}\right)\left[1 - \left(\frac{k_x v}{\omega}\right)^2\right]^{1/2} P(k_x, z, \omega), \tag{8}$$

which is the exact extrapolation equation for constant velocity. Velocity variations with respect to the depth variable are accommodated easily by simply varying the velocity with z in equation (7).

Lateral velocity variations

The simple analytic solution expressed by equation (7), which is the basis of the phase-shift method, is not valid for velocity fields with lateral variations. In this case, the square root expression in equation (8) is usually expanded into some kind of truncated series. The Taylor series expansion of equation (8) can be developed into an accurate numerical method in the (x, ω) domain (Gazdag, 1980). Another alternative is approximation by continued fractions (Hildebrand, 1956, p. 422). This approach to developing wave extrapolators was first suggested by Francis Muir (Claerbout, 1980). The idea is to expand

$$R = (1 - K^2)^{1/2} \tag{9}$$

based on the recurrence relation

$$R_{n+1} = 1 - \frac{K^2}{1 + R_n}, \tag{10}$$

in which R_n is the nth order approximation. To show that this expansion converges to equation (9), one needs only to substitute $n = \infty$ into equation (10). Starting out with $R_0 = 1$, one obtains

$$R_1 = 1 - \frac{K^2}{2} \tag{11}$$

and

$$R_2 = 1 - \frac{K^2}{2 - \frac{1}{2}K^2}. \tag{12}$$

Approximating the square root expression by R_2 and substituting it into equation (6) with

$$K = k_x v/\omega, \tag{13}$$

one obtains

$$k_z = \frac{\omega}{v} - \frac{k_x^2}{2\omega/v - k_x^2 v/2\omega}. \tag{14}$$

Similarly, the second-order approximation of equation (8) becomes

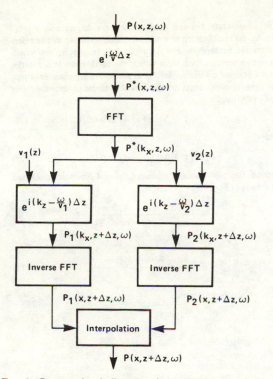

$$P(x, z, \omega)$$

$$e^{i\frac{\omega}{v}\Delta z}$$

$$P^*(x, z, \omega)$$

FFT

$$P^*(k_x, z, \omega)$$

$$v_1(z) \qquad v_2(z)$$

$$e^{i\left(k_z - \frac{\omega}{v_1}\right)\Delta z} \qquad e^{i\left(k_z - \frac{\omega}{v_2}\right)\Delta z}$$

$$P_1(k_x, z+\Delta z, \omega) \qquad P_2(k_x, z+\Delta z, \omega)$$

Inverse FFT Inverse FFT

$$P_1(x, z+\Delta z, \omega) \qquad P_2(x, z+\Delta z, \omega)$$

Interpolation

$$P(x, z+\Delta z, \omega)$$

FIG. 1. Computational diagram of the PSPI extrapolation scheme corresponding to equations (25) and (26).

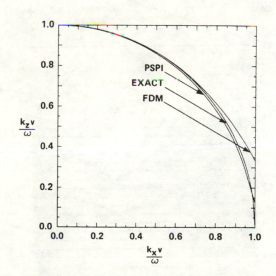

FIG. 2. Dispersion relations for the finite-difference method [equation (14)] and the PSPI method [equation (37)]. For comparison the exact dispersion relations corresponding to equation (6) are also shown. The PSPI curve refers to velocity $v = 1.25v_1$ with reference velocities v_1 and $v_2 = 1.5v_1$.

$$\frac{\partial P}{\partial z} = i\left[\frac{\omega}{v} - \frac{(v/2\omega)k_x^2}{1 - (k_x v/2\omega)^2}\right]P. \qquad (15)$$

A finite-difference extrapolator in the (x, ω) domain

A practical approach to solving extrapolation equations like equation (15) is by splitting. This is done by decomposing equation (15) into two extrapolators:

$$\frac{\partial P}{\partial z} = \frac{i\omega}{v}P \qquad (16)$$

and

$$\frac{\partial P}{\partial z} = \frac{-i(v/2\omega)k_x^2}{1 - (k_x v/2\omega)^2}P. \qquad (17)$$

Advancing to greater depths is done by applying equations (16) and (17) alternately in small Δz steps. To express equation (17) in the (x, ω) domain, it is first multiplied by the denominator of its right-hand side, and then both sides are Fourier transformed with respect to k_x. The result is

$$\left[1 + \left(\frac{v}{2\omega}\right)^2 D_{xx}\right]D_z P = \frac{iv}{2\omega}D_{xx}P, \qquad (18)$$

in which D_{xx} and D_z signify partial derivatives with respect to x and z. Migration by means of equations (16) and (18) was first reported by Kjartansson (1978). In this equation, velocity dependence need not be restricted to the depth variable alone, but it may also include lateral variations.

Equation (18) is solved numerically by means of approximating it with a set of algebraic equations involving the values of P at the grid points (j, n). If Δx is the grid spacing and Δz is the integration step, then the indices j and n refer to $x = j\Delta x$ and $z = n\Delta z$, respectively. To simplify notations in the finite-difference formulation, the wave field at grid point (j, n) will be denoted as $P_{j,n}$, where

$$P_{j,n} = P(j\Delta x, n\Delta z, \omega). \qquad (19)$$

If the differential operators are approximated as

$$D_z P_{j, n+1/2} = (P_{j, n+1} - P_{j, n})/\Delta z \qquad (20)$$

and

$$D_{xx}P_{j, n} = (P_{j-1, n} - 2P_{j, n} + P_{j+1, n})/\Delta x^2, \qquad (21)$$

the finite-difference approximation of equation (18) becomes

$$P_{j, n+1} + (\alpha - i\beta)(P_{j-1, n+1} - 2P_{j, n+1} + P_{j+1, n+1})$$

$$= P_{j, n} + (\alpha + i\beta)(P_{j-1, n} - 2P_{j, n} + P_{j+1, n}), \qquad (22)$$

where

$$\alpha = \left(\frac{v}{2\omega\Delta x}\right)^2$$

and $\qquad\qquad\qquad\qquad\qquad\qquad (23)$

$$\beta = \frac{v\Delta z}{4\omega\Delta x^2}.$$

Equation (22), for $j = 1, 2, \ldots, N_x$, represents a system of linear differential equations with complex coefficients. This system of equations is then solved for all ω values, and the migrated data at $z + \Delta z$ are obtained by

$$P(x, z + \Delta z, t - 0) = \sum_\omega P(u_i = | \Delta z, \omega), \qquad (24)$$

where the summation is carried out for all ω values. The implementation of equation (24) represents the imaging step of the migration process. In what follows, the migration technique described above will be referred to as FDM. In the Fortran program of the FDM scheme which is used in the migration examples of this paper, the system of equations (22) involving a complex tridiagonal matrix is solved by using the subroutine CJTSL of LINPACK (Dongarra et al, 1979).

The wave-extrapolation method described above has been formulated in the (x, ω) domain. Starting with equation (18), or its inverse Fourier transform with respect to ω, one can also develop finite-difference solution methods in the (x, t) domain. The main advantage of the (x, ω) domain approach is that numerical errors due to time differencing are eliminated.

MIGRATION BY PHASE SHIFT PLUS INTERPOLATION

Under the horizontally layered velocity assumption, the phase-shift method expressed by equation (7) has considerable advantage over extrapolators expressed in the (x, t) and the (x, ω) domains. Its accuracy is due to exact dispersion relations [equation (6)] and the existence of a simple analytic solution to equation (8). The numerical implementation of equation (7) is unconditionally stable in both two and three dimensions. These attractive qualities provide ample motivation for inquiring into the possibilities of generalizing the phase-shift method to laterally varying velocities.

The PSPI extrapolation method has been developed to meet this objective. This procedure is centered around the idea that lateral velocity variations can be taken into account by interpolation among wave fields that were downward-continued by phase shift, using two or more reference velocities. Since the concepts are not altered by the number of reference velocities, the PSPI extrapolation scheme will be discussed for the two-velocity case, the computational diagram of which is shown in Figure 1.

To maintain high accuracy for small dips, when $k_x v / \omega \ll 1$, the phase shift expressed in equation (4) is split into two distinct operations:

$$P^*(z) = P(z) \exp\left(i \frac{\omega}{v} \Delta z \right) \qquad (25)$$

and

$$P(z + \Delta z) = P^*(z) \exp\left[i \left(k_z - \frac{\omega}{v'} \right) \Delta z \right], \qquad (26)$$

where $v' \neq v(x, z)$ is an approximation to $v(x, z)$ as outlined below. Equation (25) is a time shift applied to each trace. This algorithm involves no numerical approximations in the (x, ω) domain. Since equation (26) cannot be computed directly when $v = v(x, z)$, its implementation is approximated in several steps shown in Figure 1. First P^* is Fourier transformed with respect to x. This is followed by the phase-shift operation expressed in equation (26) using two velocities, v_1 and v_2, which are defined as the extrema of $v(x, z)$:

$$v_1(z) = \text{Min } [v(x, z)]$$

and $\qquad\qquad\qquad\qquad\qquad\qquad\qquad (27)$

$$v_2(z) = \text{Max } [v(x, z)].$$

The phase-shifted wave fields $P_1(k_x, z + \Delta z, \omega)$ and $P_2(k_x, z + \Delta z, \omega)$ are then inverse transformed, resulting in the reference wave fields $P_1(x, z + \Delta z, \omega)$ and $P_2(x, z + \Delta z, \omega)$, which serve as reference data from which the final result $P(x, z + \Delta z, \omega)$ is obtained by interpolation in the following manner. First, the Fourier coefficients are expressed in terms of their modulus and phase angle:

$$P_1(x, z + \Delta z, \omega) = A_1 \exp(i\theta_1) \qquad (28)$$

and

$$P_2(x, z + \Delta z, \omega) = A_2 \exp(i\theta_2). \qquad (29)$$

Second, the modulus and phase of the end result are obtained by means of linear interpolation:

$$A = \frac{A_1(v_2 - v) + A_2(v - v_1)}{v_2 - v_1} \qquad (30)$$

and

$$\theta = \frac{\theta_1(v_2 - v) + \theta_2(v - v_1)}{v_2 - v_1}, \qquad (31)$$

from which one can write

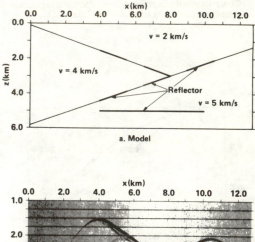

FIG. 3. (a) Schematic of model 1, representing a dipping multilayer example. The thick-line segments denote where the reflector segments have been "turned on." (b) Synthetic zero-offset time section of model 1.

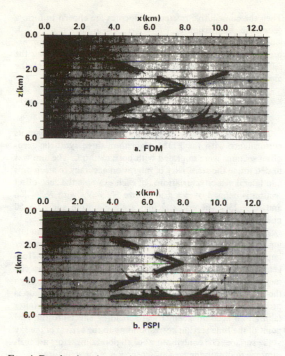

FIG. 4. Depth migration sections obtained from the zero-offset section shown in Figure 3b by means of (a) the finite-difference method (FDM) and (b) the PSPI method.

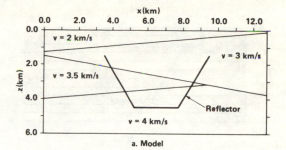

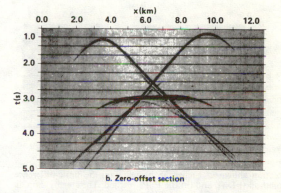

FIG. 5. (a) The aim of model 2 is to illustrate the performance of the two migration schemes for reflectors with 60-degree dips in a dipping multilayer medium. (b) Synthetic zero-offset time section of model 2.

$$P(x, z + \Delta z, \omega) = A \exp (i\theta). \qquad (32)$$

Definition of k_z

To obtain $\theta = (k_z - \omega/v)\Delta z$ in equation (26) by interpolation over the domain

$$-1 \leq k_x v/\omega \leq 1, \qquad (33)$$

the reference phase angle θ_2 must be specified over a domain that is v_2/v times greater. To accommodate any velocity v such that $v_1 \leq v \leq v_2$, the vertical wavenumber k_z should be defined over the domain

$$-v_2/v_1 \leq k_x v/\omega \leq v_2/v_1. \qquad (34)$$

It should be noted that k_z has no physical significance outside of the domain given by equation (33). The need for the extended domain is to obtain satisfactory interpolated phase angles for $k_x v/\omega$ that satisfy equation (33).

Depending upon v_2/v_1 and other considerations such as the extrema (Nyquist) of k_x, there are many ways of continuing k_z beyond the limits set in equation (33). One solution, which is by no means an optimal choice with respect to any particular criterion, to be considered in this paper is the following:

$$k_z = \frac{\omega}{v}(1 - K^2)^{1/2} \qquad \text{for} \quad 0 \leq |K| \leq 0.95,$$

$$(35)$$

$$k_z = \frac{\omega}{v}[0.312 - 3.042(|K| - 0.95)] \quad \text{for} \quad 0.95 \leq |K| \leq 1.5,$$

where $K = k_x v/\omega$ and $v_2 \leq 1.5v_1$. In this definition, k_z matches the exact square root expression up to $k_x v/\omega = 0.95$, which corresponds to 71.8 degree dip. From there it is continued along the tangent drawn at that point.

Accuracy of interpolation

The interpolation accuracy depends upon the ratio of the reference velocities (v_2/v_1) and the actual migration velocity $v(x)$. The effective phase rotation resulting from the PSPI method shown in Figure 1 is

$$\phi = \frac{\omega}{v} \Delta z + \theta, \qquad (36)$$

in which the two terms correspond to the contributions from equations (25) and (26). By setting Δz to unity and multiplying both sides by v/ω, one obtains

$$\left(\frac{k_z v}{\omega}\right) = 1 + \theta \frac{v}{\omega}. \qquad (37)$$

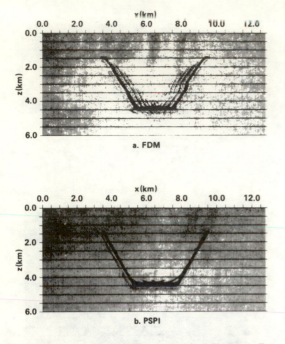

FIG. 6. Depth migration sections obtained from the zero-offset section shown in Figure 5b by using (a) the finite-difference method (FDM) and (b) the PSPI method. The steeply dipping events are smeared by the FDM migration but suffer no such distortion from the PSPI method.

By expressing θ as the function of $k_x v/\omega$, one can calculate and plot the dispersion relations, $k_z v/\omega$ versus $k_x v/\omega$, for the velocity range under consideration.

Figure 2 shows three dispersion relations corresponding to (1) the exact phase shift expressed in equation (6), (2) the FDM 45-degree algorithm characterized by equation (14), and (3) the PSPI method as described above. The latter result applies to the $v_2/v_1 = 1.5$ case. Moreover, the velocity of interpolation $v = 1.25v_1$ is in the middle of the (v_1, v_2) interval, and hence it falls in the least accurate velocity range. By contrast, as v approaches either v_1 or v_2, the PSPI dispersion relation curve approaches the exact curve up to $k_x v/\omega = 0.95$, or equivalently, 71.8-degree dips.

Choice of reference velocities

Up to now the number of reference velocities was limited to two, in order to facilitate the description of the ideas. When at some depth $z = z'$ the ratio of Max $[v(x, z')]$ and Min $[v(x, z')]$, which will be referred to as R, exceeds some upper bound, say $\rho_{max} = 1.5$, more than two reference velocities are needed. The number of reference velocities ℓ required is obtained as the smallest integer for which

$$(\rho_{max})^{\ell - 1} \geq R. \tag{38}$$

Then the consecutive reference velocities are chosen to form a geometric progression $v_{i+1}/v_i = v_i/v_{i-1} = \rho$ for $2 \leq i \leq \ell - 1$, where $(\rho)^{\ell - 1} = R$. If ℓ is greater than 2, the interpolation expressed in equations (30) and (31) is performed by using the reference velocities that are closest to the velocity of the medium $v(x, z)$.

SYNTHETIC EXAMPLES

To test the reliability of the PSPI method and compare its performance with the FDM algorithm, three synthetic zero-offset sections were migrated with both methods. The aim was to determine the sensitivity of migration accuracy to steep dips and lateral velocity variations. In each example the zero-offset section is defined over a grid of size $N_x \times N_t$, with $N_x = 256$ and $N_t = 256$. The grid spacings are $\Delta x = 50$ m and $\Delta t = 50$ msec. The integration step is $\Delta z = 50$ m.

The synthetic zero-offset sections are generated by simulating reflectors as an ensemble of point scatterers. The computer implementation of this method is essentially based upon Huygens' construction. To obtain the zero-offset section of an elementary diffractor, rays are traced in every direction through the medium up to the surface. A wavelet whose amplitude is proportional to the strength of the diffractor is assigned (after taking into account geometrical spreading effects) to the grid point of the time section corresponding to the arrival of the ray at the surface. The contributions of all point diffractors are then summed to obtain the zero-offset section of the structure.

Figure 3 shows the schematic and zero-offset section of model 1. The reflectors are indicated by means of thick lines to distinguish them from interfaces of velocity regions depicted by thin lines. Figure 4 shows the migrated results of model 1 obtained by the FDM and the PSPI methods. This model has modest velocity variations and dip angles, and therefore, it does not present much challenge to either method. The situation is quite different in the case of model 2, shown in Figures 5 and 6, which aims to test the ability of the methods to migrate steeply dipping events. The reflectors situated on the two sides form 60-degree dips with respect to the horizontal axis. While the reflectors in model 3 are horizontal, the challenge is in coping with a velocity ratio of 2.5 along the interface shown in Figure 7, part of which is directed at 60 degrees with respect to the horizontal axis.

DISCUSSION OF RESULTS

The migrated depth sections of model 1 shown in Figure 4 are of similar overall quality. There are small differences in the three top reflectors, all of which are located at the lower boundary of the $v = 2$ km/sec velocity region. These events are better migrated by the PSPI method. A noteworthy feature of the migrated sections shown in Figures 4a and 4b is a circular pattern extending down to the $z = 5$ level. The similarity between these phenomena in the two sections (FDM and PSPI) rules out the possibility that they might be of numerical origin. Detailed study of the origin of these patterns indicates that they are diffracted waves originating from the neighborhood of $(x = 8, z = 3)$, where the $v = 4$ km/sec velocity region narrows to a point.

The migrated sections of model 2 (Figure 6) demonstrate the superiority of the PSPI over the FDM for migration of events having 60 degree dips. The FDM results are dispersed, and the

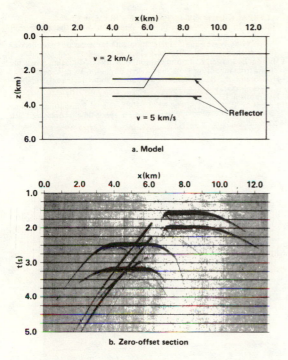

FIG. 7. (a) Model 3 is intended to test the performance of the two migration schemes in the presence of strong lateral velocity variations. (b) Synthetic zero-offset time section of model 3.

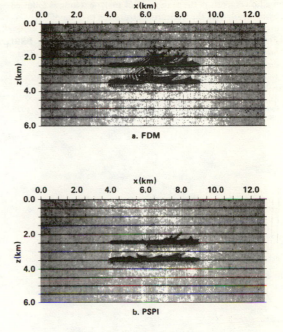

FIG. 8. Depth migration sections obtained from the zero-offset section shown in Figure 7b with (a) the finite-difference method and (b) the PSPI method. The PSPI scheme results in better images of the two horizontal reflectors.

reflector appears to have a break in the vicinity of ($x = 8.5$, $z = 3$), where it traverses the $v = 3.5$ km/sec region.

Model 3 tests the sensitivity of migration accuracy to strong velocity variations. The two horizontal reflectors are imbedded in a medium having a sudden velocity change along an interface directed 60 degrees with respect to the x axis (Figure 7a). In the zero-offset section each reflector results in three distinct recorded images (Figure 7b). Whereas the reconstruction of the horizontal reflectors shown in Figure 8 is not completely satisfactory with either case, the PSPI result is significantly better than the one obtained by FDM. The left-hand side of the upper reflector imbedded in the $v = 2$ km/sec region is weaker than its remaining part. This is due to the particular nature of the modeling program, which takes into account amplitude attenuation due to geometric spreading as a function of time $(t)^{1/2}$ rather than as a function of distance $(r)^{1/2}$. Consequently, the attenuation is exaggerated for waves traveling in low-velocity regions relative to those propagating in higher velocity media.

An ideal wave extrapolator, while restricting propagation in the reverse-time direction, is expected to simulate wave phenomena to a limited extent. Waves being extrapolated in a medium having sudden velocity changes are subject to sideways reflections, and in the case of sharp corners such as in model 1, the waves are scattered in a way that resembles diffraction. No attempt was made to remove such effects from the migrated results. In practice, however, such reflected and diffracted events are hardly useful at all. Fortunately, their effect can be easily minimized by choosing smooth velocities across interfaces. Perhaps the best strategy is simply to avoid sudden variations in the migration velocity. When the velocity variations are gradual, i.e., changes take place over several grid spacings instead of one, the undesirable "wave phenomena" are minimized without affecting the quality of migration.

CONCLUSIONS

It has been shown that the phase-shift plus interpolation method is a practical alternative solution to migration by finite-difference methods in the presence of lateral velocity variations. The dispersion relations of this migration method approach the exact ones in the neighborhood of the reference velocities. Consequently, as the ratio of two consecutive reference velocities approaches unity, which may be the case in regions of mild lateral velocity variations, the performance of this method approaches that of the phase-shift method. Similarly, the overall accuracy of migration increases with the number of reference velocities used. The use of more reference velocities means more computation, but the user has the option to determine the trade-off between performance and cost. Since the PSPI method consists of the evaluation of analytic solutions followed by interpolation, the truncation errors due to finite differencing are minimized, and the question of stability never even arises. Finally, perhaps the most significant attribute of the PSPI method is that it is generally applicable to problems of any dimensionality.

REFERENCES

Claerbout, J. F., 1980, Personal communication.

Dongarra, J. J., Moler, C. B., Bunch, J. R., and Stuart, J. W., 1979, LINPACK Users' Guide: Philadelphia, SIAM, p. 7.1–7.6.

Gazdag, J., 1978, Wave equation migration with the phase-shift method: Geophysics, v. 43, p. 1342–1351.

———— 1980, Wave equation migration with the accurate space derivative method: Geophys. Prosp., v. 28, p. 60–70.

Hatton, L., Larner, K. L., and Gibson, B. S., 1981, Migration of seismic data from inhomogeneous media: Geophysics, v. 46, p. 751–767.

Hildebrand, F. B., 1956, Introduction to numerical analysis: New York, McGraw-Hill.

Kjartansson, E., 1978, Personal communication.

Robinson, E. A., and Silvia, M. T., 1981, Digital foundations of time series analysis: v. 2—Wave-equation space-time processing: San Francisco, Holden-Day, Inc.

Reprinted from GEOPHYSICS, **44**, 1329-1344.

Wave-equation datuming

John R. Berryhill*

Wave-equation datuming is the name given to upward or downward continuation of seismic time data when the purpose is to redefine the reference surface on which the sources and receivers appear to be located. This technique differs from conventional datuming methods in the repositioning of seismic reflections laterally as well as vertically in response to observed time dips. The most interesting applications of the technique are those in which the redefined reference surface is an actual geologic interface having an irregular topography and a large velocity contrast. Wave-equation datuming can remove the deleterious effect such an interface has on seismic reflections originating below it. Wave-equation datuming also is applicable in seismic modeling.

The computations required in wave-equation datuming are related to those of migration. The Kirchhoff integral formulation of the wave equation can provide a basis for computation to deal with the irregular surfaces and variable velocities that are central to the problem. The numerical implementation of the Kirchhoff approach can be reduced to an efficient procedure involving summations and convolutions of seismic traces with short shaping and weighting operators.

INTRODUCTION

A number of different ways of accomplishing wave-equation migration have been described in the past few years; recent contributions by Stolt (1978) and Schneider (1978) include lists of references to earlier authors. In several discussions of migration, the occasion arises to distinguish migration per se from what is variously called downward or upward continuation or extrapolation. Given a zero-offset seismic section parameterized as $U(x, z = 0, t)$, the stated objective of migration is to transform this into the migrated section $U(x, z, t = 0)$. The perhaps more modest aim of extrapolation is simply to create from $U(x, z = 0, t)$ a new section $U(x, z = z_1, t)$ which appears as if the sources and receivers had been moved to a different elevation z_1. It is widely appreciated in the literature that migration may in fact be accomplished as a succession of downward continuation steps, with z moving ever deeper into the subsurface.

The term "wave-equation datuming" will be used to denote upward or downward continuation of $U(x, z = z_1, t)$ to produce $U(x, z = z_2, t)$, in cases in which z_1 and z_2 are general functions of x, and the velocities entering the problem are realistically variable. Examples will be presented to clarify this definition and to show that a wave-equation datuming process capable of accounting for irregular datum surfaces and variable velocities has useful applications in correcting and modeling seismic data.

METHOD

As an aid to understanding, we adopt the "normal incidence" view of zero-offset seismic data. We imagine that instead of reflectors, the earth contains a large number of small impulsive sources distributed uniformly over what formerly were called reflecting surfaces. We imagine that all sources are activated at once (at $t = 0$) and that the seismic section is the result of a large number of detectors recording simultaneously on separate channels or traces. In this view, at the unique time $t = 0$, a wavefront emerges parallel to each buried surface, and raypaths emerge perpendicular to the surfaces and wavefronts. The arrival times we compute by considering only these upgoing

Presented in part at the 48th Annual International SEG Meeting, October 31, 1978 in San Francisco. Manuscript received by the Editor June 15, 1978; revised manuscript received January 24, 1979.
*Exxon Production Research, P.O. Box 2189, Houston, TX 77001.

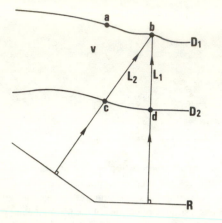

FIG. 1. Two raypaths arriving at one point on surface D_1 cross surface D_2 at separate locations.

wavefronts and raypaths will correctly represent two-way travel if we simply halve all the velocities. Some aspects of actual zero-offset amplitude behavior are not consistent with this one-way viewpoint; nevertheless, the arrival times that result are perfectly accurate.

Figure 1 suggests why a wave-equation approach is required to simulate correctly the effect of moving the detectors recording a zero-offset section from one datum surface to another. Two raypaths are shown emerging from different points on reflector R, with both being recorded at detector b on datum D_1. If the detectors had been on datum D_2 instead, the vertical raypath would have encountered detector d, and the slanting raypath, detector c. If we wish to predict what would be observed on datum D_2 from a knowledge of what is recorded on D_1, we cannot simply

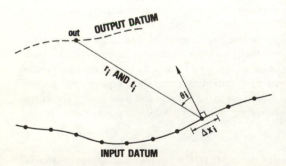

FIG. 2. Definition of the geometrical quantities appearing in equation (1).

reassign trace b to location d. Not only is the time shift $2L_1/v$ different from the time shift $2L_2/v$, but the slanting raypath actually needs to be reassigned to detector c. To be correct, any procedure for changing the datum of a seismic section from D_1 to D_2 must be sensitive to the time dip of reflections in the section (the evidence of nonvertical raypaths) and correspondingly shift both arrival times and lateral locations. The wave-equation method discussed below provides this capability by operating simultaneously on a multiplicity of traces to move laterally coherent signals in a manner consistent with wave propagation theory.

The mathematical basis of the present implementation of wave-equation datuming is described in the Appendix. It is essentially a precise and efficient computerized form of Huygens principle. Conceptually, each detector is replaced by a small loudspeaker which plays back the seismic signal previously recorded at that location; the output trace simply represents what one would hear at the output location with all the loudspeakers operating at once. The formal statement of this is equation (A–12); modified slightly to conform to the definitions in Figure 2, which is

$$U_{\text{out}}(t) = \frac{1}{\pi} \sum_i \Delta x_i \cos \theta_i \frac{t_i}{r_i} \overline{Q}(t - t_i). \quad (1)$$

In equation (1), $U_{\text{out}}(t)$ is one of presumably many traces referenced to the output datum. Equation (1) prescribes that $U_{\text{out}}(t)$ can be calculated by performing a summation of traces at locations i on the input datum. The quantities entering the summation include the separation Δx_i between adjacent trace locations on the input datum, the distance r_i between the input and output locations, the angle θ_i between r_i and the normal to the input datum, and the traveltime t_i between the input and output locations. The symbol $\overline{Q}(t - t_i)$ is defined in considerable detail in the Appendix; briefly, it represents the input trace at location i delayed by the traveltime t_i and convolved with a particular time-domain shaping operator. Although the shaping operator usually needs to be only five or ten samples long, its action is of extreme importance to the faithful preservation of waveforms and amplitudes.

A computer program designed to implement wave-equation datuming must incorporate some means of describing the input and output datums as curves in (x, z) space, together with some system for defining the wave propagation velocity in the medium bounded by the two datums. The parameterization scheme

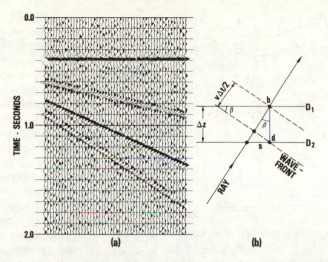

FIG. 3. A synthetic seismic section (a) showing reflections having dips of 0, 15, 30, and 45 degrees. The diagram (b) indicates how dip angle determines the vertical and lateral shifts occurring in a change of datum.

must be flexible enough to permit a faithful representation of geology. The details of the parameterization are immaterial; it is necessary only that the quantities x_i, θ_i, r_i, and t_i in equation (1) be economically and accurately calculable.

The particular computer program that produced the examples presented in this paper parameterizes datums as segments of cubic splines strung together and velocities by piecewise linear horizontal interpolation. A datum change is controlled by a single-interval geologic model, and one run of the program executes a single change of datum in either direction. The input tape is referenced to the input datum, and the output tape is referenced to the output datum. The

examples that follow generally were produced by iterated or sequential runs of the program correlated with appropriate changes in the controlling model.

SIMPLE CHANGE OF DATUM

Figures 3 and 4 illustrate in simple terms the effect of a single pass through a wave-equation datuming program. The input section shown in Figure 3a consists of synthetic data generated for subsurface reflectors having dips of 0, 15, 30, and 45 degrees. Noise was added by convolving a train of random spikes with the same Ricker wavelet used for the reflections. The depth to the flat reflector is 1400 ft, the velocity is 8400 ft/sec uniformly throughout the sec-

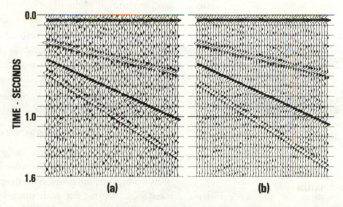

FIG. 4. The result of changing the datum of Figure 3a to coincide with the flat reflector, employing a conventional method (a) and a wave-equation method (b).

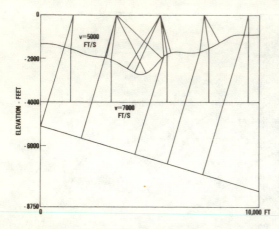

Fig. 5. A model exhibiting a water bottom with irregular topography and a large velocity contrast. The raypaths within the water layer have a wide range of dip angles.

tion, and the lateral separation between traces is 100 ft.

The change of datum to be illustrated is a downward extrapolation that moves the detectors (conceptually) from the original datum of Figure 3a to a new datum coincident with the flat reflector. Figure 3b suggests some statements that can be made concerning changes of datum between parallel surfaces in a uniform medium with velocity v. We know that the dip angle β of a wavefront (with respect to horizontal) or of a raypath (with respect to vertical) is related to the time dip $\Delta t / \Delta x$ of a reflection on the seismic section by

$$\sin \beta = (v/2) \Delta t / \Delta x.$$

Wave-equation processes (extrapolation and migration) depend upon time dip and are thus sensitive to the corresponding angle of incidence of a wavefront or raypath upon a datum. The quantity $\sin \beta / (v/2)$ is equal to the time dip $\Delta t / \Delta x$; it is also a quantity that Snell's law preserves when velocity v changes across an interface. Thus, to preserve time dip at a velocity interface is to obey Snell's law, and vice versa.

The angle β together with the change in depth Δz determines the importance of the wave-equation effects. The lateral shift s of the signal associated with a given raypath is

$$s = \Delta z \tan \beta.$$

At a given lateral x location, the difference in time Δt between the arrival of a wavefront at detector d on datum D_2 and its arrival at b on D_1 is

$$\Delta t = (2/v) \Delta z \cos \beta.$$

For the 45-degree reflection in Figure 3a, s amounts to 14 trace intervals, and Δt differs from a purely vertical time displacement by 98 msec.

Figure 4a shows the result of changing the datum of Figure 3a to coincide with the flat reflector, using what we might call the "traditional" or "static" approach—each trace has merely been shifted earlier by $(2 \times 1400)/8400 = 333$ msec. For the steeper reflections, this result is significantly incorrect. In contrast, Figure 4b shows the same change of datum executed by a wave-equation datuming program. In terms of equation (1), each output trace in Figure 4b was created by combining 39 input traces from Figure 3a using appropriate time shifts and amplitude weights. (Input traces extending left and right of those actually shown in Figure 3a were included, where necessary, to prevent "end" effects.) The wave-equation result is correct. By comparison, the 45-degree reflection is arriving 98 msec too early on the traditional section, and the other reflections are in error to an extent that diminishes with decreasing dip.

Figures 3a and 4b exemplify some fine points of terminology that need to be clearly understood. In the one-way travel concept of zero-offset seismic data, the time value $t = 0$ is unique and denotes the instant at which the sources distributed over the buried surfaces are activated. This unique time value appears as the starting time of both sections. Figure 4b suggests the interpretation that when the sources are activated, it takes no time for the signal from the flat interface to reach the detectors; this means the detectors have been moved to, and are now located upon, this particular reflector. This effect is independent of the shape of the reflector; the reflection corresponding to a new datum always occurs at zero time.

Note that the time dips of the reflections in Figure 4b are the same as the time dips of the corresponding reflections in Figure 3a. If the initial datum and the new datum had not been parallel, the time dips in Figure 4b would have changed to give a correct indication of the angles of incidence of the wavefronts with respect to the new datum.

DATA CORRECTIONS

It sometimes happens that variations in the thickness or velocity of a shallow layer can cause fluctuations in the arrival times of reflections from deeper interfaces. We regard such fluctuations as spurious, and we seek to remove them from a seismic section before it is used for geologic interpretation or depth

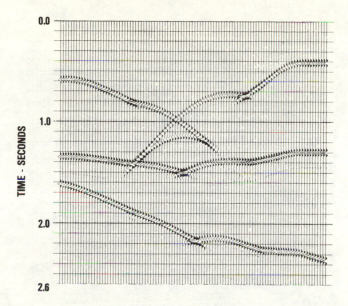

FIG. 6. A synthetic seismic section computed for the model of Figure 5. The two planar reflectors bear a distinct impression of the water bottom.

estimation. An important instance of this problem arises in marine data where the water bottom has considerable topography and the sediments just below the water bottom have a velocity significantly higher than that of water.

As a specific example, Figure 5 shows a model of an irregular water bottom overlying a flat horizon and also a reflector dipping downward to the right at 16 degrees. The sediment velocity of 7000 ft/sec contrasts sharply with the water velocity of 5000 ft/sec. The water layer ranges in depth from 900 ft at the extreme right to 2700 ft at the bottom of the channel. The raypaths show that reflections from the planar horizons undergo significant deviations at the water bottom, due to Snell's law. Figure 6 shows a synthetic seismic section computed for the model of Figure 5 (by a method described later). As expected, the two planar reflectors bear a distinct impression of the water bottom.

Wherever the raypaths are exactly vertical, we know that a change ΔW in the arrival time of the water bottom reflection induces an undesired shift ΔT in underlying reflections according to the relationship

$$\Delta T = \Delta W(1 - v_W/v), \qquad (2)$$

where v_W and v are the velocities of water and sedi-

ment, respectively. Equation (2) provides the basis of the conventional "static" correction method in which one takes for ΔW the water bottom arrival time and removes an amount ΔT of time from the beginning of each trace. The effect of this procedure is as if the velocity of water had been changed to equal the sediment velocity, and we will refer to the process as *velocity replacement*.

Velocity replacement by the conventional method is bound to fail in its objective wherever raypaths deviate significantly from the purely vertical. The model of Figures 5 and 6 is just such a case. Figure 7 shows the result of implementing velocity replacement corrections for the data of Figure 6 by making vertical ΔT time shifts. The procedure was improved over that discussed above by the use of migrated water bottom times (not shown) for ΔW. Still, this conventional method has not removed all the undesired water bottom effects from the planar reflectors. Important nonvertical raypaths are present from two sources: the inherent dip of the deepest reflector and the Snell's-law deviations along the dipping parts of the water bottom.

In comparison, velocity replacement by wave-equation datuming can achieve an essentially perfect result under these same circumstances. In two passes through a wave-equation datuming program, we can

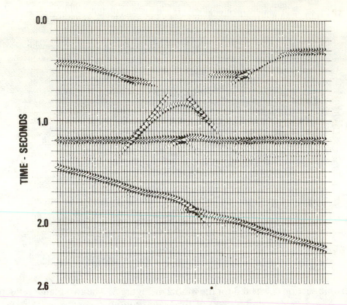

FIG. 7. An attempt to remove the effect of the water bottom from the data of Figure 6 solely by vertical time shifts. (The partial loss of the water bottom reflection is not a necessary aspect of this procedure and could be avoided.)

effectively strip off the water layer and replace it with a material having a velocity more like that of the uppermost sediments. To do this, the program must have access to an accurate representation of the water bottom (migrated and scaled according to depth) and proper initial and replacement velocities.

Figure 8 shows the result of processing the data of Figure 6 by the wave-equation method to change the datum from the surface of the water to the water bottom. We may say that Figure 8 is showing us what we would have seen if the seismic section had been recorded with detectors on the water bottom. The computational power of the process is apparent; focusing effects have been removed from both the new datum and the underlying reflections. The main point, though, is that the overlying water layer has been removed.

Figure 8 shows a seismic section datumed on the water bottom and completely independent of what lies overhead. The second step in velocity replacement by wave-equation datuming is to move the datum of this section from the water bottom back up to the water surface, using a replacement velocity instead of the velocity of water. Figure 9 shows the result of processing the data of Figure 8 by the wave-equation method to change the datum from the water bottom

to sea level, assuming a velocity of 7000 ft/sec. Figure 9 is a perfectly corrected section. The water bottom no longer has a velocity contrast across it, and so reflections from deeper horizons are not affected by it. The water bottom itself still possesses a focusing geometry, but the effects of focusing are now characterized by a higher velocity.

If we are not interested in the water bottom itself or in data from the shallowest sediments, there is an alternative to velocity replacement that is somewhat more flexible in what it can handle. Instead of restoring the datum to sea level, we can continue downward to a flat datum beneath the lowest point of the water bottom. Figure 10 shows the result of processing the data of Figure 8 by the wave-equation method to change the datum from the water bottom to a flat horizon beneath the water bottom, assuming a velocity of 7000 ft/sec. Again, the two planar reflections are perfectly corrected.

The "successive datums" approach of Figure 10 has the advantage of making specific the heretofore vague concept of a replacement velocity; we need to know the sediment velocity characterizing the interval between the water bottom and the buried datum. This velocity might very well be laterally variable.

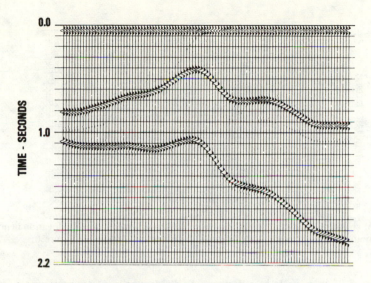

FIG. 8. The result of changing the datum of Figure 6 from sea level to coincide with the water bottom, employing a wave-equation method.

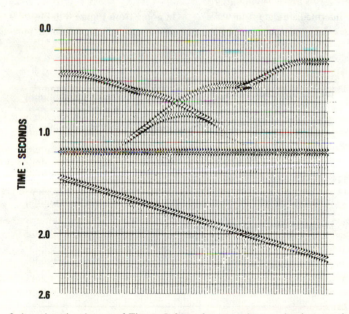

FIG. 9. The result of changing the datum of Figure 8 from the water bottom back to sea level, using a wave-equation method and substituting 7000 ft/sec in place of the velocity of water. This achieves a perfect correction of Figure 6.

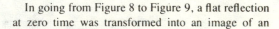

FIG. 10. The result of changing the datum of Figure 8 from the water bottom to a flat elevation beneath the water bottom. This also perfectly corrects the planar reflections in Figure 6.

In a real seismic section, there might be reflectors or unconformities below the water bottom having both significant velocity contrast and irregular topography. These would adversely affect reflections from deeper strata just as an irregular water bottom does and could be corrected by a generalization of the successive datums method—the datum of the section could be moved downward to each of the surfaces in turn until a horizontal datum is reached beneath the deepest irregular surface.

The reason for using every surface having a large velocity contrast as an intermediate datum in wave-equation datuming is that doing so takes care of the Snell's-law *refraction* at each surface, and assures us that we are always dealing with *interval* velocities. The velocities needed for the computations are then the same as the physical velocities for wave propagation. This way of proceeding seems advantageous compared to the treatment of velocities in prior descriptions of Kirchhoff migration (Schneider, 1978) and modeling (Hilterman and Larson, 1975).

MODELING

In going from Figure 8 to Figure 9, a flat reflection at zero time was transformed into an image of an

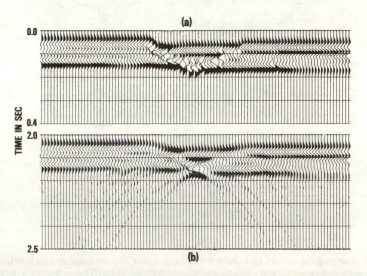

FIG. 11. The predicted seismic section for a thin, multilayer stratigraphic interval as viewed from the top of the interval itself (a), and a datum lying 5000 ft above the interval (b).

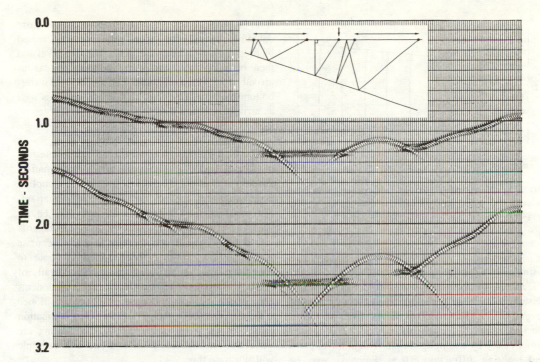

FIG. 12. The primary and the first multiple reflection for a model of a complicated water bottom. Inset: a type of multiple for which the outgoing and returning legs of the raypath become superimposed in the limit of zero source-receiver separation.

irregular water bottom. This suggests a procedure by which synthetic data consistent with the wave equation can be computed for multilayered models through repeated application of wave-equation datuming. We start by preparing a synthetic seismic tape containing identical traces with a single pulse at time zero and zeroes elsewhere. We inform our wave-equation datuming program that this is the seismic response of surface N datumed on surface N, and request the program to move the datum up to surface $N - 1$. The output tape will show how surface N appears when viewed from surface $N - 1$; we add to this a flat reflection at time zero to represent surface $N - 1$ and request the program to move the datum again, up to surface $\dot{N} - 2$. By repeating this sequence of operations a number of times, we can construct a model with many interfaces. Snell's law is obeyed at each interface because of the relationship between time dip and angle of incidence or emergence.

The synthetic seismic section shown in Figure 6 was computed by this iterated datuming method. (The same model was also computed by a ray-tracing method, and it was verified that the arrival times are identical.) To avoid aliasing effects, the lateral spacing of traces in the computations was half that shown in the display. To avoid end effects, fifty additional traces were computed past the left and right boundaries shown.

The method of iterated datuming is a convenient and inexpensive way to perform wave-equation modeling for moderately complicated models. The resulting arrival times are reliable. The behavior of amplitudes is that appropriate for the assumption of sources distributed over the buried surfaces. In some aspects, e.g., diffraction effects, the results may differ from the behavior of amplitudes assuming a point source. The method as described so far is insensitive to reflection and transmission coefficients; however, one might consider additionally manipulating wave amplitudes to simulate such factors.

One variation of this basic modeling approach is applicable to the special case of a multilayered stratigraphic unit no more than a few hundred feet thick, buried at a depth of several thousand feet. The procedure is to compute the effects of the layering within the unit by whatever means is convenient and to use

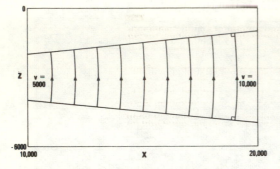

FIG. 13. A truncated wedge whose thickness and velocity both increase linearly to the right. All the raypaths shown have a common traveltime.

a single change of datum to simulate the geometrical effects due to depth of burial. Figure 11a shows the seismic section for a stratigraphic unit composed of several dozen thin layers, as viewed from the top of the unit itself. A section like this can be computed rather easily by assuming simple vertical travel paths and ignoring diffraction effects. Emphasis may be placed instead on careful handling of reflection coefficients and on accumulating the total reflected pulse due to the large number of reflecting interfaces. Figure 11b was produced from Figure 11a by shifting the datum upward from the top of the unit itself to the true surface of the earth. The result is a considerable change of appearance; diffraction effects are now present, and the extent to which they limit the resolution of details is quite apparent.

A second extension of the basic modeling method permits the computation of a certain class of multiple reflections. The simplest example of the class consists of two reflections from a surface such as a water bottom with an intervening reflection off the top of the water. The inset in Figure 12 shows three such raypaths and illustrates their critical property: as the source-detector separation approaches zero, the downgoing legs of such a raypath become exactly superimposed upon the upcoming legs in such a way that the angle of the reflection at the water surface is then 90 degrees.

The applicability of wave-equation datuming to the modeling of such multiple reflections depends on the perception that, in the inset diagram of Figure 12, the zero-offset raypath *also* describes a plane wave that propagates downward from the water surface and is reflected upward from the water bottom. The earliest reflection in Figure 12 is the primary reflection

from an irregular water bottom computed by the basic datuming method. The second reflection was created by propagating a plane wave down from sea level and then back up from the water bottom; insofar as arrival times are concerned, this produces a water-bottom multiple. One finds that lateral shifts as well as arrival times are roughly doubled.

VELOCITY VARIATION

The final example is concerned with the question of lateral velocity variation. The mathematical development of equation (1) begins with an implicit assumption that the velocity is laterally invariant. For practical purposes, this restriction may be relaxed. To produce accurate results, at least as regards arrival times, the chief requirement is that accurate individual time shifts t_i be supplied as input to equation (1), irrespective of the specific details of velocity variation. If the individual components entering the summation in equation (1) are positioned at their correct times, then the result of the summation cannot be far from its correct time (providing the waveform remains recognizable). The final example will illustrate this.

Figure 13 shows a model with a rather large lateral velocity gradient whose effects we can predict precisely. This model has the peculiarity that the lateral (x) coordinate, the thickness of the truncated wedge, and the velocity within the wedge all extrapolate to zero at the same point. The ray-tracing solution to this model tells us that all raypaths are arcs of circles centered on the line $x = 0$ and that the particular family of concentric arcs normal to both surfaces of the wedge is characterized by a single traveltime. The velocity gradient in this model is in fact stronger (in proportion to the thickness of the wedge) than the author considers geologically realistic, the raypaths having 11.5 degrees of curvature. The traveltime for the curved raypaths shown must be computed from a nontrivial formula. Nevertheless, this traveltime (399 msec) is nearly the same as the straight-line vertical traveltime (400 msec). This model was used to test the accuracy of equation (1) in the presence of a lateral velocity gradient.

Figure 14a shows the input section, consisting of a 16-Hz Ricker wavelet with onset at time zero for all traces. A datuming program incorporating equation (1) was used to change the datum of this section from the lower surface of the truncated wedge to the upper surface. Figure 14b shows the result. As the ray-tracing solution predicts, the outcome is a reflection at constant time. By visual inspection, the change of

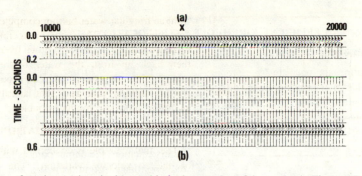

FIG. 14. The datum of section (a) is coincident with the lower edge of the wedge in Figure 13. Section (b) is the result of changing the datum to the upper edge, employing a wave-equation process.

arrival time is consistent with the expected 399 msec. No distortion of the waveform is apparent.

Figure 14 represents an output with no time dip produced from an input with no time dip through a nontrivial computation. The true-space dip and the varying velocity must both be handled properly to obtain the mutual cancellation of their effects. In evaluating equation (1), values of t_i are quite different for input traces at equal distances left and right of an output trace; the curve of t_i versus distance (a symmetric hyperbola in simpler cases) is significantly skewed. Nevertheless, the result in Figure 14b is correct, to a very good approximation. The arrival time is accurate; variants of the model with faults in the lower surface (not shown) confirm behavior consistent with curved, rather than straight, raypaths.

Obtaining accurate output from equation (1) is primarily a matter of putting in correct time delays. This apparently holds true even for the amount of raypath curvature shown in Figure 13, and most readers will agree that this is sufficient for the needs of exploration seismology. Nevertheless, we will note a systematic procedure for handling even larger horizontal velocity gradients: subdivide the interval vertically, so that raypath curvature is limited to a few degrees within each subinterval, then extrapolate the wave field across each of the subintervals in turn.

CONCLUSIONS

Wave-equation datuming provides, as an al-

ternative to velocity-replacement static corrections, a method of correcting zero-offset seismic sections for the effects of large variation in the thickness or velocity of near-surface layers. Also, wave-equation datuming provides a means of computing synthetic seismic sections for multilayered subsurface models containing irregular interfaces and variable velocities. Certain classes of multiple reflections may be included.

A specific mathematical implementation of wave-equation datuming based on a novel form of Huygens principle has been described. It has been shown by example that this method is adequate to deal with the irregular surfaces and variable velocities expected in the most interesting potential applications.

REFERENCES

Berryhill, J. R., 1977, Diffraction response for nonzero separation of source and receiver: Geophysics, v. 42, p. 1158–1176.
Hilterman, F. J., 1970, Three-dimensional seismic modeling: Geophysics, v. 35, p. 1020–1037.
——— 1975, Amplitudes of seismic waves—a quick look: Geophysics, v. 40, p. 745–762.
Hilterman, F. J., and Larson, D., 1975, Kirchhoff wave theory for multivelocity media: Presented at the 45th Annual Intl. SEG Meeting, October 14 in Denver.
Schneider, W. A., 1978, Integral formulation for migration in two and three dimensions: Geophysics, v. 43, p. 49–76.
Stolt, R. H., 1978, Migration by Fourier transform: Geophysics, v. 43, p. 23–48.
Trorey, A. W., 1970, A simple theory for seismic diffractions: Geophysics, v. 35, p. 762–784.

APPENDIX
AN EFFICIENT KIRCHHOFF EXTRAPOLATION METHOD

Convolutional form of the integral

The integral equation approach to wave propagation problems yields a computationally efficient method of extrapolating a wave field defined on a surface to points outside the surface. A convenient point of departure is the statement of the Kirchhoff theorem,

$$U(x, y, z, t)$$
$$= \frac{1}{2\pi} \int dA_0 \frac{\cos\theta}{r^2} \left[U(x_0, y_0, z_0, t - r/c) \right.$$
$$\left. + \frac{r}{c} \frac{\partial}{\partial t} U(x_0, y_0, z_0, t - r/c) \right].$$
$$(A-1)$$

In this equation and in Figure A–1, U is an upward-traveling scalar wave assumed to be known, along with its time-derivative, at all points (x_0, y_0, z_0) of an infinite planar surface A_0 for all times t. The objective is to compute U for all times at a point (x, y, z) lying above the plane. The velocity at which U propagates is taken to be c, so that r/c is the traveltime from a point (x_0, y_0, z_0) on the plane to the point (x, y, z). The equation allows the desired result to be obtained by incorporating a time delay into the known wave function, weighting the delayed wave and its derivative by some purely geometrical factors, and performing a two-dimensional (2-D) integration over the plane.

If U is known only along a line in the plane but is assumed independent of the direction perpendicular to this line, a useful simplification of the integral can be found. In Figure A–2, a coordinate system has been established with $z = 0$ in the A_0-plane and with U no longer dependent on y. Equation (A–1) becomes

$$U(0, z, t)$$
$$= \frac{1}{2\pi} \int_{-\infty}^{\infty} dx \, z \int_{-\infty}^{\infty} dy \frac{1}{r^3} \left[U(x, 0, t - r/c) \right.$$
$$\left. + \frac{r}{c} \frac{\partial}{\partial t} U(x, 0, t - r/c) \right].$$
$$(A-2)$$

In this equation, the output point is taken to lie above the coordinate origin $(x = 0)$, and the value z/r has been substituted for $\cos\theta$.

The y-integration in equation (A–2) can be converted to a time-domain convolution. With the definitions

$$\tau \equiv r/c,$$

and

$$r_x \equiv (x^2 + z^2)^{1/2},$$

it follows that

$$y = \pm(c^2\tau^2 - r_x^2)^{1/2},$$
$$(A-3)$$

and

$$dy = \pm \, c^2\tau \, d\tau / (c^2\tau^2 - r_x^2)^{1/2}.$$
$$(A-4)$$

The sign of the square root in (A–3) and (A–4) selects one or the other half-plane. The argument of the square root is kept nonnegative by the condition $c\tau \geq r_x$. Figure A–2 shows that this condition is satisfied physically. However, it requires an alteration of the limits of integration,

$$\int_{-\infty}^{\infty} dy \rightarrow 2 \int_{r_x/c}^{\infty} d\tau.$$

The same effect can be achieved by introducing the unit step function $H(t)$ defined by

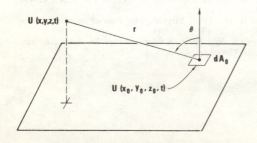

FIG. A-1. The geometry involved in the statement of the Kirchhoff theorem [equation (A–1)].

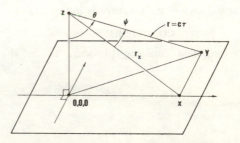

FIG. A-2. The situation of Figure A-1 specialized for the case in which U is independent of y.

$$H(t) = 0, t < 0,$$

and

$$H(t) = 1, t \geq 0.$$

This allows equation (A–2) to be written

$$
U(0, z, t)
$$
$$
= \frac{1}{\pi} \int_{-\infty}^{\infty} dx \frac{z}{c^2} \int_{-\infty}^{\infty} d\tau \, H(\tau - r_x/c) \cdot
$$
$$
\cdot \frac{1}{\tau^2} \frac{1}{(\tau^2 - r_x^2/c^2)^{1/2}} \cdot
$$
$$
\cdot \left[U(x, 0, t - \tau) + \tau \frac{\partial}{\partial t} U(x, 0, t - \tau) \right].
$$
$$(A-5)$$

It needs to be noted that

$$
f(t) * g(t) \equiv \int_{-\infty}^{\infty} f(t - \tau) g(\tau) d\tau,
$$

$$
\frac{df(t)}{dt} * g(t) = f(t) * \frac{dg(t)}{dt},
$$

$$
\frac{d}{dt} [g(t) H(t - t_0)] = H(t - t_0) \frac{dg(t)}{dt}
$$

$$
+ g(t_0) \delta(t - t_0),
$$

and

$$
\frac{1}{t^2} \frac{1}{(t^2 - r_x^2/c^2)^{1/2}} = \frac{d}{dt} \left[\frac{(t^2 - r_x^2/c^2)^{1/2}}{tr_x^2/c^2} \right].
$$

It then follows that

$$
\frac{d}{dt} \left[H(t - r_x/c) \frac{c^2(t^2 - r_x^2/c^2)^{1/2}}{tr_x^2} \right]
$$

$$
= H(t - r_x/c) \frac{1}{t^2(t^2 - r_x^2/c^2)^{1/2}},
$$

and

$$
U(x, 0, t) * \left[H(t - r_x/c) \frac{1}{t^2(t^2 - r_x^2/c^2)^{1/2}} \right]
$$

$$
= \frac{dU(x, 0, t)}{dt} * \left[H(t - r_x/c) \cdot \right.
$$

$$
\left. \cdot \frac{c^2(t^2 - r_x^2/c^2)^{1/2}}{tr_x^2} \right].
$$

There is no distinction in this case between d/dt

and $\partial/\partial t$, so equation (A–5) becomes

$$
U(0, z, t)
$$
$$
= \frac{1}{\pi} \int_{-\infty}^{\infty} dx \frac{z}{c^2} \frac{dU(x, 0, t)}{dt} * H(t - r_x/c) \cdot
$$
$$
\cdot \left[\frac{c^2(t^2 - r_x^2/c^2)^{1/2}}{tr_x^2} + \frac{1}{t(t^2 - r_x^2/c^2)^{1/2}} \right].
$$

The terms within the brackets sum to

$$
\frac{c^2}{r_x^2} \frac{t}{(t^2 - r_x^2/c^2)^{1/2}}.
$$

In Figure A–2,

$$
\tan \psi \equiv (c^2 t^2 - r_x^2)^{1/2}/r_x,
$$

so that

$$
\frac{d \tan \psi}{dt} = \frac{c}{r_x} \frac{t}{(t^2 - r_x^2/c^2)^{1/2}}.
$$

(Because the convolution is no longer expressed as an explicit integral, t replaces the τ in Figure A–2.) Thus, it can be concluded that

$$
U(0, z, t) = \frac{1}{\pi} \int_{-\infty}^{\infty} dx \frac{z}{cr_x} \frac{dU(x, 0, t)}{dt}
$$

$$
* H(t - r_x/c) \frac{d \tan \psi}{dt}.
$$
$$(A-6)$$

This completes the task of expressing the y-integration as a time-domain convolution.

Implementation for sampled data

In cases of practical importance, the convolution indicated in equation (A–6),

$$
Q(t - r_x/c) \equiv \frac{dU(x, 0, t)}{dt}
$$

$$
* \left[H(t - r_x/c) \frac{d \tan \psi(t)}{dt} \right],
$$

must be carried out using discrete time-sampled wave functions of finite length. It is worthwhile to state carefully the correspondences between continuous and sampled quantities. Assuming a time-sample interval Δt,

$$
U(x, 0, t) \rightarrow U_k \equiv U(x, 0, t_k),
$$
$$
U_{k \pm 1} \equiv U(x, 0, t_k \pm \Delta t),
$$

$$
dU(x, 0, t)/dt \rightarrow \Delta U_k/\Delta t \equiv (U_k - U_{k-1})/\Delta t,
$$

where

$$t_k = (k - 1)\Delta t;$$
$$d \tan \psi(t)/dt \rightarrow \Delta \tan \psi_j / \Delta t,$$
$$\equiv (\tan \psi_{j+1} - \tan \psi_j)/\Delta t,$$

where

$$t_j = (j - 1)\Delta t + r_x/c.$$

For the input trace U_k, the index value $k = 1$ is assumed to correspond to a zero of time which is arbitrary but shared by all input traces. The operator $\Delta \tan \psi_j$ is taken to have both its first nonzero sample and an index value $j = 1$ corresponding to a *delay* of r_x/c. This takes into account the effect of the step function H and makes it possible to write

$$Q(t - r_x/c) \rightarrow Q_j \equiv \sum_{i=1}^{L} \frac{1}{\Delta t} \Delta U_{j-i+1} \Delta \tan \psi_i.$$

$$(A-7)$$

Again, it is understood that $j = 1$ for a time-delay r_x/c which is itself x-dependent. The upper limit of summation L is equal to j (to keep the subscript of ΔU nonnegative) or N (the number of $\Delta \tan \psi$ samples actually computed), whichever is smaller.

Limiting the number N of $\Delta \tan \psi$ samples entering into equation (A–7) is equivalent to limiting the maximum value of y in equation (A–2) or Figure A–2. This is physically the same as terminating the plane of integration at some $y = \pm y_{max}$, and should give rise to a type of diffraction. [Compare the computations in Berryhill (1977).] Figure A–3 shows the result of evaluating equation (A–7) with $\Delta t =$

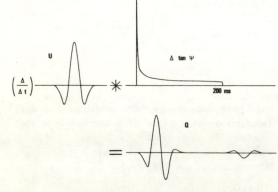

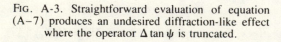

FIG. A-3. Straightforward evaluation of equation (A–7) produces an undesired diffraction-like effect where the operator $\Delta \tan \psi$ is truncated.

0.001 sec, $z = 5000$ ft, $x = 0$, $c = 5000$ ft/sec, and $N = 201$, using a 16-Hz Ricker wavelet as the input trace. The output Q_j clearly consists of a desirable contribution coincident with the onset of $\Delta \tan \psi$ plus an undesirable diffractive artifact due entirely to the finite value of N. It is found in practice that N must be made uneconomically large to make the undesired diffraction acceptably small, if equation (A–7) is evaluated directly.

Fortunately, equation (A–7) can be simplified into something more practical, without loss of accuracy, by exploiting the time separation between the signal and the diffraction apparent in Figure A–3. The calculation implied by equation (A–7) is written out in detail:

$$\Delta t Q_j = \sum_{i=1}^{N} \Delta U_{j-i+1} \Delta \tan \psi_i,$$

$$= \Delta U_j \Delta \tan \psi_1 + \Delta U_{j-1} \Delta \tan \psi_2$$
$$+ \Delta U_{j-2} \Delta \tan \psi_2 + \cdots,$$
$$+ \Delta U_{j-N+2} \cdot$$
$$\cdot \Delta \tan \psi_{N-1} + \Delta U_{j-N+1} \Delta \tan \psi_N,$$
$$= (U_j - U_{j-1}) \Delta \tan \psi_1 + (U_{j-1} - U_{j-2}) \cdot$$
$$\cdot \Delta \tan \psi_2$$
$$+ (U_{j-2} - U_{j-3}) \Delta \tan \psi_3 + \cdots,$$
$$+ (U_{j-N+2} - U_{j-N+1}) \Delta \tan \psi_{N-1}$$
$$+ (U_{j-N+1} - U_{j-N}) \Delta \tan \psi_N,$$
$$= U_j \Delta \tan \psi_1 + U_{j-1}(\Delta \tan \psi_2 - \Delta \tan \psi_1)$$
$$+ U_{j-2}(\Delta \tan \psi_3 - \Delta \tan \psi_2) + \cdots,$$
$$+ U_{j-N+1}(\Delta \tan \psi_N - \Delta \tan \psi_{N-1})$$
$$- U_{j-N} \Delta \tan \psi_N,$$
$$= \sum_{i=1}^{N} U_{j-i+1}(\Delta \tan \psi_i - \Delta \tan \psi_{i-1})$$
$$- U_{j-N} \Delta \tan \psi_N, \qquad (A-8)$$

with the added definition $\Delta \tan \psi_0 \equiv 0$. Examination of Figure A–3 shows that the difference between successive terms in the $\Delta \tan \psi$ sequence approaches zero much faster than the terms themselves, for large N. In equation (A–8) above, the series indicated by the summation converges very rapidly, leaving the separate negative term isolated and identified as the undesired diffraction effect. Dropping this term produces the useful expressions

$$Q_j = \sum_{i=1}^{L} U_{j-i+1}(\Delta \tan \psi_i - \Delta \tan \psi_{i-1})/\Delta t,$$

$$(A-9)$$

and

$$Q_j = \sum_{i=1}^{L} U_{j-i+1} \frac{\Delta^2 \tan \psi_i}{\Delta t^2} \Delta t. \qquad \text{(A–10)}$$

Equation (A–10) states that Q is essentially the result of convolving the input trace with the second derivative of the function $\tan \psi$. With derivatives calculated as shown in the preceding steps, equation (A–10) provides the basis of a comparatively inexpensive convolutional y-integration method. With typical seismic data, an operator length L of 20 to 30 msec is usually adequate.

The lateral integral

In terms of quantities defined up to this point, equation (A–6) may be restated

$$U(0, z, t) = \frac{1}{\pi} \int_{-\infty}^{\infty} dx \, \frac{z}{c} \frac{1}{r_x} Q(t - r_x/c).$$

$$\text{(A–11)}$$

In cases of practical importance, the x-integral must be carried out over a finite range of x using data sampled at certain discrete locations on a nonplanar surface in the presence of a variable velocity. A modified form of equation (A–11) which confronts these difficulties is

$$U_{\text{out}}(t) = \frac{1}{\pi} \sum_{i=-M}^{M} \Delta x_i \, \frac{z_i}{c_i} \frac{1}{r_i} \overline{Q}(t - t_i). \quad \text{(A–12)}$$

The symbol $U_{\text{out}}(t)$ identifies the desired output result without referring to any universal coordinate system. Instead, the coordinates x and z in equation (A–11) are to be defined separately at each point where an input trace is provided. Figure A–4 defines the geometry. A plane A_0 tangent to the input surface S_0 at input point i makes an angle α with the horizontal. If x' and z' are the surveyed horizontal and vertical displacements between input point i and the output point, then

$$x_i = x' \cos \alpha - z' \sin \alpha,$$
$$z_i = z' \cos \alpha + x' \sin \alpha,$$

and

$$\Delta x_i \equiv (x_{i+1} - x_{i-1})/2.$$

The distance r_i is independent of the coordinate system.

The role of velocity c in the derivation leading up to equation (A–11) is chiefly to establish a connection between r_x and the corresponding time delay. When the velocity of the medium intervening between the

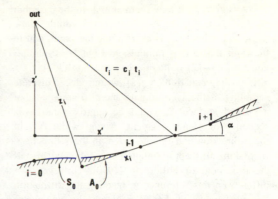

FIG. A-4. Definition of the geometrical quantities entering into equation (A–12).

input point i and the output point behaves in a way too complex to be determined by a single constant, the crucial requirement is that the time delay t_i can somehow still be calculated. Once t_i is known, it is adequate to take

$$c_i = r_i / t_i,$$

inasmuch as c_i in equation (A–12) directly affects only amplitudes, not arrival times.

When the lateral integration in equation (A–11) is approximated as the discrete sum in equation (A–12), each separate input trace i becomes accountable for all time delays between $t_i - \delta t_i/2$ and $t_i + \delta t_i/2$, where $\delta t_i \equiv (t_{i+1} - t_{i-1})/2$. When δt_i exceeds one time sample interval Δt, some form of data interpolation is called for. The simplest method is to convolve $Q(t - t_i)$ with a flat-topped averaging operator of length $1 + \delta t/\Delta t$. In equation (A–12), the symbol $\overline{Q}(t - t_i)$ represents the result of such interpolation.

The limits of the summation in equation (A–12) are taken to be $\pm M$ input traces on either side of the output trace location. The particular value of M required for a given set of data can be determined objectively from consideration of the maximum time dip present and the maximum elevation change (z) desired. An insufficient value of M results in reduced amplitudes for steep reflections and not erroneous positions. Typical values of $(2M + 1)$ are between 25 and 75, generally less than in Kirchhoff-based migration.

Equation (A–11) was derived from equation (A–1) through a series of connected mathematical steps. Equation (A–12), as defined in the preceding paragraphs, is a useful generalization of equation (A–11). The leap from equation (A–11) to equation (A–12) is not as great as it may seem at first. In the case that surface S_0 in Figure A-4 is truly an infinite plane (A_0), and velocity c_i is everywhere the same, and in the limit that Δx_i is infinitessimal but M is infinite, equation (A-12) is identical to equation (A-11).

In writing equation (A-12), we introduced discrete sampling of the x-coordinate and also assumed that a curved surface S_0 could be approximated as a sequence of planar segments having width Δx_i in the x-direction and infinite extent in the y-direction. In making this approximation, we followed in part the precedent of Hilterman (1970). However, we differ from this reference in ignoring the diffraction that must arise at the boundary of two adjoining planar segments. On the other hand, the work of Trorey (1970) and Hilterman (1975) makes it clear that this diffraction is truly negligible if the dihedral angle between adjoining segments is a few degrees or less. If the dihedral angle between adjoining planar segments (each tangent to S_0 at a field-recording location) is larger than this limit, the pertinent conclusion is really that the field-recording technique has seriously undersampled surface S_0.

In considering that the relationship between traveltime and distance may not be expressible by a single constant velocity, we have followed the precedent of a number of authors. Notably among these, Schneider (1978) has reported that root-mean-square (rms) velocity is suitable for the purposes of migration in a multilayered medium; and Hilterman and Larson (1975) have suggested that ray tracing may be necessary to find traveltimes in some situations. These authors were chiefly concerned with vertical velocity inhomogeneities due to layering; the present

author prefers to deal with layering and other discontinuous inhomogeneities by an alternative method, as outlined in the text. Still, for handling smooth lateral velocity variations (e.g., of the sort caused by differential compaction), it seems attractive to define "equivalent" velocities c_i and thereby exploit the fact that it is the time delays t_i which are of paramount importance in determining the results of equation (A-12). Further discussion of this point is contained in the text.

Equation (A-11) retains the contributions from both the near-field (U/r^2) and far-field [$(1/r)\partial U/\partial t$] terms in equation (A-1), and in this sense is unrestricted as regards the magnitude of z. In replacing the x-integration of equation (A-11) with the discrete summation of equation (A-12), we introduced the possibility of error if Δx_i is not small compared to z_i. This is not a serious limitation in the applications we envision, since z is generally large.

Equation (A-12) as it stands describes the upward continuation of upgoing waves and (with Figure A-2 held upside down) the downward continuation of downgoing waves—cases in which the traveling wave encounters the input locations before the output location. If the traveling wave encounters the output location *before* the input locations, it is necessary only to imagine a movie of the event run backwards to recover the situation previously considered. Thus, a computer program implementing equation (A-12) permits the downward continuation of upgoing waves (and the upward continuation of downgoing waves) simply by time-reversing all input traces prior to processing and all output traces afterward. The essential effect here is to reverse the relative sign of t and t_i.

Reprinted from GEOPHYSICS, *45*, 361-375.

Depth migration after stack

D. R. Judson∗, J. Lin∗, P. S. Schultz∗, and J. W. C. Sherwood∗

The conventional methods for migrating a seismic section, e.g., the finite-difference method and the Kirchhoff summation method, are inadequate in the presence of significant lateral variations in velocity. For this type of velocity distribution, the basic migration output should be in true depth, although for practical purposes it may be preferable to display it with a nonlinear depth scale.

A finite-difference method has been implemented for obtaining migrated depth sections. The concept underlying this involves all the usual assumptions of a dip line and primary reflections only, with the seismic section considered as the surface measurement of an upcoming wave field which we process with downward continuation in small increments of depth, rather than the customary increments of traveltime. The specified velocity variation laterally along a thin layer results in transmission time changes which must be corrected by a small static time shift applied to each seismic trace. This additional operation within the migration algorithm can be difficult and expensive to implement and is the main reason for its prior omission.

Results are given of depth migration applications to both synthetic and real seismic data.

INTRODUCTION

The purpose of migrating seismic data is to transform or invert it into a cross-section of the earth's subsurface, wherein geologic reflectors are accurately portrayed in terms of their horizontal position, depth, and structural dip. Accurate migration requires not only an accurate estimate of the seismic wave velocity and its spatial variations, but also an algorithm sophisticated enough to apply it in migrating the seismic data. We will not concern ourselves with the troublesome and critical problem of estimating the subsurface velocities accurately. Instead, we will simply address the problem of accurately honoring a specified velocity model within the migration process. Note that we also avoid the question of how to obtain a coherent common-depth-point (CDP) stack section in the presence of significant lateral velocity variations. It is assumed that this has already been accomplished and that we are dealing with reasonable quality data which approximate a "zero-offset" section without significant noise or long path multiple

reflections.[1] The method is applicable to three-dimensional (3-D) data organized on a rectangular grid, but at this time we have only implemented it for two-dimensional (2-D) data, assumed to be along a dip line.

We begin by noting [see Larner et al (1978)] that none of the commonly used approaches to wave-equation migration—the Kirchhoff summation method (Schneider, 1978), the finite-difference method (Claerbout and Doherty, 1972), or the frequency-wavenumber (f-k) domain method (Stolt, 1978)—migrates seismic reflections to proper positions if there is a significant lateral variation in velocity. In fact, the theory of the f-k method is based upon a simple *constant-velocity* model, and this method can even have significant accuracy problems with severe vertical velocity variations. Larner et al (1978) proposed a partial solution to the lateral veloc-

[1]In complex situations, it may be necessary to resort to depth migration before stack, as discussed in Schultz and Sherwood (1980, this issue).

Presented at the 48th Annual International SEG Meeting, October 30, 1978 in San Francisco. Manuscript received by the Editor December 22, 1978; revised manuscript received June 1, 1979.
∗ Digicon, Inc., 3701 Kirby Drive, Houston, TX 77098.

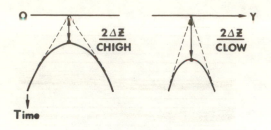

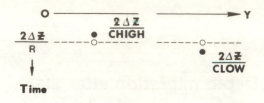

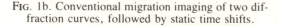

FIG. 1a. Zero-offset response due to two diffracting points at depth Δz.

FIG. 1b. Conventional migration imaging of two diffraction curves, followed by static time shifts.

ity variation problem, consisting of the depth conversion of a conventionally migrated seismic section by an image-ray method based on the work of Hubral (1977). The examples of Larner et al (1978) clearly show that this depth conversion by the image-ray technique can produce accurate depth migration in certain circumstances. However, they also noted that the division of the migration process into two stages, namely, conventional migration followed by image-ray depth migration, is somewhat artificial. In situations with rapid lateral velocity changes, this two-stage procedure is inaccurate, as will be clearly shown by the examples. A comprehensive depth migration implementation is required which will more closely honor the laws of wave propagation through an inhomogeneous medium; we present such a solution.

METHOD

It is apparent that the finite-difference migration method originated by Claerbout and Doherty (1972) is inherently better suited to dealing with rapid velocity variations than the Kirchhoff or f-k methods. It is mainly for this reason that so much effort has been applied to its development [see, for example, Loewenthal et al (1976), Quay and Whittlesey (1977)]. Throughout this development, it has been realized that a factor has been omitted from the finite-difference method which is particularly significant for lateral velocity variations. This is the "time-shifting" or "thin-lens" term discussed in Claerbout (1976) and Larner et al (1978). The reason for its delayed introduction has been two-fold. First, the term is difficult to implement into the finite-difference method and will increase the computer processing time by

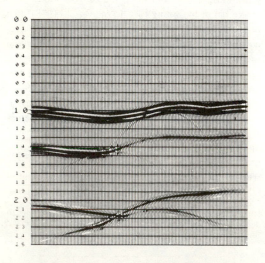

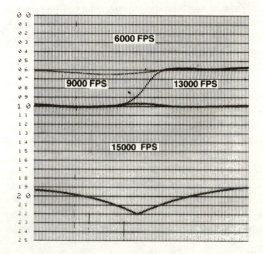

FIG. 2. Zero-offset time section obtained by ray tracing from the subsurface model in Figure 3. Reflection from 9000–13,000 ft/sec (FPS) interface is not included.

FIG. 3. Subsurface depth model expressed in vertical two-way time using a replacement velocity of 10,000 ft/sec.

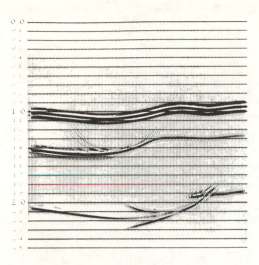

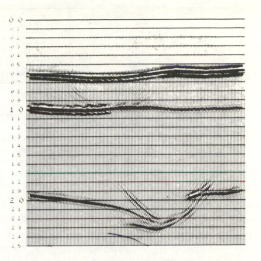

FIG. 4. Conventional migration of Figure 2.

FIG. 5. Vertical depth conversion of Figure 4 expressed in vertical two-way time using a replacement velocity of 10,000 ft/sec.

about 50 percent. Second, the associated depth migration results are apparently very sensitive to inaccuracies in the model of the lateral velocity variations, in precisely the same way as a conventional depth section formed from a time section by a transformation from vertical two-way time to depth. Consequently, considerably more attention needs to be applied to the migration velocity model than is customary in routine data processing.

A brief mathematical derivation of the depth migration procedure discussed here is provided in the Appendix. It shows that an additional time-shifting operation needs to be incorporated with conventional finite-difference time migration in order to yield a depth migration algorithm. A clear understanding of this static time-shift modification can be obtained from the following discussion and Figure 1. We consider the initial downward continuation step where we have a stacked section, or zero source-to-receiver offset section, recorded at the surface $z = 0$. Following Loewenthal et al (1976), we wish to apply downward continuation to the data to depth Δz, so that the data down to Δz are fully migrated, and the deeper data represent a "conventional" time section associated with source initiation and recording at depth Δz. The interval velocity $C(y)$ over the interval Δz varies with horizontal position y along the line. Two diffracting points exist at the depth Δz, at locations with interval velocities CHIGH and CLOW.

A schematic of the "hyperbolic" seismic response of the diffracting points is shown in Figure 1a. Con-

ventional migration processing will collapse the hyperbolas to points located at the apices of the hyperbolas, with times of $2\Delta z$/CHIGH and $2\Delta z$/CLOW. For true depth imaging, these points need to be imaged at the same depth Δz. It is convenient to retain a time scale and locate both point images at a time of $2\Delta z/R$, where R is some laterally invariant velocity, as indicated in Figure 1b. This relocating is accomplished by following the conventional small

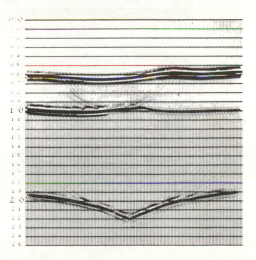

FIG. 6. Complete finite-difference depth migration of Figure 2 expressed in vertical two-way time using a replacement velocity of 10,000 ft/sec.

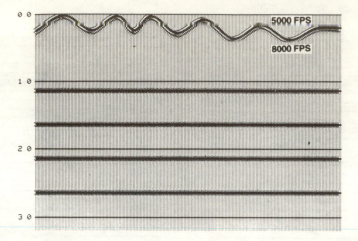

FIG. 7. Depth section with a shallow rugged water bottom. Water velocity is 5000 ft/sec (FPS) and subbottom velocity is 8000 ft/sec. Depth is expressed in vertical two-way time using a replacement velocity function $R1$ of 5000 ft/sec down to 950 ft (0.38 sec) and 8000 ft/sec thereafter.

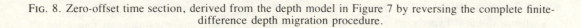

FIG. 8. Zero-offset time section, derived from the depth model in Figure 7 by reversing the complete finite-difference depth migration procedure.

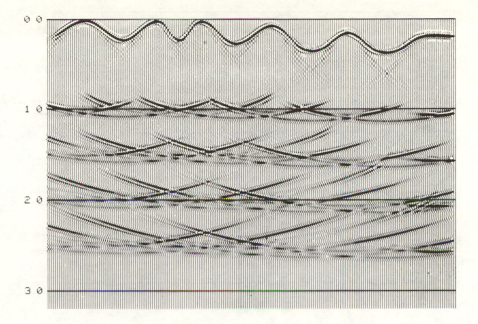

FIG. 9. Conventional migration of Figure 8.

downward continuation migration step with a static time shift of $2\Delta z\{(1/R) - [1/C(y)]\}$ on each individual trace for output times greater than $2\Delta z/R$. Hereafter, R will be referred to as the "replacement" velocity and may vary with depth, in which case the final result will be a depth section converted to a time section by the laterally invariant replacement velocity function $R(z)$. This depth-to-time transformation, which we call elasticizing, is defined in Schultz and Sherwood (1980, this issue).

It is obvious that the laterally changing static time shifts referred to above will alter the apparent dips of the deeper seismic events. This will lead to a response different from conventionally migrated data when propagated through the next layers. Ultimately, the differences will be extremely significant for large velocity variations with rapid spatial fluctuations, as will be shown later with examples.

In summary, true depth migration by the finite-difference method simply requires the additional application of a static time shift on each individual seismic trace for each downward continuation step Δz. Given input stacked seismic data with some actual velocity model, the computed output depth section is elasticized to time by a specified laterally invariant replacement velocity function. It is a simple matter to

reverse this depth migration process so it can also be used to model the zero-offset seismic data associated with an input depth cross-section and an actual velocity model. We shall investigate the capabilities of this method with several synthetic examples.

SYNTHETIC EXAMPLE 1

The test example shown in Figure 2 is a zero-offset time section derived from a ray-tracing program, with a distance between traces of 80 ft. The associated structural and interval velocity model is shown in Figure 3. It should be noted that the reflection from the 9000–13,000 ft/sec interface is not included in Figure 2. The migrated time section, which was obtained by a conventional finite-difference algorithm, and its elasticized depth section ($R = 10,000$ ft/sec) are shown in Figures 4 and 5, respectively. Note that Figure 5 is a true depth section with equal horizontal and vertical scales. It is clear that the deep reflector has not been migrated into a sensible configuration. Unfortunately, an image-ray depth migration routine was not available to apply to the result in Figure 4, but thoughtful consideration indicates that it would not remove the incorrect multiple branches on the deep reflector.

The results from the complete finite-difference

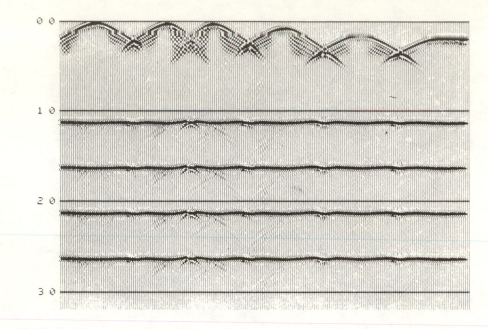

FIG. 10. Vertical depth conversion of Figure 8. Depth is expressed in vertical two-way time using the replacement velocity function $R1$.

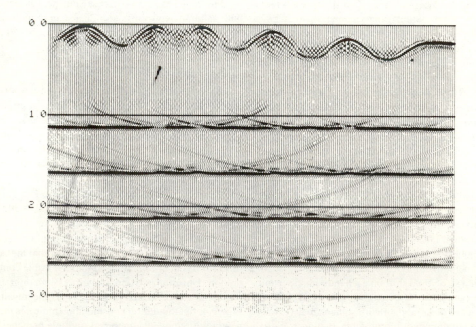

FIG. 11. Conventional migration of Figure 10 using replacement velocity function $R1$.

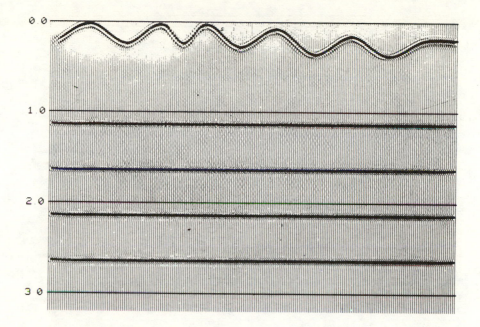

FIG. 12. Complete finite-difference depth migration of Figure 8. Replacement velocity function is $R1$.

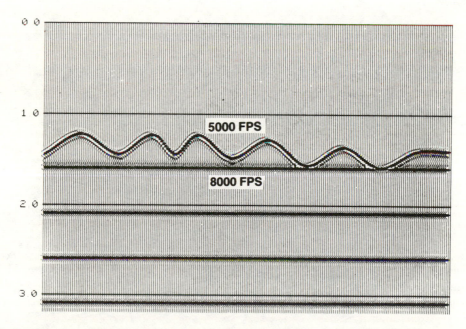

FIG. 13. Depth section with a deep rugged water bottom. Water velocity is 5000 ft/sec (FPS) and subbottom velocity is 8000 ft/sec. Depth is expressed in vertical two-way time using a replacement velocity function $R2$ of 5000 ft/sec down to 3950 ft (1.58 sec) and 8000 ft/sec thereafter.

FIG. 14. Zero-offset time section corresponding to the depth model of Figure 13.

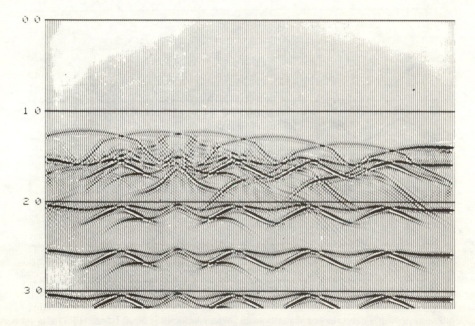

FIG. 15. Vertical depth conversion of Figure 14. Depth is expressed in vertical two-way time using the replacement velocity function $R2$.

depth migration algorithm are shown in Figure 6 ($R = 10,000$ ft/sec). These results are in excellent agreement with the original structural velocity model given in Figure 3.

The logical conclusion from this schematic example is that the finite-difference depth migration algorithm discussed here provides an accurate method for migrating an adequately stacked seismic section through an inhomogeneous velocity structure.

SYNTHETIC EXAMPLE 2

It was noted earlier that accurate depth migration requires both an accurate velocity model and a sophisticated migration algorithm. With the current advances in digital migration routines, the velocity model is undoubtedly becoming the weak link in complex geology. However, there is at least one situation where we can derive a good estimate of a severe lateral velocity change, namely, that associated with rugged water bottom topography. Hence the following synthetic example.

Shallow water bottom topography

Figure 7 depicts a rugged shallow water bottom with a maximum depth of 950 ft, where the water velocity is 5000 ft/sec. The subwater velocity is 8000 ft/sec, with flat reflectors at depths of 4000, 6000, 8000, and 10,000 feet. The depth section in Figure 7 has been elasticized to vertical two-way time by a replacement velocity function $R1$ of 5000 ft/sec down to 950 feet, and 8000 ft/sec thereafter. The separation between traces is 50 ft.

Reversing the mode of the finite-difference depth migration program and using it as a modeling program, we obtain the zero-offset time section in Figure 8. The buried foci effects associated with the water bottom valleys are noteworthy. More important is the two-way transmission effect of the variable water layer on the subwater bottom reflections. Generally, we have transmission time anomalies together with amplitude variations associated with focusing and defocusing through the near-surface anomaly.

The result of conventional migration is shown in Figure 9, and it is clearly unacceptable as far as the deep reflectors are concerned. If we are going to restrict ourselves to conventional migration, it is necessary to precorrect the data for transmission through the anomalous region. Since this region is reasonably shallow, we might use static time corrections, or some more sophisticated correction procedure. As an example, the section in Figure 10 is the result of conventional vertical depth conversion of the section in Figure 8, followed by vertical stretching to two-way

time corresponding to the replacement velocity function $R1$. The dominant transmission time anomalies evidenced on the deeper reflections in Figure 8 have been greatly diminished, and the subsequent conventional migration in Figure 11 is correspondingly improved over that in Figure 9. It should be emphasized, however, that the amplitude anomalies associated with focusing effects still persist along the deeper reflectors.

The most effective way that we know to invert the data of Figure 8 is by the complete finite-difference depth migration algorithm. The result, expressed in vertical two-way time using the replacement velocity function $R1$, is shown in Figure 12 and corresponds excellently with the original depth section in Figure 7.

The above results indicate that depth conversion followed by conventional migration produced fairly reasonable results. This was mainly due to the shallow nature of the velocity anomaly. In the next section we will investigate the effect of considerably increasing the water depth.

Deep water bottom topography

Figure 13 displays a depth section identical to that of Figure 7, except that the water depth has been increased by 3000 ft. Application of the new depth migration program in its modeling mode yields the zero-offset seismic time section in Figure 14. Conventional migration of this section will yield nonsensical results. Neither would we anticipate much improvement from operations equivalent to those used to obtain the sections in Figures 10 and 11. However, we performed this sequence of processing and show the confused results in Figures 15 and 16.

A better approximate correction procedure is first to apply conventional finite-difference migration for the layers from zero to 3000-ft depth, thereby effectively moving the seismic survey line to that depth. Subsequent vertical depth conversion (to equalize the vertical transmission times across the anomalous regions, as in Figure 15), followed by the remainder of the conventional migration from 3000 ft down (using the velocity function $R2$), will produce a result very similar in quality to that in Figure 11. In fact, this procedure is essentially a coarse version of the complete depth migration program, in which the transmission across the anomalous region is approximated by a single static time shift operation followed by conventional migration. As we saw in Figures 11 and 12, this "coarse" layering procedure is somewhat inferior to the "thin" layering procedure employed in

FIG. 16. Conventional migration of Figure 15 using velocity function $R2$.

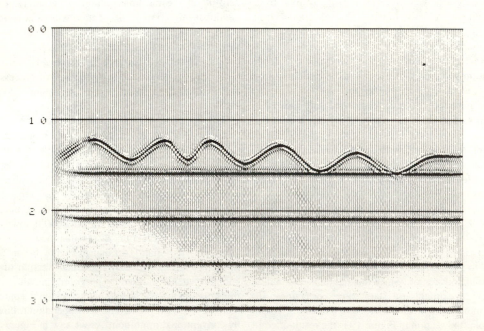

FIG. 17. Complete finite-difference depth migration of Figure 14. Replacement velocity function is $R2$.

3600 METERS

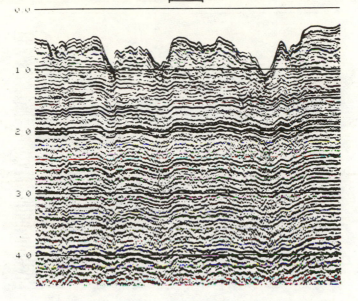

FIG. 18. CDP stack section from the East Coast of the United States.

3600 METERS

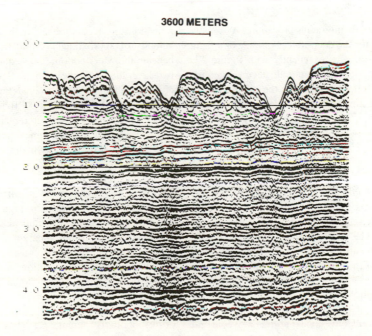

FIG. 19. Vertical depth conversion of the section in Figure 18. Depth is expressed in vertical two-way time using the replacement velocity function $R3$.

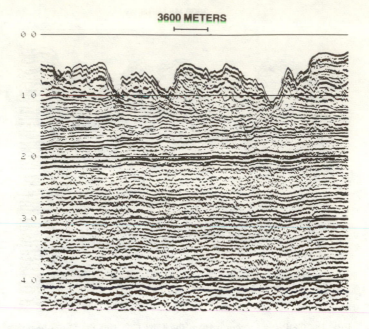

3600 METERS

FIG. 20. Conventional migration of the section in Figure 19 for a layering sequence from 0.0 to 1.1 sec using velocity function $R3$.

the complete depth migration procedure. We show the latter result for the deep water bottom example in Figure 17. It compares excellently with Figure 13, except for some localized amplitude decreases on the dipping water bottom reflector at the extremities of the line. These amplitude anomalies are associated with missing data, i.e., the water bottom reflections beyond the ends of the recorded line in Figure 14.

REAL DATA EXAMPLE

A simple example to which we can apply the preceding concepts is shown in Figure 18. This is a 36-fold CDP stack section from the East Coast of the United States. The source was a 1700-inch³ air gun array, and the recording configuration consisted of a 3600-m cable with a 100-m receiver group interval. Specialized prestack processing had considerably alleviated the problem of obtaining a coherent stack in the presence of the large transmission anomalies through the water. Conventional normal moveout and migration of the near-trace section with water velocity (4950 ft/sec) yielded a water bottom model. The effect of this rugged bottom on the deeper primary

reflections produces time anomalies consistent with a velocity of approximately 6500 ft/sec immediately beneath the water bottom. Consequently, we adopt a two-layer model for the earth of 4950 and 6500 ft/sec down to about 2720 ft. We define a replacement velocity function $R3$ of 4590 ft/sec down to this same depth, with the deeper replacement velocities being identical with the assumed earth velocities. It is to be noted that the results which follow are independent of these deeper velocities.

Corresponding to the coarse layering approach, the section in Figure 19 is the result of vertical depth conversion of the section in Figure 18 to correct approximately for vertical transmission time anomalies down to 2720 ft. It is apparent that the complexity of the deeper reflections is considerably diminished by this correction. The section in Figure 20 is the result of applying conventional finite-difference migration to the section in Figure 19, for a layering sequence from sea level down to 2720 ft (1.100 sec). Correspondingly, Figure 21 shows the result of applying the comprehensive thin layer depth migration procedure to the section in Figure 18 for the same layering sequence.

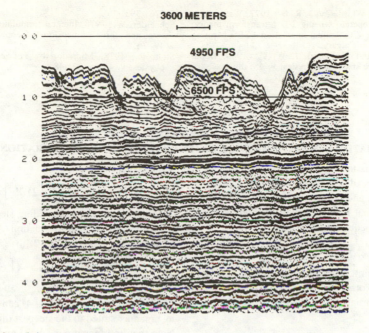

FIG. 21. Complete finite-difference depth migration of the section in Figure 18 for a layering sequence from 0.0 to 1.1 sec in the replacement medium $R3$.

The sections in Figures 19, 20, and 21 are all considered preferable to the section in Figure 18 for interpretation purposes, but the fact that they are almost identical warrants some discussion. First note that Figure 19 is unmigrated, while both Figures 20 and 21 are fully migrated down to 2720 ft (1.100 sec vertical two-way time) and thereafter correspond to a conventional unmigrated time section obtained at a depth of 2720 ft. The close similarity of the water bottom in the migrated and unmigrated sections is related to the fact that the sections have a vertical exaggeration of about 10 to 1. The gross water bottom dip is actually under 30 degrees everywhere, and the greatest lateral migration of any part of it is about 500 ft, a distance which is not easily resolved on the figures. The similarity of the deep data on the sections in Figures 19, 20, and 21 is also due to its very shallow dip and lateral velocity homogeneity. Trivial calculations show that migration of such data through the water bottom anomaly will produce lateral shifts again limited to about 500 ft. It should be emphasized that a section with complex geology at considerable depth would better illustrate the powerful capability of the complete depth migration procedure.

CONCLUSIONS

It is believed that the concepts and implementation of the complete finite-difference depth migration procedure will prove to be a valuable tool in interpreting data associated with a variety of geologic situations. Deep and variable water depths are the most obvious candidates. However, it seems mandatory that the basic ideas have to be applied whenever substantial lateral velocity changes occur above the zone of exploration interest.

REFERENCES

Claerbout, J., 1976, Fundamentals of geophysical data processing: New York, McGraw-Hill Book Co., Inc., p. 254.
Claerbout, J., and Doherty, S., 1972, Downward continuation of moveout corrected seismograms: Geophysics, v. 37, p. 741–768.
Hubral, P., 1977, Time migration—Some ray theoretical aspects: Geophys. Prosp., v. 25, p. 738–745.
Larner, K., Hatton, L., and Gibson, B., 1978, Depth migration of complex offshore profiles: Proc. OTC, Houston.
Loewenthal, D., Lu, L., Roberson, R., and Sherwood, J. W. C., 1976, The wave equation applied to migration: Geophys. Prosp., v. 24, p. 380–399.

Quay, R. G., and Whittlesey, J. R. B., 1977, Wave equation migration operators using 2-D z-transform theory. Presented at the 47th Annual International SEG Meeting September 21, in Calgary.

Schultz, P. S., and Sherwood, J. W. C., 1980, Depth migration before stack: Geophysics, v. 45, this issue, p.

376–393.

Schneider, W., 1978, Integral formulation for migration in two and three dimensions: Geophysics, v. 43, p. 49–76.

Stolt, R., 1978, Migration by Fourier transform: Geophysics, v. 43, p. 23–48.

APPENDIX
MATHEMATICAL DEVELOPMENT OF FINITE-DIFFERENCE DEPTH MIGRATION METHOD

Our purpose is to provide a brief development of the mathematics of the finite-difference scheme of depth migration and portray its differences from the more conventional time migration method. We take the scalar wave equation

$$[\partial_{zz} + \partial_{yy} - C^{-2}(y, z)\, \partial_{tt}]P = 0 \quad (A-1)$$

as a starting point, where the notation ∂ associated with a subscript or superscript of y, z, or t indicates, respectively, partial differentiation or definite integration e.g.,

$$\partial^t P = \int_{\pm\infty}^{t} P\, dt$$

with respect to that variable. The symbols t, y, and z represent time, horizontal coordinate, and depth coordinate, respectively. The instantaneous wave velocity $C(y, z)$ can vary with spatial position. To invert or migrate the seismic data, we assume an upcoming scalar wave field $P(y, z, t)$ which has been recorded as $P_0(y, t)$ at $z = 0$. For regions where the velocity C is constant, we can factorize equation (A–1) and select the factor representing upcoming waves

$$[\partial_z - (C^{-2}\partial_{tt} - \partial_{yy})^{1/2}]P = 0,$$

or

$$[\partial_z - C^{-1}\partial_t(1 - C^2\partial_{yy}^{tt})^{1/2}]P = 0. \quad (A-2)$$

Equation (A–2) will now be used to approximate upward propagation through a medium with variable velocity $C(y, z)$. Note that it is accurate for upward-traveling data within constant-velocity regions, the nature of the approximation being to neglect reflections and transmission coefficient effects. Claerbout (1976, section 5.2) gives an explanatory discussion of the square root term in equation (A–2) and details its relationship to the square root of a finite-difference matrix operator. Adequate accuracy for events with shallow dips may be obtained by approximating the square root factor with the first two terms in its Taylor

series expansion,

$$[\partial_z - C^{-1}\partial_t + (C/2)\partial_{yy}^{t}]P = 0. \quad (A-3)$$

Note that an equation providing steep dip accuracy, but with a form similar to equation (A–3), is provided by modifying equation (A–2) to

$$\{\partial_z - C^{-1}\partial_t + C^{-1}\partial_t[1 - (1 - C^2\partial_{yy}^{tt})^{1/2}]\}\cdot$$
$$\cdot P = 0. \quad (A-4)$$

The last term in equation (A–4) can be approximated in a manner suited to the steepest dip of significance. Note that both equations (A–3) and (A–4) have the form

$$[\partial_z - C^{-1}\partial_t + M(C, \partial_t, \partial_y)]P = 0, \quad (A-5)$$

with the analytical solution to the initial value problem being

$$P(y, z, t) = \exp\left\{\int_0^z dz\,[C^{-1}\partial_t - M(C, \partial_t, \partial_y)]\right\}P_0(y, t). \quad (A-6)$$

This solution cannot be easily evaluated due to the complexity of the exponent. The procedure adopted for the complete finite-difference depth migration procedure is to use small increments in z and evaluate the scalar wave field $P(y, z, t)$ at depths $z_1, z_2, \ldots z_i, \ldots$ using the recursive adaptation of equation (A–6)

$$P(y, z_{i+1}, t)$$
$$= \exp\{(z_{i+1} - z_i)\,[C^{-1}\partial_t - M(C, \partial_t, \partial_y)]\}P(y, z_i, t). \quad (A-7)$$

To conform with the text and the actual implementation of depth migration, it is necessary to introduce a change of variables,

$$y' = y, \ \tau = \int_0^z R^{-1}(z)\, dz, \ t' = t + \tau. \quad (A-8)$$

$R(z)$ is the replacement velocity, whose only role is to

transform depth to a vertically elasticized time scale τ. The displacement of the time variable from t to t' stems from consideration of the "exploding" reflector model of migration introduced by Loewenthal et al (1976) (see Figures 2 and 3). At an intermediate depth z, this allows a data storage area to contain the fully migrated results from $t' = 0$ to τ, while the region $t' \geq \tau$ contains the upcoming wave field for times $t \geq 0$.

The change of variables leads to replacements for the differential operators,

$$\partial_y \to \partial_{y'}, \ \partial_t \to \partial_{t'}, \ \partial_z \to R^{-1}(z)(\partial_\tau + \partial_{t'}).$$
(A–9)

Substituting these into equation (A–5) and then dropping the primes leads to equations (A–10) and (A–11), replacements for equations (A–5) and (A–7),

$$[\partial_\tau - R(C^{-1} - R^{-1})\partial_t + RM(C, \partial_t, \partial_y)]P = 0, \quad (A–10)$$

$$P(y, \tau_{i+1}, t)$$
$$= \exp\{\Delta z[(C^{-1} - R^{-1})\partial_t - M(C, \partial_t, \partial_y)]\}P(y, \tau_i, t), \quad (A–11)$$

where the layer thickness Δz is

$$\Delta z = R(\tau_{i+1} - \tau_i). \quad (A–12)$$

For a thin layer, it is permissible to split the exponential in equation (A–11) into two separate multiplicative factors and consider them as operating sequentially upon the upcoming wave field $P(y, \tau_i, t)$. The first operator $\exp[-\Delta z M(C, \partial_t, \partial_y)]$ corresponds to that used in conventional finite-difference migration algorithms and will not be discussed further here. The

second operator $\exp[\Delta z(C^{-1} - R^{-1})\partial_t]$ is equivalent to inserting a perfect static time shift,

$$\Delta t = [\Delta z(R^{-1} - C^{-1})], \quad (A–13)$$

which may vary with y, into the upcoming wave field $P(y, \tau_i, t)$. This corresponds exactly[2] to the additional operation discussed in the main text that modifies conventional time migration to depth migration. A heuristic derivation of this term is included in the "Method" section and in Figure 1.

Some comments concerning the omission of the time-shifting term from conventional time migration are appropriate here. First, we note that the time shift can only be made zero for all y, z by assuming the velocity $C(y, z)$ to be laterally invariant and having $R(z) = C(z)$. In this situation, the output migrated section will be a true depth section elasticized to time using the replacement velocity function $C(z)$. If $C(y, z)$ varies only slowly with y, then omission of the time shift term will still result in a good approximation to a depth section elasticized to time using a replacement velocity $C(y, z)$. Thus, conventional time migration followed by transformation from time to depth is clearly a reasonable method for forming depth sections when the lateral velocity variations are slow. For rapid lateral velocity changes, it is clear that the time shift operator does not commute with the conventional migration operator [see equation (11)]; hence the time-shifting term must remain included as an integral part of the true migration procedure.

[2]A factor of 2 is apparently missing from the expression for Δt here. This is because the velocities $R(z)$ and $C(y,z)$ need to be halved throughout the Appendix in order to migrate data with two-way rather than one-way travel paths [see Loewenthal et al (1976) for a more detailed explanation].

Reprinted from GEOPHYSICS, **45**, 376-393.

Depth migration before stack

Philip S. Schultz* and John W. C. Sherwood*

When seismic data are migrated using operators derived from the scalar wave equation, an assumption is normally made that the seismic velocity in the propagating medium is locally laterally invariant. This simplifying assumption causes reflectors to be imaged incorrectly when lateral velocity gradients exist, irrespective of the degree of accuracy to which the subsurface velocity structure is known.

A finite-difference method has been implemented for migration of unstacked data in the presence of lateral velocity gradients, where the operation of wave field extrapolation is done in increments of depth rather than time. Performing this depth migration on unstacked data results in the imaging of reflectors on the zero-offset trace, whereupon a zero-offset section becomes a fully imaged-in-depth seismic section. Such a section, in addition to being a correctly migrated depth section, shows the same order of signal amplitude enhancement as in a normal stacking process.

INTRODUCTION

The task of imaging seismic reflectors in the presence of complex subsurface structure and sufficiently slowly laterally varying seismic velocities is generally accomplished by one of the three most common methods of wave-equation migration: the finite-difference method (Claerbout and Doherty, 1972), the Kirchhoff summation method (for example, Schneider, 1978), or by Fourier transform (F-K) migration (Stolt, 1978).

Each of these imaging methods is most accurate in a constant-velocity medium. However, as velocity gradients increase, these techniques experience varying degrees of success in maintaining accuracy. Of the three, the finite-difference and Kirchhoff methods are most amenable to vertical velocity gradients. However, when lateral velocity inhomogeneities are considered, particularly when velocity changes are rapid, the finite-difference method becomes the most convenient and natural way to deal with these complications. An important reason for this is that each step in the migration operation using finite-difference can be interpreted as a downward continuation operation, in which the generalization to severe lateral velocity variations becomes conceptually simple. Between each step in the downward continuation, an operation similar to a static time correction is included. This operation is the "time-shifting" term described by Claerbout (1976), and generally is neglected in derivations of finite-difference migrating equations. A discussion of the time-shifting term is given in a paper by Judson et al (1980, this issue).

Another advantage of the finite-difference method is that its operators tend to be more localized than those of the Kirchhoff summation method, which migrates using global operators (i.e., those which are applied once to the data for every output point). In a rapidly varying velocity medium, such global operators can become awkward to construct and implement. [Several researchers (Berryhill, 1978; Berkhout et al, 1978) have shown that the Kirchhoff operators can be convenient to construct when extrapolating a wave field from one irregular velocity interface to another, provided, however, that the velocity function between these interfaces has a sufficiently small lateral gradient.]

We have chosen the finite-difference method to develop a scheme that will maintain the same high

Presented at the 48th Annual International SEG Meeting October 30, 1978, in San Francisco. Manuscript received by the Editor December 22, 1978; revised manuscript received June 1, 1979.
*Digicon Inc., 3701 Kirby Drive, Houston, TX 77098.

fidelity in regions of severe lateral velocity variations as is achieved by current methods when the velocity is better behaved.

We use the term "depth migration" to mean a migration which includes the time-shifting term; it implies that the downward continuation is performed in successive increments of *depth*. This is in contrast to conventional migration, which can be called "time migration" because the downward continuation is performed in increments of time.

We will not address the difficult but important problem of the estimation of the velocity model to be used in the migration. It is conceivable that one may be able to develop a scheme for making a local velocity estimate at each downward continuation step, thereby constructing a velocity model concurrently with the depth migration. This concept, however, will have to be addressed in future investigations. There are indeed many cases in which severe lateral velocity variations exist and can be estimated fairly easily. The most obvious example is a deep sea floor with severe topography, such as off the Atlantic coast of North America.

In a region of very complex structure, even when we assume a constant velocity, conventional time migration after stack may not be given a fair opportunity to image steeply dipping reflectors. The stacking process, when occurring before migration, may discriminate against steeply dipping events because they have higher apparent stacking velocities than if they were of zero dip. In such a case, it may become desirable to perform a migration process before stack. Similarly, when lateral velocity variations are deep enough and severe enough, an attempt to stack before depth migration may result in the summation being discriminatory against some events. This paper,

Table 1. Minimum processing required to image reflectors in depth.

	Constant Velocity	Depth varying velocity	Slowly laterally varying velocity	Rapidly laterally varying velocity
Simple structure		Vertical raypath depth conversion	Image ray depth conversion	True depth migration
Complex structure	Time migration	Time migration + Vertical raypath depth conversion	Time migration + Image ray depth conversion	True depth migration

therefore, will deal with depth migration on unstacked data.

A perspective on depth migration is given by Table 1 which shows minimum processing required to image reflectors in *depth*. In a constant-velocity medium, no special imaging procedure is needed when reflectors have little structure, but in complex structure, simple time migration is needed. If we have only a depth variable velocity, we must include a vertical raypath depth conversion in the above processing. If we have a laterally variable velocity, then a true depth migration is required.

It has been demonstrated (Larner et al, 1977) that in some cases where the lateral velocity gradient is sufficiently small, an approximate depth migration of stacked data can be effective. This method is a depth conversion of time-migrated data along an image ray-

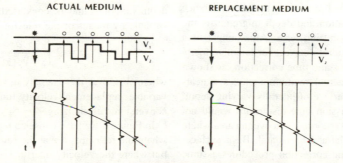

FIG. 1. An example of simple time shifting to correct for transmission through lateral velocity inhomogeneities. Static corrections apply a time shift equal to the difference in traveltime between the actual medium and some laterally invariant replacement medium.

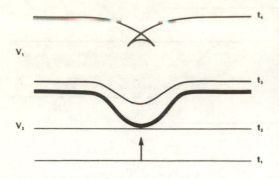

FIG. 2. A schematic diagram showing wave propagation effects. A planar wavefront is traveling upward at times t_1 and t_2 approaching a synclinal velocity interface. At time t_3, the wavefront has passed through the interface and has been distorted by it in a simple manner. On further propagation, as at time t_4, focusing effects can appear.

path (Hubral, 1977). When this procedure is acceptable, it is convenient and economical to use.

A more accurate depth migration procedure is described here (and in Judson et al, 1980). Such a procedure can be expected to image reflectors properly in the presence of an arbitrary lateral velocity gradient, provided that the velocity function is known. The last column in Table 1 indicates that the need for depth migration is not controlled by the reflector geometry, but only by the velocity structure. This implies that in certain situations we may not be able to image a perfectly horizontal reflector without some type of depth migration. The two synthetic data examples given later will demonstrate that this is indeed the case.

THE METHOD

We have mentioned previously that the difference between time migration and depth migration is the addition of a time-shifting term. An example of a time-shifting operation to correct for lateral velocity variations is simple static time correction, as shown in Figure 1. The actual medium includes a near-surface velocity interface of laterally varying depth. Anomalous transmission time distortions would not be present if we had recorded data over an area where that interface was flat and horizontal. Basic surface-consistent static time correction procedures assume that the only difference between the actual and desired recording is a unique time shift for each trace. The magnitude of the shift is given by the difference

in vertical traveltimes in the two media. Thus, static corrections substitute a "replacement medium," whose near-surface velocity is laterally invariant, for the actual medium.

Static time corrections work well when the lateral velocity inhomogeneities are at a shallow depth. When the anomalous zone is at a greater depth, however, transmission effects involve more than a simple static time shift. In Figure 2, an upward-traveling planar wavefront at times t_1 and t_2 is approaching a buried synclinal feature which is also a velocity interface. At time t_3, the wavefront has completed transmission through the feature and has been distorted by it. At this point, a static time shift will return the distorted wavefront to its original shape. If, however, we permit the distorted wavefront to propagate farther, such as shown at time t_4, the effects of the propagation are such that a simple time-shifting operation will no longer return the wavefront to its original shape.

Figure 3 shows a schematic display of an actual medium consisting of a buried channel, acting also as a velocity interface, overlying a flat reflector. Also shown is the zero-offset section we expect to receive over such a structure. The buried channel reflection is diffracted and shows an effect due to a buried focus. The reflection from the flat horizon is also somewhat diffracted due to the transmission distortion and subsequent propagation effects which the wavefront had suffered. Clearly, time migration would drastically misrepresent the flat reflector. To obtain a true structural picture, it is necessary to have an accurate velocity model and perform a depth migration operation designed for zero-offset sections, as discussed by Judson et al (1980). Within this procedure, it is convenient to convert the depth section to a vertical two-way traveltime section using some suitably chosen replacement velocity function which is laterally invariant, but which may vary with depth. That is, the depth variable z may be converted to a two-way time variable by means of the transformation

$$t = 2 \int_0^z \frac{dz}{V_r(z)} ,$$

where $V_r(z)$ is the laterally invariant, but depth-variable, replacement velocity function. We call this conversion "elasticizing."

In Figure 3 we have chosen a replacement medium where the reflectors are at the same depth locations, but where the velocity interface is everywhere horizontal; its continuation across the synclinal feature is shown as a dotted line. Had the data been recorded over this medium, the reflection from the underlying

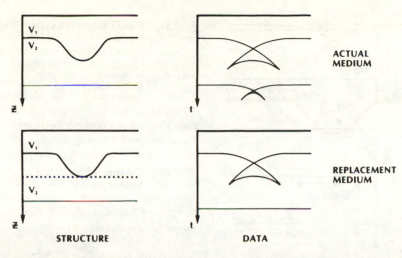

FIG. 3. Simulated transmission focusing in seismic data. When data are recorded over a lens-shaped velocity interface, focusing effects can appear on underlying reflections as in the upper two figures. If the velocity interface were everywhere horizontal, as in the lower figures (the dotted line marks the velocity interface), no transmission effects would distort underlying reflections.

flat horizon would show no anomalous transmission effect.

The removal of these transmission effects on *unstacked* data is accomplished by downward continuation of the sources and receivers. Since unstacked data are not exclusively zero source-receiver offset, we cannot use the "exploding reflector" formulation [as in Judson et al (1980)] to accomplish a simultaneous downward continuation of both sources and receivers. Instead, we must extrapolate the receivers and sources separately.

Assuming the usual case of weak reflectors, we have upward- and downward-traveling waves which are decoupled except for the free surface reflection. If the corresponding ghosting can be considered part of the source pulse, we then may regard a common-source gather (field record) as the sampling of an upcoming wave field. Thus, we may use a one-way wave equation to extrapolate this wave field in depth, say by an amount Δz. By so doing, we have, in effect, downward continued the receivers and left the source at the surface. If we perform this downward continuation operation on all such common-source gathers in the unstacked data, we will have downward continued all receivers by an increment Δz and will have left all sources at the surface.

We regather the data such that each gather contains data with a common-receiver coordinate; we call it a common-receiver gather. We now invoke the principle of reciprocity (for example, Claerbout, 1976, p. 174) and assume that there exist sources, in exchange with our receivers, and receivers, in exchange with our sources, which will produce an identical data set, trace for trace. Therefore, a common-receiver gather is the sampling of a physical wave field; as a result, the sources may be downward continued by wave field extrapolation. As we have done with the common-source gathers, we perform the wave field extrapolation in depth on all common-receiver gathers in the data set. Both sources and receivers will then have been downward continued by an amount Δz.

Figure 4 shows the physical interpretation of the process described above. As implied in the figure, the process may be repeated for downward continuation of the data to any depth.

Suppose the sources and receivers are at the same depth, say $z = z_1$. Then, within the context of a primaries-only model, the data above this depth would be fully migrated and imaged on the zero-offset trace, and they would not be included in subsequent downward continuation. Below this depth, the data would appear as if with a new zero time, equal to the vertical two-way traveltime to the depth mentioned above through the laterally invariant replacement medium.

To illustrate this point, Figure 5 shows a synthetic

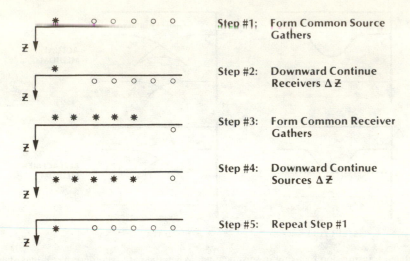

FIG. 4. The downward continuation sequence. The time shifting is done after step 2 for the receivers, and after step 4 for the sources. The sequence may be repeated for downward continuation of data to any desired depth.

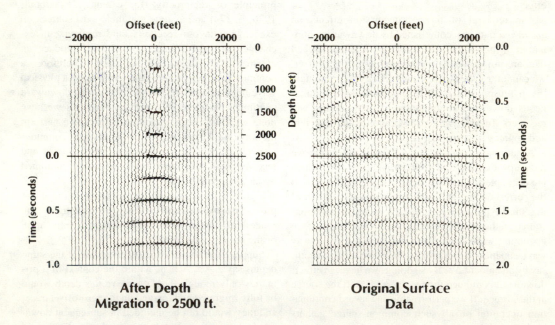

FIG. 5. Synthetic common-source field record before and after depth migration down to 2500 ft. The medium velocity is everywhere 5000 ft/sec, the group interval is 100 ft, and horizontal reflectors exist every 500 ft. After depth migration, data from reflectors above 2500 ft are focused on the zero-offset trace and can be displayed on a depth scale, while data from reflectors below 2500 ft are displayed on a time scale with a redefined time origin.

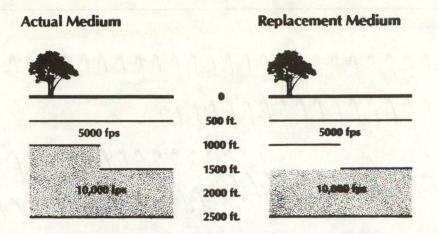

FIG. 6. Model for synthetic data of Figures 7 through 13. In the actual medium, the velocity interface occurs along the faulted reflector. In the replacement medium, the velocity interface occurs uniformly at 1500 ft.

common-source split-spread record over a constant-velocity medium (5000 ft/sec) where horizontal reflectors exist every 500 ft (200 msec two-way traveltime). Random noise has been added. Since the velocity in the "actual" medium is laterally invariant, the replacement medium will be the same as the actual medium, and no time-shifting need be done by the depth migration process. The frame on the left shows the data after depth migration, where we have downward continued the sources and receivers to 2500 ft (one second of two-way time).

Above 1 sec, the data have been imaged onto the zero-offset trace. The data from 1 to 2 sec are partially imaged, and appear as if there were a new zero time, located at 1 sec on the original time scale. The new vertical scales on the migrated record show that above the new datum we have a depth section, while below it we have a time section with the time origin redefined.

The zero-offset trace for the imaged data (i.e., for reflectors above 2500 ft) is the trace which we would use on the output section. No further stacking is required. We demonstrate in the Appendix that we expect the signal to random noise ratio on the zero-offset trace of the imaged data to be effectively identical to that of a common-depth-point (CDP) stacked trace.

The data *below* 2500 ft are partially focused, but the zero-offset trace here would have a poorer signal-to-noise ratio than a stacked trace. These data need to be further imaged or stacked.

At this stage in the processing, three options appear. First, we may stack the data below the new zero-time in the conventional manner (but redefine zero-time) and append the stack to the imaged data. Second, we may reverse the algorithm and upward propagate the shots and groups back to the surface, but through the laterally invariant replacement medium. All the data can then be processed as if they were recorded in a laterally homogeneous velocity medium. Third, we may continue the depth migration process all the way to the deepest reflector.

Economy and convenience would usually dictate which of the three options to use in a particular situation. For example, if the anomaly is at a fairly shallow depth, we may wish to upward propagate the partially migrated data back to the surface to facilitate subsequent processing or to make a more meaningful comparison with the original input data.

FAULT MODEL

Figure 6 shows a vertical cross-section of a synthetic earth model. The actual medium shows a flat reflector at 500 ft, a reflector at 1000 ft dropping to 1500 ft and acting as a velocity interface (5000 ft/sec over 10,000 ft/sec), and a horizontal reflector at 2500 ft. The replacement medium which we have chosen for this model also is shown. In it, the velocity interface occurs at a uniform depth of 1500 ft.

Figure 7 shows synthetic common-source trace gathers over the model. The common-source gathers are 24 traces with a 164-ft receiver interval arranged

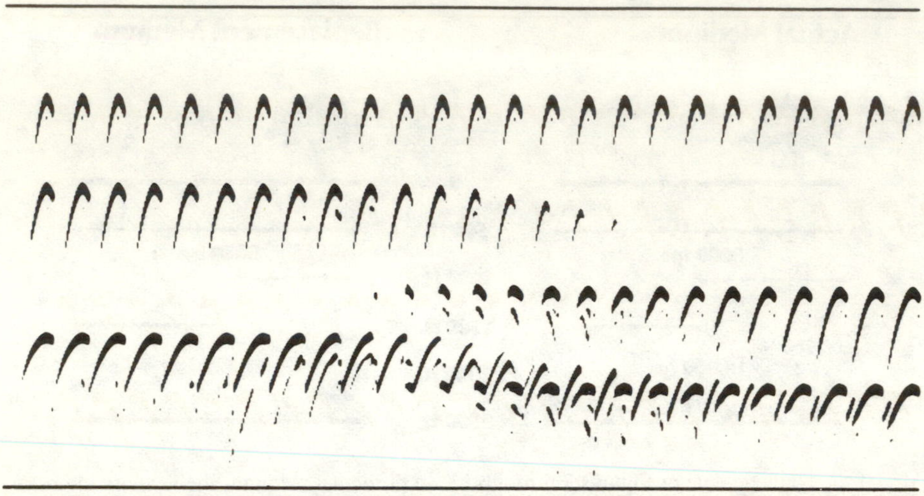

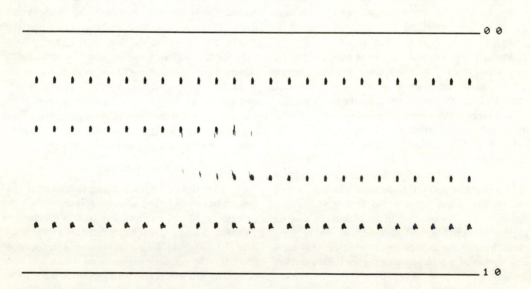

FIG. 7. Synthetic common-source field records generated by the program in the modeling mode. Each record is 24-trace, with a 50-m source and receiver interval. The flat reflector at 2500 ft shows complex transmission effects.

FIG. 8. Imaging of the reflectors onto the zero-offset trace of the common-source records by depth migration of the full 1 sec of data. The apparent overlap of the second reflection is caused by the absence of very high-spatial wavenumbers.

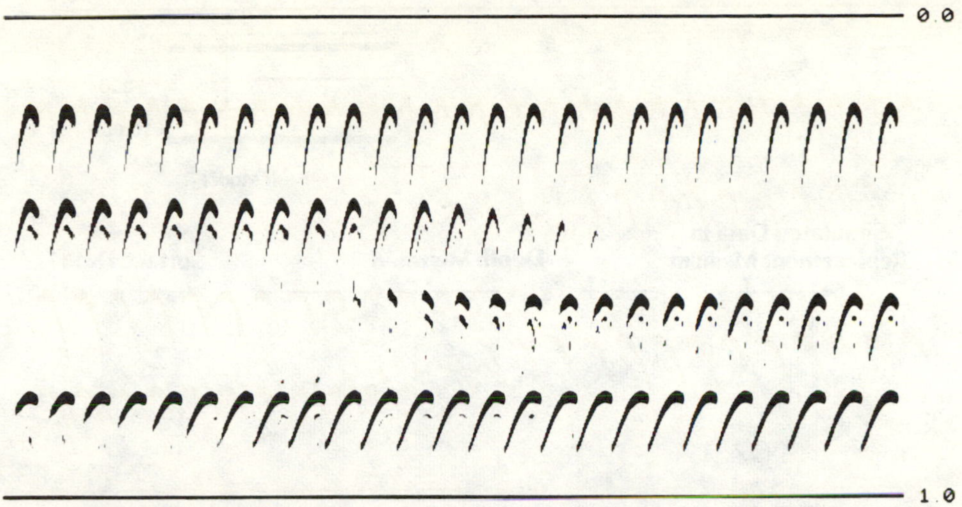

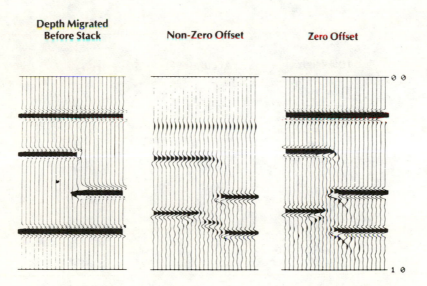

FIG. 9. Simulated data in replacement medium. Starting with the imaged data shown in Figure 8, the sources and receivers were upward continued through the replacement medium to the surface. The deep reflection shows a simple signature with no anomalous transmission distortions. The missing tails of some of the hyperbolas on the deep reflection represent seismic energy which would have originated outside the finite extent of the data, as shown in Figure 8.

FIG. 10. Zero-offset sections of the data of Figures 7 and 8. The two sections of surface data (i.e., synthetic field data) are plotted with the same amplitude scaling; the depth migrated zero-offset section is on a smaller amplitude scale.

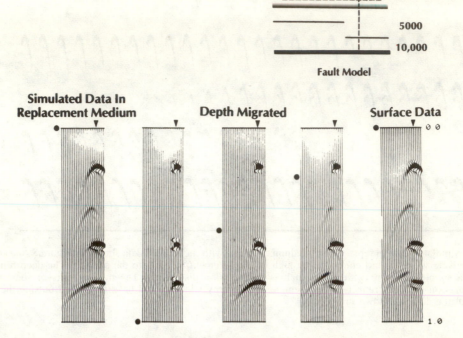

Fault Model

FIG. 11. Common-source record taken from the data of Figure 7. The inset shows the position of the source (asterisk) and receiver spread (dashed horizontal line) for this record relative to the fault. The three middle records show various stages of downward continuation. Record on far left is data as it would appear if recorded in replacement medium. Markers show downward continuation level, and arrowheads indicate zero-offset trace.

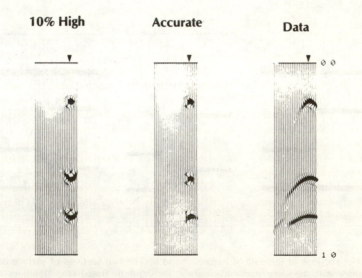

FIG. 12. The effect of using migration velocities everywhere 10 percent too high.

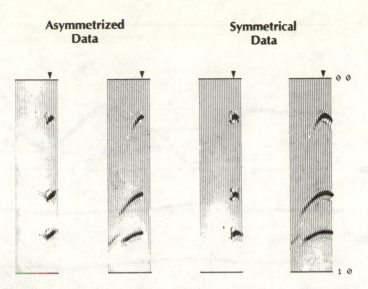

FIG. 13. The effect of asymmetry. The foci have noticeable dip, but the energy is still centered on the zero-offset trace.

asymmetrically about the shot, simulating a possible land acquisition geometry. Figure 8 shows the depth migration before stack done for the full 1 sec of data. The replacement medium shown in Figure 6 was used for the depth migration, so the imaged reflectors are at a time equal to the two-way vertical traveltime through the replacement medium. The deep reflector is therefore seen to be flat. The data generation is the result of the program running in the modeling mode, whereas the depth-migrated data result from running the program in the migrating mode. While we have demonstrated only that the process is reversible, there are some instructive observations which can be made.

The migrated data have the focus appearing smeared over perhaps three traces. This is also the reason the vertical escarpment appears to be overlapping. Spatial wavelengths shorter than twice the group interval are generally attenuated by field-recording geometries and group array responses; we filtered these components out of the synthetic data. Even if we had not filtered them out, the finite-difference algorithm would be accurate up to, but not past, the spatial aliasing wavenumber. We are, therefore, observing an effect due to a finite bandwidth of spatial wavenumbers.

As expected, the shallow event is unaffected by the deeper velocity anomaly, and maintains a simple structure along the line. The reflector at 1000 ft is

more complex, but the data can be understood as the response of a truncated reflector. The diffraction from the truncation of the reflector at 1500 ft has raypaths through the complex velocity medium of the step. However, the effects of the velocity anomaly can be seen most readily on the deepest reflection. It has raypaths through both the high and low parts of the step and shows both the time shift and its associated diffraction effects.

Figure 9 shows the result of upward propagating the imaged data of Figure 8 to the surface through the laterally invariant replacement medium. This figure shows the data as they would have appeared had we actually recorded data over the replacement medium shown in Figure 6. Since this exercise has simulated data in a medium with no lateral velocity inhomogeneity, depth migration would not be needed to process the data further.

Figure 10 shows the data of Figures 7 and 8 in common-offset section form. Both a zero and a non-zero offset section are shown for the surface data, and in each of them the complex transmission effects can easily be seen on the deepest reflection. Both offset sections of the surface data are plotted with the same amplitude level and can be compared directly. The shallow event on the nonzero offset section has a lower amplitude because it is a sampling along the slope of the hyperbolic signature, where the amplitude is lower. An interesting feature of the nonzero

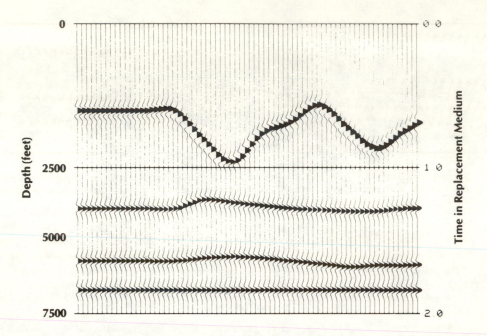

FIG. 14. Second synthetic model. Zero-offset section shows irregular sea floor which is a 5000-to-10,000 ft/sec velocity interface (the "actual" medium), and is underlain by three reflectors of gentle structure, the deepest being perfectly flat. These three events are portrayed at a two-way traveltime in the replacement medium, which has the velocity interface at a uniform depth of 2500 feet. A nonlinear depth scale is also defined as shown.

offset section can be seen on the deep reflection, where it appears that there is a double step. The intermediate step occurs when the source is over the lower part of the escarpment and the receiver is over the upper part. Complex diffraction effects can also be seen on this feature.

The section in Figure 10 labeled "Depth migrated before stack" is the zero-offset section taken from the data of Figure 8. Because the migration has the effect of stacking the data onto the zero-offset trace, it is plotted at a much lower gain than the other sections. As we would reasonably expect, depth migration represents the deep reflector as flat, at a time equal to the two-way traveltime through the replacement medium.

Figure 11 shows a common-source record taken from the data set of Figure 7. The inset shows the position of both the recording line and the zero-offset imaged trace with reference to the fault. Depth migration of the complete data set was performed here in three steps: first, downward continuation to 224 msec, then to 616 msec, then all the way to 1000

msec. The display shows (from right to left) the same record as recorded at the surface (surface data), after each of the three downward continuation steps (depth migrated) and also as the simulated recording in the replacement medium. The markers on each record show the zero-offset trace (arrow at top of each record) and the current downward continuation datum (solid circle). Notice that as we apply downward continuation, the data above the new zero-time (solid circle) are imaged on the zero-offset trace, and data below appear in reference to the new zero-time. Note also that by the time we have reached 616 msec, most of the lateral inhomogeneity is above the current zero time, and there are no apparent anomalous transmission effects remaining on the deep event. On the simulated recording in the replacement medium (first record on left), diffractions from the escarpment are apparent, but no anomalous transmission effects are seen on the deep reflection.

Figure 12 shows the effect of the depth migration using a velocity that is 10 percent too high everywhere. Particularly on the deeper events, the effects

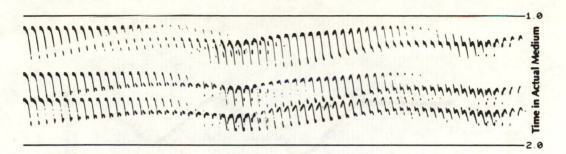

Fig. 15. Synthetic field records generated from the model of Figure 14. Records are 36-trace with three negative offsets and a 50-m source and receiver interval. Only data between 1.0 and 2.0 sec were included to avoid visual contamination by diffractions from the sea-floor reflector itself.

of overmigration are immediately evident. It is conceivable that this phenomenon could be used as a tool to improve the accuracy of the velocity model.

In acquisition geometries, such as marine "end-on shooting," where there are no zero-offset and nearest-offset traces, or where gaps exist in the receiver spread, special data preparation must be made. Since all uniformly spaced offsets down to zero are necessary (and several "negative-offset" traces are generally desirable), any nonexistent traces at offset positions therefore must be supplied either as zeroed traces or as extrapolations or interpolations of exist-

ing traces. In Figure 13, a comparison is made between a data set which has actual traces over all offsets ("symmetrical data") and a data set containing some zeroed traces ("asymmetrized data"). The symmetrical data result in a collapse to a symmetrical focus, while the focus in the asymmetrical data has a noticeable dip. The zero-offset trace, however, is through the center of the focused energy in either case. The data labeled symmetrical data were modified to a symmetrical split-spread by a program option which invoked the reciprocity principle. This option was not used on the asymmetrized data. If it

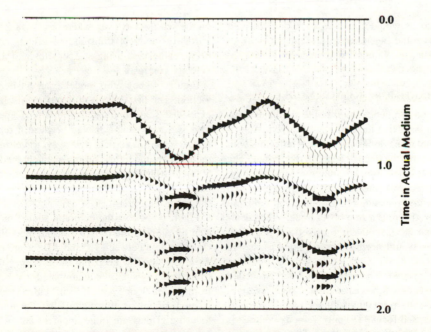

Fig. 16. The surface data as a zero-offset section. Shown are zero-offset traces from Figure 15 appended to the first second of the data of Figure 14. The three deep reflections are depicted here with the velocity interface following the sea-floor event (i.e., the time scale refers to the actual medium).

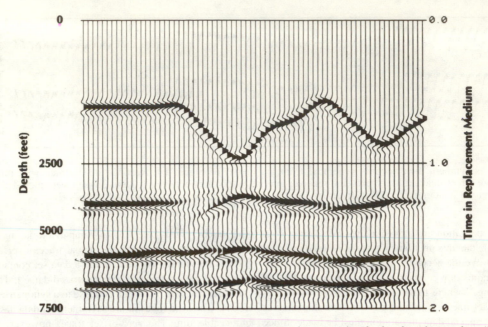

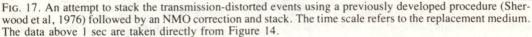

FIG. 17. An attempt to stack the transmission-distorted events using a previously developed procedure (Sherwood et al, 1976) followed by an NMO correction and stack. The time scale refers to the replacement medium. The data above 1 sec are taken directly from Figure 14.

were, the focus would appear as an X-pattern, with both positive and negative dip present but with no zero dip because of the missing inner trace.

BALTIMORE CANYON MODEL

The second synthetic data example simulates a situation in an area like the Baltimore Canyon. A simulated irregular sea floor overlies three reflectors with gentle structure, the deepest being perfectly flat. The data are in 36 trace common-source records with the source and receiver interval both 164 ft. Thirty-two receivers are on one side of the source and three are on the other. There is a zero-offset trace. When dealing with marine data, the zero-offset and negative-offset traces would normally be supplied as zeroed traces prior to depth migration (as in the following real data example). The zero-offset trace is needed because the foci will be centered on it. The negative-offset traces, when present, facilitate identification of a fully collapsed focus. The only velocity contrast is across the sea floor reflector, and is 5000 ft/sec over 10,000 ft/sec. This fairly severe contrast was used to emphasize the transmission effects. The structural dip

on the sea floor is everywhere less than 27 degrees. Virtually identical transmission effects could have been synthesized using steeper dips and less severe velocity contrasts.

Figure 14 shows this model as an elasticized depth section. The first 2500 ft of depth are displayed at an expanded scale compared with the deeper region. This is a convenient display because it represents a linear scale in vertical two-way traveltime in the replacement medium. Our replacement medium has a 5000- to 10,000-ft/sec velocity interface at a constant depth of 2500 ft. The corresponding time scale is also shown on Figure 14.

Figure 15 shows the common source records created by running the program in the modeling mode. Only the data between 1.0 and 2.0 sec are portrayed. This was done to keep the three deeper reflections from being visually contaminated by diffractions from the sea floor event. It is difficult to see anything in great detail on this figure, but the transmission focusing effects are quite apparent and will clearly frustrate any attempt at a simple stacking of the data.

Figure 16 shows the simulated surface data in a

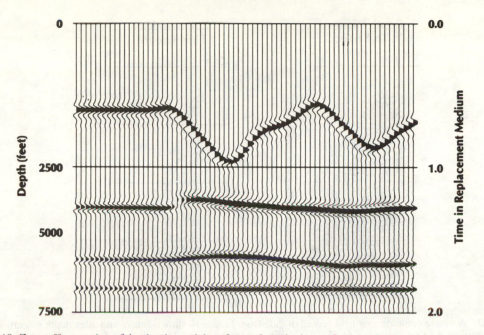

FIG. 18. Zero-offset section of depth migrated data. Loss of continuity on the second event is due to energy lost off the end of the spread. The loss of amplitude on the left is due to energy loss off the end of the line (similar to CDP taper-on and taper-off).

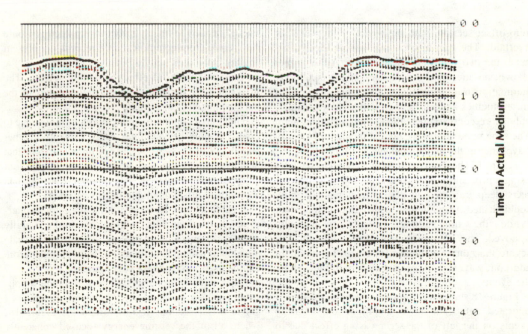

FIG. 19. Nearest-offset section of NMO-corrected data from offshore United States. The sea floor is a 4950 to 6500 ft/sec velocity interface. The replacement medium fills the valleys with 6500 ft/sec velocity material.

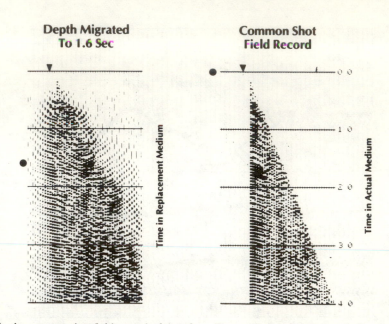

Depth Migrated To 1.6 Sec

Common Shot Field Record

Time in Replacement Medium

Time in Actual Medium

FIG. 20. A single common shot field record of data from Figure 19 shown before and after depth migration of the complete data set down to 1.6 sec (in the replacement medium). Data above 1.6 sec are focused onto the zero-offset trace, while data below are noticeably undermigrated. The low dominant frequency on migrated events above 1.0 sec is due to spatial aliasing of the higher frequencies caused by the large group interval (100 m) and also by the absence of traces with small offset.

zero-offset section, with the model of the sea floor overlaid. The transmission foci appear clearly beneath the two valleys of the sea floor. Although not shown, an attempted simple stack of the data along summation trajectories calculated from the vertical rms velocities at each location will yield a considerably more confused result than that portrayed here. The time scale here is the normal incidence two-way traveltime in the actual medium.

Figure 17 shows our attempt at a coherent stack by using a procedure (Sherwood et al, 1976) which corrects deep transmission effects in a manner more sophisticated than simple static time corrections. While the time-shifting anomaly is virtually removed, the amplitude anomalies, although less severe than those of Figure 16, are still present. In addition, waveform inconsistency is noticeable along the events.

Figure 18 shows a zero-offset section of the depth-migrated data. The lower amplitude on the deep events on the left of the section is an effect due to energy being lost off the end of the survey line when in the modeling mode. That is, some of the seismic

energy needed to image fully those reflectors would have come from common source records not present in this finite data set. It is an effect similar to the taper-on and taper-off effect on a CDP section. The effect is more noticeable on the left because of the data asymmetry.

The loss of continuity on the second event is due to energy lost off the end of the receiver spread. In other words, all of the seismic energy reflected from that segment of the second reflector which was not imaged would be present only on offsets greater than the farthest offset available in this geometry.

Figures 17 and 18 show a time scale to be interpreted as two-way vertical traveltime in the laterally invariant replacement medium. The depth scale which was used for the model in Figure 14 therefore can be used also in these two figures.

Figures 16–18 were not plotted at the same amplitude. The gain was highest for Figure 16, since it is a single offset section. It was lowest for Figure 18, where all of the seismic energy focused coherently onto the zero-offset trace (except that energy which was lost off the end of the line or off the end of the spread). Thus, if random noise had been a factor, Figure 18 would have shown the highest signal-to-noise ratio.

REAL DATA EXAMPLE

Figure 19 shows a zero-offset section of data from the East Coast of the United States. The cable contained 36 receiver groups. Both source and receiver intervals are 100 m. The interval velocity contrast at the water bottom is 4950 ft/sec to approximately 6500 ft/sec. Figure 20 shows a common source record and the same record after all data were depth-migrated to 1.6 sec (solid circle). The record was expanded to 48 traces with the addition of 12 dead traces, 9 of which had offsets on the opposite side of the source. The zero-offset trace is indicated by an arrow. The data were migrated without the symmetrizing option, so that the foci appear with dip, as in the synthetic data example (Figure 13).

The reflection amplitude is greatest on the inner traces, and, therefore, it is difficult to detect the remnant hyperbolic tendency (c.f., Figure 5) on the events below 1.6 sec. It is clear, however, that the events at and above 1.6 sec are focused approximately on the zero-offset trace, while those deeper are undermigrated, as expected. All of the energy which is not focused onto the zero-offset trace above 1.6 sec was interpreted by the algorithm as nonsignal in nature. An example of this effect was seen in Figure 5, where the nonsignal was specifically random noise.

The sea floor reflection itself is subject to fairly severe spatial aliasing because of the large group interval and the absence of very small-offset traces. The migration has therefore recovered only the low frequencies. This low-frequency feature is quite apparent on Figure 21, the zero-offset section of the depth migration to 1.6 sec. In the replacement medium, we have filled the valleys of the sea floor with the subbottom velocity of 6500 ft/sec. Three aspects of this section contrast it to the input section, Figure 19. First, the anomalous time-shifting due to the sea-floor transmission effect has been removed. Secondly, continuity appears enhanced on many of the events between 1.0 and 2.0 sec. Third, the focusing by the depth migration process has resulted in higher

FIG. 21. Zero-offset section of depth-migrated data. The low-frequency character above 1.0 sec is due to spatial aliasing. Data below 1.6 sec are not fully focused, and further focusing or stacking is required. Data above 1.6 sec have been fully imaged by depth migration.

reflection amplitudes in the 1.0- to 2.0-sec region relative to the background noise level.

If the problems causing loss of high-frequency information above 1.0 sec had not been present, the data of Figure 21 down to 1.6 sec could be considered as data of the final output section. Data below 1.6 sec are not fully focused and either must be stacked using an NMO function based on the new time origin, or the depth migration must be continued to the full 4.0 sec. The appearance of the data subjectively verifies this. Data above approximately 2.0 sec show improved coherence and signal-to-noise ratio, while below 2.0 sec, the data do not appear to have been significantly improved.

CONCLUSIONS

Depth migration before stack has been implemented using a finite-difference migration method which includes the time-shifting term important in laterally varying velocity media. In cases where lateral velocity variations are sufficiently severe and deep, previously available approximations to wave-equation migration will not maintain the fidelity which may be desired. In these cases, true depth migration before stack, such as implemented here,

can provide a more accurate imaging process when the velocity model can be provided.

REFERENCES

Berkhout, A. J., Larson, D. E., and Rockwell, D., 1978, A prestack migration system employing downward continuation and optimal concurrent interval velocity estimation: Presented at the 48th Annual International SEG Meeting October 30, in San Francisco.

Berryhill, J. R., 1979, Wave equation datuming: Geophysics, v. 44, p. 1329–1351.

Claerbout, J., 1976, Fundamentals of geophysical data processing: New York, McGraw-Hill Book Co., Inc.

Claerbout, J., and Doherty, S. M., 1972, Downward continuation of moveout corrected seismograms: Geophysics, v. 37, p. 741–768.

Hubral, P., 1977, Time migration—Some ray theoretical aspects: Geophys. Prosp., v. 25, p. 738–745.

Judson, D. R., Lin, J., Schultz, P. S., and Sherwood, J. W. C., 1980, Depth migration after stack: Geophysics, v. 45, this issue, p. 361–375.

Larner, K., Hatton, L., and Forshaw, R., 1977, Depth migration of imaged time sections: Presented at the 39th Annual EAEG Meeting, Zagreb.

Schneider, W., 1978, Integral formulation for migration in two and three dimensions: Geophysics, v. 43, p. 49–76.

Sherwood, J. W. C., Adams, H., Blum, C., Judson, D. R., Lin, J., and Meadours, B., 1976, Developments in filtering seismic data: Presented at the 46th Annual International SEG Meeting October 25, in Houston.

Stolt, R., 1978, Migration by Fourier transform: Geophysics, v. 43, p. 23–48.

APPENDIX
SIGNAL-TO-RANDOM NOISE RATIO ON THE IMAGED ZERO-OFFSET TRACE AND CDP-STACKED TRACE

We give a reasonability argument to show that the signal-to-random noise ratio on the imaged zero-offset trace is essentially the same as that on a CDP-stacked trace.

We consider the hyperbolic signature on a common-source field record of a single reflecting horizon in the presence of noise. The hyperbolic signature may be collapsed into a focus by a wave-equation propagation operator of the form

$$e^{i\phi(f,k)z},\qquad (A-1)$$

which is an all-pass filter in both frequency and wavenumber (Claerbout, 1976, p. 190). The total energy in the $f-k$ spectrum is the same before and after application of the all-pass focusing filter, and by Parseval's theorem, the same is true in the $x-t$ domain.

Suppose that we have M traces over which the

hyperbolic signature is present, and that the amplitude of the signature is s_i on trace i. Suppose also that after focusing the signature onto the zero-offset trace, its amplitude is S_f. Then, a statement that energy has been conserved in the $x-t$ domain is

$$\sum_{i=1}^{M} s_i^2 = S_f^2. \qquad (A-2)$$

If we make the simplifying assumption that the amplitude s_i is the same on all traces, then equation (A-2) can be written

$$Ms^2 = S_f^2,$$

or

$$S_f = \sqrt{M}\,s. \qquad (A-3)$$

At the same time, the random noise amplitude has stayed the same everywhere (as shown in Figure 5).

So, if the noise amplitude on each input trace is n, the input signal-to-random noise ratio was s/n. If the output noise amplitude is also n, then the output signal-to-random noise ratio is

$$S_f/n = \sqrt{M}\, s/n, \qquad (A-4)$$

showing that we have improved the signal to random noise ratio by $\sqrt{M}$.

If, instead, we summed along the proper hyperbolic trajectory, we would add the individual trace amplitudes

$$S_{\text{stack}} = \sum_{i=1}^{M} s_i,$$

or

$$S_{\text{stack}} = Ms, \qquad (A-5)$$

but the random noise would add in quadrature

$$\sum_{i=1}^{M} n_i \rightarrow \sqrt{M}\, n, \qquad (A-6)$$

giving a final signal-to-random noise ratio of

$$S_{\text{stack}}/\sqrt{M}\, n = Ms/\sqrt{M}\, n$$
$$= \sqrt{M}\, s/n, \qquad (A-7)$$

which is the same as that achieved by wave-equation focusing.

The above argument is not meant to be a rigorous proof. We attempt only to show that we may reasonably expect to use the focused zero-offset trace in place of the CDP-stacked trace without degrading the signal-to-random noise ratio.

Reprinted from GEOPHYSICS, **46**, 734-750.

Depth migration of imaged time sections

Kenneth L. Larner*, Leslie Hatton‡, Bruce S. Gibson*, and
I-Chi Hsu§

ABSTRACT

None of the leading approaches to the migration of seismic sections—the Kirchhoff-summation method, the finite-difference method, or the frequency-domain method—readily migrates seismic reflections to their proper positions when overburden velocities vary laterally. For in-homogeneous media, the diffraction curve for a localized, buried scatterer is no longer hyperbolic and its apex is displaced laterally from the position directly above the scatterer. Hubral observed that the Kirchhoff-summation method images seismic data at emergent "image ray" locations rather than at the desired positions vertically above scatterers. In addition, distortions in diffraction shapes lead to incorrect imaging (i.e., incomplete diffraction collapse) and, hence, to further displacement errors for dipping reflections.

The finite-difference method has been believed to continue waves downward correctly through inhomogeneous media. In conventional implementations, however, both the finite-difference method and frequency-domain approach commit the same error that the Kirchhoff method does. Synthetic examples demonstrate how conventional migration fails to image events completely.

Hubral's solution to this migration problem is two- (or three-) dimensional mapping of imaged time sections into depth. This mapping, "depth migration," replaces simple vertical conversion from time to depth. Such depth migration can be postponed until after efficient image-ray modeling has been performed to (1) support the final choice of velocity model, and (2) determine whether depth migration is necessary.

Comparisons between depth-migrated and conventionally depth-converted sections of both synthetic and field data properly show that significant lateral displacement is often required to position reflectors properly. Monte Carlo studies show that the lateral corrections can be important not only in absolute terms but also in relation to errors expected from an inaccurate velocity model.

INTRODUCTION

Migration of seismic data is performed to position reflections correctly on seismic sections. Accurate migration, by any method, requires not only a high-quality estimate of velocity but also an algorithm that can properly apply detailed velocity information to the unmigrated data. Unfortunately, none of the commonly used approaches to wave-equation migration—the Kirchhoff-summation method, the finite-difference method, or frequency-domain methods—migrates reflections to their proper positions if subsurface structure is complex and is characterized by significant lateral variation of velocity.

Today, all valid seismic migration processes are known to be firmly rooted in the wave equation description of seismic wave propagation. We owe this new (or renewed) enlightenment largely to the contributions of Jon Claerbout, particularly his description of the finite-difference approach to migration (Claerbout and Doherty, 1972; Claerbout, 1976). French (1975) showed that, with some refinement, the heuristic migration approach of summing amplitudes along diffraction hyperbolas is equivalent to the Kirchhoff-integral solution of the wave equation. To Larner and Hatton (1976), the two approaches—finite-difference and Kirchhoff summation—simply involve different approximations for the solution of the same wave equation. Schneider (1978) further demonstrated that the Kirchhoff-integral method can be formulated to simulate the downward continuation of the finite-difference method. Stolt (1978) and Gazdag (1978) presented other migration approaches based upon solutions to the wave equation in frequency-wavenumber space. Implementation of any one of these approaches involves approximations that have counterparts in the others; consequently, the quality of migration results depends largely upon the validity of these approximations in the presence of data characteristics such as signal-to-noise (S/N) ratio, maximum dip, trace spacing, sampling interval, and frequency content.

Once current debates on the relative merits of these and other methods of migration are over, we will find that the proper treatment of velocity is the most important issue determining the accuracy of any particular migration result. When one analyzes results of migrating field data, a most appropriate question is: Were the velocities used in the migration correct? (We know, for

Presented at the 47th Annual International SEG Meeting, September 20, 1977, in Calgary. Manuscript received by the Editor November 29, 1979; revised manuscript received September 4, 1980.
*Western Geophysical Company, P.O. Box 2469, Houston, TX 77001.
‡Formerly Western Geophysical Company, Houston; presently Merlin Geophysical Company, Woking, Surrey, England.
§Formerly Western Geophysical Company, Houston; presently Occidental Eastern, Inc., Hong Kong.

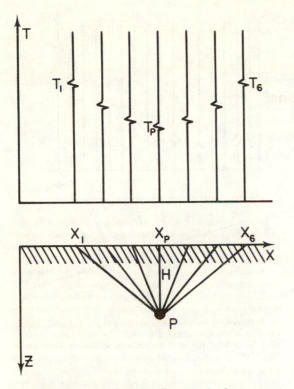

FIG. 1. Bottom: raypaths from a small scatterer in a homogeneous medium. Top: resulting diffraction curve on a time section.

example, that sections "overmigrate" where migration velocities are erroneously high and "undermigrate" where they are low.) Moreover, when geologic structures are complex, the question could become more involved: Were the detailed velocity variations (and resulting complicated raypaths) in the subsurface treated properly in the migration process? For the interpretation of geologically complex sections, the proper treatment of migration velocity is as important as migration itself. If we fail to take account of detailed velocity variation in an inhomogeneous overburden, we can miss structural highs, penetrate unexpected faults,

and make seismic interpretations that contradict dip measurements in wells.

Wherever velocities vary laterally, the diffraction patterns from subsurface scatterers can depart significantly from the normal hyperbolic shape associated with the Kirchhoff solution. The apexes of these patterns will be displaced laterally from the positions of the scatterers; moreover, the patterns will generally be asymmetrical and might well exhibit more severe distortions such as triplications. These departures from the simple hyperbola are attributable to complicated propagation paths in the inhomogeneous structure.

Where a medium is laterally inhomogeneous, the Kirchhoff approach is known to migrate seismic events erroneously. In studying this problem, Hubral (1977) concluded that Kirchhoff summation along hyperbolas is, nevertheless, a particularly appropriate first step toward the correct positioning of seismic events. He reasoned that, in moving energy to the apexes of hyperbolas, the conventional Kirchhoff approach transforms an unmigrated time section into an imaged time section that can then be mapped into a fully migrated depth section.

This reasoning holds as well for other conventional forms of wave-equation migration. Thus, conventional finite-difference methods and frequency-domain methods also effectively move seismic energy to the apexes of hyperbolas. Following Hubral, we shall refer to all such approaches to migration as forms of "time migration." We shall refer to the depth mapping of a time-migrated section as "ray-theoretical depth migration."

The central idea in this two-step approach to creating a migrated depth section is that the seismic energy on a conventionally migrated time section reaches the surface along raypaths that are computable—Hubral's *image* raypaths. The ray-theoretical depth migration itself is accomplished in two stages: First, given a velocity model (in the form of velocity versus time on a time-migrated section), these raypaths are computed with a simple ray-tracing algorithm. In the second stage, the time-migrated section is mapped into depth along the computed paths. Ray-theoretical depth migration is, in this sense, a generalization of conventional conversion to depth.

While time migration followed by ray-theoretical depth migration is an appealing approach, it involves an artificial separation of what is truly a single process: the transformation of unmigrated seismic time data directly to an image of the subsurface in depth. That single-step process, which we shall call "wave-theoretical" depth migration, more accurately honors the scalar wave equation appropriate to an inhomogeneous medium. We leave to another

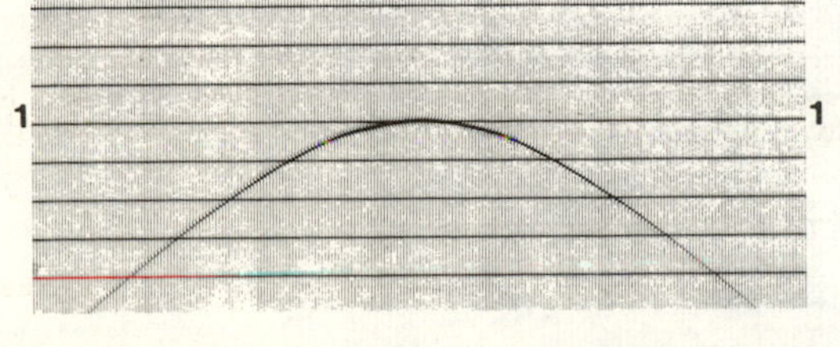

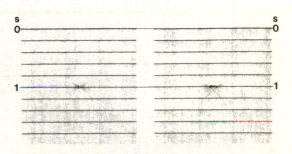

FIG. 2a. Hyperbolic diffraction pattern on a synthetic time section. The lateral position of the scatterer coincides with the minimum-time location.

FIG. 2b. Results of applying Kirchhoff-summation migration (left) and finite-difference migration (right) to the data shown in Figure 2a.

paper (Hatton et al, 1981, this issue) a discussion and demonstrations of the wave-theoretical approach. For another perspective on this method, see also Judson et al (1980).

We use synthetic examples to discuss the problem of wave-equation migration in inhomogeneous media and consider the similar actions of the various conventional migration methods. We shall demonstrate the ray-theoretical depth migration approach using synthetic data and demonstrate its importance in applications to two field-data examples.

THE PROBLEM

Before discussing problems arising in inhomogeneous media, let us first review how time migration works properly when the medium is homogeneous. For convenience, we shall use two-dimensional (2-D) examples although, as Hubral points out, the ideas readily generalize to three dimensions.

The bottom portion of Figure 1 shows a depth section with a localized scatterer at P embedded in a homogeneous medium. Suppose we conduct a zero-offset survey (i.e., shots and receivers are coincident) over the scatterer. Since the velocity is constant, all raypaths between the surface and P are straight. Hence, the diffraction curve in the resulting time section (upper part of Figure 1) has the familiar hyperbolic shape with its apex (i.e., minimum-time point) directly above the scatterer. Figure 2a is a synthetic time section containing the hyperbolic diffraction pattern from such a small scatterer. Figure 2b shows results for two forms of time migration, Kirchhoff-summation and 15-degree finite-difference. Although the migrated results differ in detail (because of different approximations underlying the two methods), both accomplish the desired action of concentrating amplitudes near the apex position.

The successful collapse of the diffracted energy in these migrations depends directly on the choice of migration velocity. For the Kirchhoff-summation approach, choosing the correct velocity ensures that the summation path is the hyperbola that matches the diffraction pattern. For the finite-difference technique, choosing the correct velocity dictates that the surface-recorded energy will be downward-continued to a common focus in the subsurface.

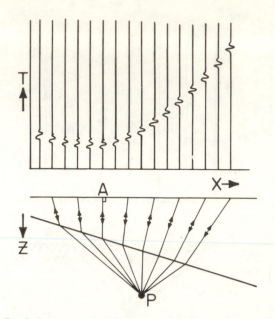

FIG. 3. Bottom: raypaths from a scatterer below a dipping interface. The interface separates a low-velocity layer above from a high-velocity layer below. Top: resulting diffraction curve on a time section.

The results in Figure 2b confirm the basic equivalence of the two types of migration; because of its intuitive appeal, most of our subsequent points on time migration will be introduced in the language of Kirchhoff summation. Our conclusions, however, apply to all forms of time migration.

Now let us see what can go wrong when velocity within the medium varies laterally. Figure 3 shows perhaps the simplest

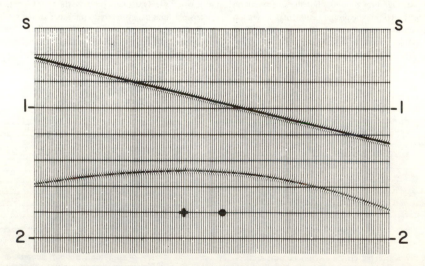

FIG. 4a. Time section computed for depth model in Figure 3. Lateral position of the diffraction-pattern apex is marked by the heavy cross. The true position of the scatterer is shown as the dot.

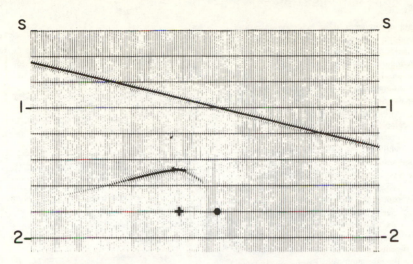

FIG 4b. Result of Kirchhoff migration of the data in Figure 4a.

model with lateral velocity variation. The localized scatterer now lies beneath a dipping interface that separates a low-velocity layer above from a higher velocity layer below. For this model, the raypaths between the surface and the scatterer at P are no longer straight; they now refract at the dipping interface in accordance with Snell's law. Because of the ray bending, the diffraction curve in the resulting time section (upper part of Figure 3) will be distorted from the usual hyperbolic shape. It will be asymmetric, having greater curvature on the downdip side and less curvature updip. More significantly, however, the minimum-time location of the diffraction curve (the apex) is displaced laterally from the scatterer P.

Figure 4a shows a zero-offset time section corresponding to the depth model in Figure 3. The dot indicates the lateral position of the subsurface scatterer, and the cross is directly below the apex of the (asymmetrical) diffraction curve. Figure 4b is the result of applying Kirchhoff migration to the section in Figure 4a. Although the dipping reflection has been migrated properly, the diffraction pattern beneath has been improperly imaged. First, the pattern has not collapsed as completely as when velocity was constant. The reason is that Kirchhoff migration uses amplitudes over hyperbolic paths, whereas, in this case, the diffraction pattern was not a hyperbola. Second, the energy that did collapse is positioned at the apex of the input diffraction pattern rather than at the correct position of the scatterer. Since the output point in the Kirchhoff process is the apex of the hyperbola, diffraction patterns tend to collapse around their minimum-time points.

Figure 4c shows the result of migrating the section in Figure 4a

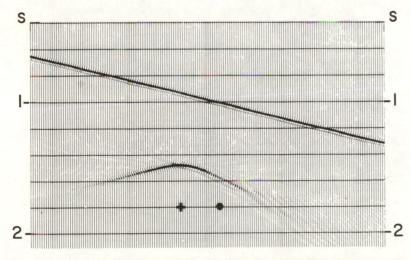

FIG. 4c. Result of finite-difference migration of the data in Figure 4a.

with a conventional finite-difference method; in this case we used an explicit algorithm and the so called 15-degree approximation. Although it differs in detail from the Kirchhoff migration result, the finite-difference section also shows incomplete imaging about the incorrect apex position. This erroneous result cannot be attributed to limitations in accuracy of the 15-degree approximation because that approximation was seen to perform well in the example where the medium was homogeneous (Figure 2).

In considering the incomplete collapse of the diffraction pattern in Figures 4b and 4c, one may first think that the migration velocities were chosen slightly too low—the data look a bit undermigrated. The migration velocities used here were calculated in accordance with Hubral's description of diffraction in a heterogeneous medium (Hubral, 1975). (Hubral's expression for migration velocity is recast in Appendix A.) In this approach, migration velocities are chosen so that hyperbolic summation paths match the diffraction patterns near their apexes. When velocity varies laterally, the quality of match deteriorates along the flanks of the diffraction curves; the incomplete collapse here can be attributed to this fact. No choice for migration velocity, however, could provide a match over more than some portion of the diffraction pattern. Of the possible choices, Hubral's provides the best collapse (see Appendix A).

It could be argued that when velocity varies laterally within a medium, the summation path should not be hyperbolic nor should it be symmetric about the apex. For example, the velocity used to describe the summation path could have been (1) a function of the output migration location and, hence, constant over the summation path, (2) a function of the input unmigrated positions and, thus, variable over the summation path, or (3) a function of position between the input and output locations (this actually was the choice that produced the result in Figure 4b). We tried all three possibilities. For each, the diffraction pattern was collapsed incompletely, with the dominant amplitude collecting about the (incorrect) apex location.

Clearly, the choice of migration velocity has no effect on the fundamental mispositioning problem. Moreover, where lateral variation is severe, time migration simply cannot collapse the diffraction pattern of a point scatterer to a well-confined region,

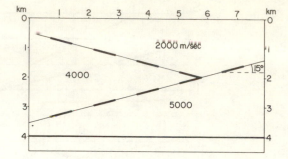

FIG. 5. Dipping multilayer model. The thick-line portions of the interfaces denote where the reflector segments have been "turned on."

regardless of the velocity used. We conclude that, strictly speaking, there is no "correct" velocity choice for time migration when the medium varies laterally.

The conclusion, based on the results for the Kirchhoff-summation and finite-difference methods given here and for the frequency-wavenumber approaches as well, is that *time migration* fails to migrate data correctly when the velocity of the medium varies laterally. Time migration fails because inherently it is founded on the assumption that, locally, the medium is horizontally layered.

We have been aware of distortion in the shapes of diffraction curves and how the Kirchhoff summation method might be modified for complex media: Instead of migrating by summing amplitudes along hyperbolic curves, one could perform ray tracing to determine the distorted shapes of the diffraction curves and the positions (away from the apexes) at which summed values should be placed. That approach, however, has two drawbacks:

(1) It introduces a time-consuming ray-tracing procedure

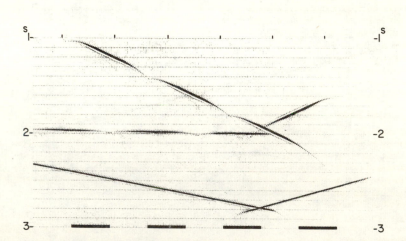

FIG. 6a. Synthetic zero-offset time section for the model shown in Figure 5. The horizontal bars at the bottom denote the lateral positions of the reflector segments shown in Figure 5.

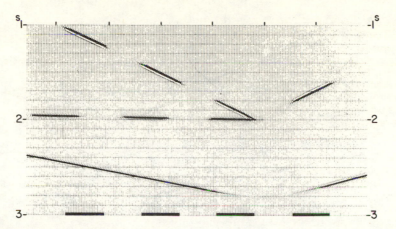

FIG. 6b. Imaged time section generated by applying the Kirchhoff-summation method of migration to the synthetic section in Figure 6a. Horizontal bars at the bottom denote the lateral positions of the reflector segments shown in Figure 5.

into the already long migration process; but more significantly, •

(2) it requires a commitment to a detailed velocity model too early in processing—before migration has clarified geologic structure.

The shortcomings of the finite-difference result in Figure 4c are equally easy to explain. In conventional implementations of the method, an important part of the wave equation is ignored. That part has been called the "thin-lens term" by Claerbout. Hatton et al (1981) show that the thin-lens term governs Snell's law refraction at dipping interfaces or wherever lateral velocity gradients exist. When the thin-lens equation is excluded from the finite-difference system, Snell's law is violated and, consequently, migration results are erroneous. The various frequency-domain approaches likewise fail to honor Snell's law, since they too are based on a wave equation without the thin-lens term.

Judson et al (1980) and Hatton et al (1981) demonstrate that full wave-theoretical depth migration can be implemented with a finite-difference algorithm based on a wave equation that includes the thin-lens term. The central issue in performing such wave-theoretical depth migration is again the choice of velocity, which must be specified as a function of depth before the data are migrated. By contrast, the ray-theoretical approach (described below) refines the velocity model after the time migration step to take advantage of any clarity gained in the initial imaging.

IMAGE RAYS

Suppose that we have time migrated a section conventionally and wish to correct the result by a conversion to depth. The key to the depth migration solution to this problem is the simple, yet profound observation of Hubral (1977): The minimum time or apex position of the diffraction curve lies at a known and readily computable location relative to the position of the scatterer. In Figure 3 the apex is located at the surface position A where the ray from point P emerges *vertically*. This ray Hubral calls the "image ray."

The significance of Hubral's observation is that conventional migration positions events below the surface points where their image rays emerge. Thus, apart from amplitude considerations, migrated sections simulate what one would have recorded had each source and receiver been unidirectional and pointed vertically downward into the earth. (Actual sources and receivers are more nearly omnidirectional; they send and receive energy in all directions.) Reflection events in such a unidirectional survey would be displayed vertically below the source-receiver locations, although we know that Snell's law refraction must deflect the beam paths laterally whenever they encounter dipping interfaces (i.e., lateral inhomogeneities). Simply stated, the beam paths follow image rays. The migration process, however, is not complete unless reflectors have been imaged where they would be encountered if we drilled wells *vertically* into the earth.

RAY-THEORETICAL DEPTH MIGRATION

The conversion of an unmigrated time section to a fully migrated depth section proceeds as follows: We start by performing conventional wave-equation migration to image the data. That is, we use wave-equation migration to collapse diffraction patterns and move events to image ray locations. Migration velocities derived from velocity analyses or well log information, although imperfect, are usually adequate for the purpose of this imaging.

Next we interpret and digitize key horizons on the imaged time section. Although their positions and shapes are displaced and distorted compared to the final locations, these horizons are more readily interpretable than were those on the unmigrated sections (e.g., crossing features have been imaged into synclines).

We must now commit ourselves to selecting an interval velocity model. It is in complex structural geologic settings—where we most expect depth migration to be a correction of first-order importance—that we least trust the velocity values obtained from surface velocity analyses. We recommend that interval velocities be specified on the basis of geologically plausible extrapolation of well log data or from velocity analyses in neighboring areas of less complicated structure.

If we already know the model of velocity as a function of depth (as we might for synthetic data), it is an easy matter to construct the image rays. We merely trace the image rays downward, starting perpendicular to the surface and applying Snell's law at each

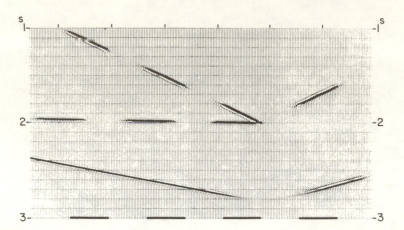

FIG. 6c. Imaged time section generated by applying a finite-difference migration method to the synthetic section in Figure 6a.

interface. No ray searching is required in such forward modeling since the image ray from a surface location to any given interface is just a continuation, from the interface immediately above, of the ray that started vertically at the surface of the model.

Under normal circumstances, however, we do not initially have velocity specified as a function of depth. Rather, we have estimates of velocity for intervals bounded by the horizons interpreted from the conventionally migrated time section. Given interval velocity as a function of (migrated) time, we then do an *inverse* (as opposed to forward) image ray tracing. Appendix B gives mathematical details of the inverse problem. In inverse image-ray tracing, a ray is started vertically downward. It intersects the first interface at a depth computed from the given horizon time and interval velocity; the spatial dip of the interface is computed from the time gradient observed for the horizon; Snell's law then governs refraction of the ray into the next layer (interval velocity given). Intersection points with deeper interfaces are determined in a similar manner. The intersections of image rays and interfaces define the fully migrated depths of the digitized horizons. On a depth-section plot, the lateral deflections of the image rays from vertical give an estimate of the mispositioning of events observed on the conventionally migrated section.

In all the examples shown here, whether for synthetic or field data, we use this inverse ray-tracing approach. Thus we treat the synthetic cases just as we would for field data, as though the velocity information were not available as a function of depth.

Obviously, the image ray deflections are controlled by the chosen velocity model, and in practice the velocity model for field data cases inevitably will be inaccurate to some extent. The amount of uncertainty this inaccuracy causes in the position of any image raypath can be estimated by performing a simple Monte Carlo experiment. We do such an error analysis for the second field data example (below).

Because image-ray tracing is inherently more efficient than conventional ray tracing, we can readily repeat the process using progressively more refined velocity models until we select a model for use in the final depth-mapping step.

This last step—depth mapping—may be considered as a generalization of the simplistic, conventional conversion from time to depth. In the conventional approach, where traces are simply

stretched vertically, the conversion from time to depth cannot correct for lateral position errors in imaged sections. Rather than stretching traces vertically, however, we map amplitudes along the computed image-ray paths. These two steps, image-ray tracing and depth mapping, together constitute ray-theoretical depth migration. The following synthetic example further demonstrates the time migration problem and the importance of depth migration when the geology is complex.

SYNTHETIC EXAMPLE

In the depth model shown in Figure 5, an interface dipping at 15 degrees is truncated at an unconformity having opposite 15-degree dip. Both overlie a layer bounded below by a horizontal interface. The velocity contrasts are large. Densities in the model have been manipulated to turn off reflections from portions of the dipping interfaces. We shall see how the separate reflecting segments are treated by various processes.

Figure 6a is a synthetic zero-offset section generated by use of the wave-theoretical modeling approach of Trorey (1970) as modified by Larson and Hilterman (1976) to accommodate multilayered models. The model was designed so that the deeper dipping reflector appears to be almost horizontal on the time section, a

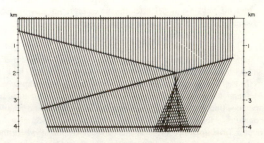

FIG. 7. Image-ray plot showing image rays and depth migrated horizons. The image ray tracing used the known layer velocities and a digitized interpretation of the imaged time section in Figure 6b.

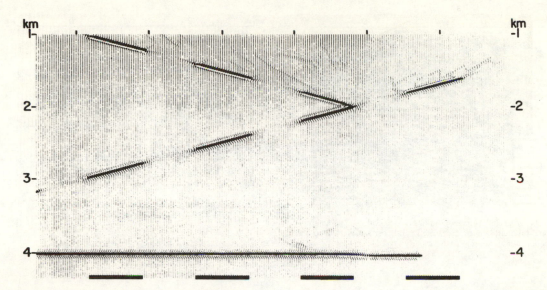

FIG. 8a. Depth migration section obtained by mapping the imaged section (Figure 6b) with the mapping parameters determined by image-ray tracing. The reflector segments are now directly above the horizontal bars. The deep, horizontal interface is truncated at the right of the section because we ignored the part of the reflection that had migrated out of the imaged section.

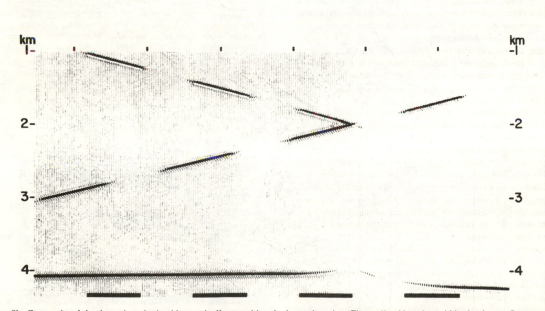

FIG. 8b. Conventional depth section obtained by vertically stretching the imaged section (Figure 6b). Note the void in the deep reflector and the incorrect positions of the reflector segments of the second dipping interface. Horizontal scale equals vertical scale in all sections shown in this paper.

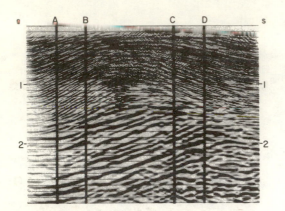

FIG. 9a. Migrated (imaged) marine time section. The interpretational problem here involves the lateral positioning of deep features beneath an arching high-velocity layer.

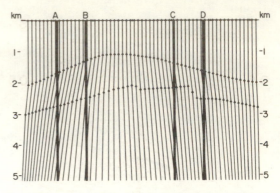

FIG. 10a. Image-ray plot for the model in Figure 9b, plotted on the resulting depth model.

common occurrence when horizons are overlain by dipping overburden. Also, note that the deep horizontal interface has broken into two dipping reflections connected by a phantom diffraction. That diffraction is caused by wavefront distortion during transit through the truncated-layer region above. Complete migration is required to correct for lateral position errors on all interfaces.

Figure 6b is the result of Kirchhoff-summation migration of the section in Figure 6a. Note that the reflector segments of the first interface have been migrated excellently (the medium is homogeneous down to the first interface). Beneath the first interface, however, horizons are mispositioned (too far to the left on the left side of the truncation zone and too far to the right on the other side). In particular, note that migration did not move the horizontal reflections. Conventional finite-difference migration of these data (Figure 6c) produces a substantially identical imaged section. Symmetry in both migration algorithms prevents their displacing horizontal reflections.

Next, as would be done for field data, we digitized the horizons seen on the imaged time section (Figure 6b) and supplied the known interval velocities. Figure 7 shows the image-ray plot

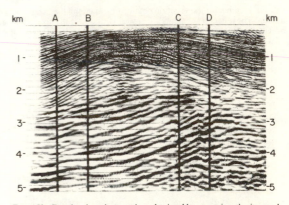

FIG. 10b. Depth migration section obtained by mapping the imaged section (Figure 9a) along the image rays shown in Figure 10a.

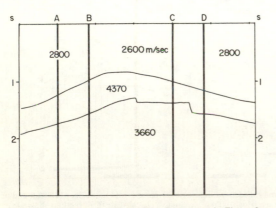

FIG. 9b. Velocity model for the time data shown in Figure 9a.

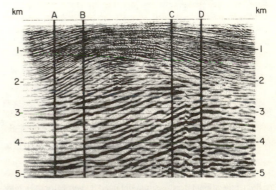

FIG. 10c. Conventional depth section obtained by vertically stretching the imaged section (Figure 9a).

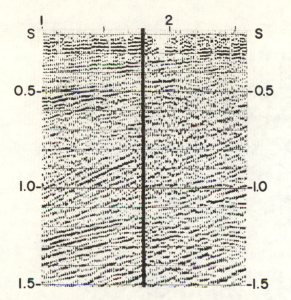

FIG. 11a. Conventional stacked time section from the Rocky Mountains.

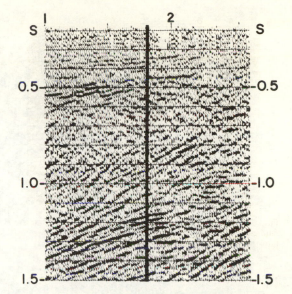

FIG. 11b. Imaged time section obtained by finite-difference migration of the Rocky Mountains data.

generated by inverse image ray tracing. The horizons, shown by the open triangles on Figure 7, are fully depth-migrated reflector positions. Consider the image of the horizontal interface beneath the unconformity. The image-ray plot tells us that the two segments (Figure 6b) must be moved toward each other in order to "heal" the intervening void region.

The final step of depth migration uses the image ray tracing results of Figure 7 as parameters for a depth mapping of the imaged time section into the depth-migration section shown in Figure 8a. This final section overlies the known model very well. For comparison, Figure 8b shows the depth section resulting from conventional (vertical) depth conversion of the imaged time section. Conventional conversion to depth simply cannot correct for erroneously positioned events on imaged sections.

In particular, note the failure of the conventional approach to heal the gap in the deep reflection, whereas depth migration has closed this gap quite well. We attribute the slight weakening in amplitude within the healed zone to the fact that this approach involves a ray-theoretical correction to what is truly a wave-theoretical phenomenon. In general, amplitude treatment in ray theory only grossly approximates the more correct treatment in wave theory.

ARCHING LAYER EXAMPLE

The interpretational problem for the imaged section in Figure 9a involves determining the proper spatial positions of reflections and fault boundaries beneath the arching high-velocity layer seen in the center of the section.

Using well velocities and velocity analyses from neighboring regions, we chose a simple velocity model. In this model, only the horizons bounding the arching high-velocity layer were digitized from Figure 9a and are shown in Figure 9b. Had we included more detail within the layers of the model, the resulting

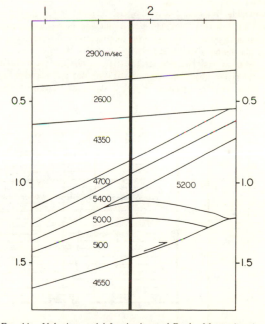

FIG. 11c. Velocity model for the imaged Rocky Mountains time section.

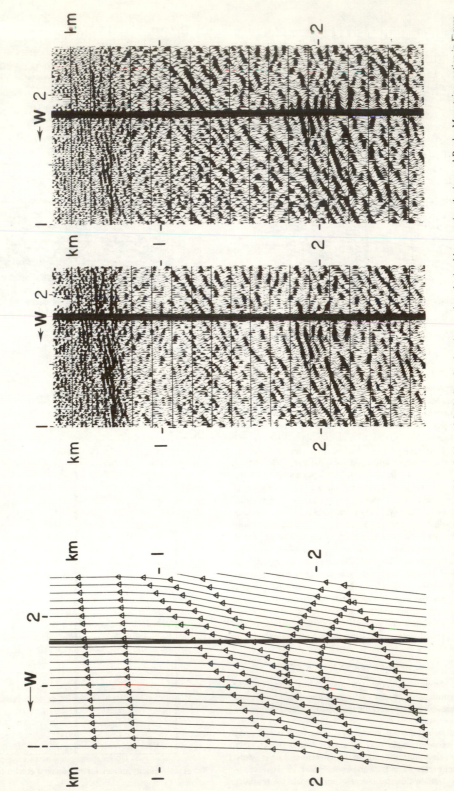

FIG. 12(a). Image-ray plot for the Rocky Mountains data plotted on the resulting depth model. (b) Depth migration section obtained by mapping the imaged Rocky Mountains section in Figure 11b along the image rays shown in Figure 12a. (c) Conventional depth section for the Rocky Mountains data.

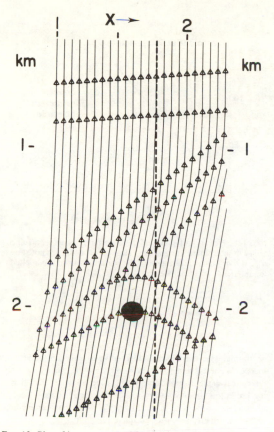

FIG. 13. Plot of image rays used for the depth migration shown in Figure 12b. The black oval near the apex of the anticline is the zone of uncertainty defined in a velocity-model perturbation study. The dashed vertical line marks the surface origin of the image.

image ray picture would not have changed significantly. That is, had the velocity transition across the arching interface been gradual rather than abrupt, as here, the deflections of the image rays would be little different from those in Figure 10a. The velocities shown in Figure 9b are representative values; actually, the model velocities varied slowly within the layers.

Figures 10a, b, and c compare the image ray plot, the ray-theoretical depth migration, and the conventional depth section. At large depths, image rays on the left side of the arch deflect almost 0.5 km. The deflections toward the right side are not so large. These deflections account for all the differences between reflector positions seen on the depth migration section in Figure 10b and the conventional depth section in Figure 10c. The most extensive lateral movement is the 500-m displacement for the deep event at 5 km, across vertical reference line A. This displacement is of major significance when we consider that a lateral positioning accuracy of 100 m or better is essential in most current exploration problems.

It may be noted that the combination of large-throw faults and a large velocity contrast at the base of the high-velocity layer in Figure 9b has introduced discontinuities into deeper events in both

Figures 10b and 10c. Without accurate information about the shallower section, one might not wish to introduce such a severely faulted model into our process. If the base of that layer had been chosen to be smoother, these discontinuities would be removed, but the image-ray picture would show essentially the same deflections as those seen in these examples.

OVERTHRUST EXAMPLE

Figure 11a is a stacked section from a structurally complex region in the Rocky Mountains. It typifies common problems in choosing well locations.

Figure 11b is a section imaged with finite-difference migration. Above the unconformity at 0.6 sec, the section is gently dipping toward the west. Beneath the unconformity, dips are as large as 35 degrees toward the west down to 1.1 sec where a tight anticline is encountered above an overthrust plane. The dipmeter log of a well drilled into the anticline below 2 km measured gentle dip to the southeast rather than the westerly dip interpreted from the conventionally migrated section.

We can anticipate that the steep dips and known large-velocity contrasts in the dipping strata beneath the unconformity would be sufficient to cause substantial deflections of image rays. The velocity model shown in Figure 11c was drawn from an interpretation of Figure 11b in conjunction with sonic log information from the well shown. For this complex structure, the computed image rays (Figure 12a) are seen to deflect significantly, tightening the anticline and displacing its apex about 150 m to the west.

The same result is displayed on the depth-migration section shown in Figure 12b. At a depth below 2 km, the dip on this section is easterly, consistent with the dipmeter readings. In contrast, the conventional depth section in Figure 12c shows westerly dip with the apex of the anticline at an erroneous position east of the well.

Since the velocities of Figure 11c were taken from well logs and were assumed constant within each layer, they are certainly inaccurate in some parts of the model. An appropriate question is: How sensitive is the lateral position of the anticline apex to our particular choice of velocity model? To answer this question, we performed a Monte Carlo investigation of the behavior of a particular image ray (that ray which departs the surface at the well location and intersects the anticline near the apex). In particular, we perturbed all layer velocities randomly and independently from their presumed values and traced that image ray for each of the perturbed models. The velocity perturbations had a normal probability distribution with standard deviation of 500 m/sec. For some layers, ±500 m/sec corresponds to as much as 20 percent uncertainty in velocity (see Figure 11c). The elliptical zone of uncertainty in Figure 13 was produced by plotting the various computed image-ray intersections with the anticline for 100 different perturbations of the velocity model.

The shape and size of the zone of uncertainty are revealing. The elliptical shape indicates that both the depth and the lateral position of the feature are sensitive to velocity variation. The direct dependence of depth on velocity is well known in conventional conversion from time to depth. It is the lateral position error, however, that most concerns us here. Note that even for a 500 m/sec uncertainty in velocity, the zone of uncertainty for the intersection does not include the vertical line from the surface origin of the image ray; at this level of uncertainty, none of the image raypaths includes the vertical, the path for conventional depth conversion. Under these circumstances depth migration is a desirable procedure since it produces a correction significant in relation to its expected errors.

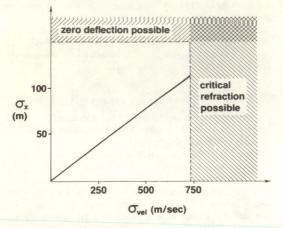

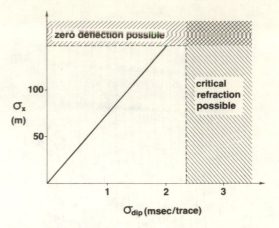

FIG. 14a. Plot of the half-width of the uncertainty zone versus the magnitude of velocity uncertainty. The actual data points are close to the plotted line and are not shown. The half-width of the uncertainty zone is described by σ_x, the standard deviation of the x positions of the image ray for 100 trials. The standard deviation of the applied velocity perturbations is denoted by σ_{vel}.

FIG. 14b. Plot of half-width of the uncertainty zone versus uncertainty in interpreted horizon dip. Data points again are not shown. σ_{dip} is the standard deviation of the perturbations to dip applied to the interfaces of the model.

This experiment was repeated for several levels of velocity uncertainty, and the results are summarized in Figure 14a. The width of the zone of uncertainty was approximately proportional to the uncertainty in velocity. For the velocity model of Figure 11c, the center of the zone of uncertainty was displaced from the vertical by about 150 m. Thus, for that ray, the zone in Figure 14a marked "zero deflection possible" indicates width of the uncertainty zone large enough to include the vertical. Note that, for this example, the result of vertical conversion to depth is never a possible solution at any level of uncertainty. The reason is that as the velocity uncertainty increases, there comes a point where the incidence angle of the (downward traveling) image ray exceeds the critical refraction angle for at least one interface. Critical refraction of an image ray is unacceptable for it would imply a shadow zone, a phenomenon not evidenced on the unmigrated or migrated sections; thus any proposed velocity model that dictates the critical refraction of image rays will be rejected.

The velocity model can contain another type of inaccuracy; the interface shapes may have been improperly interpreted. In another series of Monte Carlo experiments, velocity values were held constant (at their original values), but the dips of all layer interfaces were perturbed independently by various amounts. The results, shown in Figure 14b, again demonstrate a linear relationship, this time between the width of the uncertainty zone and the uncertainty in dip.

With this second type of perturbation, however, the zone of uncertainty includes the vertical conversion to depth as a possible solution before critical refractions are encountered. This result reinforces the importance of time migration to the ray-theoretical technique. In clarifying the seismic section, time migration greatly improves the prospects for accurately interpreting the subsurface structure. In this example, errors in dip that admit vertical depth conversion as a solution are of such large magnitude that they could usually be avoided in practice.

Although confined to a specific data example, this uncertainty study shows that lateral positioning of events by ray-theoretical depth migration is reasonably insensitive to errors in the velocity model. We must remember, however, that the quality of a ray-theoretical depth migration section can be no better than that of the preceding time-migration result. The key horizons must be identifiable on the time-migrated section. Only then can we hope to have a velocity model of adequate reliability.

CONCLUSIONS

Conventional migration owes its immense success to the fact that it typically images seismic data well (i.e., it collapses diffraction patterns) despite uncertainties in the migration velocities used. Ray-theoretical depth migration is a subsequent step required when the overburden has large dips and large velocity contrasts. Depth migration is likely important in more data environments than previously suspected.

The greatest difficulty in migration practice remains the limited accuracy of velocity determinations. Depth migration is, however, an important processing step even in the presence of uncertainties in estimated velocities. In the two-stage method of conventional wave-equation migration followed by ray-theoretical depth migration, a detailed velocity model is constructed *after* the time migration (imaging) has been performed. At this stage we may know more about our sections and can more readily interpret them.

We saw that, from the viewpoint of the conventional finite-difference method, the need to perform ray-theoretical depth migration when geologic structure is complex arises from neglect of the thin-lens equation. Judson et al (1980) and Hatton et al (1981) show that to implement complete wave-theoretical migration into depth, the thin-lens equation and the so-called diffracting equation must be solved alternately at each depth step. Ray-theoretical depth migration may be viewed as an implementation of the thin-lens equation *after* we have completed time migration using the diffracting equation (Hatton et al, 1981). Strictly, we would be justified in postponing depth migration to this final step only if the thin-lens and diffraction operations were commutative throughout the migration process. As noted by Judson et al (1980), this is only approximately the case.

Thus, dividing the migration process into the two stages—imaging followed by depth migration—is somewhat artificial. Whenever the near-surface is so complex that lensing effects cause severe focusing and crossing of image rays, the imaging will be incomplete (i.e., diffraction patterns will not collapse completely). Even subsequent depth mapping will then leave events mispositioned. In such cases, however, image-ray plots can reveal the likelihood that the data will be imaged imperfectly. Moreover, the image-ray computations can serve a vital modeling function in determining a plausible velocity model that has the detail required for performing the wave-theoretical depth migration. Once a velocity model is developed, we recommend that the original data be remigrated, this time using the more complete, wave-theoretical approach so as to image and position reflected energy concurrently.

ACKNOWLEDGMENTS

We particularly appreciate Ron Chambers's leadership in coordinating the computer program design, and thank Ying Huang, Jakob Shapiro, Roy Forshaw, and Linda Elliot for their programming contributions in image ray tracing and depth migration. Our thanks go also to Albert Ng who developed the forward modeling program, and to Dolores Meeks for typing the manuscript.

REFERENCES

Claerbout, J., 1976, Fundamentals of geophysical data processing: New York, McGraw-Hill Book Co., Inc., 274 p.
Claerbout, J., and Doherty, S. 1972, Downward continuation of moveout-corrected seismograms: Geophysics, v. 37, p. 741–768.
French, W., 1975, Computer migration of oblique seismic reflection profiles: Geophysics, v. 40, p. 961–980.
Gazdag, J., 1978, Wave equation migration with the phase shift method: Geophysics v. 43, p. 1342–1351.
Hatton L., Larner, K., and Gibson, B., 1981, Migration of seismic data from inhomogeneous media: Geophysics, v. 46, this issue, p. 735–755.
Hubral, P., 1975, Locating a diffractor below plane layers of constant interval velocity and varying dip: Geophys. Prosp., v. 23, p. 313–322.
——— 1977, Time migration—Some ray theoretical aspects: Geophys. Prosp., v. 25, p. 738–745.
Hubral, P., and Krey, T., 1980, Interval velocities from seismic reflection time measurements: SEG, Tulsa, 203 p.
Judson, D., Lin, J., Schultz, P., and Sherwood, J., 1980, Depth migration after stack: Geophysics, v. 45, p. 361–375.
Larner, K., and Hatton, L., 1976, Wave equation migration: Two approaches: Presented at 8th Annual OTC, May, Houston.
Larson, D., and Hilterman, F., 1976, Diffractions: Their generation and interpretation use: Presented at 29th Annual Midwestern SEG Meeting, March 7–9, Dallas.
Schneider, W., 1978, Integral formulation for migration in two and three dimensions: Geophysics, v. 43, p. 49–76.
Sorrells, G., Crowley, J., and Veith, K., 1971, Methods for computing ray paths in complex geologic structures: SSA Bull., v. 61, p. 27–54.
Trorey, A., 1970, A simple theory for seismic diffractions: Geophysics, v. 35, p. 762–784.

APPENDIX A
MIGRATION VELOCITY

From the perspective of the Kirchhoff approach to migration, the best velocity to use in time migration is the one that provides the best match between hyperbolic summation paths and diffraction patterns in the seismic data. Only when the medium is homogeneous, however, can the match be exact. There, the migration velocity is just the medium velocity.

For horizontally layered media, diffraction patterns can be approximated by hyperboloids characterized by the root-mean-square (rms) velocity function,

$$V_{\text{rms}, N} = \left[\frac{\sum\limits_{1}^{N} V_j^2 t_j}{\sum\limits_{1}^{N} t_j} \right]^{1/2}, \qquad (A-1)$$

where V_j is the (constant) velocity of layer j; t_j is the one-way traveltime vertically through layer j; and the layer index j varies from 1 at the top of the medium to N for the layer having the diffractor at its base. Where velocities vary substantially among layers, the match can be poor along the flanks of diffraction patterns. Consequently, $V_{\text{rms}, N}$ would serve well as the migration velocity only for reflections that are not too steeply dipping. In such circumstances, stacking velocities computed by fitting hyperbolas to moveout patterns most often suffice as migration velocities and are suitable for treating a broad range of dips.

Recall from the text that where velocity varies laterally, time migration can do no better than collapse diffraction patterns around their apexes, and these apexes will be displaced from the correct diffractor positions. At that, the only portion of an asymmetric diffraction pattern that we can expect to match with a hyperbola whose apex is coincident with that of the diffraction curve is that region near the apex, the region of less steep dip on the unmigrated time section. For the special case of a 2-D medium consisting of arbitrarily plane-dipping isovelocity layers, Hubral (1975) derived an expression for the migration velocity that provides the required fit near the apex. The migration velocities used in our examples are based on Hubral's result. Hubral and Krey (1980) generalized this result to include layers with curved interfaces in three dimensions.

A recasting of Hubral's (1975) result provides several insights. Consider an image ray through a point diffractor at depth. The ray surfaces at the lateral position of the diffraction apex. Using a relationship between the shape of the diffraction pattern and the curvature of the wavefront of a diffracted wave as it arrives at the earth's surface, Hubral obtained the following result for migration velocity:

$$V_{\text{mig}, N}^2 = \frac{1}{T_N} \sum_{j=1}^{N} \prod_{k=0}^{j-1} \frac{\cos^2 \alpha_k}{\cos^2 \beta_k} V_j^2 t_j, \qquad (A-2)$$

where

$T_N = \sum\limits_{j=1}^{N} t_j$ is the total traveltime along the image raypath from the surface to the diffractor,

t_j = the traveltime through layer j along the image raypath,

α_k = the angle of incidence of the downgoing image ray at the kth interface, and

β_k = the angle of refraction of the image ray as it proceeds downward from the kth interface.

In equation (A–2), we set $\alpha_0 = \beta_0 = 0$. (β_0 is the angle of emergence of the image ray at the surface.)

Let us simplify the form of equation (A–2) by defining a new layer velocity quantity σ_j:

$$\sigma_j \equiv \prod_{k=0}^{j-1} \frac{\cos \alpha_k}{\cos \beta_k} V_j. \qquad (A-3)$$

Then V_{mig} takes on the familiar form of an rms velocity [see equation (A–1)],

$$V_{\mathrm{mig},N} = \left[\frac{\sum\limits_{1}^{N} \sigma_i^2 t_i}{\sum\limits_{1}^{N} t_j} \right]^{1/2}. \qquad (A-4)$$

The only difference, then, between $V_{\mathrm{mig},N}$ and $V_{\mathrm{rms},N}$ is that the true layer velocities in the expression for $V_{\mathrm{rms},N}$ have been replaced by the modified velocities σ_j. We note also that t_j in equation (A–1) pertains to the vertical path through layer j, whereas in equation (A–4) it pertains to the slanting, image raypath. This distinction, however, vanishes in a horizontally layered medium; the image raypath is vertical, and expression (A–1) yields the rms velocity.

Let us take a closer look at the velocities σ_j. From equation (A–3), the ratio σ_{j+1}/σ_j is simply

$$\sigma_{j+1}/\sigma_j = \frac{\cos \alpha_j}{\cos \beta_j} \frac{V_{j+1}}{V_j}. \qquad (A-5)$$

Equation (A–5) provides a means of computing the new velocities recursively. Given that $\alpha_0 = \beta_0 = 0$, we have

$$\sigma_1 = V_1$$

and

$$\sigma_{j+1} = \sigma_j (V_{j+1}/\cos \beta_j)/(V_j/\cos \alpha_j), \qquad (A-6)$$

for $j = 1, 2, \ldots N - 1$.

Results (A–6) and (A–4) have an interesting interpretation. The migration velocity for a 2-D plane dipping, layered medium is a generalized form of rms velocity. In it, the *pseudo*-layer velocity σ_j is the actual layer velocity V_j modified by the geometry of the image raypath. Note that the quantity $V_j/\cos \alpha_j$ is just the *apparent velocity, normal* to jth interface, of a plane wave incident at the image ray angle α_j. Likewise $V_{j+1}/\cos \beta_j$ is the apparent velocity in the normal direction, for the refracted wave. Thus, in the recursive computation, we find that $\sigma_1 = V_1$, and thereafter the ratio of the pseudo-layer velocity in a given layer to the pseudo-layer velocity in the layer above is given by the ratio of apparent velocity normal to the interface for the refracted wave to that for the incident wave. Note that for horizontal layering, $\alpha_j = \beta_j = 0$ for all j, and hence $\sigma_j = V_j$ for all j. Stated differently, for horizontal layering the migration velocity equals the rms velocity.

In using the term *migration velocity* here, we have been taking the perspective of the Kirchhoff method wherein the migration velocity represents an average velocity from the surface to the depth of the diffractor. The finite-difference implementation, on the other hand, uses *interval velocity* as the governing parameter for doing migration. Where the medium consists of dipping layers, the quantities σ_j, rather than the actual layer velocities, would be used.

APPENDIX B
INVERSE IMAGE RAY TRACING IN THREE DIMENSIONS

Image raypaths obey the same laws of geometric optics and acoustics as do other rays in an acoustic medium. They are unique only in that they depart vertically downward from the horizontal surface of the medium. Thereafter, the paths are governed solely by Snell's law of refraction.

The forward problem of computing traveltimes to interfaces in a given velocity-depth model is a familiar ray-tracing problem.

Suppose that the model consists of homogeneous layers separated by straight or curved interfaces. Then, within layers, raypaths are straight. Ray tracing involves the usual alternating steps: (1) compute the point of intersection of the ray in a given layer and the underlying interface, and (2) using Snell's law at the interface, compute the direction of the raypath in the next layer down. The ray tracing is one-directional (downward), and computed traveltimes are doubled to match reflection times observed on time-migrated sections.

In image ray tracing for depth migration, an *inverse* ray tracing procedure is required; the depth model is computed by use of observed (migrated) traveltimes and interval velocities. We shall treat the inverse ray-tracing problem for models in which velocities are constant within layers, and layers are bounded by planar interfaces that dip in 3-D space.

The goal is to compute the coordinates of intersections between image rays and interfaces in the depth model. We have three basic tasks to perform:

(1) Use the observed traveltime to a given interface and the computed raypath in the layer above to determine the coordinates of the intersection with that interface;

(2) use Snell's law to determine the direction of the ray refracted past a given interface; and

(3) use observed horizontal gradients of traveltime to determine the spatial orientation of an interface.

Figure B–1 depicts an image ray in a layered, 3-D depth model. In the model, any layer j has constant velocity V_j and is bounded below by interface j. The orientation of interface j is specified in terms of the unit (downward) normal vector $\mathbf{n}_j$. The image ray starts at surface position $\mathbf{r}_0$ and intersects interface j at $\mathbf{r}_j = x_j\mathbf{i} + y_j\mathbf{j} + z_j\mathbf{k}$. ($\mathbf{i}$, $\mathbf{j}$, and $\mathbf{k}$ are unit vectors in the x, y, and z directions, respectively.) Within layer j, the ray parallels the unit vector $\mathbf{p}_j$.

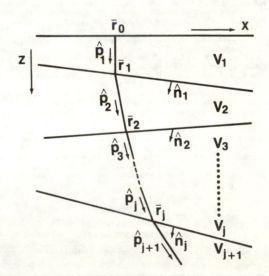

FIG. B–1. Schematic depth section (projected in two dimensions) showing an image raypath. The ray intersects interface j at $\mathbf{r}_j = x_j\mathbf{i} + y_j\mathbf{j} + z_j\mathbf{k}$. Within layer j it parallels the unit vector $\mathbf{p}_j$. Note: In the figure, the overbar denotes a general vector and the circumflex denotes a unit direction vector.

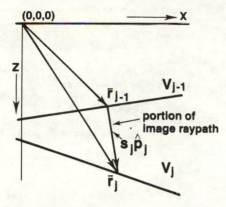

FIG. B–2. Schematic (projected in *two* dimensions) showing that the vector $\mathbf{r}_j$ is merely the vector sum of $\mathbf{r}_{j-1}$ and the vector $s_j\mathbf{p}_j$ connecting the intersection points.

Intersection of a ray and interface.—Suppose we have already traced the ray past interface $j-1$ and into layer j. Hence, both $\mathbf{r}_{j-1}$ and $\mathbf{p}_j$ already have been computed. It is a straightforward matter to compute the intersection point $\mathbf{r}_j$ with the next interface. As Figure B–2 illustrates,

$$\mathbf{r}_j = \mathbf{r}_{j-1} + s_j\mathbf{p}_j, \tag{B–1}$$

where s_j is the distance between intersection points on interfaces $j-1$ and j. The distance s_j is just

$$s_j = \frac{1}{2} V_j (T_j - T_{j-1}), \tag{B–2}$$

where T_j is the two-way (migrated) traveltime to the jth interface.

Snell's law for a refracted ray.—In 2-D problems, Snell's law is used to obtain the angle of refraction for a ray refracted across an interface in terms of (1) the angle of incidence of the ray at the interface, (2) the dip of the interface, and (3) the velocities of the two layers separated by the interface. In 3-D problems, Snell's law is used to determine the unit ray vector $\mathbf{p}_{j+1}$ in terms of (1) the unit vector $\mathbf{p}_j$ for the incident ray, (2) the orientation $\mathbf{n}_j$ of the interface, and (3) layer velocities V_j and V_{j+1}. Snell's law in three dimensions can be expressed in the form (Sorrells et al, 1971),

$$\mathbf{p}_{j+1} = \frac{V_{j+1}}{V_j} \left\{ \mathbf{p}_j - \left[\mathbf{p}_j \cdot \mathbf{n}_j - \left([\mathbf{p}_j \cdot \mathbf{n}_j]^2 \right. \right. \right.$$
$$\left. \left. \left. - \frac{V_{j+1}^2 - V_j^2}{V_{j+1}^2} \right)^{1/2} \right] \mathbf{n}_j \right\}. \tag{B–3}$$

According to this equation, computation of the ray direction $\mathbf{p}_{j+1}$ requires that we know the orientation $\mathbf{n}_j$ of the jth interface. The determination of $\mathbf{n}_j$ is the primary task remaining. A derivation of expressions for the components of $\mathbf{n}_j$ follows.

Orientation of an interface.—Three points determine a plane. Suppose that, in addition to following the image ray that starts at $\mathbf{r}_0$ on the surface of the medium, we also follow image rays starting at neighboring positions $\mathbf{r}_{0x}$ and $\mathbf{r}_{0y}$, given by

$$\mathbf{r}_{0x} = \mathbf{r}_0 + \Delta X \mathbf{i},$$

and

$$\mathbf{r}_{0y} = \mathbf{r}_0 + \Delta Y \mathbf{j}. \tag{B–4}$$

In equation (B–4), ΔX and ΔY are arbitrary small distances.

Under the assumption that all interfaces are locally planar, the three closely spaced image rays will be parallel in every layer. Then the intersection points $\mathbf{r}_j$, $\mathbf{r}_{jx}$, and $\mathbf{r}_{jy}$ for the three parallel image rays define the orientation of the jth interface. According to equation (B–1), we have

$$\mathbf{r}_{jx} = \mathbf{r}_{j-1,x} + s_{jx}\mathbf{p}_j,$$

and

$$\mathbf{r}_{jy} = \mathbf{r}_{j-1,y} + s_{jy}\mathbf{p}_j, \tag{B–5}$$

where the distances s_{jx} and s_{jy} are given by

$$s_{jx} = \frac{1}{2} V_j \left[\left(T_j + \frac{\partial T_j}{\partial x} \Delta X \right) - \left(T_{j-1} + \frac{\partial T_{j-1}}{\partial x} \Delta X \right) \right],$$

and

$$s_{jy} = \frac{1}{2} V_j \left[\left(T_j + \frac{\partial T_j}{\partial y} \Delta Y \right) - \left(T_{j-1} + \frac{\partial T_{j-1}}{\partial y} \Delta Y \right) \right]. \tag{B–6}$$

In expression (B–6), $\partial T_j/\partial x$ and $\partial T_j/\partial y$ are the observed gradients of (migrated) traveltime in the x and y directions, respectively.

The equation for the jth plane interface is

$$\mathbf{n}_j \cdot \mathbf{r}_j = h_j, \tag{B–7}$$

where h_j is the normal distance from the origin of coordinates to the interface. Our immediate goal is to determine the three components (n_{jx}, n_{jy}, n_{jz}) of the unit vector $\mathbf{n}_j$. Given the intersections $\mathbf{r}_j$, $\mathbf{r}_{jx}$, and $\mathbf{r}_{jy}$, we can use equation (B–7) to provide two equations,

$$\mathbf{n}_j \cdot (\mathbf{r}_j - \mathbf{r}_{jx}) = 0,$$

and

$$\mathbf{n}_j \cdot (\mathbf{r}_j - \mathbf{r}_{jy}) = 0 \tag{B–8}$$

for the three unknowns. The third equation follows from the requirement that $\mathbf{n}_j$ be a unit vector, i.e.,

$$|\mathbf{n}_j|^2 = 1. \tag{B–9}$$

Let us solve equations (B–8) and (B–9) explicitly for n_{jx}, n_{jy}, and n_{jz}. Let the components of the vector $\mathbf{r}_j - \mathbf{r}_{jx}$ be given by a_x, a_y, and a_z and the components of the vector $\mathbf{r}_j - \mathbf{r}_{jy}$ be given by b_x, b_y, and b_z. Then equations (B–8) become

$$n_{jx}a_x + n_{jy}a_y = -n_{jz}a_z,$$

and

$$n_{jx}b_x + n_{jy}b_y = -n_{jz}b_z.$$

Solution of these two equations gives

$$n_{jx} = An_{jz},$$

and

$$n_{jy} = Bn_{jz}, \tag{B–10}$$

where

$$A = \frac{-a_z b_y + a_y b_z}{a_x b_y - a_y b_x},$$

and

$$B = \frac{-a_x b_z + a_z b_x}{a_x b_y - a_y b_x}. \qquad (B-11)$$

Now, combining equations (B–9) and (B–10), we have

$$(A^2 + B^2 + 1) n_{jz}^2 = 1,$$

from which we get

$$n_{jx} = A[A^2 + B^2 + 1]^{-1/2},$$

$$n_{jy} = B[A^2 + B^2 + 1]^{-1/2},$$

and

$$n_{jz} = [A^2 + B^2 + 1]^{-1/2}. \qquad (B-12)$$

We can now list the sequence of computational steps for inverse-image ray tracing in three dimensions.

(1) Start at positions r_0, r_{0x}, and r_{0y} on the surface [see equation (B–4)].

(2) For image rays, p_1 is the unit vector $\mathbf{k}$ in the z-direction.

(3) Intersect the ray with the first interface. From equations (B–1) and (B–2), $r_1 = r_0 + 1/2\, V_1 T_1 p_1$. r_{1x} and r_{1y} are similarly computed using equations (B–5) and (B–6).

(4) Use equations (B–11) and (B–12) to compute the orientation of the first interface n_1.

(5) Use Snell's law, equation (B–3), to compute the ray direction p_j in layer $j = 2$.

(6) The intersection points with the second interface are given by setting $j = 2$ in equations (B–1), (B–2), (B–5), and (B–6).

(7) Set $j = 2$ in equations (B–11) and (B–12) to compute n_2.

Steps (5), (6), and (7) are repeated for successive layers $j = 2$, 3, ...

Reprinted from GEOPHYSICS, **46**, 751-767.

Migration of seismic data from inhomogeneous media

Les Hatton*, Ken Larner‡, and Bruce S. Gibson‡

ABSTRACT

Because conventional time-migration algorithms are founded on the implicit assumption of locally lateral homogeneity, they leave events mispositioned when overburden velocity varies laterally. The ray-theoretical depth migration procedure of Hubral often can provide adequate first-order corrections for such position errors. Complex geologic structure, however, can so severely distort wavefronts that resulting time-migrated sections may be barely interpretable and thus not readily correctable. A more accurate, wave-theoretical approach to depth migration then becomes essential to image the subsurface properly. This approach, which transforms an unmigrated time section directly into migrated depth, more completely honors the wave equation for a medium in which variations in interval velocity and details of structural shape govern wave propagation. Where geologic structure is complicated, however, we usually lack an accurate velocity model. It is important, therefore, to understand the sensitivity of depth migration to velocity errors and, in particular, to assess whether it is justified to go to the added effort of doing depth migration.

We show a synthetic data example in which the wave-theoretical approach to depth migration properly images deep reflections that are poorly resolved and left distorted by either time migration or ray-theoretical depth migration. These imaging results are, moreover, surprisingly insensitive to errors introduced into the velocity model. Application to one field data example demonstrates the superior treatment of amplitude and waveform by wave-theoretical depth migration. In a second data example, deep reflections are so influenced by anomalous overburden structure that the only valid alternative to performing wave-theoretical depth migration is simply to convert the unmigrated data to depth.

When the overburden is laterally variable, conventional time migration of unstacked data can be as destructive to steeply dipping reflections as is CDP stacking prior to migration. A schematic example illustrates that when migration of unstacked data is judged necessary, it should normally be performed as a depth migration.

INTRODUCTION

In many areas of interest, seismic data are collected over geologic structures that have substantial lateral velocity variation. Very often, lateral inhomogeneity is directly related to steeply dipping beds and, in processing the seismic data, migration will almost always be performed.

The importance of velocity estimates to the success of migration has been well documented (e.g., use of a migration velocity that is too low results in the incomplete collapse of diffractions and the insufficient movement of dipping reflections). When the medium is inhomogeneous, the proper specification of velocity is of even greater importance. In that case, the geophysicist must make a commitment to a detailed model of overburden velocities in order to migrate properly. Furthermore, details of the migration algorithm itself must be carefully considered because a sophisticated migration algorithm is required to honor the detailed velocity information.

Hubral (1977) showed that the Kirchhoff summation migration of data from laterally inhomogeneous media fails to position reflected events properly. Larner et al (1981, this issue) show further that migration by *any* conventional technique cannot properly position subsurface features when the overburden has substantial lateral variation in velocity. Errors in position result from approximations made to the scalar wave equation, the foundation of all migration techniques commonly used today (finite-difference, Kirchhoff summation, or frequency domain). Approximations can be identified in each technique to explain why each approach misplaces reflected events in about the same way.

Even though they provide for gross lateral variation in velocity, conventional migration techniques fail to position reflections properly because all techniques include the implicit assumption that, locally, the velocity in the medium does not change in the horizontal direction. Following Hubral, we shall refer to migration algorithms that assume such local homogeneity as *time-migration* algorithms. A common characteristic of time-migration techniques is that their direct output is a seismic section in time.

Hubral's observation was that time-migration algorithms position reflected events at locations that have a simple relationship to their true locations. He identified those locations as the surface terminations of *image raypaths*. Exploiting Hubral's work, Larner

Presented at the 48th Annual International SEG Meeting October 31, 1978, in San Francisco. Manuscript received by the Editor November 29, 1979; revised manuscript received September 4, 1980.
*Formerly Western Geophysical Co., Houston; presently Merlin Geophysical Co., Morris House, Commercial Way, Woking, Surrey, England.
‡Western Geophysical Co., P. O. Box 2469, Houston, TX 77001.

et al (1981) present a ray-theoretical solution to the mispositioning problem.

In the ray-theoretical approach, the first step is a standard time migration. Next, velocities are specified for intervals between prominent horizons on the time-migrated section. Such a velocity model provides sufficient information to trace image raypaths through the medium. (Image raypaths leave the surface vertically and obey Snell's law at layer interfaces, velocity contrasts across dipping horizons thus cause these raypaths to be deflected laterally.) In the second step, the time-migrated section is converted to depth by stretching traces along the image raypaths. This process is called *depth migration* because the result is a depth section with properly positioned reflections.

An important aspect of ray-theoretical depth migration is that one need not adopt a final, detailed interval-velocity model until time migration has clarified the subsurface structure. Nevertheless, any velocity model derived for field data will be imperfect because typically it is constructed either from isolated well logs or from seismic velocity analyses that are themselves dependent upon a model. Larner et al (1981) use a Monte Carlo numerical experiment to demonstrate that where the subsurface is complex and a reasonably accurate velocity model can be constructed, ray-theoretical depth migration is a process of primary importance, i.e., corrections in lateral position can be significant and are subject to a relatively small probable error.

That two-step approach is, however, an artificial division of the full migration process. Although the ray-theoretical technique has been successful for field cases of moderate complexity, we shall exhibit a simple but severely inhomogeneous velocity model for which time migration cannot image the subsurface correctly, and thus ray-theoretical depth migration is inappropriate.

We shall then examine the scalar wave equation to reveal the source of positioning and imaging errors in time migration, and we suggest a cure. The proper treatment of the wave equation yields an algorithm that properly images an unmigrated time section and maps it directly into depth in one step. This technique we call *wave-theoretical depth migration* because of its direct development from the scalar wave equation. Superior treatment of both synthetic and field data by this wave-theoretical approach demonstrates that the technique honors the essentials of the scalar wave equation for inhomogeneous media. As in the ray-theoretical approach, however, we must ask again whether the added effort and cost of this more sophisticated approach are justified in the presence of imperfect velocity information. Applications of the technique with intentionally incorrect velocity models demonstrate a fortunate insensitivity to velocity uncertainty.

However, this extended migration algorithm does not properly account for all the effects of velocity inhomogeneity. In complex media, for example, raypath distortion can be so severe that conventional CDP stacking destroys much reflection information. Subsequent migration then cannot reconstruct an image of the subsurface. The conventional solution to such a problem is to migrate the unstacked data. We shall, however, demonstrate schematically that if the data are merely *time* migrated and then stacked, reflections can be damaged as much as in conventional CDP stacking. The proper solution is to *depth* migrate the unstacked, variable-offset data.

We observe that where the subsurface is inhomogeneous and structurally complex, it is usually complex in all three dimensions. The migration of data from inhomogeneous media is thus a problem of three-dimensional (3-D) migration. While we shall not address 3-D depth migration here, we see it as important and inevitable future work.

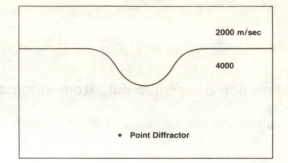

FIG. 1a. Depth model depicting a subsurface scatterer beneath an anomaly in the overburden.

SHORTCOMINGS IN RAY-THEORETICAL DEPTH MIGRATION

The success of ray-theoretical depth migration is predicated on the ability of time migration to form an adequate image of subsurface reflectors. When the medium is homogeneous, time migration is accurate both in positioning features and in imaging them. In addition, Larner et al (1981) indicate that for certain types of lateral heterogeneity, time migration mispositions reflected energy but does not severely degrade the quality of the image. We now show an example of inhomogeneity in which the issue of mispositioning is subordinate; the principal failure is the construction of the image.

Figure 1a shows a point diffractor in a two-layer medium. The interface between the two layers has a smooth depression with steep flanks. By ray tracing from the point diffractor, we obtain the traveltimes of its diffraction pattern at the surface (Figure 1b). The complex pattern in Figure 1b would be approximately that recorded by a coincident source-receiver survey over this model. For this example, the diffraction pattern is split into two hyperbola-like features. Perhaps worse yet, these features exhibit such large curvatures that their proper migration would seem to require

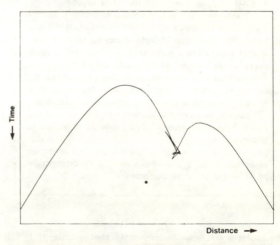

FIG. 1b. Unmigrated time section displaying the diffraction pattern, the curve of two-way traveltime between source-receiver points at the earth's surface, and the scatterer in Figure 1a. The dot marks the correct lateral position of the point scatterer.

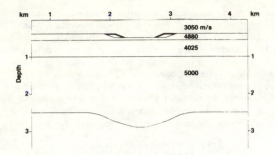

FIG. 2a. Depth model. The heavy lines on the upper interface indicate a perturbation of the original model used in the velocity sensitivity tests. The reflection coefficient of the interface between the 4880 and 4025-m/sec layers is zero. Reflection coefficients at all other interfaces are identical to one another.

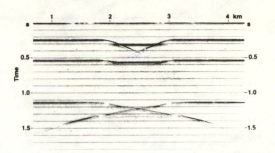

FIG. 2b. Normal-incidence time section for the depth model in Figure 2a. This zero-offset section was generated by the diffraction response technique of Trorey (1970) as modified by Larson and Hilterman (1976). Note the phantom diffractions on the second and third horizons.

velocities that are ridiculously low relative to the average velocity for the medium.

Imagine time migrating the intricate pattern of Figure 1b by the Kirchhoff technique of summation along hyperbolas. (The results obtained by any other method would be quite similar.) The best we might hope for is a partial collapse of energy near the two peaks of the pattern. In fact, the result will be more confused than that because of the complex of events between the peaks; the image after time migration will not remotely resemble that of a point scatterer. Subsequent depth conversion (whether conventional or image ray) will not remedy this situation. Clearly, what is required is a proper incorporation of the detailed velocity model into the initial migration step.

WAVE-THEORETICAL DEPTH MIGRATION

To understand the failure of time migration (and, consequently, ray-theoretical depth migration), we must make a brief examination of the scalar wave equation. The scalar wave equation in two dimensions can be written

$$\frac{\partial^2 P}{\partial x^2} + \frac{\partial^2 P}{\partial z^2} = \frac{4}{v^2(x, z)} \frac{\partial^2 P}{\partial t^2}, \qquad (1)$$

where $P(x, z, t)$ is the observed pressure wave field, $v(x, z)$ is the velocity of the medium, and t is the *two-way* reflection time appropriate for zero-offset data (hence, the factor 4 on the right side of the equation). In order to make the migration computation tractable, Claerbout (1976, p. 234) proposed a family of coordinate transformations. The key element in any of Claerbout's transformations is that the frame of reference for the migration algorithm moves along with the wave field being migrated. Consider the transformation:

$$x' = x,$$
$$z' = z,$$

and

$$t' = t + 2 \int_0^z \frac{d\sigma}{\bar{v}(x, \sigma)}.$$

The motion of the reference frame is specified by a parameter called the frame velocity $\bar{v}$.

If we choose the frame velocity to be a constant, the scalar wave equation can be rewritten exactly (see Appendices A and B) as

$$P_{t'z'} + \frac{\bar{v}}{4} P_{z'z'} = A(\bar{v})P_{x'x'} + B(v, \bar{v})P_{t't'}, \qquad (2)$$

where P and v are now functions of the new primed coordinates, and $P_{t'z'}$ denotes $\partial^2 P/\partial t' \partial z'$, etc. Here A and B are used to denote more complicated expressions that involve the parameters indicated [see equation (B–1), Appendix B]. If the frame velocity is not constant, other terms involving its derivatives must be added to the right side of equation (2).

In implementing migration by means of equation (2), we are free to choose the frame velocity as we wish, providing we choose it to be constant. The usual practice in time migration is to bend the rules and choose the frame velocity to be the velocity of the medium, *in spite of the spatial variance* of that velocity. This choice simplifies equation (2) by forcing term B on the right side to be identically zero, leaving only the so-called diffraction term A. Time migration, then, is performed with the equation

$$P_{t'z'} + \frac{v}{4} P_{z'z'} = A(v)P_{x'x'}. \qquad (3)$$

Another common approximation is to set $P_{z'z'} = 0$ (c.f., Appendix B), yielding

$$P_{t'z'} = A(v)P_{x'x'}. \qquad (4)$$

Since v is a function of x and z, equation (4) *apparently* allows the use of a laterally variable velocity, and one would think that it properly describes the propagation of waves in an inhomogeneous medium. In reality, this equation propagates waves locally as though the medium were horizontally layered. In bending the assumptions, we have, among other things, discarded the gradients of $\bar{v}$ which embody Snell's law for a laterally inhomogeneous medium (see Appendix A). In wave-theoretical depth migration, we assume a constant $\bar{v}$ and honor Snell's law by including the so-called shifting term B of wave equation (2). When we do so, the term $P_{z'z'}$ cannot be so casually dropped (see Appendix B). Our studies with synthetic data indicate that *so long as the lateral velocity variation is not too severe*, neglect of the gradients of $\bar{v}$ is not highly detrimental to the *imaging* ability of time migration. As will be seen in the next example, however, for severe inhomogeneity such neglect is patently inappropriate.

Although our discussion of the wave equation has been along the lines used by Claerbout in developing his finite-difference migration techniques, the same (and some even more restrictive)

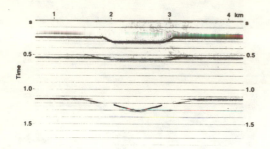

FIG. 2c. Time migration of the data in Figure 2b by the Kirchhoff summation technique.

approximations are made when migration is performed in the frequency-wavenumber domain (Stolt, 1978; Gazdag, 1978). In addition, Larner and Hatton (1976) showed that approximations made in conventional Kirchhoff summation migration are equivalent to those discussed here. Thus, our remarks on the positioning and imaging behavior of time migration apply regardless of the particular implementation.

Clearly, resolution of imaging and mispositioning problems requires incorporation of the shifting term in the migration algorithm. Consequently, $\bar{v}$ can be assumed constant and the conditions on its derivatives satisfied. Because the shifting term embodies the proper refraction of energy in laterally inhomogeneous media (Snell's law), this more complete form of the wave equation will solve both imaging and lateral positioning problems. For ease in implementation, we chose to perform our migrations by the finite-difference method.

The input to this extended technique is an unmigrated (stacked) time section. The output, obtained directly in one step, is a *fully migrated* depth section. We refer to this method as wave-theoretical depth migration to distinguish it from its ray-theoretical counterpart. [Judson et al (1980) and Gazdag (1980) present additional perspectives on depth migration.]

As in the ray-theoretical depth migration process, specification of migration velocity is again of central importance. Wave-theoretical depth migration requires that interval velocity be specified as a function of depth and horizontal position. Because the velocity function is not available in advance, we start with a reasonably close estimate of velocity as a function of those coordinates. The depth migration process will then provide information for improving the estimated structure. The relative insensitivity of the technique to imperfect velocity modeling usually allows the solution for structure to converge adequately after only one iteration.

SYNTHETIC DATA EXAMPLE

Consider the depth model in Figure 2a. The layers have constant velocities, and interfaces are basically horizontal except for a syncline at depth. The near-surface anomaly of this model constitutes the significant inhomogeneity whose disruptive effects will be evident. (If the top layer were less thick relative to the depths of deeper interfaces, time distortions caused by the near-surface anomaly could be corrected by conventional static time adjustments.) Dips in the near-surface exceed 30 degrees, and those in the syncline exceed 40 degrees. Figure 2b shows a synthetic, zero-offset section for this model, generated by a wave-theoretical, diffraction response technique.

The most important features in the unmigrated section are direct results of the near-surface inhomogeneity. Both the flat-lying event (just below 0.5 sec) and the buried focus from the syncline exhibit phantom diffractions and gaps that are the result of distortion of the wavefront during its transit through the anomalous near-surface.

Figure 2c displays the Kirchhoff time migration of the data in Figure 2b based on the known velocities in the medium. Note that the shallowest interface is correctly migrated (the medium is homogeneous above it), but the time migration has failed to treat the phantom diffractions on deeper horizons. Residual diffraction effects remain in the second reflection, and large gaps remain evident in the syncline. Note also the double events at the base of the syncline and at its edge. No form of depth conversion that involves stretching of traces—either vertically or along image raypaths—can convert these double events into the simple wavelet that should mark the interface.

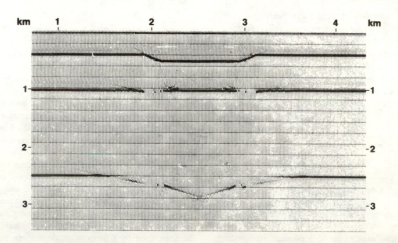

FIG. 3a. Ray-theoretical depth migration of the section in Figure 2b. Note that this is a depth section.

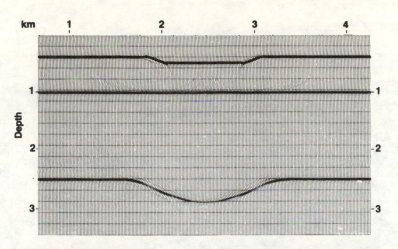

FIG. 3b. Wave-theoretical depth migration of the data in Figure 2b. Note that this is a depth section.

The double events are not the only problems confronting the image ray conversion from time to depth, as the ray-theoretical depth migration in Figure 3a vividly demonstrates. The strong velocity contrast and appreciable dip at the upper interface dictate large deflections for the image rays. With so pronounced an inhomogeneity, image-ray theory breaks down just as the imaging of time migration did. The subsequent redistribution of amplitudes into depth along the image rays actually creates more problems than it solves; large gaps are opened in the second reflection, and some diffracted energy is left essentially untreated.

Figure 3b is a depth section displaying the result of wave-theoretical depth migration of the data in Figure 2b. The process has yielded a section in which the positions and shapes of imaged interfaces compare well with their known positions (compare with the depth model in Figure 2a). Moreover, it has both properly imaged the syncline and correctly collapsed phantom diffractions.

In this example, the only velocity inhomogeneity of concern is that in the near-surface. A useful way to visualize migration is as a downward continuation process: the technique proceeds step by step to predict the seismic sections that would have been recorded had the sources and receivers of the survey been buried deeper and deeper. In finite-difference migration, the observation plane is in effect progressively lowered to the reflectors. At any particular stage of downward continuation, the seismic data above the observation plane are fully migrated, while the data below represent the (unmigrated) time section that would have been recorded at that depth of burial.

In Figure 3c, the data have been downward continued to a depth of just over 750 m (the dashed line). The resulting section is a hybrid; above the dashed line it is a (fully migrated) depth section of the near-surface, while below it is the time section that would have been recorded 750 m below the surface.

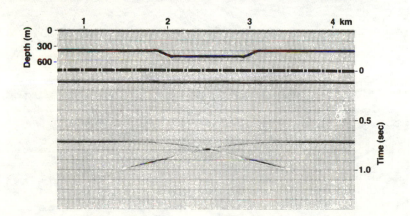

FIG. 3c. Partial downward continuation. The wave-theoretical depth migration was stopped with the source-receiver level at approximately 750-m depth (dashed line). Above the dashed line, the section is fully migrated in depth; the dashed line also corresponds to time zero on the time section (below the line) that would have been recorded had sources and receivers been buried at 750 m.

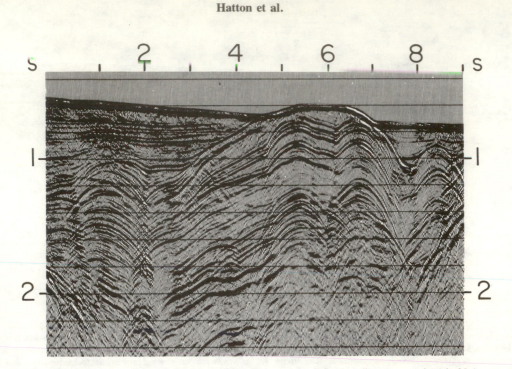

FIG. 5a. CDP stack section of data recorded in the Santa Barbara channel. Steepest dips are approximately 25 degrees.

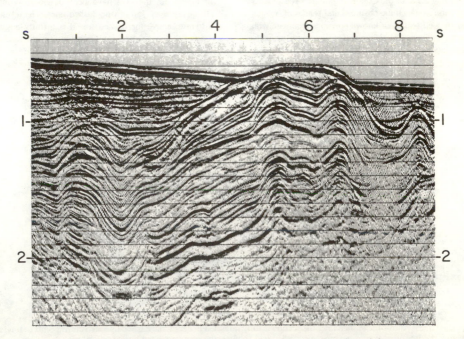

FIG. 5b. Finite-difference (time) migration of the Santa Barbara channel data.

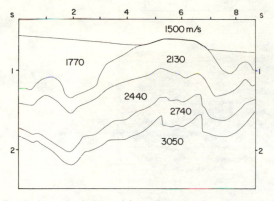

FIG. 5c. Interval velocity model used for performing depth migration on the data shown in Figure 5a.

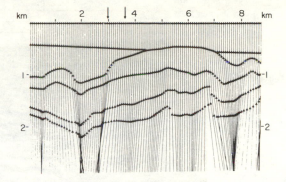

FIG. 5d. Image raypaths (plotted in depth) obtained by ray tracing a velocity model interpreted from the time-migrated section in Figure 5b.

slightly shallower than they should be. These depth errors are simply the familiar errors in converting from time to depth with the wrong velocity. Results of the 10 percent increase (not shown) were similar in character; the error in depth was naturally in the opposite direction.

Other types of errors can also contaminate the velocity model; in particular, interpreted shapes of layer interfaces could be inaccurate. To examine the effects of such an error, we perturbed the shape of the first interface as shown by the heavy lines in Figure 2a. Here, interpretation of the first interface was made from the unmigrated time section, certainly a crude thing to do for this model. In practice, the migrated time section yields a much more accurate estimate of the shallow structure. In this experiment, correct velocities were used for all layers.

Results of the wave-theoretical depth migration are shown in Figure 4b. The gross error in layer shape has caused considerable distortion in the deeper section and again introduced some spurious low-amplitude diffractions. Nevertheless, even for this severe error the reflectors are imaged rather coherently across the section. One can interpret the deeper section here better than on any of the sections that involved time migration as an intermediate step.

The results of this small sensitivity study parallel those of the image-ray uncertainty study of Larner et al (1981). Depth migration is less sensitive to the specification of velocity values than to interpreted shapes of layer interfaces. We consider the proper interpretation of layer shape as the keystone in approximating horizontal velocity gradients for a layered model. Horizontal gradients constitute the central issue in laterally inhomogeneous media.

FIELD DATA EXAMPLE: OFFSHORE CALIFORNIA

Figure 5a is a CDP stack of data from the Santa Barbara Channel, offshore California. The (finite-difference) time migration in Figure 5b shows the expected changes; crossing diffraction patterns are collapsed to synclines, and anticlines are constricted. In this profile, dips do not exceed about 25 degrees. We observe that the imaging has been reasonably successful in the sense that the time-migrated section presents a more plausible picture of the subsurface than does the original stack. Plausibility, however, is no guarantor of truth.

The unconformity surface between the gently dipping Quaternary section and the prominently folded Tertiary section could

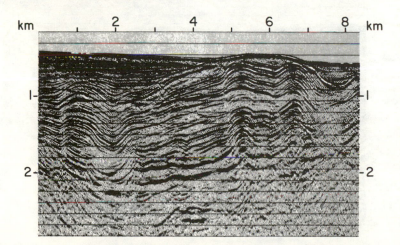

FIG. 5e. A conventional depth section for the time-migrated data recorded in the Santa Barbara channel. Note the amplitude weakness in the synclinal features at lateral positions 2 and 7.5 km.

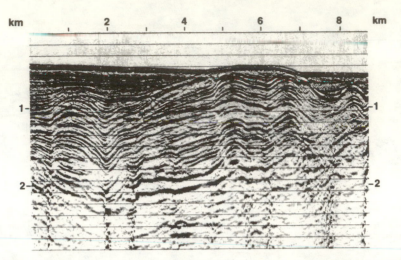

FIG. 5f. Depth section generated by stretching time-migrated data (Figure 5b) into depth along the image raypaths of Figure 5d. These steps constitute ray-theoretical depth migration.

provide the circumstances that lead to significant deflection of image rays from the vertical. If that is the case, we expect that (although plausible) the time-migrated section in Figure 5b is distorted. Using independently known velocities for the area and a digitized interpretation of several major horizons (including the unconformity, Figure 5c), we computed the paths of the image rays plotted in Figure 5d. This plot is in depth, and the triangular symbols denote intersections of the image raypaths with the (now fully migrated) horizons. In the model, velocities increased from about 1770 m/sec in the Quaternary section to about 2100 m/sec across the unconformity and then to about 2750 m/sec in the deep layer at 2 sec.

In the image-ray plot for this section (Figure 5d), rays tend to converge downward into synclines and to diverge beneath anticlines. Another model of this line (not shown here) differed from this one only in that velocities increased more rapidly with depth, attaining values 50 percent higher for the layer immediately above basement. For that quite different velocity model, the pattern of convergence and divergence of image rays was much like that shown in Figure 5d. As noted earlier, computed depths are highly sensitive to the model velocities, but the lateral position errors requiring correction by depth migration are more directly related to the interpreted horizon shapes. For this reason, it is important to understand whether the velocities of a particular section are

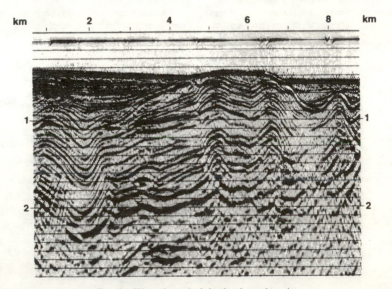

FIG. 5g. Wave-theoretical depth migrated section.

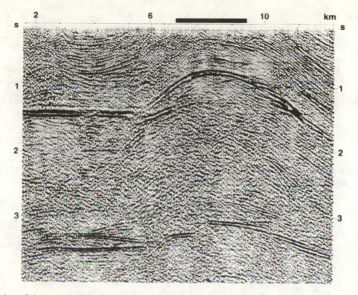

FIG. 6a. CDP-stacked section of data recorded in Central America. The prominent reflector at about 1.4 sec on the left side is the top of a thick, high-velocity carbonate/anhydrite sequence. The horizontal bar denotes the length of the spread.

controlled primarily by depth or by lithology, i.e., it is important to know whether or not velocity gradients trend normal to interpreted horizons.

Figure 5e is a conventional depth section obtained by stretching the traces in Figure 5b vertically from time to depth. Note the generally weaker amplitudes below about 1.5 km in depth at lateral positions of 2 km and 7.5 km. These zones coincide with the greatest degree of folding in the section.

Conversion to depth along image rays will tend to boost amplitudes in places where the rays converge, thereby compensating

approximately for amplitude weakness. The image-ray depth conversion of the traces in Figure 5b is shown in Figure 5f. Inadequacies in amplitude treatment attributable to ray theory are evident, but the result is encouraging; amplitudes tend to be restored over those zones where they were originally weak.

In spite of substantial uncertainty in the available velocity information, the wave-theoretical depth migration shown in Figure 5g is superior to either the conventional or the image-ray depth section. In particular, note the generally improved continuity and more uniform treatment of amplitude throughout the syncline at

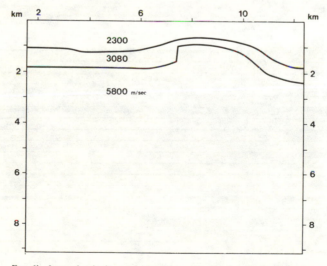

FIG. 6b. Interval velocity model used in processing the data in Figure 6a.

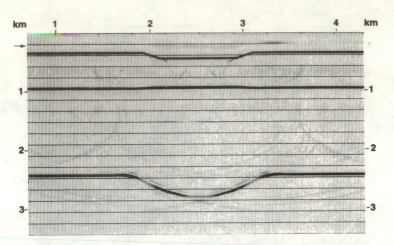

Fig. 4a. Wave-theoretical depth migration with the velocity of the uppermost layer specified as 2745 m/sec (10 percent lower than the known velocity for the model).

Note that the time section shows none of the distortion attributable to the near-surface anomaly. The middle reflector (just below the dashed line) is flat, and the phantom diffractions have healed completely. The buried focus from the syncline is also free of phantom diffractions. This time section could now be successfully treated with a conventional time-migration algorithm.

The advantage of such partial migration is clear. Often, we might expect to have a detailed velocity model for the near-surface from high-resolution seismic work, refraction shooting, or well surveys, whereas the velocity information for the deep section is less precisely known. Partial migration can remove the distorting influence of an inhomogeneous near-surface zone from deep data without the need for a full depth migration. In a sense, partial depth migration is thus a wave-theoretical extension of approaches to statics corrections for near-surface time anomalies. After performing partial depth migration to correct for near-surface variations, one can finish with time migration (based on whatever velocity information is available for the deeper section) and a final direct conversion to depth.

Again, the essential issue is the sensitivity of migration quality to inaccuracies in velocity. To study sensitivity, we performed wave-theoretical depth migration on the data of Figure 2b using velocity information that was known to be incorrect. The first two tests were based on the use of wrong velocities for the uppermost layer (correct velocity = 3050 m/sec); specifically, we performed depth migrations with that velocity 10 percent too low and then 10 percent too high.

Only the results of the 10 percent decrease are shown (Figure 4a). Surprisingly, the images of both the syncline and the flat reflector are still coherent and continuous across the section. Although the section exhibits some faint spurious diffraction effects, the syncline shape, wavelet character, and even amplitude treatment are quite good—far superior to the results of either time migration or ray-theoretical depth migration. The resulting reflectors are all

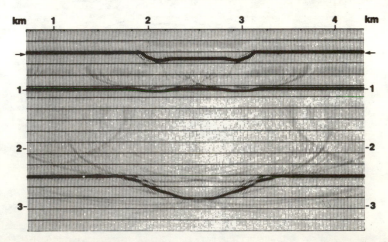

Fig. 4b. Wave-theoretical depth migration with the velocity model perturbed as shown by heavy lines in Figure 2a.

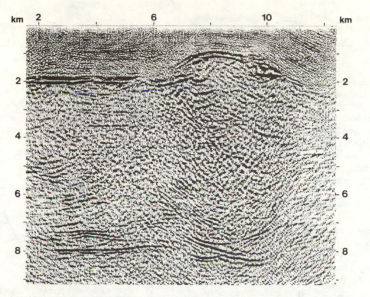

FIG. 6c. Depth section produced by time migration of the section in Figure 6a and simple conversion to depth. Both the time migration and conversion to depth used the velocity model in Figure 6c.

2 km. Again, the imaging of this depth migration process is relatively insensitive to inaccuracies in the velocities used for the computation.

Reviewing the sequence of sections from Figures 5a through 5g, we see an interesting progression near the lateral position of 2 km —from the implausible crossing of events before migration to a broad syncline in the conventional depth section and then back

toward a tighter syncline in the depth-migrated sections. Note that differences exist in the velocities used to generate the two forms of depth-migrated data, and these differences account in part for the different shapes obtained for the syncline at 2 km. The basic conclusions, however, are unaffected: (1) realistic models for velocity lead to convergence of image rays beneath the synclines and resultant tightening of the synclines relative to the shapes on

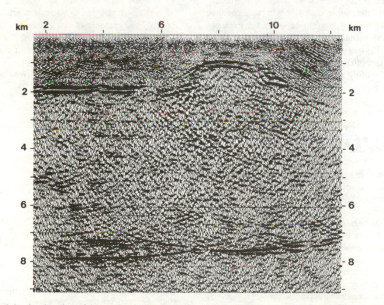

FIG. 6d. Wave-theoretical depth migration of the section in Figure 6a based on the velocity model in Figure 6b. Note the simplified image of the deep reflector.

the conventional depth section, and (2) from comparison of results obtained using different velocity models, we can put realistic bounds on the range of possible structural interpretations.

FIELD DATA EXAMPLE: CENTRAL AMERICA

Figure 6a shows an unmigrated stacked section of data collected in Central America. Regional tectonic forces have produced substantial folding and faulting in the upper part of the section. Above the prominent horizon (at 1.4 sec on the left side), the geologic section is comprised of Tertiary clastics with velocities in the neighborhood of 2500 to 3000 m/sec. The strong reflection marks the boundary between the Tertiary section above and a substantially higher-velocity Cretaceous carbonate/anhydrite sequence below. Velocities in the Cretaceous section are approximately 6000 m/sec.

This two-to-one velocity contrast, combined with the rollover and steeply dipping fault, result in greatly distorted deeper reflections. Specifically, regional information suggests that the deep reflector appearing at about 3.1 to 3.6 sec is essentially flat and continuous; the complexity it exhibits in this stacked section results from the distorting influence of the velocity variations in the overburden.

As noted earlier, the most critical step in imaging such a distorted reflector is the specification of a velocity model for the overburden. For the data here, the key issues are (1) the shape of the Tertiary-Cretaceous boundary, and (2) the interval velocities above and below. Well log information indicates that the Tertiary section is relatively homogeneous; hence, a preliminary time migration provides a good image of the Tertiary-Cretaceous horizon. Interpretation of the migrated data and well-log information suggests that the Tertiary section is acceptably described by a two-layer model and that the Cretaceous section is approximately homogeneous. The selected model (Figure 6b) has a slight, vertical velocity gradient in the uppermost layer and constant velocities in the other two.

For this section, the velocity model was developed with relative ease because the interval velocities were assumed reasonably well known and varied little within each layer. In more complex regions, interval velocities may have to be estimated from observed moveout reflection times by use of techniques such as those of Larner and Rooney (1972) or Hubral (1976). Such estimates, however, are often crude and may require careful interpretation (namely, editing and smoothing). The uncertainties in interval velocity estimates are related directly to the quality of surface-moveout analyses; such surface-velocity analyses will often be ambiguous when conducted over complex structures. For this reason, the use of independent velocity information, such as that from well logs, is always desirable and often essential. As indicated in the synthetic-data studies, depth migration is fairly insensitive to errors in the velocity model. We therefore expect that a preliminary time migration can clarify layer structure and thus be used to improve the velocity model.

Figure 6c is a conventional depth section obtained by vertically stretching a (finite-difference) time-migration result for the data in Figure 6a. Both the time migration and depth conversion used the velocity model of Figure 6b. As expected, the data down to the Cretaceous horizon are generally clarified; the deep reflection, however, is still quite complex, actually more confused than it was on the unmigrated data. A better image would have been gained by simply depth converting the original unmigrated section.

In contrast, the wave-theoretical depth migration (Figure 6d) produces a much simplified image of the deep reflector. The image here is essentially continuous and flat across the extent of

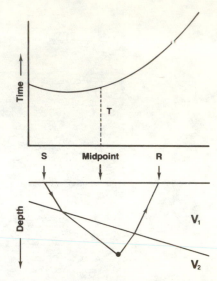

FIG. 7a. Depth model of a point diffractor beneath a plane dipping interface. A diffraction pattern for some nonzero source-receiver offset is also shown. Traveltime from source to scatterer to receiver is plotted at the source-receiver midpoint.

the section. The slight pull-up of that reflection near the center of the section is likely attributable to some imperfection in the model for the overburden and could be corrected by an adjustment to the velocity or shape of the upper layers. Most importantly, though, wave-theoretical depth migration has effectively compensated for the severe lensing caused by the shallow velocity structure.

While the deep reflection in Figure 6d is now approximately flat and continuous, a bowl-shaped artifact has appeared just above it. The presence of this pattern can be traced to the stacked section (Figure 6a). In that figure note that segments of the deep reflection lack diffraction tails that must exist in a recorded wave field. The tails likely have been destroyed in CDP stacking because complex propagation paths across the folded and faulted

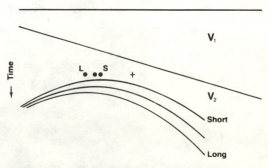

FIG. 7b. Diffraction patterns for the scatterer of Figure 7a as they would be seen on three common-offset time sections. L and S denote long and short offsets, respectively. The cross indicates the correct lateral position of the scatterer.

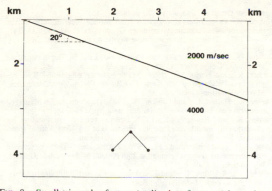

FIG. 8a. Small triangular feature (stylized reef or trap) beneath a dipping interface.

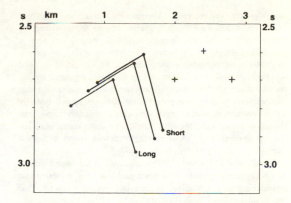

FIG. 8b. Results of time migration before stack for the feature of Figure 8a as seen on three different common-offset sections. Crosses mark the correct lateral positions of the three edges of the small feature.

structure have severely disrupted normal-moveout (NMO) relationships. The bowl-shaped pattern in Figure 6d then is a direct result of the migration of those truncated reflections. This artifact reminds us that a CDP-stacked section is not equivalent to a zero-offset section.

DEPTH MIGRATION BEFORE STACK

Since we are addressing various aspects of the migration of data from inhomogeneous media, it is fair to ask whether a poor migration for a complex section might be attributable to prior degradation by the CDP stacking process. Certainly, raypaths and traveltimes are likely to be complex and not well described by the horizontally layered model that underlies the stacking process.

Where a CDP stack is suspect, the conventional thinking has been that the data ought to be migrated before being stacked. Migration of unstacked data involves considerably more computation than migration applied after stack; hence, a close look at the assumed benefits of this more extensive effort is in order.

Consider first the simple case of a point diffractor beneath a dipping interface, as depicted in Figure 7a. Regroup traces from a conventional CDP survey so that they appear as though we had conducted a number of common-offset surveys. In each common-offset section, we would observe the diffraction pattern of traveltimes (from source to diffractor and back to receiver) plotted as a function of the point midway between the source and receiver. Figure 7b shows schematically three such diffraction patterns for small, medium, and large offsets. As in the normal-incidence case, the diffraction patterns are asymmetric with their apexes displaced laterally from the correct diffractor position.

Whereas for zero offset the position of the minimum time corresponds to the image-ray location, no such simple relationship exists for the minimum-time points on common-offset diffraction patterns. For the dipping interface model in Figure 7a, the apexes are displaced progressively farther updip as offset increases.

Now, suppose we performed conventional time migration on each offset section. Then, despite a failure to collapse the diffraction patterns completely, the process would condense amplitudes to a *different minimum point* for each offset section. With this lateral spread of minima, the time-migrated offset sections cannot be stacked properly.

Common-offset migration is not the only way to accomplish migration of unstacked data. An alternative approach is to downward continue, over each depth step, first the receivers associated with each shot and then the shots associated with each receiver (Clayton, 1978). As long as this process is performed without consideration of refraction through the overburden, it cannot resolve the lateral mispositioning problem any better than common-offset migration can.

We can illustrate this problem differently with a schematic model of a small structural feature beneath a dipping interface. In the depth section of Figure 8a, a small "trap" (the triangularly shaped feature) is embedded in an otherwise gently dipping section. Figure 8b shows the time-migrated positions (in close-up) of the trap as it would be seen on sections for three different offsets. The left dipping segments nearly coincide, but these features are not yet NMO-corrected. The processing steps that would follow the migration are velocity analysis and then stacking of the data. What velocities are required to correct the left and right segments of this feature for normal moveout? Along the left segment, the differential time from short offset to long is so small that the required stacking velocity might be as high as 6000 m/sec or more. In contrast, on the right segment, the required stacking velocity could be as low as 1400 m/sec or less. Between the two, the normal moveout may not be even approximately hyperbolic.

How could one hope to stack this small feature properly? Recall, our trap is a subtle structure in an otherwise gently dipping medium. In practice we would smooth variations in the stacking velocity over a distance of a kilometer or more. We would have little choice other than to correct for normal moveout with a velocity of around 3200 m/sec, not at all close to the velocities required to stack the time-migrated limbs of our small feature.

As this example illustrates, time migration before stack can be as damaging to reflections, particularly those from subtle features, as conventional CDP stacking. Thus, where the subsurface is sufficiently complex to warrant depth migration, the most appropriate solution often may be depth migration of the unstacked data. Conventional time migration of unstacked data, for all the added computational effort, may yield a result no better than migration after stack.

Schultz and Sherwood (1980) showed how depth migration of unstacked traces can be performed by the downward continuation of sources and receivers. Alternatively, Yilmaz and Claerbout (1980) discussed how depth migration can be accomplished for common-offset sections. In their technique, each offset section is partially migrated to convert it (approximately) to an unmigrated normal-incidence time section. The resulting sections are then stacked, and the stack can be depth migrated by the method discussed here. During each downward continuation step of the prestack migration process, the data are shifted laterally by an amount dependent upon the offset, reflection time, and degree of lateral heterogeneity. The lateral shifts compensate for the offset-dependent displacement of reflected and diffracted events.

Schultz and Sherwood (1980) demonstrated the application of depth migration to another common problem—the distortion and misstacking of deep reflection data by irregular, water-bottom topography.

CONCLUSIONS

Although virtually every seismic section ought to be migrated, not all need to be depth migrated. Where the overburden is known to be laterally inhomogeneous, however, depth migration can be a refinement of primary importance. Given a reasonably accurate estimate of velocity, the relatively efficient process of image-ray tracing (and velocity modeling) can be applied to the time-migrated data to determine whether the more complete, more accurate, and more costly wave-theoretical depth migration is warranted.

Information about detailed velocity structure will be imperfect, sometimes grossly inaccurate. Fortunately, like time migration, both ray-theoretical and wave-theoretical depth migration are reasonably insensitive to errors in the velocity model; nevertheless, the quality of depth migration is ultimately limited by the accuracy of velocity information available. We have demonstrated an algorithm that migrates data correctly *if* we know the velocity model. The primary task must be to improve our ability to estimate velocity.

Geophysicists are now squarely confronting seismic problems attributable to complex geologic structures and are finding that the issue of velocity is fundamental. We caution against applying too much effort to any single problem at the expense of others. At present, we artificially categorize depth migration, migration before stack, and 3-D migration as three *separate* tools for migrating data from inhomogeneous media. The problems they address are, however, deeply interconnected—they cannot be considered apart from one another. Thus, we must pursue a task that could not have been considered seriously only a short time ago—implementing 3-D depth migration before stack. Executed correctly, it will be based on the elastic wave equation.

ACKNOWLEDGMENTS

We gratefully acknowledge Ron Chambers for his expert computer programming, and Carl Savit for reviewing the manuscript. Stew Levin made significant contributions to the analysis shown in Appendix B. Our thanks go also to Brenda Edwards and Rhonda Boone for their creative drafting of the figures and Evelyn Fulford, Grace Bonaventura, and Dolores Meeks for their patient attention to the typing of the manuscript.

REFERENCES

Claerbout, J. F., 1976, Fundamentals of geophysical data processing: New York, McGraw-Hill Book Co., Inc., 274 p.
Clayton, R., 1978, Common midpoint migration: Stanford Expl. Proj. rep. no. 14, Stanford, p. 21–36.
Gazdag, J., 1978, Wave equation migration with the phase shift method: Geophysics, v. 43, p. 1342–1351.
——— 1980, Wave equation migration with the accurate space derivative

method: Geophys. Prosp., v. 28, p. 60–70.
Hubral, P., 1976, Interval velocities from surface measurements in the three dimensional plane layer case: Geophysics v. 41, p. 233–242.
——— 1977, Time migration—Some ray-theoretical aspects: Geophys. Prosp., v. 25, p. 738–745.
Judson, D., Lin, J., Schultz, P., and Sherwood, J., 1980, Depth migration after stack: Geophysics, v. 45, p. 361–375.
Larner, K., and Hatton, L., 1976, Wave equation migration—Two approaches: Presented at 8th Annual Offshore Technology Conference, May, in Houston.
Larner, K., Hatton, L., and Gibson, B., 1981, Depth migration of imaged time sections: Geophysics, v. 46, this issue, p. 724–738.
Larner, K., and Rooney, M., 1972, Interval velocity computation for plane dipping multilayered media: Presented at the 42nd International SEG Meeting, November 30, in Anaheim.
Larson, D., and Hilterman, F., 1976, Diffractions—Their generation and interpretation use: Presented at 29th Annual Midwestern SEG Meeting, March, in Dallas.
Mitchell, A. R., 1969, Computational methods in partial differential equations: New York, J. Wiley and Sons.
Schultz, P., and Sherwood, J., 1980, Depth migration before stack: Geophysics, v. 45, p. 376–393.
Stolt, R. H., 1978, Migration by Fourier transform: Geophysics, v. 43, p. 23–48.
Trorey, A., 1970, A simple theory for seismic diffractions: Geophysics, v. 35, p. 762–784.
Yilmaz, O., and Claerbout, J., 1980, Prestack partial migration: Geophysics, v. 45, p. 1753–1779.

APPENDIX A
IMAGE RAYS IN FINITE-DIFFERENCE ALGORITHMS

Larner et al (1981) show that both Kirchhoff summation and finite-difference time-migration algorithms fail to position events correctly in the presence of laterally varying velocity and indeed fail in similar ways. Hubral's (1977) elegant analysis introduces the concept of the image ray and its relationship to Kirchhoff summation. It is of interest, therefore, to identify image rays in the formalism of finite-difference algorithms.

An appropriate coordinate frame is useful when trying to identify a particular phenomenon. We start with the scalar wave equation

$$\frac{\partial^2 P}{\partial x^2} + \frac{\partial^2 P}{\partial z^2} = \frac{4}{v^2(x, z)} \frac{\partial^2 P}{\partial t^2}, \qquad (A-1)$$

where $P = P(x, z, t)$ is the disturbance as a function of x and z (the horizontal and vertical coordinates, respectively) and of two-way time t.

In the spirit of the coordinate transformations introduced by Claerbout (1976), we introduce the transformation:

$$x' = x,$$
$$z' = z,$$

and

$$t' = t + 2 \int_0^z \frac{d\sigma}{\bar{v}(x, \sigma)}. \qquad (A-2)$$

The rationale behind this choice is its intimate connection with ray tracing. Equation (A–1) can then be written with suffix notation for partial derivatives as

$$P'_{z't'} + \frac{\bar{v}'}{4} P'_{z'z'} = -\frac{v'}{4} P'_{x'x'} + \bar{v}' \, a \, P'_{x't'}$$

$$+ v' \left(\frac{1}{v'^2} - \frac{1}{\bar{v}'^2} - b \right) P'_{t't'} + \frac{\bar{v}'}{2} c \, P'_{t'}, \qquad (A-3)$$

where

$$P'(x', z', t') = P(x, z, t),$$
$$\bar{v}'(x', z') = \bar{v}(x, z),$$
$$v'(x', z') = v(x, z)$$

$$a(x', z') \equiv \int_0^z \frac{\bar{v}_x}{\bar{v}^2}\, d\sigma, \qquad (A-4)$$

$$b(x', z') \equiv a^2(x', z'),$$

and

$$c(x', z') \equiv \frac{\bar{v}_z}{2\bar{v}^2} + \int_0^z \left[\frac{\bar{v}_{xx}}{\bar{v}^2} - \frac{2\bar{v}_x^2}{\bar{v}^3} \right] d\sigma.$$

Note that equation (A–3) is a more general form of equation (2) in the text (i.e., $\bar{v}$ is not assumed constant). Now neglect the term involving $P'_{z'z'}$, a common approximation in finite-difference migration. Equation (A–3) is therefore only approximate, but it will suffice to illustrate the derivation of image rays.

At this stage we will also set the frame velocity $\bar{v}'$ equal to the medium velocity v'.

A popular technique for solving equations like (A–3) is that of Marchuk splitting (Mitchell, 1969). In this method the initial equation is split into separate equations which are solved in parallel. Equation (A–3) can be written as

$$P'_{z't'} = -\frac{v'}{4} P'_{x'x'}, \qquad (A-5)$$

and

$$P'_{z't'} = v'\, a\, P'_{x't'} - v'\, b\, P'_{t't'} + \frac{v'c}{2} P'_{t'}. \qquad (A-6)$$

It can be shown that equation (A–5) describes the process of diffraction. As we shall see, the ability to refract at dipping interfaces is contained solely in equation (A–6).

The wave equation alluded to in conventional time migration is (A–5) or an equation with similar properties. Equation (A–6) or its equivalent is essential if propagation in a laterally heterogeneous medium is to be treated correctly.

Let us analyze equation (A–6): Further splitting reveals that the first term on the right is the image ray term, the second term on the right is a time-shifting term, and the third term on the right is

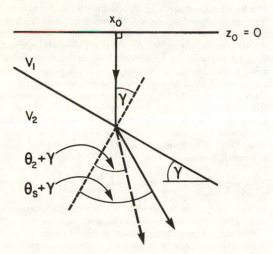

FIG. A–1. The raypaths resulting from a single dipping layer with velocity contrast as shown. $\theta_s + \gamma$ is the Snell's law angle, $\theta_2 + \gamma$ is the refraction angle associated with equation (A–10), and γ is the dip.

a residual-amplitude term. We shall discuss the image ray term first.

With one t'-integration, the first term on the right of equation (A–6) behaves in split form as

$$\left(v'\, a\, \frac{\partial}{\partial x'} - \frac{\partial}{\partial z'} \right) P' = 0. \qquad (A-7)$$

In the nontrivial case $v'\, a \neq 0$, equation (A–7) can be written

$$\frac{DP'}{Dx'} = 0, \qquad (A-8)$$

where

$$\frac{D}{Dx'} = \frac{\partial}{\partial x'} + \frac{dz'}{dx'} \frac{\partial}{\partial z'}$$

is the Lagrangean derivative used frequently in describing the dynamics of fluids, and

$$\frac{dz'}{dx'} = -\frac{1}{v'\, a}. \qquad (A-9)$$

Equations (A–8) and (A–9) indicate that P' should be kept constant along the trajectory given by (A–9). This is exactly the behavior we would expect along a raypath in the time-shifted frame described by equation (A–2). Let us further analyze the path described by equation (A–9).

Using equation (A–4), we write (A–9) as

$$\frac{dx'}{dz'} = -v' \int_{z_0'}^{z'} \frac{\partial v'/\partial x'}{v'^2}\, d\sigma, \qquad (A-10)$$

and assume (x_0', z_0') is the starting point of the trajectory of interest. We suspect immediately that equation (A–10) is closely analogous to an image ray by observing that for $z_0' = 0$, and within an underlying homogeneous layer (where $\partial v'/\partial x' = 0$), $dx'/dz' = 0$ and remains so until the depth is reached at which $\partial v'/\partial x' = 0$. Now, $dx'/dz' = 0$ implies that the corresponding trajectory departs z_0' (=0) normal to that surface. Furthermore, dx'/dz' is always constant across a homogeneous layer.

Consider the refraction of such a ray across the dipping interface in the model of Figure A–1. Initially, suppose that a transition zone centered at the dipping interface separates the homogeneous layers above and below. Also, let the velocity vary linearly across the transition zone (from V_1, the velocity in the upper layer, to V_2, the velocity in the lower layer). That is,

$$v'(x, z) = \begin{cases} V_1 & \text{for } z < z_b - \delta \\ V_0 + \lambda_1 x + \lambda_2 z & \text{for } |z - z_b| \leq \delta, \\ V_2 & \text{for } z > z_b + \delta \end{cases}$$

where $z_b(x)$ is the depth of the center line of the transition zone, δ is the vertical thickness of the zone, and V_0, λ_1, and λ_2 are constants. This model requires that

$$\lambda_2/\lambda_1 = -\tan \gamma,$$

where γ is the dip of the transition zone.

According to equation (A–10), the derivative dx'/dz' for the image ray trajectory at and beneath the transition zone is

$$\frac{dx'}{dz'} = -V_2 \int_{z_b - \partial}^{z_b + \partial} \frac{\lambda_1\, d\sigma}{(V_0 + \lambda_1 x + \lambda_2 z)^2},$$

$$= (\lambda_1/\lambda_2)\, V_2 \left(\frac{1}{V_2} - \frac{1}{V_1} \right),$$

$$= (V_2 - V_1) \tan \gamma / V_1. \qquad (A-11)$$

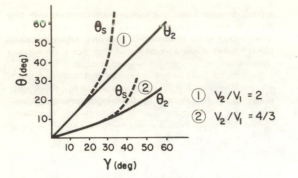

FIG. A–2. A graph of θ_2 and θ_s as functions of dip for two values of the ratio V_2/V_1.

Note that beneath the transition layer, the slope of the image raypath is constant (because V_2 is constant) and independent of the thickness of the transition layer. Letting δ go to zero gives the two-layer model shown in Figure (A–1). The slope $dx'/dz' = \tan\theta_2$, where θ_2 is the angle that the image raypath makes with the vertical. Some manipulation of (A–11) then gives

$$\frac{V_2}{V_1} = \frac{\sin(\theta_2 + \gamma)}{\sin\gamma\cos\theta_2}. \tag{A–12}$$

We may compare this result with Snell's law for the same model, which can be written

$$\frac{V_2}{V_1} = \frac{\sin(\theta_s + \gamma)}{\sin\gamma}. \tag{A–13}$$

Equations (A–12) and (A–13) then give the following relation between the Snell angle, $\theta_s + \gamma$, and the refraction angle of equation (A–10), $\theta_2 + \gamma$:

$$\sin(\theta_s + \gamma) = \frac{\sin(\theta_2 + \gamma)}{\cos\theta_2}.$$

Figure A–2 compares θ_2 and θ_s as a function of the dip γ for two choices of V_2/V_1. We see immediately that θ_2 is always less than θ_s; for small dip, however, θ_2 and θ_s are very close. That they differ at all is, in large part, a legacy of the paraxial approximation $P'_{z'z'} = 0$. For practical usage, however, the difference is slight. Should greater accuracy be required, it can be achieved by upgrading the quality of the paraxial approximation made to the wave equation.

In Appendix B, we approach the problem from a slightly different viewpoint. Instead of setting $\bar{v}' = v'$, we make $\bar{v}'$ constant (hence, not equal to v'). While that choice offers certain computational advantages, it requires special care in the treatment of the $P'_{z'z'}$ term.

For completeness, we discuss briefly the remaining two terms in equation (A–6). These can be written in split form as:

$$P'_{z't'} = -v' b P'_{t't'},$$

and

$$P'_{z't'} = \frac{v' c}{2} P'_{t'}.$$

One t'-integration of each yields

$$P'_{z'} = -v' b P'_{t'}, \tag{A–14}$$

and

$$P'_{z'} = \frac{v' c}{2} P'. \tag{A–15}$$

Equation (A–14) is equivalent in form to equation (A–7). The trajectories are now in the (z', t') plane and hence determine residual shifting in t'-time. The amount of shifting is different from that due to the term B of equation (2) in the text; it is termed residual because it is closely connected with the relatively small differences between integrating along z in equation (A–2) and integrating along the image raypath. Note that in the absence of lateral velocity variation, $b = 0$. Residual shifting most likely plays a role in the differences found between θ_2 and θ_s above.

Equation (A–15) has an obvious exponential solution that corresponds to small changes of amplitude with depth, providing of course that v' is not too heterogeneous [c.f., equation (A–4)].

APPENDIX B
THE PARAXIAL APPROXIMATION: $P_{z'z'} = 0$

In deriving equation (4) of the text, a key step was specifying $P_{z'z'} = 0$, a choice often called the paraxial approximation. This approximation is of considerable importance since P_{zz} cannot be obtained from surface seismic data [i.e., the wave field $P(x, z = 0, t)$]. Indeed, migration is not well posed with that term explicitly included in the wave equation. To make the migration problem tractable, it is necessary to reduce the wave equation from second order in z to first order. In so doing, one obtains a one-way wave equation that propagates waves in only the positive or negative z-direction but not both.

The neglect of $P_{z'z'}$ is justified on the grounds that upward traveling waves, as seen in the moving reference frame (x', z', t'), change very slowly. This approximate invariance is, in fact, true only for waves traveling near the vertical (the paraxial approximation is also called the 15-degree approximation). This restriction to small angles from vertical limits the quality of time migration attainable with equation (4).

Is there some better approximation than that $P_{z'z'} = 0$? For time migration, so-called 45- and 55-degree approximations have been proposed (see, e.g., Claerbout, 1976, p. 202) to extend the range of acceptable accuracy. The important point here, however, is that for *some* range of angles, accurate time migration is achievable even though the $P_{z'z'}$ term is *completely* neglected. For depth migration, on the other hand, we shall show that proper approximation of $P_{z'z'}$ is absolutely necessary if we are to obtain correct results, even for waves traveling vertically upward.

Let us return to the full, scalar wave equation in the moving coordinate frame [equation (A–3) in Appendix A]. Instead of forcing the frame velocity to equal the medium velocity as we did in Appendix A, let us now fix the frame velocity at some constant value. Then, all the terms in equation (A–3) involving derivatives of $\bar{v}$ vanish, leaving

$$P_{z't'} + \frac{\bar{v}}{4} P_{z'z'} = -\frac{\bar{v}}{4} P_{x'x'} + \hat{v}\left(\frac{1}{v^2} - \frac{1}{\bar{v}^2}\right) P_{t't'}. \tag{B–1}$$

We have dropped the primes on P, v, and $\bar{v}$ for clarity but retain them in the independent coordinates to avoid confusion.

If the $P_{z'z'}$, term could be ignored, the resulting equation could be treated conveniently by splitting it into two more simple equations (Marchuk splitting; Mitchell, 1969),

$$P_{z't'} = -\frac{\bar{v}}{4} P_{x'x'}, \tag{B–2}$$

and

$$P_{z't'} = \bar{v}\left(\frac{1}{v^2} - \frac{1}{\bar{v}^2}\right)P_{t't}. \qquad (B-3)$$

In time migration, $\bar{v} = v$; consequently, equation (B–3) vanishes leaving equation (B–2) to perform the required diffraction of waves. For depth migration, in which $\bar{v}$ does not everywhere equal v, the two equations would be solved alternately, once for each step of integration in z'. Equation (B–2) performs diffraction and equation (B–3), a shift in time, thereby allowing refraction to occur. But here difficulties arise. Equation (B–2) (a 15-degree equation) diffracts waves as though the medium velocity were the frame velocity $\bar{v}$, instead of the actual velocity v. Moreover, as we shall show, the time shifts specified by equation (B–3) are incorrect for waves propagating at any angle. The source of these difficulties is the dropping of the $P_{z'z'}$ term.

To see this, we first perform a t'-integration of equation (B–3), yielding an equation equivalent in form to equation (A–8):

$$\frac{DP}{Dz'} = P_{z'} - \bar{v}\left(\frac{1}{v^2} - \frac{1}{\bar{v}^2}\right)P_{t'} = 0. \qquad (B-4)$$

Equation (B–4) indicates that P is kept constant along the trajectory given by

$$\frac{dt'}{dz'} = -\bar{v}\left(\frac{1}{v^2} - \frac{1}{\bar{v}^2}\right). \qquad (B-5)$$

Stated differently, for each integration step $\Delta z'$, equation (B–3) shifts the time t' by an amount

$$\Delta t' = -\bar{v}\left(\frac{1}{v^2} - \frac{1}{\bar{v}^2}\right)\Delta z'. \qquad (B-6)$$

To see that the time shift indicated by equation (B–5) is erroneous, consider migration of a horizontal reflection occurring on a surface-recorded seismic trace at time

$$t = t_0 = 2\sum_{i=1}^{N}\frac{\Delta z}{v_i}.$$

Here, we have assumed the earth consists of N horizontal layers (down to the reflector of interest), each of thickness Δz and having depth-varying velocity v_i. After one integration step (i.e., one downward continuation step), the time of the reflection in the moving coordinate system governed by equation (A–2) should be

$$t' = \left(t_0 - \frac{2\Delta z}{v_1}\right) + \frac{2\Delta z'}{\bar{v}}$$

$$= t_0 + \Delta t_1', \qquad (B-7)$$

where, using the identity $\Delta z = \Delta z'$, we have

$$\Delta t_1' = 2\left(\frac{1}{\bar{v}} - \frac{1}{v_1}\right)\Delta z'.$$

In general, the time shift at the ith integration step will be given by

$$\Delta t_i' = 2\left(\frac{1}{\bar{v}} - \frac{1}{v_i}\right)\Delta z'. \qquad (B-8)$$

At any step, let us downward continue the wave field using equations (B–2) and (B–3) alternately. For horizontal reflections, the diffraction equation (B–2) has no effect on timing, i.e., the choice of $\bar{v}$ is immaterial, and equation (B–2) need not even be applied. In other words, the shifting equation (B–3) must bear the entire burden of migrating horizontal reflections. On comparing equations (B–6) and (B–8), however, we see that equation (B–3) will not provide the correct shift; hence, the migrated reflection will be incorrectly positioned in depth.

Equation (B–1), however, is *exact*. The discrepancy lies in the simple neglect of $P_{z'z'}$. How then should $P_{z'z'}$ be approximated? Of the several ways, the following is particularly illuminating. For convenience, Fourier-transform time in equation (B–1) to obtain the monochromatic wave equation:

$$i\,\omega P_{z'} + \frac{\bar{v}}{4}P_{z'z'} = -\frac{\bar{v}}{4}P_{x'x'} - \omega^2\bar{v}\left(\frac{1}{v^2} - \frac{1}{\bar{v}^2}\right)P. \qquad (B-9)$$

Now make the one-way wave assumption,

$$P_{z'} = \left(\sum_{k=0}^{\infty}\alpha_k\frac{\partial^k}{\partial x'^k}\right)P, \qquad (B-10)$$

where the α_k are constants. (More generally, the α_k can be considered spatially variable but this line will not be pursued here.) Equation (B–10) can be used to express $P_{z'z'}$ in terms of P. Making that substitution into equation (B–9) replaces the explicit presence of $P_{z'z'}$. This approach is analogous to that employed in the development of higher-order approximations (45-degree, etc.) in time migration.

Equating coefficients of $\partial^j/\partial x^j$ results in the one-way equation

$$P_{z'} = 2i\,\omega\left(\frac{1}{v} - \frac{1}{\bar{v}}\right)P + \left(\left[\frac{-4\,\omega^2}{v^2} - \frac{\partial^2}{\partial x'^2}\right]^{1/2} - \frac{2i\,\omega}{v}\right)P. \qquad (B-11)$$

We can now apply Marchuk splitting to equation (B–11) to obtain separate equations for diffraction and shifting. Doing so and transforming back into the time domain, we get

$$P_{z't'} + \frac{v}{4}P_{z'z'} = -\frac{v}{4}P_{x'x'} \qquad \text{(diffraction)}, \qquad (B-12)$$

and

$$P_{z'} = 2\left(\frac{1}{v} - \frac{1}{\bar{v}}\right)P_{t'} \qquad \text{(shifting)}. \qquad (B-13)$$

From a comparison of these equations with equations (B–2) and (B–3), we draw several conclusions. The diffraction equation, now expressed as a two-way wave equation, contains the $P_{z'z'}$ term explicitly. Note, however, it differs from equation (B–2) in that it contains the actual medium velocity v rather than the frame velocity $\bar{v}$. This result is more satisfying; equation (B–2) will not, in general, diffract properly since $\bar{v}$ can be chosen arbitrarily. The shifting equation (B–13) agrees with the simple argument for a horizontal reflector that led to equation (B–8). The influence of $P_{z'z'}$ is embedded implicitly in this equation.

Equations (B–12) and (B–13) could also have been derived by taking a correct one-way wave equation first in a fixed coordinate frame and then transforming to the moving frame.

To conclude, wave-theoretical depth migration must be implemented with a form of equations (B–12) and (B–13) in which $P_{z'z'}$ has been correctly approximated rather than with equations (B–2) and (B–3) in which the simplest assumption is made, viz., $P_{z'z'} = 0$. Although that simple assumption previously proved adequate for wave-theoretical *time* migration, more care must be exercised in developing *depth* migration.

Reprinted from GEOPHYSICS, **46**, 1638-1656.

Wave field extrapolation techniques in seismic migration, a tutorial

A. J. Berkhout*

ABSTRACT

The objective of this paper is to provide a general view on methods of wave field extrapolation as used in seismic modeling and seismic migration, i.e., the Kirchhoff-summation approach, the plane-wave method (k-f method), and the finite-difference technique.

Particular emphasis is given to the relationship between the different methods. By formulating the problem in the space-frequency domain (x, y, ω-domain), a systems approach can be adopted which results in simple and concise expressions. These expressions clearly show that forward extrapolation is described by a spatial *convolution* procedure and inverse extrapolation is described by a spatial *deconvolution* procedure. In the situation of lateral velocity variations, the (de)convolution procedure becomes space-variant. The space-frequency domain is most suitable for *recursive* depth migration. In addition, frequency dependent properties such as absorption, dispersion, and spatial bandwidth can be handled easily.

It is shown that all extrapolation methods are based on two equations: Taylor series and wave equation. In the Kirchhoff-summation approach all terms of the Taylor series are summed to an exact analytical expression—the *Kirchhoff-integral* for plane surfaces. It formulates the extrapolation procedure in terms of a spatial convolution integral which must be discretized in practical applications. The Fourier-transformed version of the Kirchhoff-integral is used in the plane wave method (k-f method). This actually

means that spatial (de)convolution in the x, y, ω-domain is translated into multiplication in the k_x, k_y, ω-domain. Of course, this is not allowed if the extrapolation operators are space-variant.

In explicit finite-difference techniques a truncated version of the Taylor series is used with some optimum adjustments of the coefficients. For only one or two terms in the Taylor series, a spatial low-pass filter must be applied to compensate for the amplitude errors at high tilt angles. Explicit methods are simple and most suitable for three-dimensional (3-D) applications.

In implicit finite-difference schemes the wave field extrapolator is written in terms of an explicit *forward* extrapolator and an explicit *inverse* extrapolator. Properly designed implicit schemes do not show amplitude errors and, therefore, amplitude correction filters need not be applied. In comparison with explicit schemes, implicit schemes are more sensitive to improper boundary conditions at both ends of the data file.

It is shown that the forward seismic model can be elegantly described by a matrix equation, using separate operators for *downward* and *upward* traveling waves. Using this model, inverse extrapolation involves one matrix inversion procedure to compensate for the downward propagation effects and one matrix inversion procedure to compensate for the upward propagation effects.

INTRODUCTION

The objective of most echo techniques is the collection of information on the internal structure of a medium without destructive penetration. Experience has taught that raw echo recordings need extensive processing before reliable information can be extracted from it. The most fundamental processing technique for echo recordings is generally referred to as "image reconstruction." Image reconstruction is applied in ultrasonic echo techniques (Johnson et al, 1979), tomography (Brooks and DiChiro, 1976),

radio astronomy (Fomalont, 1979) and particularly in seismology (Claerbout, 1976; Berkhout, 1980). Advanced acoustic image reconstruction techniques combine different echo measurements (multichannel processing) such that diffractions are collapsed, the distortion (in position and amplitude) of reflections is corrected for, and noise is attenuated.

Seismic migration is an *acoustic* image reconstruction technique based on the scalar wave equation. Application involves two steps (Figure 1):

Manuscript received by the Editor October 20, 1980; revised manuscript received April 28, 1981.
*Technische Hogeschool Delft, Laboratorium voor Technische Natuurkunde, Postbus 5046, 2600 GA Delft, The Netherlands.

(1) Wave field extrapolation

Using the scalar wave equation, the recorded data are transformed into a series of new recordings which represent simulated registrations at new positions of the recording plane.

(2) Imaging

The imaging principle formulates that the upper parts (i.e., the data around zero traveltime) of the simulated recordings, related to recording planes inside the medium of investigation, together form the migrated result.

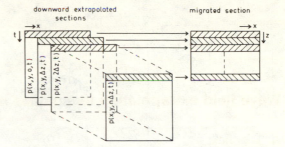

FIG. 1. A migrated seismic section may be synthesized by a number of data strips, each strip being defined by the upper part (i.e., the data around zero traveltime) of a seismic section which is simulated for a recording plane below the surface.

The wave field extrapolator is derived from the Kirchhoff integral for plane surfaces (summation approach) or directly from a simplified version of the wave equation (finite-difference approach). For the wavenumber-frequency approach, the wave field extrapolator can either be derived from the Kirchhoff integral or directly from the Fourier-transformed wave equation. Wave field extrapolation can be applied *recursively*, i.e., the extrapolated output is used as input for the next extrapolation step (Figure 2). Recursive extrapolation has the significant advantage that local velocities can be used and, therefore, spatial velocity variations can be handled properly. It is important to realize that in complicated geologic situations a reliable solution can only be obtained if, in addition to recursive extrapolation, the velocity input and the migration output are defined in the *depth* domain (depth migration). If, as in complicated geologic situations, the initial velocity-depth model is not accurately known, an *interpretive* approach is called for (model verification). We may expect that the interest in interpretive migration will increase considerably.

Modeling and migration are techniques which are closely related: modeling involves forward extrapolation and simulates the effects of wave field propagation; migration involves inverse extrapolation and removes the effects of wave field propagation. Of course, the success of inverse extrapolation will greatly depend upon the validity of the forward model being used. In the space-frequency domain it can be easily seen that for each frequency component, forward extrapolation is realized by convolution along the spatial axes.

Inverse extrapolation is realized by deconvolution along the spatial axes, the deconvolution operator being derived from the forward model. This is a very important observation because fundamental problems such as stability of the inverse extrapolator, optimum spatial bandwidth, and optimum spatial resolution can be approached fruitfully from the theory of inverse filtering. In the plane-wave method, the (de)convolution procedure along the spatial axes is replaced by a simple multiplication in the wavenumber domain.

Seismic modeling and migration methods are based on the wave equation for compressional waves (*P*-waves):

$$\nabla^2 P + k^2 (1 - j\eta)^2 P = \nabla \ln \rho \cdot \nabla P, \qquad (1)$$

where

$$\nabla = \mathbf{i}_x \frac{\partial}{\partial x} + \mathbf{i}_y \frac{\partial}{\partial y} + \mathbf{i}_z \frac{\partial}{\partial z},$$

$$\nabla^2 = \frac{\partial^2}{\partial x^2} + \frac{\partial^2}{\partial y^2} + \frac{\partial^2}{\partial z^2},$$

and

$P = P(x, y, z, \omega)$ represents pressure (in fluids) or a scaled version of the average of the principal stresses (in solids),

$k = k(x, y, z, \omega) = \omega / c(x, y, z, \omega),$

$c = c(x, y, z, \omega)$ represents the frequency-dependent phase velocity; if $\eta = 0$ (no absorption), then $c = c(x, y, z)$ represents the propagation velocity,

$\rho = \rho(x, y, z)$ represents the density distribution of the medium, and

$\eta = \eta(x, y, z, \omega)$ represents the loss angle which quantifies the absorption of the medium: $\eta \geq 0$. In practical seismic situations $\eta \ll 1$.

In principle, the propagation of *P*-waves in an inhomogeneous subsurface cannot be considered without simultaneously taking into account wave conversion. However, if we consider individual geologic layers to be weakly inhomogeneous, then inside each layer the influence of wave conversion ($P \leftrightarrow S$) on the propagation of compressional waves can be ignored. This does not hold at the boundary of two significantly different geologic layers. There, application of proper boundary conditions should take wave conversion into account by means of angle-dependent transmission and reflection coefficients. According to current practice, transmission losses due to reflection and wave conversion will be ignored. Moreover, multiply reflected and/or multiply diffracted energy will not be considered. However, angle-dependent reflection coefficients are included with the aid of the scattering matrix.

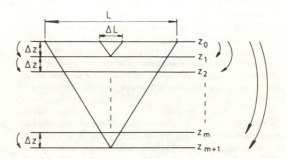

FIG. 2. In recursive extrapolation the output of the previous extrapolation step is used as input for the next step; in nonrecursive applications the starting data are used as input for all extrapolation steps.

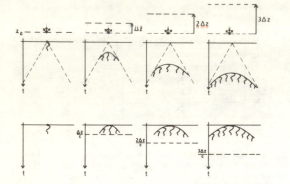

FIG. 3. Schematic illustration of the influence of forward extrapolation. The upper part shows $p(x, z, t)$ for four different recording planes. The lower part shows the time-advanced versions: $p(x, z, t + n\Delta z/c)$.

In the following we will derive expressions for the *forward* extrapolator first. From these expressions the *inverse* extrapolator will be defined.

TIME-ADVANCED PRESSURES

In the upper part of Figure 3 the influence of forward extrapolation on the upward traveling wave field of a buried point source (dipole at depth level z_0) is shown schematically. Note that after each extrapolation step, two changes can be seen: (1) The seismic energy shifts away from the apex of the response curve; and (2) the apex is shifted toward larger traveltimes ($\Delta z/c$ for each step). If the time delay $\Delta z/c$ is deleted, the results are shown in the lower part of Figure 3. We will call each result a *time-advanced* pressure section p' since it represents the original pressure at the reference planes $z = z_0 - n\Delta z$ advanced by the vertical traveltime $n\Delta z/c$:

$$p'(x, y, z_0 - n\Delta z, t) = p\left(x, y, z_0 - n\Delta z, t + \frac{n\Delta z}{c}\right).$$

In Figure 3 the result is shown for $n = 0, 1, 2, 3$.

For each extrapolation step forward extrapolation can be carried out in two phases. First, the advanced pressure section (advanced by the vertical traveltime $\Delta z/c)p'$ is computed. Second, the advanced pressure section is delayed by $\Delta z/c$.

From Figure 3 it can be easily seen that the change of the time-advanced pressure p' during extrapolation is significantly smaller than the change of the original pressure p during extrapolation, particularly for small emergence angles. Actually, for a horizontal plane wave in loss-free homogeneous media, the change of p' would be zero. Use of time-advanced pressures is identical to use of the floating time reference (Claerbout, 1976):

$$t' = t - (n\Delta z)/c.$$

We will see that, by splitting the forward extrapolator into two suboperators—(1) forward extrapolator which does not apply the vertical traveltime, and (2) time-delay operator which applies the vertical traveltime—an important computational advantage is obtained, particularly in finite-difference methods.

In the following we will derive the wave equation for time-advanced pressures. First let us define for the loss-free situation

$$p'(x, y, z, t) = p(x, y, z, t \pm \tau) \tag{2a}$$

or

$$P'(x, y, z, \omega) = P(x, y, z, \omega) e^{\pm j\omega\tau}, \tag{2b}$$

where

$$\tau = \int_{z_0}^{z} \frac{dz}{\bar{c}(x, y, z)},$$

or

$$j\omega\tau = j \int_{z_0}^{z} \bar{k}\, dz, \tag{2c}$$

and where $\bar{c}(x, y, z)$ represents the spatially averaged (local) propagation velocity, the spatial averaging being such that the derivatives of $\bar{c}$ with respect to x and y may be neglected. Of course, $\bar{c}$ may be chosen laterally constant: $\bar{c} = \bar{c}(z)$. If $z > z_0$, then we want to apply a forward extrapolation procedure to waves traveling in the positive z-direction, and the positive sign should be used. If $z < z_0$, then we want to apply a forward extrapolation procedure to waves traveling in the negative z-direction, and the negative sign should be used. Using equation (2b), we may write

$$\frac{\partial P}{\partial x} = \frac{\partial P'}{\partial x} e^{\mp j\omega\tau}, \frac{\partial P}{\partial y} = \frac{\partial P'}{\partial y} e^{\mp j\omega\tau},$$

$$\frac{\partial^2 P}{\partial x^2} = \frac{\partial^2 P'}{\partial x^2} e^{\mp j\omega\tau}, \frac{\partial^2 P}{\partial y^2} = \frac{\partial^2 P'}{\partial y^2} e^{\mp j\omega\tau},$$

$$\frac{\partial P}{\partial z} = \frac{\partial P'}{\partial z} e^{\mp j\omega\tau} \mp j\bar{k} P' e^{\mp j\omega\tau},$$

$$\frac{\partial^2 P}{\partial z^2} = \frac{\partial^2 P'}{\partial z^2} e^{\mp j\omega\tau} \mp 2j\bar{k} \frac{\partial P'}{\partial z} e^{\mp j\omega\tau}$$

$$+ \left[-\bar{k}^2 \pm j\bar{k} \frac{\partial \ln \bar{c}}{\partial z} \right] P' e^{\mp j\omega\tau}.$$

Substitution of these results in wave equation (1) yields, deleting absorption for the moment ($\eta = 0$),

(1) for forward extrapolation in the *positive* z-direction:

$$\nabla^2 P' - 2j\bar{k}\frac{\partial P'}{\partial z} = -(\gamma + 2jr)\bar{k}^2 P'$$

(2) for forward extrapolation in the *negative* z-direction:

$$\nabla^2 P' + 2j\bar{k}\frac{\partial P'}{\partial z} = -(\gamma - 2jr)\bar{k}^2 P'.$$

In the above equations we have neglected $\nabla P' \cdot \nabla \ln \rho$ and

$$\bar{k} = \omega/\bar{c}, r = \frac{1}{2\bar{k}} \frac{\partial \ln \rho\bar{c}}{\partial z},$$

$$\gamma = \left(\frac{\bar{c}}{c}\right)^2 - 1.$$

If $\Delta c = c - \bar{c}$ is small, then $\gamma \approx -2\Delta c/\bar{c}$.

For horizontally stratified media, $\gamma = 0$. For homogeneous media, $\gamma = 0$ and $r = 0$. If $\gamma = 0$, then τ represents true vertical traveltime. Note that for r we may write:

$$(\bar{k}\Delta z)r = \frac{1}{2} \frac{\partial \ln \rho\bar{c}}{\partial z} \Delta z, \text{ with } \Delta z = z_{i+1} - z_i$$

$$= \frac{1}{2} \Delta \ln \rho\bar{c}, \text{ for } \Delta z \text{ sufficiently small}$$

$$= \frac{\rho \bar{c}(z_{i+1}) - \rho \bar{c}(z_i)}{\rho \bar{c}(z_{i+1}) + \rho \bar{c}(z_i)}.$$

Hence r represents a scaled version of the normal-incidence reflection coefficient if $c = c(z)$.

The above derivation can also be given taking absorption into account. Then equation (2b) must be replaced by

$$P'(x, y, z, \omega) = P(x, y, z, \omega) e^{\pm s\tau},$$

where

$$s\tau = j\omega \int_{z_0}^{z} \frac{1 - j\eta}{\bar{c}(x, y, z, \omega)} dz$$

$$= j \int_{z_0}^{z} \bar{k}(1 - j\eta) dz$$

$$= j \int_{z_0}^{z} \bar{k}_1 dz.$$

If we compare this result with equation (2c), we may conclude that absorption can be included in the foregoing results if we replace $\bar{k}$ by $\bar{k}_1$.

As before, in the following we will neglect transmission losses due to reflection. This might not be correct for reflectors with large reflection coefficients, but one should realize that for strong reflectors a more realistic approach (taking wave conversion into account) should be used anyway. Hence, our discussions on forward extrapolation will be based on the wave equation for compressional waves *without transmission losses but with absorption*:

(1) for the original waves P:

$$\nabla^2 P + k^2 (1 - j\eta)^2 P = 0$$

or

$$\nabla^2 P + \bar{k}_1^2 P = -\gamma \bar{k}_1^2 P; \tag{3a}$$

(2) for the time-advanced waves P':

$$\nabla^2 P' - 2j\bar{k}_1 \frac{\partial P'}{\partial z} = -\gamma \bar{k}_1^2 P' \text{ (traveling in the } +z\text{-direction)} \tag{3b}$$

$$\nabla^2 P' + 2j\bar{k}_1 \frac{\partial P'}{\partial z} = -\gamma \bar{k}_1^2 P' \text{ (traveling in the } -z\text{-direction),} \tag{3c}$$

where

$$\bar{k}_1 = \frac{\omega}{\bar{c}} (1 - j\eta), \gamma = \left(\frac{\bar{c}}{c}\right)^2 - 1 \text{ and } c = c(x, y, z, \omega).$$

Note that

$$\bar{k}_1 = \left(1 + \frac{\Delta c}{\bar{c}}\right) k_1,$$

or

$$\bar{k}_1 = (1 + \gamma)^{-1/2} k_1. \tag{3d}$$

In equations (3b) and (3c) spatially averaged propagation velocity $\bar{c}$ may be laterally space-variant but its derivatives with respect to x and y should be small. The term $-\gamma \bar{k}_1^2 P$ can be considered as a source term, representing the distributed secondary sources in the subsurface due to spatial velocity variations. If $\bar{c}$ is taken as constant, then all lateral velocity variations are included in the source term $-\gamma \bar{k}_1^2 P$. If $\bar{c}$ is laterally space-variant, then only the fast lateral velocity variations are included in source

term $-\gamma \bar{k}_1^2 P$. This subdivision plays an important role in statistical scattering.

UNIFIED APPROACH TO FORWARD EXTRAPOLATION

In seismic techniques forward traveling P-waves occur in two ways: (1) Waves traveling *downward* from the surface to the inhomogeneities of interest; (2) waves traveling *upward* from the inhomogeneities back to the surface.

In the forward problem both types of waves have to be extrapolated. We will review the basic principles now.

Forward wave field extrapolation can be quantified by two basic equations:

(1) **The Taylor series**

$$P(z_0 \pm \Delta z) = P(z_0) \pm \frac{\Delta z}{1!} \frac{\partial P}{\partial z_0} + \frac{\Delta z^2}{2!} \frac{\partial^2 P}{\partial z_0^2} \pm \ldots \tag{4a}$$

Hence, given the data itself *and* the derivatives toward z in the data-plane $z = z_0$, then the data in the plane $z = z_0 \pm \Delta z$ can be computed with the aid of equation (4a).

In equation (4a) the positive sign must be used for downward traveling waves and the negative sign must be used for upward traveling waves (remember that downward means increasing z).

(2) **The wave equation**

$$\nabla^2 P + k^2 (1 - j\eta)^2 P = 0, \tag{4b}$$

where $k = \omega / c(x, y, z, \omega)$.

If time-advanced pressures are used, then equation (4b) should be replaced by equation (3b) for downward traveling waves and by equation (3c) for upward traveling waves. Standard seismic acquisition techniques do not measure derivatives and, therefore, the derivatives with respect to z as needed in the Taylor series must be computed from the data themselves. We will see that the wave equation facilitates this computation:

$$\frac{\partial P}{\partial z_0} = \pm jkH(x, y, \omega) * P(x, y, z_0, \omega); \tag{5a}$$

$$\frac{\partial^2 P}{\partial z_0^2} = (\pm jk)^2 H(x, y, \omega) * H(x, y, \omega) * P(x, y, z_0, \omega); \text{ etc.,} \tag{5b}$$

H being a function to be derived later and $*$ denoting convolution.

In expressions (5a) and (5b) the negative sign relates to downward traveling waves and the positive sign relates to upward traveling waves. This is in agreement with the difference between equations (3b) and (3c). Accurate and efficient computation of the derivatives with respect to z is one of the most important problems in finite-difference techniques.

In the situation of lateral velocity variations, the operator H is space-variant and, therefore, for fast lateral variations a matrix formulation of expressions (5a) and (5b) is more appropriate. Note that in expression (5b) c is assumed locally independent of z. This is acceptable unless accurate modeling of transmission effects is desired. However, as before, this refinement is only meaningful if wave conversion at inhomogeneities (particularly at major reflectors) is included as well.

If we combine expressions (4a) and (5), the basic expression for forward extrapolation is obtained:

$$P(z_0 \pm \Delta z) = P(z_0) + \frac{(-jk\Delta z)}{1!} H * P(z_0)$$

$$+ \frac{(-jk\Delta z)^2}{2!} H * H * P(z_0) + \ldots,$$

or

$$P(z_0 \pm \Delta z) = \left[\delta(x)\delta(y) + \frac{(-jk\Delta z)}{1!} H \right.$$
$$\left. + \frac{(-jk\Delta z)^2}{2!} H * H + \dots \right] * P(z_0),$$

or

$$P(z_0 \pm \Delta z) = [\delta(x)\delta(y) + G(x, y, \Delta z, \omega)] * P(z_0), \quad (6a)$$

or

$$P(x, y, z_1, \omega) = W(x, y, \Delta z, \omega) * P(x, y, z_0, \omega), \quad (6b)$$

where $W(x, y, \Delta z, \omega)$ represents the forward extrapolator for the extrapolation distance $\Delta z = |z_1 - z_0|$. Note that $G(x, y, \Delta z, \omega)$ plays the role of a spatial prediction-error filter. Again, in the situation of lateral velocity variations the operator W is space-variant and, therefore, for fast lateral variations a matrix formulation of expressions (6a) and (6b) is more appropriate. If we assume in the plane $z = z_0$ a pressure point source, i.e., a dipole

$$P(x, y, z_0, \omega) = \delta(x)\delta(y), \quad (7a)$$

then the pressure in the plane $z = z_1$ is given by

$$P(x, y, z_1, \omega) = W(x, y, \Delta z, \omega). \quad (7b)$$

Therefore $W(x, y, \Delta z, \omega)$ may be considered as a *spatial impulse response*. We will see that in the summation approach, W is given by the Kirchhoff integral for plane surfaces, i.e., the Rayleigh integral. Hence, in the summation approach summation of all terms of the Taylor series has already been carried out, resulting in an analytical expression for W. In finite-difference techniques, a truncated version of the Taylor series expansion of W is used with some optimum adjustment of the coefficients.

To illustrate the foregoing, let us consider expressions (5), (6), and (7) for loss-free, homogeneous media ($\bar{c} = c = $ constant). For this simplified situation the wave equation may be written as

$$\frac{\partial^2 P}{\partial x^2} + \frac{\partial^2 P}{\partial y^2} + \frac{\partial^2 P}{\partial z^2} + k^2 P = 0, \quad (8a)$$

where k is independent of the spatial coordinates.

Fourier transformation of equation (8a) toward x and y yields

$$\frac{\partial^2 \bar{P}}{\partial z^2} + (k^2 - k_x^2 - k_y^2)\bar{P} = 0, \quad (8b)$$

where $\bar{P} = \bar{P}(k_x, k_y, z, \omega)$.

Equation (8b) can also be written as

$$\frac{\partial P}{\partial z} = \pm j\sqrt{k^2 - k_x^2 - k_y^2}\,\bar{P}. \quad (8c)$$

Bearing in mind that the expression of a plane wave traveling in the positive z-direction is given by

$$A\,e^{-j(\omega/c_z)z}$$

and the expression of a plane wave traveling in the negative z-direction is given by

$$A\,e^{+j(\omega/c_z)z},$$

c_z being the phase velocity along the z-axis ($c_z = c/\cos\alpha$), then it can be easily seen that in equation (8c) we must choose the negative sign when we are dealing with downward traveling waves and we must choose the positive sign when we are dealing with upward traveling waves. Hence, if we compare equation (8c) with

equation (5a) we may conclude

$$\bar{H}(k_x, k_y, \omega) = \sqrt{1 - (k_x^2 + k_y^2)/k^2}, \quad (9a)$$

or, in terms of the tilt angle α,

$$\bar{H}(k_x, k_y, \omega) = \cos\alpha \quad (9b)$$

or, according to Appendix B,

$$H(x, y, \omega) = \frac{k}{2\pi} j_1(kr_0)/r_0, \quad (9c)$$

where $r_0 = \sqrt{x^2 + y^2}$ and j_1 represents the first-order spherical Bessel function. If we consider the Taylor series in the (k_x, k_y, z, ω) domain and use equation (8c), then we may write

$$\bar{P}(z_1) = \left[1 + (-j\sqrt{k^2 - (k_x^2 + k_y^2)})\frac{\Delta z}{1!} \right.$$
$$\left. + (-j\sqrt{k^2 - (k_x^2 + k_y^2)})^2 \frac{\Delta z^2}{2!} + \dots \right] \bar{P}(z_0), \quad (10a)$$

or, bearing in mind the series expansion of the exponential function,

$$\bar{P}(z_1) = e^{-j\sqrt{k^2 - (k_x^2 + k_y^2)}\,\Delta z}\,\bar{P}(z_0). \quad (10b)$$

For the special situation that $k_x^2 + k_y^2 > k^2$, it follows from equation (10b)

$$\bar{P}(z_1) = e^{-\sqrt{(k_x^2 + k_y^2) - k^2}\,\Delta z}\,\bar{P}(z_0). \quad (10c)$$

The negative exponential has to be chosen in equation (10c) since no physical wave will increase without limit in its direction of propagation.

If we compare equations (10b) and (10c) with equation (6b), we obtain

$$\bar{W}(k_x, k_y, \Delta z, \omega) = e^{-j\sqrt{k^2 - (k_x^2 + k_y^2)}\,\Delta z} \text{ for } k_x^2 + k_y^2 \le k^2; \quad (11a)$$

$$\bar{W}(k_x, k_y, \Delta z, \omega) = e^{-\sqrt{(k_x^2 + k_y^2) - k^2}\,\Delta z} \text{ for } k_x^2 + k_y^2 > k^2. \quad (11b)$$

The part at $k_x^2 + k_y^2 > k^2$ is called the *evanescent* wave field of spatial impulse response W. Note that the evanescent wave field attenuates exponentially with distance.

Application of the inverse Fourier transformation to equation (11), as derived by Berkhout (1980, Appendix D), yields

$$W(x, y, \Delta z, \omega) = \frac{\Delta z}{2\pi} \frac{1 + jkr}{r^3} e^{-jkr}, \quad (11c)$$

where $r = \sqrt{x^2 + y^2 + \Delta z^2}$. The time domain version of equation (11c) was used by French (1975) in a discussion on computer migration of oblique profiles. Expression (11c) will be further discussed when the "summation approach" is considered.

In the foregoing we have heuristically shown the advantage of using time-advanced pressure data. We now examine it more rigorously using expression (11a). We have seen that using time-advanced pressures actually means splitting the extrapolation operator into two suboperators:

$$W(x, y, \Delta z, \omega) = W'(x, y, \Delta z, \omega)e^{-j\omega(\Delta z/\bar{c})}, \quad (12)$$

the operator W' producing a time-advanced section and $\exp(-j\omega\Delta z/\bar{c})$ being a time-delay operator.

According to equation (12) we may write for the loss-free, homogeneous situation

$$\bar{W}'(k_x, k_y, \Delta z, \omega) = \bar{W}(k_x, k_y, \Delta z, \omega)e^{+jk\Delta z}$$

or, using equation (11a)

$$\bar{W}'(k_x, k_y, \Delta z, \omega) = e^{-j(\sqrt{k^2 - k_x^2 - k_y^2} - k)\Delta z} \tag{13a}$$

or, using spherical coordinates in the Fourier space,

$$\bar{W}'(k_x, k_y, \Delta z, \omega) = e^{-jk(\cos \alpha - 1)\Delta z}, \tag{13b}$$

where $\cos \alpha = \sqrt{k^2 - k_x^2 - k_y^2}/k = k_z/k$ [see also equation (9b)].

From equation (13b) it follows

$$\tilde{P}'(z_1) = e^{-jk(\cos \alpha - 1)\Delta z} \tilde{P}'(z_0),$$

or, using the series expansion of the exponential function,

$$\tilde{P}'(z_1) = \left[1 + (\cos \alpha - 1)\frac{(-jk\Delta z)}{1!} \right.$$
$$\left. + (\cos \alpha - 1)^2 \frac{(-jk\Delta z)^2}{2!} + \ldots \right] \tilde{P}'(z_0). \tag{14a}$$

If we compare this expression with equation (10a),

$$\tilde{P}(z_1) = \left[1 + \cos \alpha \frac{(-jk\Delta z)}{1!} \right.$$
$$\left. + \cos^2 \alpha \frac{(-jk\Delta z)^2}{2!} + \ldots \right] \tilde{P}(z_0), \tag{14b}$$

then we may draw the following conclusions:

(1) For small tilt angles $|\alpha|$ the change of time-advanced pressures as a function of z is significantly smaller than the change of the original pressures as a function of z: If we write for small Δz

$$\Delta \tilde{P}' = \frac{\partial \tilde{P}'}{\partial z}\Delta z = (-jk\Delta z)(\cos \alpha - 1)\tilde{P}' \tag{15a}$$

and

$$\Delta \tilde{P} = \frac{\partial \tilde{P}}{\partial z}\Delta z = (-jk\Delta z)\cos \alpha \,\tilde{P}, \tag{15b}$$

then it can be easily seen that $|\Delta \tilde{P}'| \ll |\Delta \tilde{P}|$ for small tilt angles $|\alpha|$. From equations (15a) and (15b) it also follows that use of suboperator W' is advantageous only if $|\cos \alpha - 1| < |\cos \alpha|$ or $|\alpha| < 60°$.

(2) For small tilt angles $|\alpha|$ the Taylor series of the total operator W converges fast, only if the extrapolation distance Δz is very small ($k\Delta z \ll 1$).

(3) For small tilt angles $|\alpha|$ the Taylor series of the suboperator W' converges fast, even for considerable extrapolation distances ($k\Delta z \approx 1$).

In the foregoing we have seen that in situations with laterally changing velocities forward extrapolation can be formulated in terms of space-variant convolution. From systems theory we know that the relation between input and output of nonstationary systems can be beautifully described by means of matrices. For the two-dimensional (2-D) situation the matrix formulation of equation (6b) becomes:

(1) In terms of column vectors

$$\mathbf{P}(z_1) = \mathbf{V}(z_1, z_0)\mathbf{P}(z_0). \tag{16a}$$

(2) In terms of row vectors

$$\mathbf{P}^T(z_0)\mathbf{W}(z_0, z_1) = \mathbf{P}^T(z_1), \tag{16b}$$

where the symbol T means interchanging rows and columns and $\mathbf{W}(z_0, z_1) = \mathbf{V}^T(z_1, z_0)$.

In expression (16b) we have defined

$$\mathbf{P}^T(z_0) = [P(0, z_0, \omega), P(\Delta x, z_0, \omega), \ldots P(n\Delta x, z_0, \omega),$$
$$\ldots P(N\Delta x, z_0, \omega)]$$

= data vector (row vector) containing the pressure values in the data plane $z = z_0$. If z_0 defines the surface ($z_0 = 0$), then $\mathbf{P}^T(z_0)$ defines a source array (of course for one frequency component only).

$\mathbf{W}(z_0, z_1)$ = forward propagation matrix between the depth levels z_0 and z_1. If all elements of $\mathbf{P}^T(z_0)$ are zero except at $x = n\Delta x$, where $P(n\Delta x, z_0, \omega) = 1$, then $\mathbf{P}^T(z_1)$ represents the response of a spatial impulse (= dipole) at $(n\Delta x, z_0)$, i.e., $\mathbf{P}^T(z_1) = = \mathbf{W}_n^T(z_0, z_1)$. Hence, symbolically,

$$\mathbf{W}(z_0, z_1) = \begin{pmatrix} \mathbf{W}_0^T(z_0, z_1) \\ \mathbf{W}_1^T(z_0, z_1) \\ \mathbf{W}_n^T(z_0, z_1) \\ \mathbf{W}_N^T(z_0, z_1) \end{pmatrix}.$$

Note that propagation vector $\mathbf{W}_n^T(z_0, z_1)$ is determined by the velocity distribution in layer (z_0, z_1) around $x = n\Delta x$; its main amplitudes occur around $n\Delta x$. Note also that without lateral velocity variations, $\mathbf{W}$ represents a Toeplitz matrix: $\mathbf{W}_n^T$ is a shifted version of $\mathbf{W}_m^T$, the shift being $(m - n)\Delta x$. This property is indicated by the slanted presentation of $\mathbf{W}$.

In the following we have chosen for the row vector presentation; thus the data of one seismic experiment, measured on a horizontal plane at some given depth level, are defined by row vectors.

In Figure 4 the total data flow is shown for a response from depth level $z = z_m$ due to one source (or one source array) at the surface $z = z_0$. If we use the symbol S for the *downgoing* pressure field and P for the *upgoing* pressure field, then we may write in terms of vectors and matrices:

$$\mathbf{P}_n^T(z_0) = \mathbf{S}_n^T(z_0)[\mathbf{W}(z_0, z_m)\mathbf{R}(z_m)\mathbf{W}(z_m, z_0)]\mathbf{D}(z_0), \tag{17}$$

where

$\mathbf{S}_n^T(z_0)$ = row vector defining the source array on the surface around $x = n\Delta x$;

$\mathbf{W}(z_0, z_m)$ = forward propagation matrix between the depth levels z_0 and z_m;

$\mathbf{W}(z_m, z_0)$ = forward propagation matrix between the depth levels z_m and z_0; assuming that the principle of reciprocity holds for seismic media, then $\mathbf{W}(z_m, z_0) = \mathbf{W}^T(z_0, z_m)$;

$\mathbf{R}(z_m)$ = scattering matrix at depth level z_m; if there exists a reflector at z_m then the nth row defines the angle-dependent reflection coefficient at $x = n\Delta x$; if the reflectivity is not angle-dependent (locally reacting assumption), then $\mathbf{R}$ is a diagonal matrix and, finally, if the reflectivity is not depending on x then $\mathbf{R}$ is a Toeplitz matrix;

$\mathbf{D}(z_0)$ = detector matrix, the nth column representing the detector array on the surface around $x = n\Delta x$: $\mathbf{D}_n(z_0)$. Hence,

$$\mathbf{D}(z_0) = \begin{pmatrix} \mathbf{D}_0(z_0) \\ \quad \mathbf{D}_1(z_0) \\ \qquad \diagdown \\ \qquad \mathbf{D}_n(z_0) \\ \qquad\quad \diagdown \\ \qquad\qquad \mathbf{D}_N(z_0) \end{pmatrix}.$$

Note that in practical situations seismic data are *always* collected with detector arrays;

$\mathbf{P}_n^T(z_0)$ = row vector defining the pressure values as recorded by the detector arrays on the surface and generated by one source array on the surface around $n\,\Delta x$ [i.e., one seismic recording due to $\mathbf{S}_n^T(z_0)$].

If we consider many seismic experiments ($n = 1, 2, \ldots, N$) and combine all the row vectors $\mathbf{S}_n^T(z_0)$ in a source matrix $\mathbf{S}(z_0)$ and all the row vectors $\mathbf{P}_n^T(z_0)$ in a data matrix $\mathbf{P}(z_0)$, then we obtain the matrix formulation of a multirecord data set:

$$\mathbf{P}(z_0) = \mathbf{S}(z_0)[\mathbf{W}(z_0, z_m)\mathbf{R}(z_m)\mathbf{W}(z_m, z_0)]\mathbf{D}(z_0), \quad (18a)$$

or, taking the responses from all depth levels of interest,

$$\mathbf{P}(z_0) = \mathbf{S}(z_0)\left[\sum_{m=1}^{M} \mathbf{W}(z_0, z_m)\mathbf{R}(z_m)\mathbf{W}(z_m, z_0)\right]\mathbf{D}(z_0). \quad (18b)$$

Note that for recursive applications

$$\mathbf{W}(z_0, z_N) = \mathbf{W}(z_0, z_1)\mathbf{W}(z_1, z_2) \ldots \mathbf{W}(z_{M-1}, z_M) \quad (18c)$$

From equations (18b) and (18c) it follows that multirecord data sets can also be formulated as

$$\mathbf{P}(z_m) = \mathbf{W}(z_m, z_{m+1})\mathbf{P}(z_{m+1})\mathbf{W}(z_{m+1}, z_m) + \mathbf{R}(z_m) \quad (18d)$$

for $m = 1, 2, \ldots$ and

$$\mathbf{P}(z_0) = \mathbf{S}(z_0)[\mathbf{W}(z_0, z_1)\mathbf{P}(z_1)\mathbf{W}(z_1, z_0)]\mathbf{D}(z_0). \quad (18e)$$

Expressions (18) define the basic equations for prestack seismic modeling. In the summation approach, $\mathbf{W}$ is defined by the Rayleigh integral. In finite-difference techniques, W is approximated by a series expansion. Note that in the plane-wave method the matrix manipulations are replaced by multiplications in the wavenumber-frequency domain with the time delay operator $\mathbf{W} = \exp(-jk_z\,\Delta z)$.

One of the most difficult tasks in the forward problem is the specification of the scattering matrices $\mathbf{R}(z_m)$. In the practice of seismic modeling, $\mathbf{R}(z_m)$ is generally specified as a *diagonal* matrix (locally reacting assumption). This actually means that with most current modeling software, the propagation effects are evaluated only.

Finally, it follows from equation (18b) that a zero-offset section is given by

$$\mathbf{P}_n(n\,\Delta x, z_0, \omega) =$$
$$\mathbf{S}_n^T(z_0)\left[\sum_{m=1}^{M} \mathbf{W}(z_0, z_m)\mathbf{R}(z_m)\mathbf{W}(z_m, z_0)\right]\mathbf{D}_n(z_0), \quad (19)$$

$n = 1, 2, \ldots, N$. From equations (18) and (19) interesting conclusions can be drawn, e.g., with respect to plane-wave methods, zero-offset methods, the extrapolation of common-midpoint gathers, and the validity of the exploding reflector model as introduced by Loewenthal et al (1976).

UNIFIED APPROACH TO INVERSE EXTRAPOLATION

In the foregoing we have derived that forward extrapolation can be described by spatial convolution for each temporal frequency component (Figure 5a):

$$P(z_1) = W(z_1, z_0) * P(z_0) + N(z_1), \quad (20a)$$

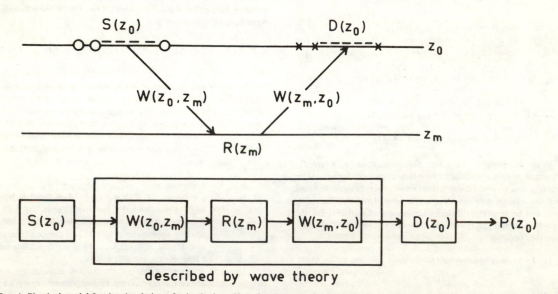

FIG. 4. Physical model for the simulation of seismic data. Here the responses from seismic discontinuities at depth level z_m are considered.

where $W(z_1, z_0)$ is symbolic notation for $W(x, y, \Delta z, \omega)$ with $\Delta z = |z_1 - z_0|$. Equation (20a) quantifies the effect of wave propagation from depth level z_0 to depth level z_1. The term $N(z_1)$ specifies the noise present on depth level z_1. Note that $N(z_1)$ may be spatially coherent.

If we want to compensate the propagation effect in layer (z_0, z_1), we need an operator such that (Figure 5b)

$$\langle P(z_0) \rangle = F(z_0, z_1) * P(z_1), \tag{20b}$$

$\langle P(z_0) \rangle$ representing an estimate of the exact pressure at z_0. We will call F the *inverse* extrapolator. If we substitute equation (20a) in equation (20b), we obtain

$$\langle P(z_0) \rangle = F(z_0, z_1) * W(z_1, z_0) * P(z_0) + N'(z_0), \tag{21}$$

where $N'(z_0)$ represents the filtered noise.

From equation (21) it can be seen that $\langle P(z_0) \rangle = P(z_0)$ if $N'(z_0) = 0$ (noise free situation) and if

$$F(x, y, \Delta z, \omega) * W(x, y, \Delta z, \omega) = \delta(x)\delta(y), \tag{22a}$$

or, after Fourier transformation,

$$\tilde{F}(k_x, k_y, \Delta z, \omega) = 1/\tilde{W}(k_x, k_y, \Delta z, \omega). \tag{22b}$$

Operator (22b) is unstable and cannot be used as it stands. This important fact can be easily understood if we consider the loss-free homogeneous situation. Substitution of equation (11) into equation (22b) yields

$$\tilde{F}(k_x, k_y, \Delta z, \omega) = e^{+j\sqrt{k^2 - (k_x^2 + k_y^2)}\,\Delta z} \text{ for } k_x^2 + k_y^2 \leq k^2, \tag{23a}$$

$$\tilde{F}(k_x, k_y, \Delta z, \omega) = e^{+\sqrt{(k_x^2 + k_y^2) - k^2}\,\Delta z} \text{ for } k_x^2 + k_y^2 > k^2. \tag{23b}$$

Expression (23b) defines an exponentially increasing operator which is unacceptable in practical situations. This also means that result (22a) can never be realized. We will see that in situations with absorption the instability is even more pronounced. In practice several alternatives are of interest:

(1) Band-limited inversion

$$\tilde{F} = \tilde{W}_0 / \tilde{W}, \tag{24a}$$

where $\tilde{W}_0$ represents a spatial low-pass filter. For one temporal frequency component and a given velocity distribution the passband of $\tilde{W}_0$ can be specified in terms of the tilt angle α. This is a valuable option, particularly in finite-difference techniques, where W has been properly described up to a maximum tilt angle (α_m) only.

(2) Least-squares inversion

$$\tilde{F} = \frac{\tilde{W}^*}{|\tilde{W}|^2 + |\tilde{N}|^2}. \tag{24b}$$

Note the interesting similarity with two-sided, least-squares temporal deconvolution, where the seismic time wavelet and the spatially incoherent noise spectrum should be specified [see, e.g., Berkhout (1977)]. Here we are dealing with the spatial wavelet and the spatially coherent noise spectrum; temporal frequency plays the role of a parameter. Hence, if we carry out inverse extrapolation with the aid of equation (24b), then shot-generated noise can be optimally taken into account. Note that spatial wavelet estimation means velocity analysis.

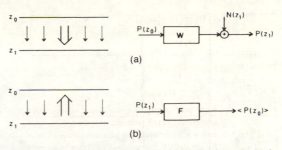

FIG. 5. (a) The physical model of wave propagation between depth levels z_0 and z_1 together with additive noise. The convolution operator W quantifies the forward propagation effects in layer (z_0, z_1). (b) Inverse extrapolation can be realized by spatial deconvolution. Application of deconvolution operator F compensates for the forward propagation effects in layer (z_0, z_1).

(3) Matched filtering

$$\tilde{F} = \tilde{W}^*. \tag{24c}$$

It is interesting to realize that if we do not consider absorption and evanescent waves, then equation (24c) defines a pure spatially zero-phasing procedure.

Note that alternative (3) formulates the simplest inversion. According to expressions (12), (24a), (24b) and (24c), if we split the forward operator into two suboperators,

$$W = W' e^{-j\omega(\Delta z/\bar{c})}, \tag{25a}$$

then the inverse operator can also be written in terms of two suboperators:

$$F = F' e^{+j\omega(\Delta z/\bar{c})}. \tag{25b}$$

Hence, the use of time-advanced pressures in the forward model means the use of time-delayed pressures in the inverse extrapolation process:

$$p'(x, y, z_0 \pm n\Delta z, t) = p\left(x, y, z_0 \pm n\Delta z, t - \frac{n\Delta z}{\bar{c}}\right),$$

where $\Delta z > 0$ and $n = 1, 2, \ldots$. From Figure 6 it can be seen that the change of the time-delayed pressure p' during inverse extrapolation is significantly smaller than the change of the actual pressure p, particularly for small emergence angles. Actually, for a horizontal plane wave in a loss-free homogeneous medium the change of p' would be zero. Use of time-delayed pressures is identical to the use of the floating time reference (Claerbout, 1976):

$$t' = t + (n\Delta z)/\bar{c}.$$

In the existing literature the inverse extrapolation problem has generally been approached from the matched-filtering point of view, although this aspect may not always be realized. Note that the well-known inverse summation operator (Schneider, 1978),

$$F = \frac{\Delta z}{2\pi} \frac{1 - jkr}{r^3} e^{+jkr}, \tag{26}$$

is obtained by applying the matched-filter concept to the summation operator for loss-free homogeneous media [equation (11c)].

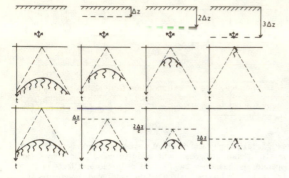

FIG. 6. Schematic illustration of the influence of inverse extrapolation. The upper part shows $p(x, z, t)$ for four different recording planes. The lower part shows the time-delayed versions: $p(x, z, t - n\Delta z/c)$.

Now formulate the inverse problem for both downward and upward propagation in terms of matrices by making use of the forward model as given by expressions (18). First we will rewrite equation (18a):

$$\mathbf{P}(z_0) = \mathbf{W}_s(z_0, z_m)\mathbf{R}(z_m)\mathbf{W}_d(z_m, z_0), \quad (27a)$$

where

$$\mathbf{S}(z_0)\mathbf{W}(z_0, z_m) = \mathbf{W}_s(z_0, z_m),$$
$$\mathbf{W}(z_m, z_0)\mathbf{D}(z_0) = \mathbf{W}_d(z_m, z_0). \quad (27b)$$

Generally, $\mathbf{S}(z_0) \neq \mathbf{D}(z_0)$ and therefore the theorem of reciprocity may not generally be applied to seismic data:

$$\mathbf{W}_s(z_0, z_m) \neq \mathbf{W}_d^T(z_m, z_0).$$

It is important to realize that application of lengthy arrays involves a serious spatial band limitation, particularly at high temporal frequencies.

The objective of inverse extrapolation is compensation for downward propagation matrix $\mathbf{W}_s(z_0, z_m)$ and upward propagation matrix $\mathbf{W}_d(z_m, z_0)$ such that an estimate of $\mathbf{R}$ is obtained. Hence, inverse extrapolation involves matrix inversion [see also equation (18a)]:

$$\langle \mathbf{R}(z_m) \rangle = \mathbf{F}_s(z_m, z_0)\mathbf{P}(z_0)\mathbf{F}_d(z_0, z_m)$$
$$= [\mathbf{F}_s(z_m, z_0)\mathbf{W}_s(z_0, z_m)] \cdot$$
$$\cdot \mathbf{R}(z_m)[\mathbf{W}_d(z_m, z_0)\mathbf{F}_d(z_0, z_m)]$$
$$= \mathbf{W}_s(z_m)\mathbf{R}(z_m)\mathbf{W}_d(z_m), \quad (28a)$$

where

$$\mathbf{W}_s(z_m) = \mathbf{F}_s(z_m, z_0)\mathbf{W}_s(z_0, z_m),$$

and   (28b)

$$\mathbf{W}_d(z_m) = \mathbf{W}_d(z_m, z_0)\mathbf{F}_d(z_0, z_m).$$

If $\mathbf{W}_s(z_m)$ and $\mathbf{W}_d(z_m)$ represent unity matrices, we obtain after inversion $\langle \mathbf{R}(z_m) \rangle = \mathbf{R}(z_m)$. However, due to the band limitation of the spatial wavelets $\mathbf{W}_s$ and $\mathbf{W}_d$, this result will never be reached [see also expressions (24)].

From the foregoing we may conclude that inverse extrapolation (prestack) involves two matrix multiplications with inverse $\mathbf{F}$: (1) To compensate for the propagation effect of upward traveling waves

$$\mathbf{Q}(z_0) = \mathbf{P}(z_0)\mathbf{F}_d(z_0, z_m); \quad (29a)$$

(2) To compensate for the propagation effect of downward traveling waves

$$\langle \mathbf{R}(z_m) \rangle = \mathbf{F}_s(z_m, z_0)\mathbf{Q}(z_0). \quad (29b)$$

It can be easily verified that equation (29a) defines inverse extrapolation of common source gathers (rows of $\mathbf{P}$) and equation (29b) defines inverse extrapolation of common detector gathers (columns of $\mathbf{Q}$).

Note that in the inversion of equation (29a) some of the undesirable effects of the detector arrays may be compensated for as well. Similarly, in the inversion of equation (29b) some of the effects of the source arrays may be compensated for. However, noise problems will always determine to what degree array compensation can be carried out. If influences of arrays are neglected during inversion then equation (28b) can be replaced by

$$\mathbf{W}_0(z_m) = \mathbf{F}(z_m, z_0)\mathbf{W}(z_0, z_m)$$
$$= \mathbf{W}(z_m, z_0)\mathbf{F}(z_0, z_m).$$

In addition, in the situation of matched filtering,

$$\mathbf{F} = \mathbf{W}^*.$$

Finally, one problem remains to be discussed. In order to obtain $\langle \mathbf{R}(z_m) \rangle$, the influence of

$$\sum_{i=m+1}^{M} \mathbf{W}(z_m, z_i)\mathbf{R}(z_i)\mathbf{W}(z_i, z_m)$$

has to be eliminated from the inverse extrapolation result, starting with $m = 0$. Bearing in mind that $\mathbf{W}(z_m, z_i)$ involves traveltimes of at least $(i - m)\Delta z/c$, $\langle \mathbf{R}(z_m) \rangle$ can be obtained (after inverse extrapolation up to depth level z_m) in the time domain at zero traveltime (imaging principle):

$$\langle \mathbf{r}(z_n) \rangle = \frac{1}{\pi} \int_{\omega_{\min}}^{\omega_{\max}} \mathbf{P}(z_m)\,d\omega \quad \text{(multichannel imaging)}.$$

Generally, only zero-offset traces are considered for imaging (Claerbout, 1976). Hence, in the situation of zero-offset imaging the diagonal elements of matrix $\mathbf{P}(z_m)$, i.e., $P_n(n\Delta x, z_m, \omega)$, are used only:

$$\langle r_n(n\Delta x, z_m) \rangle$$
$$= \frac{1}{\pi} \int_{\omega_{\min}}^{\omega_{\max}} P_n(n\Delta x, z_m, \omega)\,d\omega \quad \text{(single-channel imaging)}.$$

EXPLICIT AND IMPLICIT SCHEMES

In explicit forward formulations the relationship between the known pressure distribution $P(z_0)$ and the unknown pressure distribution $P(z_1)$ is given such that $P(z_1)$ is specified as a weighted sum of the *known* quantities $P(z_0)$, i.e., $P(z_1)$ is explicitly expressed in $P(z_0)$:

$$P(z_1) = W(z_1, z_0) * P(z_0). \quad (30a)$$

Expression (30a) is computationally very convenient. So far, our expressions were formulated explicitly.

Bearing the inverse problem in mind, the relationship between the unknown distribution $P(z_1)$ and the known distribution $P(z_0)$ can also be formulated with the aid of the inverse extrapolator:

$$F(z_0, z_1) * P(z_1) = P(z_0). \quad (30b)$$

Expression (30b) is computationally not a very convenient formulation of the forward problem because we are dealing now with a weighted sum of the unknown quantities $P(z_1)$, i.e., $P(z_1)$ is implicitly expressed in $P(z_0)$.

To generalize expressions (30a) and (30b), consider depth level z_θ such that (Figure 7):

$$z_0 \le z_\theta \le z_1.$$

Then $P(z_\theta)$ can be computed from $P(z_0)$ by forward extrapolation,

$$P(z_\theta) = W(z_\theta, z_0) * P(z_0), \qquad (31a)$$

and $P(z_\theta)$ can be computed from $P(z_1)$ by inverse extrapolation,

$$P(z_\theta) = F(z_\theta, z_1) * P(z_1). \qquad (31b)$$

Combination of equations (31a) and (31b) yields the generalized implicit formulation for wave field extrapolation:

$$F(z_\theta, z_1) * P(z_1) = W(z_\theta, z_0) * P(z_0), \qquad (32)$$

or, after Fourier transformation,

$$\tilde{P}(z_1) = \frac{\tilde{W}(z_\theta, z_0)}{\tilde{F}(z_\theta, z_1)} \tilde{P}(z_0),$$

and, for the inverse problem,

$$\tilde{P}(z_0) = \frac{\tilde{F}(z_\theta, z_1)}{\tilde{W}(z_\theta, z_0)} \tilde{P}(z_1).$$

Of course, expression (32) can also be formulated in terms of matrices:

$$\mathbf{F}(z_\theta, z_1)\mathbf{P}(z_1) = \mathbf{W}(z_\theta, z_0)\mathbf{P}(z_0),$$

or

$$\mathbf{P}(z_1) = \mathbf{F}^{-1}(z_\theta, z_1)\mathbf{W}(z_\theta, z_0)\mathbf{P}(z_0)$$
$$\text{(forward problem),} \qquad (33a)$$

$$\mathbf{P}(z_0) = \mathbf{W}^{-1}(z_\theta, z_0)\mathbf{F}(z_\theta, z_1)\mathbf{P}(z_1)$$
$$\text{(inverse problem).} \qquad (33b)$$

Similarly, all previously derived matrix equations can be formulated accordingly.

Note that for (1) $z_\theta = z_1$, expression (33a) formulates the explicit forward problem; (2) $z_\theta = z_0$, expression (33b) formulates the explicit inverse problem; (3) $z_0 < z_\theta < z_1$, expressions (33a) and (33b) formulate the forward and inverse problems implicitly. In finite-difference techniques, approximated versions of W and F are used, which means that during extrapolation amplitude and phase errors are involved. In the following we will see that within layer (z_0, z_1) the value of z_θ can be chosen such that simplified versions of the extrapolation operators $F^{-1}W$ and $W^{-1}F$ do not contain amplitude errors. This is a most attractive property of implicit finite-difference operators.

SUMMATION APPROACH

In the summation approach to wave field extrapolation, the spatial wavelet W is obtained from the Kirchhoff integral for plane surfaces, i.e., the Rayleigh integral. The derivation of the Rayleigh integral assumes homogeneous media and can be carried out in two ways.

(1) Starting with the theorem of Gauss, Green's theorems are derived. If we choose for the two scalar fields in Green's second theorem: (1) the pressure field (P) we are interested in, P having no singularities inside a closed surface S, (2) a pressure field (G)

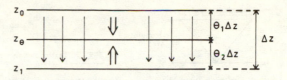

FIG. 7. Implicit, forward extrapolation from depth level z_0 to z_1 can be formulated in terms of explicit forward extrapolation from depth level z_0 to z_θ and explicit inverse extrapolation from depth level z_1 to z_θ.

with one singularity inside closed surface S, then, by substituting the wave equation for both wave fields in Green's second theorem, the Kirchhoff integral is obtained:

$$P_A = \frac{1}{4\pi} \oiint_S \left(P \frac{\partial G}{\partial n} - \frac{\partial P}{\partial n} G \right) dS, \qquad (34)$$

where

P = pressure field of interest without singularities inside S;

$\dfrac{\partial P}{\partial n}$ = component of the pressure field along the normal $\mathbf{n}$ of closed surface S ($\mathbf{n}$ pointing inward);

G = pressure field with one singularity inside S, the singularity being situated at point A.

If we choose for S a plane surface and we choose G such that $G = 0$ on S, then the Kirchhoff integral can be rewritten in terms of the Rayleigh integral:

$$P_A = \frac{1}{2\pi} \iint_S P \frac{\partial G}{\partial n} dS,$$

or, substituting the expression for G and using for S the horizontal plane $z = z_0$,

$$P_A = \frac{|z_A - z_0|}{2\pi} \iint_S P \frac{1 + jk_1 \Delta r}{\Delta r^3} e^{-jk_1 \Delta r} dxdy, \qquad (35)$$

where $k_1 = k(1 - j\eta)$, $P = P(x, y, z_0, \omega)$ and

$$\Delta r = \sqrt{(x_A - x)^2 + (y_A - y)^2 + (z_A - z_0)^2}.$$

In expression (35) it is essential that P represents a wave field which travels across surface S toward the observation point A. Note that if $p_A = p(x_A, y_A, z_A, t)$ need be computed up to a finite recording time only, integration area S becomes finite as well (finite aperture area).

(2) Starting with the wave equation,

$$\frac{\partial^2 P}{\partial x^2} + \frac{\partial^2 P}{\partial y^2} + \frac{\partial^2 P}{\partial z^2} + k^2(1 - j\eta)^2 P = 0 \text{ with } 0 \le \eta \ll 1,$$
$$(36a)$$

a Fourier transformation in the domain of the spatial variables x and y is carried out [see also equation (8b)]:

$$\frac{\partial^2 \tilde{P}}{\partial z^2} + (k_1^2 - k_x^2 - k_y^2)\tilde{P} = 0, \qquad (36b)$$

where $k_1 = k(1 - j\eta)$. Equation (36b) formulates the one-dimensional (1-D) Helmholtz equation along the z-axis; its forward solution is well-known [see also equation (10b)]:

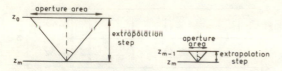

FIG. 8. For small extrapolation steps, small volumes are involved and, therefore, local propagation velocities can be used.

$$\tilde{P}(k_x, k_y, z, \omega) = \tilde{P}(k_x, k_y, z_0, \omega) \, e^{-j \sqrt{k_1^2 - (k_x^2 + k_y^2)} |z - z_0|}. \quad (37)$$

On physical grounds the imaginary part of the square root in equation (37) must always be chosen negative [see also equation (10c)]. Now if we take the inverse Fourier transform of equation (37) with respect to the spatial Fourier variables k_x and k_y, then Rayleigh integral (35) is obtained again [see also equation (11c)].

From convolution integral (35) and the definition of the spatial wavelet W according to equation (6b), it follows that in the summation approach spatial wavelet W is given by

$$W(x_A - x, y_A - y, \Delta z, \omega) = \frac{\Delta z}{2\pi} \frac{1 + jk_1 \Delta r}{\Delta r^3} e^{-jk_1 \Delta r}, \quad (38a)$$

or

$$W(x, y, \Delta z, \omega) = \frac{\Delta z}{2\pi} \frac{1 + jk_1 r}{r^3} e^{-jk_1 r}, \quad (38b)$$

where $r = \sqrt{x^2 + y^2 + \Delta z^2}$ and $\Delta z = |z_A - z_0|$.

To obtain an expression for W in the 2-D situation, we have to integrate with respect to y:

$$W(x, \Delta z, \omega) = \int_{-\infty}^{+\infty} W(x, y, \Delta z, \omega) \, dy$$

$$= \frac{\Delta z}{2\pi} \int_{-\infty}^{+\infty} \frac{1 + jk_1 r}{r^3} e^{-jk_1 r} dy$$

$$= \frac{-jk_1}{2} \cos \phi \, H_1^{(2)}(k_1 r), \quad (39)$$

where $\cos \phi = \Delta z / r$ and $H_1^{(2)}$ represents the first-order Hankel function of the second kind. In the far field ($kr \gg 1$), the expressions for W become somewhat simpler:

$$W(x, y, \Delta z, \omega) \approx \frac{jk_1}{2\pi} \cos \phi \, \frac{e^{-jk_1 r}}{r} \quad (40a)$$

$$W(x, \Delta z, \omega) \approx \sqrt{\frac{jk_1}{2\pi}} \cos \phi \, \frac{e^{-jk_1 r}}{\sqrt{r}}, \quad (40b)$$

where $\sqrt{j} = e^{j\pi/4}$. For an extensive discussion on the Kirchhoff and Rayleigh integral and the derivation of 2-D and 3-D summation operators, the reader is referred to Berkhout (1980, chapters 5 and 6).

Wave field extrapolation with summation operators can be applied both recursively and nonrecursively. In recursive applications, Δz is small ($\Delta z < \lambda / 2$). Consequently, for a given aperture angle the conical volume for which one velocity value has to be specified to compute one extrapolated pressure value will be small (Figure 8). This means that a local velocity value can be used, which may be taken differently for each extrapolated pressure value. Hence, we may conclude that, although the Rayleigh

integral was derived for homogeneous situations, moderate vertical and lateral velocity variations can be properly handled by summation operators if the wave field extrapolation procedure is carried out recursively.

In the situation of fast lateral velocity variations the velocity may change within the summation operator by assigning different velocity values to the different samples of the operator, i.e., for fast lateral velocity variations we may take, according to equation (3d) in expressions (38) and (39):

$$k_1 = \sqrt{1 + \gamma} \, \bar{k}_1$$

$$\approx \left(1 - \frac{\Delta c}{\bar{c}}\right) \bar{k}_1 \text{ for small } |\Delta c / \bar{c}|,$$

where $\bar{c}$ is constant within the operator and $\Delta c = c - \bar{c}$ may be different for each ray. However, we will see that in the situation of fast lateral velocity variations the finite-difference technique is more suitable.

In nonrecursive applications the final extrapolation result is computed by one step only. Hence, for deep data very large extrapolation steps occur and, therefore, very large volumes are involved for which one velocity value has to be specified. Of course, this procedure will never lead to satisfactory results in situations with significant lateral velocity variations. Still, in the practice of migration the summation approach is generally used nonrecursively in the time domain for 2- and 3-D data (economic reasons), the absorption being neglected. Hence, expressions (40a) and (40b) are used with the assumption $\eta = 0$:

$$w(x, y, \Delta z, t) \approx \frac{\cos \phi}{2\pi r} d_1 \left(t - \frac{r}{c}\right), \quad (41a)$$

where $r = \sqrt{x^2 + y^2 + \Delta z^2}$;

$$w(x, \Delta z, t) \approx \frac{\cos \phi}{\sqrt{2\pi r}} d_{1/2} \left(t - \frac{r}{c}\right), \quad (41b)$$

where $r = \sqrt{x^2 + \Delta z^2}$. In expression (41a), $d_1(t)$ represents the inverse Fourier transform of a band-limited version of $j\omega / c$ and $d_{1/2}(t)$ represents the inverse Fourier transform of a band-limited version of $\sqrt{j\omega / c}$.

Using for the inverse problem a spatial matched filter, it follows from equations (40a) and (40b) that inverse extrapolation in the summation approach means convolution in the space-time domain with

$$f(x, y, \Delta z, t) = \frac{\cos \phi}{2\pi r} d_{-1} \left(t + \frac{r}{c}\right),$$

$$f(x, \Delta z, t) = \frac{\cos \phi}{\sqrt{2\pi r}} d_{-1/2} \left(t + \frac{r}{c}\right),$$

where $d_{-1}(t)$ is related to $-j\omega / c$ and $d_{-1/2}(t)$ is related to $\sqrt{-j\omega / c}$.

It is interesting to realize that the conventional diffraction stack involves a nonrecursive summation operator in the time domain with floating time reference (time-delayed pressure output), without weighting factors and without time differentiation:

$$f_{st}(x, z, t) = \delta \left(t + \frac{r - z}{c}\right).$$

Note that in the summation approach the extrapolation operator is applied explicitly. Therefore, the summation approach is not very sensitive to improper boundary conditions at both ends of the data file.

We have seen, by using the Taylor series, that the forward extrapolation operator for advanced pressures can be expressed in terms of a series expansion:

$$W'(z_0, z_1) = \delta(x, y) + \frac{(-jk\,\Delta z)}{1!}H'$$

$$+ \frac{(-jk\,\Delta z)^2}{2!}H' * H' + \ldots, \quad (42a)$$

where, according to Appendix A,

$$H = \frac{k}{2\pi}\left[\frac{j_1(kr_0)}{r_0} - (jk\eta)j_0(kr_0)\right]$$

$$\text{for } \eta \ll \cos\alpha_m, \quad (42b)$$

and

$$H' = H - (1 - j\eta)\,\delta(x, y);$$

$$r_0 = \sqrt{x^2 + y^2}, \quad \Delta z = |z_1 - z_0|;$$

$$\delta(x, y) = \delta(x)\,\delta(y).$$

In the homogeneous situation

$$k\bar{H}' = \sqrt{k^2(1 - j\eta)^2 - (k_x^2 + k_y^2)} - k(1 - j\eta),$$
$$(42c)$$

and, therefore,

$$\bar{W}'(z_0, z_1) = 1 + \frac{\Delta z}{1!}(-jk\bar{H}') + \frac{\Delta z^2}{2!}(-jk\bar{H}')^2 + \ldots,$$

$$= \exp(-jk\bar{H}'\,\Delta z), \quad (43a)$$

$k\bar{H}'$ being given by equation (42c). Compare equation (43a) with expression (11a) for the loss-free situation. As was mentioned before, inverse Fourier transformation of equation (43a) yields the summation operator in the space-frequency domain:

$$W'(z_0, z_1) = \frac{\Delta z}{2\pi}\frac{1 + jk_1 r}{r^3}e^{-jk_1(r - \Delta z)}. \quad (43b)$$

Hence, expression (42a) represents the Taylor series expansion of the summation operator, H being given by equation (42b).

From equation (43a) it can be seen that for $\Delta z \to 0$, the spatial bandwidth of the summation operator increases without limit. Therefore, expression (43b) is not suitable for recursive applications and Taylor series expansion (42a) should be used. We will see that recursive summation extrapolation with the aid of equations (42a) and (42b) is closely related to the finite-difference technique.

WAVENUMBER-FREQUENCY APPROACH

In the foregoing we have seen that band-limited inverse extrapolation can be formulated in two ways: (1) In terms of matrix inversion,

$$\mathbf{F}(z_0, z_1)\mathbf{W}(z_1, z_0) = \mathbf{W}_0(z_0). \quad (44a)$$

(2) In terms of deconvolution,

$$F(x, y, \Delta z, \omega) * W(x, y, \Delta z, \omega) = W_0(x, y, \Delta z, \omega), \quad (44b)$$

where $\Delta z = |z_1 - z_0|$. If we assume no lateral velocity and absorption variations, then F is a stationary operator and equation (44b) can be translated to the spatial Fourier domain

$$\bar{F}(k_x, k_y, \Delta z, \omega)\,\bar{W}(k_x, k_y, \Delta z, \omega) = \bar{W}_0(k_x, k_y, \Delta z, \omega). \quad (44c)$$

Using expression (37), it follows from equation (44c) that the band-limited inversion operator in the wavenumber-frequency domain is given by

$$\bar{F} = \bar{W}_0 e^{+j\sqrt{k^2(1 - j\eta)^2 - (k_x^2 + k_y^2)}\,\Delta z}. \quad (45)$$

In equation (45) the real part of the square root should be taken positive and W_0 represents a spatial band-pass filter to stabilize F.

To evaluate the influence of absorption on the inverse extrapolation operator, consider the spatial transfer function,

$$\bar{W} = e^{-j\sqrt{k^2(1 - j\eta)^2 - (k_x^2 + k_y^2)}\,\Delta z}, \quad (46a)$$

for two situations: (1) If $k_x^2 + k_y^2 \ll k^2$, then

$$\bar{W} = e^{-(k\,\Delta z)\eta}e^{-j\sqrt{k^2 - (k_x^2 + k_y^2)}\,\Delta z}. \quad (46b)$$

(2) If $k_x^2 + k_y^2 = k^2$, then

$$\bar{W} = e^{-(k\,\Delta z)\sqrt{\eta}}e^{-j(k\,\Delta z)\sqrt{\eta}}. \quad (46c)$$

From expressions (46) it follows that the attenuation due to absorption increases with increasing temporal frequency ω and increasing tilt angle α. Hence, at high frequencies and high tilt angles the data will be enhanced considerably by inverse extrapolator $\bar{W}^{-1}$. Depending upon the noise spectrum in the wavenumber-frequency domain, this might be unacceptable and proper band-limitation should be applied according to equations (24a), (24b), or (24c). For further details on inverse extrapolation in the presence of noise, the reader is referred to Berkhout and Van Wulfften Palthe (1979).

In the foregoing we have also seen that inverse extrapolation of prestack data can be formulated in two ways: (1) In terms of matrix inversion,

$$\mathbf{F}(z_0, z_1)\mathbf{P}(z_1)\mathbf{F}(z_1, z_0) = \langle\mathbf{P}(z_0)\rangle. \quad (47a)$$

(2) In terms of deconvolution,

$$F(x_s, \Delta z, \omega) * P(x_s, x_d, z_1, \omega) * F(x_d, \Delta z, \omega)$$
$$= \langle P(x_s, x_d, z_0, \omega)\rangle, \quad (47b)$$

where x_s and x_d represent the source and detector coordinates, respectively. Note that for common-offset data equation (47b) yields the interesting expression

$$F^2(x_s = x_d, \Delta z, \omega) * P(x_s = x_d, z_1, \omega)$$
$$= \langle P(x_s = x_d, z_0, \omega)\rangle$$

[see also expression (19)].

If we assume again no lateral velocity and absorption variations, then F is a stationary operator and equation (47b) can be written as a multiplication procedure in the wavenumber-frequency domain:

$$\bar{F}(k_{x_s}, \Delta z, \omega)\bar{P}(k_{x_s}, k_{x_d}, z_1, \omega)\bar{F}(k_{x_d}, \Delta z, \omega)$$
$$= \langle\bar{P}(k_{x_s}, k_{x_d}, z_0, \omega)\rangle. \quad (47c)$$

Finally, if we substitute equation (45) into equation (47c), we obtain

$$[\bar{W}_0 e^{+j\sqrt{k_1^2 - k_{x_s}^2}\,\Delta z}]\bar{P}(k_{x_s}, k_{x_d}, z_1, \omega)[\bar{W}_0 e^{+j\sqrt{k_1^2 - k_{x_d}^2}\,\Delta z}]$$
$$= \langle\bar{P}(k_{x_s}, k_{x_d}, z_0, \omega)\rangle, \quad (48a)$$

where $k_1 = k(1 - j\eta)$. Note that equation (48a) can easily be extended to the 3-D situation. Expression (48a) was also derived by Yilmaz (1979) for loss-free media. The prestack extrapolation procedure, as described by equation (48a), is simple, but the method is not suitable in situations with lateral velocity variations. This is a very serious drawback. A more realistic and practical procedure would consist of transforming the individual common source records to the wavenumber-frequency domain only. In that situation equation (47b) can be rewritten as a mixed convolution-multiplication procedure:

$$F(x_s, \Delta z, \omega) * \tilde{P}(x_s, k_{x_d}, z_1, \omega) \cdot$$
$$\cdot [\bar{W}_0 e^{+j\sqrt{\bar{k}_1^2 - k_{x_d}^2}\Delta z}] = \langle \tilde{P}(x_s, k_{x_d}, z_0, \omega)\rangle. \tag{48b}$$

The original proposal of Stolt (1978) uses a mapping procedure from the $k_x - k$ to the $k_x - k_z$ domain. This method is elegant and fast, but it applies only for constant-velocity media. Although many extensions to Stolt's method have been proposed (see, e.g., Gazdag, 1978), the reader should be aware that pure wavenumber-frequency methods can only provide *approximate* solutions in laterally inhomogeneous situations. This important constraint can also be well appreciated by considering Fourier transformation of $P(x, y, z, \omega)$ in the domain of the spatial variables x and y as a decomposition procedure into plane waves (Berkhout, 1980, chapters 3, 4). Multiplication with $\bar{F}$ means compensation of the plane-wave traveltime in layer (z_0, z_1). In laterally varying layers, a plane wave does *not* remain planar and, therefore, equation (48a) does not apply.

From the foregoing it follows that it would be more elegant to refer to equations (44c) and (48a) as plane-wave methods instead of wavenumber-frequency $(k - f)$ methods.

FINITE-DIFFERENCE TECHNIQUES— EXPLICIT SCHEMES

We have seen by using the Taylor series that the forward extrapolation operator for advanced pressures can be expressed in terms of a series expansion:

$$W'(z_0, z_1) = \delta(x, y) + \frac{(-j\bar{k}_1\Delta z)}{1!} H'$$
$$+ \frac{(-j\bar{k}_1\Delta z)^2}{2!} H' * H' + \dots. \tag{49}$$

In explicit finite-difference schemes a truncated version of equation (49) is used and an explicit expression for H' is derived from the wave equation. Hence, analytic expression (42b) is not used. [For a discussion of the application of analytical expression (42b) the reader is referred to Berkhout, 1980.] To derive 3-D explicit schemes, we have to determine explicit expressions for H' with the aid of wave equation

$$\frac{\partial^2 P'}{\partial x^2} + \frac{\partial^2 P'}{\partial y^2} + \frac{\partial^2 P'}{\partial z^2} - 2j\bar{k}_1\frac{\partial P'}{\partial z} + \gamma\bar{k}_1^2 P' = 0, \tag{50a}$$

or

$$d_2(x) * P' + d_2(y) * P' + \frac{\partial^2 P'}{\partial z^2} - 2j\bar{k}_1\frac{\partial P'}{\partial z} + \gamma k_1^2 P' = 0, \tag{50b}$$

where

$$\bar{k}_1 = \frac{\omega}{\bar{c}}(1 - j\eta), \gamma = \left[\left(\frac{\bar{c}}{c}\right)^2 - 1\right].$$

In finite-difference techniques $d_2(x)$ and $d_2(y)$ are approximated by simple filters. For instance, in terms of z-transforms,

$$d_2 = \frac{\beta}{\Delta^2}\frac{z^{-1} - 2 + z}{z^{-1} + (\beta - 2) + z}, \tag{50c}$$

where z represents the unit delay operator in the x- or y-direction

and Δ is the spatial sampling interval. Ristow (1980) proposes to take β such that the error is minimized according to a least-squares criterion for a given dip range $|\alpha| < \alpha_{max}$. Generally $\beta \approx 12$.

If we write

$$d_2 = \frac{\beta}{\Delta^2}\frac{z^{-1} - 2 + z}{(A + Bz^{-1})(A + Bz)},$$

then application of d_2 involves convolution with a three-point filter, forward feedback filtering with a two-point filter, and backward feedback filtering with a two-point filter. This procedure is computationally very attractive. Of course, more accurate approximations of d_2 can be introduced at the cost of the efficiency of the algorithm. In the most extreme situation, $d_2(x)$ and $d_2(y)$ are applied in the wavenumber-frequency domain by multiplication with $(jk_x)^2$ and $(jk_y)^2$.

If the derivatives of $\bar{k}_1$ and γ with respect to z may be neglected, then it follows from equation (50b):

$$\frac{\partial^{n+1} P'}{\partial z^{n+1}} - 2j\bar{k}_1\frac{\partial^n P'}{\partial z^n} + \bar{k}_1^2 a(x, y) * \frac{\partial^{n-1} P'}{\partial z^{n-1}}$$
$$= 0 \text{ for } n = 1, 2, \dots, \tag{51}$$

where $a(x, y) = \frac{1}{\bar{k}_1^2}[d_2(x) + d_2(y)] + \gamma\delta(x, y)$.

(1) First-order explicit scheme

In the first-order scheme, H' is computed with the assumption that $\partial^2 P'/\partial z^2 = 0$. Then it follows from equation (51) for $n = 1$:

$$\frac{\partial P'}{\partial z}\bigg|_1 = -\frac{1}{2}j\bar{k}_1 a(x, y) * P'. \tag{52a}$$

In equation (52a) the total contribution to $\partial P'/\partial z$ can be subdivided into two parts, i.e., the *diffraction* term, and the so-called *thin lens* term:

$$\frac{\partial P'}{\partial z}\bigg|_1 = \frac{1}{2j\bar{k}_1}[d_2(x) + d_2(y)] * P' - j\bar{k}_1\left(\frac{\gamma}{2}\right)P'$$
$$= \frac{1}{2j\bar{k}_1}[d_2(x) + d_2(y)] * P' - j\bar{k}_1\frac{(\bar{c}/c)^2 - 1}{2}P'$$
$$\approx \underbrace{\frac{1}{2j\bar{k}_1}[d_2(x) + d_2(y)] * P'}_{\text{diffraction term}} - \underbrace{j\omega\Delta\tau_1 P'}_{\substack{\text{thin-lens} \\ \text{term}}}, \tag{52b}$$

where $\Delta\tau_1 = 1/c - 1/\bar{c}$. Note that in the first-order scheme the deviation between the actual velocity and the reference velocity is taken into account by a time shift only (thin-lens term). Recently it was proposed (Judson et al, 1980) to improve migration schemes for situations with very fast lateral velocity variations by means of time shifts.

From equation (52a) and the definition of H' we may conclude that in the first-order scheme H' is approximated by

$$H' = \frac{1}{2}a(x, y),$$

or

$$H' = \frac{1}{2\bar{k}_1^2}[d_2(x) + d_2(y)] + \frac{1}{2}\gamma\delta(x, y). \tag{53}$$

Substitution in a truncated version of equation (49) yields the first-order explicit expression for W'.

(2) Second-order explicit scheme

In the second-order scheme H' is computed with the assumption that $\partial^3 P'/\partial z^3 = 0$. Then it follows from equation (51) for $n = 2$:

$$\left.\frac{\partial^2 P'}{\partial z^2}\right|_1 = -\frac{1}{2}\,j\bar{k}_1 a * \left.\frac{\partial P'}{\partial z}\right|_1$$

$$= -\frac{1}{4}\,\bar{k}_1^2[a * a * P'],$$

where $a = a(x, y)$. Using this result in equation (51) for $n = 1$, we obtain:

$$\left.\frac{\partial P'}{\partial z}\right|_2 = -\frac{1}{2}\,j\bar{k}_1[a * P'] + \left(\frac{1}{2j\bar{k}}\right)\left.\frac{\partial^2 P'}{\partial z^2}\right|_1$$

$$= -\frac{1}{2}\,j\bar{k}_1[a * P'] + \frac{1}{8}\,j\bar{k}_1[a * a * P'].$$

Hence, in the second-order scheme we make use of the approximation

$$H' = \frac{1}{2}\,a - \frac{1}{8}\,a * a. \tag{54}$$

Substitution in a truncated version of equation (49) yields the second-order explicit expression for W'.

(3) Third-order explicit scheme

In the third-order scheme H' is computed with the assumption that $\partial^4 P'/\partial z^4 = 0$. Then it follows from equation (51) for $n = 3$:

$$\left.\frac{\partial^3 P'}{\partial z^3}\right|_1 = \left(-\frac{1}{2}\,j\bar{k}_1\right) a * \left.\frac{\partial^2 P'}{\partial z^2}\right|_1$$

$$= \frac{1}{8}\,j\bar{k}_1^3[a * a * a * P'],$$

and, using this result in equation (51) for $n = 2$,

$$\left.\frac{\partial^2 P'}{\partial z^2}\right|_2 = \left(-\frac{1}{2}\,j\bar{k}_1\right) a * \left.\frac{\partial P'}{\partial z}\right|_2 + \left(\frac{1}{2j\bar{k}}\right)\left.\frac{\partial^3 P'}{\partial z^3}\right|_1$$

$$= -\frac{1}{4}\,\bar{k}_1^2[a * a * P'] + \frac{1}{8}\,\bar{k}_1^2[a * a * a * P'].$$

Using this result in equation (51) for $n = 1$,

$$\left.\frac{\partial P'}{\partial z}\right|_3 = -\frac{1}{2}\,j\bar{k}_1[a * P'] + \left(\frac{1}{2j\bar{k}}\right)\left.\frac{\partial^2 P'}{\partial z^2}\right|_2$$

$$= -\frac{1}{2}\,j\bar{k}_1[a * P'] + \frac{1}{8}\,j\bar{k}_1[a * a * P']$$

$$- \frac{1}{16}\,j\bar{k}_1[a * a * a * P'].$$

Hence, in the third-order scheme we make use of the approximation

$$H' = \frac{1}{2}\,a - \frac{1}{8}\,a * a + \frac{1}{16}\,a * a * a. \tag{55}$$

Substitution in a truncated version of equation (49) yields the third-order explicit expression for W'.

In Figure 9a the computational diagram is given for 3-D, explicit inverse extrapolation using three terms of the Taylor series. It is important to realize that the scheme enables 3-D extrapolation by 1-D convolution with the operators $d_2(x)$ and $d_2(y)$. Therefore, explicit schemes are most advantageous for 3-D finite-difference migration. Figures 9b and 9c show the amplitude and phase errors of the scheme of Figure 9a, using for d_2 expression (50c). Of course, these results can be significantly improved if the coefficients of the truncated Taylor expansion are optimized. Note that the extrapolation operator has the property of a spatial high-cut

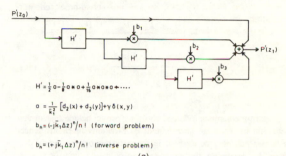

$$H' = \frac{1}{2}\,o - \frac{1}{8}\,o * o + \frac{1}{16}\,o * o * o + \cdots.$$

$$o = \frac{1}{\bar{k}_1^2}\,[d_2(x) + d_2(y)] + \gamma\,\delta(x, y)$$

$$b_n = (-j\bar{k}_1\Delta z)^n/n! \quad \text{(forward problem)}$$

$$b_n = (+j\bar{k}_1\Delta z)^n/n! \quad \text{(inverse problem)}$$

(a)

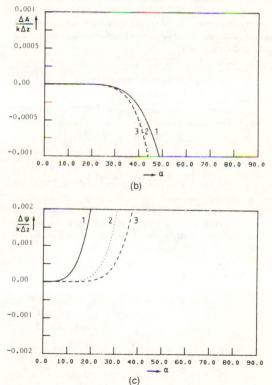

(b)

(c)

FIG. 9. (a) Full explicit finite-difference scheme with three terms of the Taylor series. Generally, the coefficients γ and b are space-variant. (b) Amplitude error of the full explicit extrapolation operator of Figure 9a, H being represented by the first-, second- and third-order approximation. (c) Phase error of the full explicit extrapolation operator of Figure 9a, H being represented by the first-, second- and third-order approximation.

filter; at high dip angles where the phase errors are unacceptably large, the amplitude becomes increasingly smaller than unity. Finally, it is worthwhile mentioning that for the simple situation of loss-free homogeneous media, i.e., $\bar{k}_1 = k$ and $\gamma = 0$, the above explicit schemes can be derived easily with the well-known series expansion of the square root (Claerbout, 1976):

$$\sqrt{1 - (k_x^2 + k_y^2)/k^2} \approx 1 - \frac{1}{2}\frac{k_x^2 + k_y^2}{k^2} \quad \text{(first-order scheme)}$$

$$\approx 1 - \frac{1}{2}\frac{k_x^2 + k_y^2}{k^2} - \frac{1}{8}\frac{(k_x^2 + k_y^2)^2}{k^4}$$
$$\text{(second-order scheme)}$$

$$\approx 1 - \frac{1}{2}\frac{k_x^2 + k_y^2}{k^2} - \frac{1}{8}\frac{(k_x^2 + k_y^2)^2}{k^4}$$
$$- \frac{1}{16}\frac{(k_x^2 + k_y^2)^3}{k^6} \quad (56)$$
$$\text{(third-order scheme)}.$$

etc.

Comparing equations (55) and (56), we may conclude that in the situation of moderate lateral velocity variations finite-difference schemes can also be derived in the wavenumber-frequency domain by assuming the homogeneous situation first ($\bar{k} = k =$ constant, $\gamma = 0$) and, afterward, taking $\bar{k} = k$ space-variant in the final expression for W'. As we have seen, this procedure was followed in the derivation of the space-variant summation operator. However, with fast lateral velocity variations the factor γ should be included *during the derivation*.

FINITE-DIFFERENCE TECHNIQUES—
IMPLICIT SCHEMES

In the foregoing we have seen that implicit forward extrapolation can be formulated according to

$$F'(z_\theta, z_1) * P'(z_1) = W'(z_\theta, z_0) * P'(z_0),$$

or, using a symbolic notation,

$$P'(z_1) = \frac{W'(z_\theta, z_0)}{F'(z_\theta, z_1)} * P'(z_0).$$

Note that in the equations above, symbolic notation division means convolution with the inverse. If we substitute in this expression the series expansion of W' and F' as used in explicit schemes,

$$W'(z_\theta, z_0) = \delta(x, y) + \sum_{n=1}^{N_1} \theta_1^n \frac{(-j\bar{k}_1\Delta z)^n}{n!} H'_n$$
$$= \delta(x, y) + G'_{N_1}(x, y, \theta_1\Delta z, \omega); \quad (57a)$$

and

$$F'(z_\theta, z_1) = \delta(x, y) + \sum_{n=1}^{N_2} \theta_2^n \frac{(+j\bar{k}_1\Delta z)^n}{n!} H'_n$$
$$= \delta(x, y) + G'_{N_2}(x, y, -\theta_2\Delta z, \omega), \quad (57b)$$

where $H'_1 = H'$, $H'_2 = H' * H'$, etc., $0 < \theta_1 < 1$, $0 < \theta_2 < 1$, $\theta_1 + \theta_2 = 1$, then we obtain

$$\langle P'(z_1)\rangle = \frac{\delta(x, y) + G'_{N_1}(x, y, \theta_1\Delta z, \omega)}{\delta(x, y) + G'_{N_2}(x, y, -\theta_2\Delta z, \omega)} * P'(z_0). \quad (58a)$$

In the literature of implicit finite-difference migration, expression (58a) is often used for $N_1 = N_2 = 1$ and $\theta_1 = \theta_2 = 1/2$, yielding

a central-difference or Crank-Nicolson formulation

$$P'(z_1) = \frac{\delta(x, y) - \dfrac{j\bar{k}_1\Delta z}{2}H'}{\delta(x, y) + \dfrac{j\bar{k}_1\Delta z}{2}H'} * P'(z_0). \quad (58b)$$

Pann et al (1979) proposed a collocation approach to equation (58b).

Note that if we choose $N_1 = N_2 = N$ and $\theta_1 = \theta_2 = 1/2$, then in the situation of loss-free media expression (57b) is the conjugate complex version of equation (57a). Therefore we may write for the spatial amplitude spectrum

$$A(k_x, k_y, \Delta z, \omega) = \left| \frac{1 + \bar{G}'_N\left(k_x, k_y, \dfrac{\Delta z}{2}, \omega\right)}{1 + \bar{G}'_N\left(k_x, k_y, \dfrac{-\Delta z}{2}, \omega\right)} \right| = 1$$

for any N. Consequently, in loss-free situations wave-field extrapolation with the aid of equation (58a) will yield a unit spatial amplitude spectrum of the extrapolation operator: $A = 1$ if $N_1 = N_2$ and $\theta_1 = \theta_2$. For $\theta_1 \neq \theta_2$, A will be close to unity for small tilt angles only. For $\theta_1 > \theta_2$, A will *increase* for increasing tilt angle; for $\theta_1 < \theta_2$, A will *decrease* for increasing tilt angle.

In the following the implicit forward extrapolator will be approximated by

$$W'(z_0, z_1) = \frac{\delta(x, y) - \theta_1\dfrac{j\bar{k}_1\Delta z}{1!}H'}{\delta(x, y) + \theta_2\dfrac{j\bar{k}_1\Delta z}{1!}H'}. \quad (59a)$$

Note that with the aid of expressions (57) and (58a) more accurate versions can easily be formulated.

From equation (59a) it can be seen that by using an implicit expression of H',

$$\bar{H}' = \bar{H}_z/\bar{H}_p, \quad (59b)$$

the expression for W' does not become significantly more complicated,

$$\bar{W}'(z_0, z_1) = \frac{\bar{H}'_p - \theta_1\dfrac{j\bar{k}_1\Delta z}{1!}\bar{H}'_z}{\bar{H}'_p + \theta_2\dfrac{j\bar{k}_1\Delta z}{1!}\bar{H}'_z}, \quad (59c)$$

while the accuracy of H', and therefore also of W', will be improved significantly. Hence, the important problem in implicit finite-difference techniques is to find simple expressions for H'_z and H'_p such that an accurate estimate of H' is obtained (Ristow, 1980). As we saw in explicit finite-difference schemes, the exact expression for H', as given by equation (42b), is also not used in implicit finite-difference schemes.

(1) First-order scheme

In the first-order scheme it is assumed in the estimation procedure for H' that $\partial^2 P'/\partial z^2 = 0$. Then it follows from equation (51) for $n = 1$:

$$2j\bar{k}_1\delta(x, y) * \frac{\partial P'}{\partial z} = \bar{k}_1^2 a(x, y) * P'. \quad (60a)$$

Hence, in the first-order scheme we make use of the approximations

$$H_z' = \frac{1}{2} a(x, y), \quad H_p' = \delta(x, y). \qquad (60b)$$

Substitution in equation (59c) yields the first-order implicit expression for W'.

(2) Second-order scheme

In the second-order scheme it is assumed in the estimation procedure for H' that $\partial^3 P'/\partial z^3 = 0$. Then, if we differentiate equation (60a) with respect to z, we obtain

$$2j\bar{k}_1\delta(x, y) * \frac{\partial^2 P'}{\partial z^2} = \bar{k}_1^2 a(x, y) * \frac{\partial P'}{\partial z}. \qquad (61a)$$

Substitution of the original wave equation,

$$\frac{\partial^2 P'}{\partial z^2} = 2j\bar{k}_1 \frac{\partial P'}{\partial z} - \bar{k}_1^2 a(x, y) * P', \qquad (61b)$$

in equation (61a) yields

$$[a(x, y) + 4\delta(x, y)] * \frac{\partial P'}{\partial z} = (-2j\bar{k})a(x, y) * P'. \qquad (61c)$$

Hence, in the second-order scheme we make use of the approximations

$$H_z' = \frac{1}{2} a(x, y), \quad H_p' = \delta(x, y) + \frac{1}{4} a(x, y). \qquad (62)$$

Substitution in equation (59c) yields the second-order implicit expression for W'.

(3) Third-order scheme

In the third-order scheme it is assumed in the estimation procedure for H' that $\partial^4 P'/\partial z^4 = 0$. Then, if we differentiate equation (61c) with respect to z we obtain

$$[a(x, y) + 4\delta(x, y)] * \frac{\partial^2 P'}{\partial z^2} = (-2j\bar{k}_1)a(x, y) * \frac{\partial P'}{\partial z}. \qquad (63a)$$

Substitution of the original wave equation,

$$\frac{\partial^2 P}{\partial z^2} = 2j\bar{k} \frac{\partial P}{\partial z} - \bar{k}_1^2 a(x, y) * P',$$

in equation (63a) yields

$$2j\bar{k}[2a(x, y) + 4\delta(x, y)] * \frac{\partial P'}{\partial z}$$
$$= \bar{k}_1^2 a(x, y) * [a(x, y) + 4\delta(x, y)] * P'. \qquad (63b)$$

Hence, in the third-order scheme we make use of the approximations

$$H_z' = \frac{1}{2} a(x, y) + \frac{1}{8} a(x, y) * a(x, y),$$
$$\qquad\qquad\qquad\qquad\qquad (64)$$
$$H_p' = \delta(x, y) + \frac{1}{2} a(x, y).$$

Substitution in equation (59c) yields the third-order implicit expression for W'.

Figures 10a and 10b show the amplitude and phase errors of the first-, second- and third-order implicit scheme. Note that the amplitude error equals zero.

Finally, for implicit finite-difference migration in the space-

time domain the operators W' and F' should be transformed to the time domain. This means that the coefficient $\overline{k}_1$ will be replaced by a time function.

For the simple situation of loss-free homogeneous media, i.e., $\overline{k}_1 = k$ and $\gamma = 0$, implicit expressions for H' can be very easily derived with the continued-fraction expansion of the square root (Engquist and Majda, 1979; Muir and Claerbout, 1980):

$$\sqrt{1 - \frac{k_x^2 + k_y^2}{k^2}} \approx 1 - \frac{1}{2} \frac{k_x^2 + k_y^2}{k^2},$$

(first-order scheme)

$$\approx 1 - \frac{(k_x^2 + k_y^2)/k^2}{2 - \frac{1}{2}(k_x^2 + k_y^2)/k^2},$$

(second-order scheme)

$$\approx 1 - \frac{(k_x^2 + k_y^2)/k^2}{2 - \dfrac{(k_x^2 + k_y^2)/k^2}{2 - \frac{1}{2}(k_x^2 + k_y^2)/k^2}}.$$

$$\vdots$$

(third-order scheme) (65)

etc.

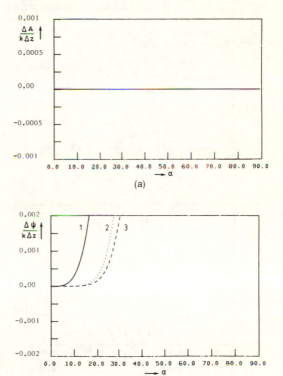

FIG. 10. (a) Amplitude error of the full implicit finite-difference extrapolation operator (Crank-Nicolson formulation) for any approximation of H. (b) Phase error of the full implicit, finite-difference extrapolation operator (Crank-Nicolson formulation), H being represented by the first-, second- and third-order approximation.

COMBINATION OF EXPLICIT AND IMPLICIT SCHEMES

In explicit schemes the expressions for H' and W' were both formulated explicitly. In implicit schemes we have formulated both the expressions for H' and W' in an implicit way.

From a practical point of view it is important to consider a third possibility as well: H' is formulated in an implicit way according to equation (59b) and W' is formulated explicitly according to equation (49). In this procedure we have to compute the derivatives $\partial^n P' / \partial z^n$ first according to

$$H_p' * \frac{\partial^n P'}{\partial z^n} = jkH_z' * \frac{\partial^{n-1} P'}{\partial z^{n-1}} , \qquad (66)$$

H_z' and H_p' being given by the previous section. Then we have to substitute the computed derivatives in a truncated version of equation (49).

The mixed procedure, as described here, is less complicated than the full implicit technique but gives slightly better results than the more simple, full explicit schemes.

To illustrate the above, consider again the finite-difference diagram of Figure 9a, but now H being defined implicitly by equation (66) and the previous section. Figures 11a and 11b show

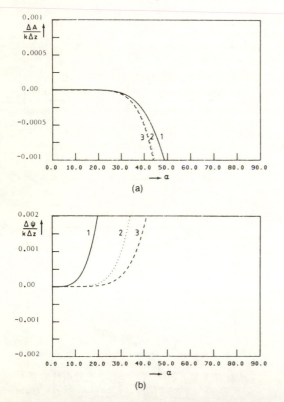

FIG. 11. (a) Amplitude error of the mixed explicit-implicit, finite-difference extrapolation operator with three terms of the Taylor series, H being represented by the first-, second- and third-order approximation. (b) Phase error of the mixed explicit-implicit, finite-difference extrapolation operator with three terms of the Taylor series, H being represented by the first-, second- and third-order approximation.

the amplitude and phase errors of the first-, second- and third-order mixed explicit-implicit scheme. Note the great similarity to the full explicit scheme. Of course this result can be significantly improved if the coefficients of equation (49) are optimally adjusted for the truncation error.

CONCLUSIONS

Extrapolation techniques as used in migration have been subdivided according to (1) the type of extrapolation operator (Figure 12a), (2) the computational method (Figure 12b), and (3) the domain of application (Figure 12c). It has been shown that explicit, recursive methods in the space-frequency domain are flexible and simple. Implicit techniques are computationally more complicated, particularly in the space-time domain. Nonrecursive methods are inflexible, particularly in the $k - f$ domain, and application is not advised in complex geologic situations.

In all of the above methods the transmission losses due to reflection and wave conversion were ignored. This is justified within geologic layers, but at major reflectors the assumption is not correct. Therefore future refinements should concentrate on handling the transmission properties at major boundaries correctly.

In all discussions multiple reflections were not considered.

A quantitative description of prestack seismic data is given by the matrix equation

$$\mathbf{P}(z_0) = \sum_{m=1}^{M} \mathbf{W}_s(z_0, z_m) \mathbf{R}(z_m) \mathbf{W}_d(z_m, z_0). \qquad (67)$$

In equation (67) the upward and downward traveling wave fields are treated separately. Generally $\mathbf{W}_s \neq \mathbf{W}_d$, which means that the principal of reprocity may not be applied. Equation (67) determines how forward and inverse extrapolation schemes have to be designed for any data configuration (common offset, common midpoint, common source, common receiver).

In the situation of significant lateral velocity variations, wave field extrapolation should be carried out recursively with small extrapolation steps. The summation approach can be applied both recursively and nonrecursively. Finite-difference methods are recursive by necessity.

The space-frequency domain is preeminently suitable for recursive wave field extrapolation:

(1) **Simplicity.**—In the 2-D situation the operator is 1-D, in the 3-D situation the operator is 2-D.
(2) **Flexibility.**—Spatial variations in the physical parameters—velocity, density and absorption—can be easily included. Frequency-dependent properties such as absorption, dispersion, and scattering can be taken into account. The extrapolation step Δz can be chosen differently for each frequency component (say $\Delta z \approx 1/3\lambda$). This choice will suppress the numerical dispersion, typically for space-time finite-difference techniques.
(3) The spatial bandwidth, as defined by W_0, can be optimized per temporal frequency component; this is of particular importance for the minimization of spatial-aliasing noise due to operator discretization (summation method), amplitude and/or phase distortion due to operator approximation (finite-difference method), and spatially coherent noise.
(4) During imaging, the temporal bandwidth can be optimized for each depth level. Filtering, such as deconvolution and band-pass filtering, should *not* be applied on the migrated

result $p(x, y, z, t = 0)$. If any filtering is desired, application should occur on the extrapolated data $P(x, y, z_i, \omega)$ as part of the imaging process. In this way a true depth-variant filter output can be realized.

The original finite-difference formulation of Claerbout (1970) was in the space-frequency domain, but very soon preference was given to the space-time domain. From a historical point of view it is interesting to see that modeling and migration software were realized in two domains: space-time and wavenumber-frequency. Surprisingly, the mixed domain, i.e., the space-frequency domain, was considered to be not very attractive. Recently, Kjartansson (1979) used a space-frequency algorithm for his absorption study.

In the summation approach the series expansion of W is represented by an explicit analytical expression. In the derivation of the operator a homogeneous medium is assumed, but if the application occurs recursively then moderate velocity variations can be properly handled.

For $\Delta z \to 0$, the spatial bandwidth of the summation operator increases without limit. For practical situations this inevitably will lead to serious spatial aliasing problems, particularly for high temporal frequencies. Therefore, for recursive applications with small Δz values it is advisable to stay away from expressions (38) and (39). Instead, a finite number of terms of the Taylor series expansion could be used, H represented by analytical expression (42b).

Typically, the summation operator is most suitable for the more simple geologic situations (i.e., simple in terms of propagation velocity). Nonrecursive applications in the space-time domain are economically very attractive.

Wave field extrapolation, applied entirely in the wavenumber-frequency domain, makes use of a plane-wave decomposition of the data. It extrapolates each plane wave by a simple multiplication with phase shift operator $\exp(\pm jk_z\Delta z)$, assuming that during propagation a plane wave stays a plane wave. This makes the method economically very attractive. However, plane-wave methods are significantly less flexible for situations with lateral velocity variations than the summation method.

Finite-difference techniques are most suitable for situations with fast lateral velocity variations. Implicit finite-difference schemes are more sensitive to improper boundary conditions at both ends of the data file than explicit ones. This forms a problem, particularly in prestack migration. Moreover, full implicit schemes are very complicated in 3-D applications. Explicit finite-difference schemes are simple and very attractive for recursive 3-D migration.

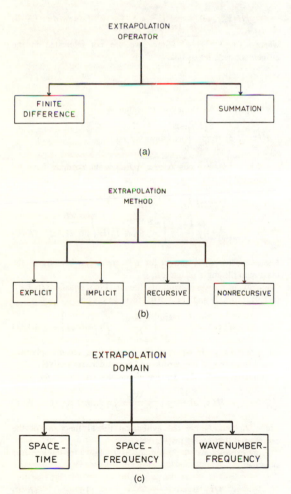

(a)

(b)

(c)

FIG. 12. (a) Subdivision of wave field extrapolation techniques according to the type of extrapolation operator. (b) Subdivision of wave field extrapolation techniques according to the computational method. (c) Subdivision of wave field extrapolation techniques according to the domain of application.

REFERENCES

Albramowitz, M. A., and Stegun, I. A., 1965, Handbook of mathematical functions.
Berkhout, A. J., 1977, Least-squares inverse filtering and wavelet deconvolution: Geophysics, v. 42, p. 1369–1383.
———— 1979, Steep dip finite-difference migration: Geophys. Prosp., v. 27, p. 196–213.
———— 1980, Seismic migration—Imaging of acoustic energy by wave field extrapolation: Amsterdam/New York, Elsevier/North Holland Publ. Co.
Berkhout, A. J., and Van Wulfften Palthe, D. W., 1979, Migration in terms of spatial deconvolution: Geophys. Prosp., v. 27, p. 261–291.
Brooks, R. A., and DiChiro, G., 1976, Principles of computer assisted tomography in radiographic and radio-isotopic imaging, a review article: Phys. Med. Biol., v. 21, p. 689–732.
Claerbout, J. F., 1970, Coarse grid calculations of waves in inhomogeneous media with application to delineation of complicated seismic structure: Geophysics, v. 35, p. 407–418.
———— 1976, Fundamentals of geophysical data processing: New York, McGraw-Hill Book Co., Inc.
Engquist, B., and Majda, A., 1979, Radiation boundary conditions for acoustic and elastic waves: Comm. Pure and Appl. Math., v. 32, p. 313–320.
Formalont, E. B., 1979, Fundamentals and deficiencies of aperture synthesis: Image formation from coherence functions in astronomy: Reidel Publ. Co., p. 3–18.
French, W. S., 1975, Computer migration of oblique seismic reflection profiles: Geophysics, v. 40, p. 961–980.
Gazdag, J., 1978, Wave equation migration with the phase-shift method: Geophysics, v. 43, p. 1342–1351.
Johnson, S. A., Greenleaf, J. F., and Rajagopalan, B., 1979, High spatial resolution ultrasonic measurement techniques for the characterization of static and moving tissues: Ultrasonic tissue characterization II: NBS Spec. Publ. 525.
Judson, D. R., Lin, J., Schultz, P. S., and Sherwood, J. W. C., 1980, Depth migration after stack: Geophysics, v. 45, p. 361–375.
Kjartansson, E., 1979, Attenuation of seismic waves in rocks and application in energy exploration: Ph.D. thesis, Stanford Univ.
Loewenthal, D., Lu, L., Roberson, R., and Sherwood, J., 1976, The wave equation applied to migration: Geophys. Prosp., v. 24, p. 380–399.
Mitchell, A. R., 1969, Computational methods in partial differential equations: New York, John Wiley.
Muir, F., and Claerbout, J. F., 1980, Impedance and wave extrapolation:

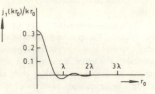

FIG. A–1. The first-order spherical Bessel function j_1 divided by its argument.

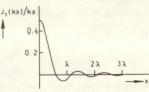

FIG. A–2. The first-order cylindrical Bessel function J_1 divided by its argument.

Presented at the 42nd EAEG meeting, Istanbul.

Pann, K., Eisner, E., and Shin, Y., 1979, A collocation formulation of wave equation migration: Geophysics, v. 44, p. 712–721.

Ristow, D., 1980, Three-dimensional finite-difference migration: Ph.D. thesis, Univ. of Utrecht, The Netherlands.

Schneider, W. A., 1978, Integral formulation in two and three dimensions: Geophysics, v. 43, p. 49–76.

Stewart, G. W., 1973, Introduction to matrix computations: New York, Academic Press.

Stolt, R., 1978, Migration by Fourier transform: Geophysics, v. 43, p. 23–48.

Yilmaz, Ö., 1979, Pre-Stack partial migration: Ph.D. thesis, Stanford Univ.

APPENDIX A

AN ANALYTICAL EXPRESSION FOR THE $\partial/\partial z$ OPERATOR

Start with the wave equation in the space-frequency domain (Helmholtz equation):

$$\nabla^2 P + k_1^2 P = 0, \qquad (A-1)$$

where $k_1 = k(1 - j\eta)$. In the homogeneous situation we may write for waves traveling in the $+ z$-direction:

$$\tilde{P}(k_x, k_y, z, \omega) = \tilde{P}(k_x, k_y, 0, \omega) e^{-jk\tilde{H}z}, \qquad (A-2a)$$

where

$$k\tilde{H}(k_x, k_y, \omega) = \sqrt{k_1^2 - (k_x^2 + k_y^2)}. \qquad (A-2b)$$

On physical grounds the imaginary part of equation (A–2b) should be chosen negative. From expression (A–2a) it follows

$$\frac{\partial}{\partial z}\tilde{P}(k_x, k_y, z, \omega) = -jk\tilde{H}(k_x, k_y, \omega)\tilde{P}(k_x, k_y, z, \omega), \qquad (A-3a)$$

or, after inverse Fourier transformation,

$$\frac{\partial}{\partial z}P(x, y, z, \omega) = -jkH(x, y, \omega) * P(x, y, z, \omega), \qquad (A-3b)$$

where

$$kH(x, y, \omega) = \frac{1}{4\pi^2}\int_{-\infty}^{+\infty}e^{-jk_x x}dk_x\int_{-\infty}^{+\infty}k\tilde{H}(k_x, k_y, \omega)\cdot$$
$$\cdot e^{-jk_y y}dk_y$$
$$= \frac{1}{4\pi^2}\int_{-\infty}^{+\infty}e^{-jk_x x}dk_x\int_{-\infty}^{+\infty}\sqrt{k_1^2 - (k_x^2 + k_y^2)}\cdot$$
$$\cdot e^{-jk_y y}dk_y. \qquad (A-4)$$

Now let us introduce the new integration variables κ and ψ:

$$k_x = \kappa\cos\psi,$$
$$k_y = \kappa\sin\psi.$$

$dk_x dk_y = \kappa\,d\kappa\,d\psi$ and equation (A–4) can be rewritten as

$$kH(x, y, \omega) = \frac{1}{4\pi^2}\int_{-\infty}^{+\infty}\kappa\sqrt{k_1^2 - \kappa^2}.$$

$$\cdot d\kappa\int_{-\infty}^{+\infty}e^{-j\kappa(x\cos\psi + y\sin\psi)}d\psi,$$

or, using the definition of the *zeroth*-order cylindrical Bessel function J_0,

$$kH(x, y, \omega) = \frac{1}{2\pi}\int_{-\infty}^{+\infty}\kappa\sqrt{k_1^2 - \kappa^2}\,J_0(\kappa r_0)\,d\kappa, \qquad (A-5)$$

where $r_0 = \sqrt{x^2 + y^2}$. Since we are not interested in the evanescent part, let us take

$$\kappa = k\sin\alpha \quad \text{with} \quad 0 \leq \alpha \leq \frac{\pi}{2}.$$

Then equation (A–5) may be rewritten as

$$H(x, y, \omega) = \frac{k^2}{2\pi}\int_0^{\pi/2}(\sin\alpha\cos\alpha)\cdot$$
$$\cdot\sqrt{\cos^2\alpha - 2j\eta}\,J_0(kr_0\sin\alpha)\,d\alpha$$
$$\approx \frac{k^2}{2\pi}\int_0^{\pi/2}\sin\alpha\cos^2\alpha\,J_0(kr_0\sin\alpha)\,d\alpha$$
$$-j\frac{k^2}{2\pi}\eta\int_0^{\pi/2}\sin\alpha\,J_0(kr_0\sin\alpha)\,d\alpha. \qquad (A-6)$$

Expression (A–6) is correct for $\eta \ll \cos\alpha_m$, where α_m is the maximum tilt angle of interest.

Finally, using Abramowitz and Stegun [1965, equations (6.1.8), (6.1.9), (10.1.1) and (11.4.10)], then we obtain

$$H(x, y, \omega) = \frac{k^2}{2\pi}\left[\frac{j_1(kr_0)}{kr_0} - j\eta\,j_0(kr_0)\right] \qquad (A-7)$$

for $\eta \ll \cos\alpha_m$. In equation (A–7) j_0 and j_1 are the spherical Bessel functions of the *zeroth*- and first-order, respectively.

Similarly, for the 2-D version of equation (A–7) we may write

$$H(x, \omega) = \frac{k}{2}\left[\frac{J_1(kx)}{kx} - j\eta\,J_0(kx)\right], \qquad (A-8)$$

where J_0 and J_1 are the cylindrical Bessel functions of the *zeroth*- and first-order, respectively.

From expressions (A–7) and (A–8) it follows for the loss-free situation:

$$H(x, y, \omega) = \frac{k^2}{2\pi}\frac{j_1(kr_0)}{kr_0} \qquad (3\text{-D}); \qquad (A-9)$$

$$H(x, \omega) = \frac{k}{2}\frac{J_1(kx)}{kx} \qquad (2\text{-D}). \qquad (A-10)$$

Figures A–1 and A–2 show j_1 and J_1.

Reprinted from GEOPHYSICS, **48**, 677-687.

Migration with the full acoustic wave equation

Dan D. Kosloff* and Edip Baysal‡

ABSTRACT

Conventional finite-difference migration has relied on one-way wave equations which allow energy to propagate only downward. Although generally reliable, such equations may not give accurate migration when the structures have strong lateral velocity variations or steep dips. The present study examined an alternative approach based on the full acoustic wave equation. The migration algorithm which developed from this equation was tested against synthetic data and against physical model data. The results indicated that such a scheme gives accurate migration for complicated structures.

INTRODUCTION

During the last decade, finite-difference migration has been dominated by the approach of Claerbout (1972). His method uses one-way wave equations which allow energy to propagate only downward. Although successful in many situations, the method is limited by the assumptions made in deriving the one-way wave equations. In particular, it is assumed that spatial derivatives of the velocity field can be ignored (Claerbout, 1972, 1976; Stolt, 1978; Gazdag, 1980; Berkhout and Palthe, 1979). However, such terms are significant in the presence of strong velocity contrasts. Furthermore, most finite-difference migration schemes which use the one-way wave equation contain a limit on the maximum dip of events which can be migrated properly. One-way wave equations are also incapable by nature of producing correct amplitudes (Berkhout and Palthe, 1979; Larner et al, 1981). The change in amplitude for smooth and discontinuous velocity variations is discussed in the Appendix for the one-dimensional case. This becomes important in the construction of before stack migration methods.

The present study tackled the problem of finding a migration scheme that would eliminate the deficiencies outlined above, offering an improvement over conventional finite-difference migration algorithms. A migration scheme for stacked sections based on the full acoustic wave equation is introduced. Because such an equation is directly derivable from continuum mechanics, it seemed likely to give accurate migration in areas with high velocity contrasts and steep dips. The main limitations which would remain in

the scheme were those associated with the conceptual model on which migration is based. For migration of stacked or zero-offset sections, this model is the so-called "exploding reflector model" (Loewenthal et al, 1976) which cannot correctly account for multiples or for complicated wave propagation such as, for example, a diffraction followed by a reflection.

Our study found that in order to implement migration based on the full acoustic wave equation, a number of obstacles had to be overcome. First, the acoustic wave equation is of second order in the spatial coordinates, and therefore requires two boundary conditions to initiate the depth extrapolation. On the other hand, the seismic data contain only one recorded field which is related either to the pressure or to the vertical pressure gradients. Second, a means must be supplied for eliminating evanescent energy which, if not removed, can cause numerical solutions to grow exponentially out of bounds. These topics are dealt with in later sections.

In the following sections we describe the numerical solution method developed for the migration algorithm. The scheme is then tested against simple synthetic examples, and against physical modeling data collected in the modeling tank at the Seismic Acoustics Laboratory at the Univ. of Houston.

BASIC EQUATIONS

The migration algorithm is derived in the space-frequency domain, with the acoustic wave equation serving as the basis. In an acoustic medium with variable density and velocity, the acoustic wave equation reads

$$\frac{\partial}{\partial x}\left(\frac{1}{\rho}\frac{\partial P}{\partial x}\right) + \frac{\partial}{\partial z}\left(\frac{1}{\rho}\frac{\partial P}{\partial z}\right) = \frac{1}{C^2}\frac{\partial^2 P}{\partial t^2}, \quad (1)$$

where x and z are the horizontal and vertical coordinates, $P(x, z, t)$ is the pressure field at time t, $C(x, z)$ is the acoustic velocity, and $\rho(x, z)$ is the density.

For the migration algorithm, equation (1) is Fourier transformed with respect to time and rewritten as a set of two coupled equations:

$$\frac{\partial}{\partial z}\begin{bmatrix} \tilde{P} \\ \frac{1}{\rho}\frac{\partial \tilde{P}}{\partial z} \end{bmatrix} = \begin{bmatrix} 0 & \rho \\ -\frac{\omega^2}{\rho C^2} - \frac{\partial}{\partial x}\left(\frac{1}{\rho}\frac{\partial}{\partial x}\right) & 0 \end{bmatrix}\begin{bmatrix} \tilde{P} \\ \frac{1}{\rho}\frac{\partial \tilde{P}}{\partial z} \end{bmatrix}, \quad (2)$$

Manuscript received by the Editor January 12, 1982; revised manuscript received August 23, 1982.
*Formerly Seismic Acoustics Laboratory, University of Houston; presently Department of Geophysics and Planetary Sciences, Tel Aviv University, Ramat Aviv, Tel Aviv 69978, Israel.
‡Formerly Seismic Acoustics Laboratory, University of Houston; presently Geophysical Development Corporation, 8401 Westheimer, Ste. 150, Houston, TX 77063.

where ω is the temporal frequency, and $\tilde{P}$ and $\partial\tilde{P}/\partial z$ represent the transformed pressure and vertical pressure gradient, respectively.

Equation (2) is written consistently with the continuity conditions of continuum mechanics which require that both the tractions and the displacements remain continuous across all possible interfaces in the medium. This follows because in an acoustic medium the tractions are equal to the pressures, and the accelerations are equal to $(1/\rho)(\partial P/\partial x)$ and $(1/\rho)(\partial P/\partial z)$, respectively.

When the density is assumed constant throughout the medium, equation (2) simplifies to give

$$\frac{\partial}{\partial z}\begin{bmatrix}\tilde{P}\\[2mm]\dfrac{\partial\tilde{P}}{\partial z}\end{bmatrix}=\begin{bmatrix}0 & 1\\[2mm]\dfrac{-\omega^2}{C^2}-\dfrac{\partial^2}{\partial x^2} & 0\end{bmatrix}\begin{bmatrix}\tilde{P}\\[2mm]\dfrac{\partial\tilde{P}}{\partial z}\end{bmatrix}. \tag{3}$$

In the following, equation (3) will be used for simplicity. It is useful in practice since the density usually varies less than the velocity in geologic structures, and is often not known accurately. However, the constant density assumption is not fundamental to the present migration scheme.

The depth extrapolation in the migration is based on equation (3). For stacked or zero-offset time sections, x and z represent midpoint coordinates and $C(x, z)$ is equal to half the true acoustic velocity in the medium (for further discussion, see Loewenthal et al, 1976; Stolt, 1978). The input section for the migration is given by $P(x, z = 0, t)$ and the migrated section is given by $P(x, z, t = 0)$ (Loewenthal et al, 1976). The integration of equation (3) is carried out for each frequency component separately. At each depth level z, components of the vector

$$\begin{bmatrix}\tilde{P}\,(x, z, \omega)\\[2mm]\dfrac{\partial\tilde{P}}{\partial z}(x, z, \omega)\end{bmatrix}$$

are stepped down to the next depth level $z + \Delta z$ for all x. The process is initiated at the surface after a specification of $\tilde{P}(x, 0, \omega)$ and $(\partial\tilde{P}/\partial z)(x, 0, \omega)$. The final migrated section $P(x, z, 0)$ is obtained after $\tilde{P}(x, z, \omega)$ has been calculated for all ω by summing over all ω (Gazdag, 1980):

$$P(x, z, 0) = \sum_\omega \tilde{P}(x, z, \omega). \tag{4}$$

METHOD OF EXTRAPOLATION IN DEPTH

For the depth extrapolation, equation (3) is spatially discretized. Let N_x be the number of seismic traces and Δx the trace spacing. We denote by $\tilde{P}(i, z, \omega)$ and $(\partial\tilde{P}/\partial z)(i, z, \omega)$ the respective values of the transformed pressure and transformed vertical pressure gradient at depth z, and at horizontal location $x = x_0 + (i - 1)\Delta x$. With this discretization in x, and with an appropriate approximation to $\partial^2 P/\partial x^2$, equation (3) becomes a set of $2N_x$ coupled ordinary differential equations in z for $P(i, z, \omega)$ and $(\partial\tilde{P}/\partial z)(i, z, \omega)$, $i = 1, \ldots, N_x$. The integration in depth can then be carried out with standard solution techniques for ordinary differential equations. We used a fourth-order Runge-Kutta method because it is accurate and easy to implement on an array processor.

Using the discretized version of equation (3) requires an approximation for $\partial^2\tilde{P}/\partial x^2$. The present scheme uses a Fourier

method to calculate this term (Kosloff and Baysal, 1982; Gazdag, 1980). Accordingly, the pressure transforms $\tilde{P}(i, z, \omega)$ are spatially transformed by a fast Fourier transform (FFT) to yield $\tilde{P}(K_x, z, \omega)$, then multiplied by $-K_x^2$ and inverse transformed to give $(\partial^2\tilde{P}/\partial x^2)(i, z, \omega)$. This derivative operator, unlike finite-differences, is accurate to the spatial Nyquist frequency. It also allows easy removal of undesirable spatial wavenumber components such as evanescent waves.

GENERATION OF SURFACE VALUES OF $\tilde{P}$ and $\partial\tilde{P}/\partial z$

The initiation of migration based on equation (3) requires the specification of both $\tilde{P}$ and $\partial\tilde{P}/\partial z$ on the surface. Since only one of these fields is recorded in practice, the remaining field must be generated from mathematical assumptions.

We assume that the seismic data are given by the pressure field $P(x, 0, t)$. $\tilde{P}(x, 0, \omega)$ can then be calculated from $P(x, 0, t)$ by Fourier transformation in time. The values of $(\partial\tilde{P}/\partial z)(x, 0, \omega)$ need to be generated from $\tilde{P}(x, 0, \omega)$ mathematically (other cases in which the recorded field is not the pressure can be treated by the same method outlined in this section). For this process we assume that the acoustic velocity is laterally uniform in the vicinity of the surface (but can be completely nonuniform everywhere else), and that the seismic time section consists of upgoing energy only. The latter assumption is also used in migration schemes based on one-way equations (Claerbout, 1976).

In a region in which the acoustic velocity C is constant, equation (1) can be doubly transformed in x and t to give

$$\frac{\partial^2\tilde{P}}{\partial z^2} = -\left(\frac{\omega^2}{C^2} - K_x^2\right)\tilde{P}(K_x, z, \omega), \tag{5}$$

where $\tilde{P}$ is the twice transformed pressure and K_x is the horizontal wavenumber. The solutions to equation (5) are given by

$$\tilde{P}(K_x, z, \omega) = e^{i\eta z}\tilde{P}(K_x, 0, \omega), \tag{6}$$

with $\eta = \sqrt{\omega^2/C^2 - K_x^2}$. The solution (6) includes only upgoing waves under the convention that z increases with depth. This study uses only nonevanescent energy components for which $\omega^2/C^2 > K_x^2$.

The doubly transformed pressure gradients $(\partial\tilde{P}/\partial z)(K_x, z, \omega)$ can be obtained from equation (6) by differentiation,

$$\frac{\partial\tilde{P}}{\partial z} = i\eta\, e^{i\eta z}\tilde{P}(K_x, 0, \omega) = i\eta\,\tilde{P}(K_x, z, \omega). \tag{7}$$

The generated vertical pressure gradients $(\partial\tilde{P}/\partial z)(x, 0, \omega)$ on the surface are obtained from equation (7) by setting $z = 0$ and by an inverse Fourier transformation with respect to x.

The procedure for generation of $\partial\tilde{P}/\partial z$ on the surface is compatible with assumptions used in most migration schemes. However, other alternatives may fit the reality of the field configuration better. For example, it may be more appropriate for data from land surveys to set the pressures $P(x, 0, t)$ equal to zero and to assume that the recorded data are proportional to $(\partial P/\partial z)(x, 0, t)$. However, we did not attempt to pursue these alternatives in the present study.

ELIMINATION OF EVANESCENT ENERGY

Evanescent waves are given by the exponentially varying solutions of the wave equation. Although in physical reality only solutions which exponentially decrease with depth are present, in numerical algorithms the exponentially increasing solutions can also be generated and cause the numerical results to grow out

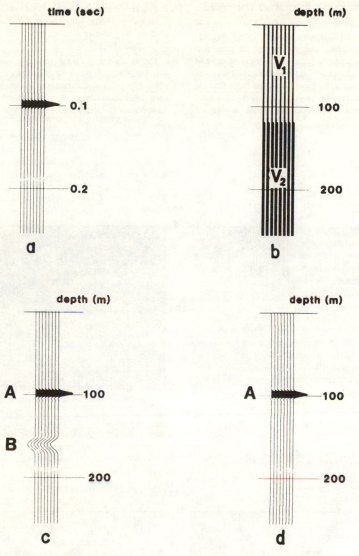

FIG. 1. (a) A one-dimensional time section containing a single event. (b) Velocity model $V_1 = 2000$ m/sec, $V_2 = 1000$ m/sec. (c) Migrated depth section based on equation (3). (d) Migrated section after the elimination of downgoing energy at time $t = 0$.

of bounds. Therefore, it is important to eliminate evanescent energy in implementing migration with the full acoustic wave equation.

For a laterally uniform medium with a depth dependent velocity $C(z)$, the evanescent solutions are defined by the relation

$$K_x > \frac{\omega}{C(z)},$$

where K_x is the horizontal wavenumber and ω is the temporal frequency. When the velocity field varies laterally, the identification

of the evanescent field becomes less clear cut. In this work we chose the definition

$$K_x > \frac{\omega}{C_{max}}, \qquad (8)$$

where C_{max} is the highest velocity at the depth z. Our experience indicates that criterion (8) assures numerical stability. However, in some cases it may cause the elimination of steeply dipping events in low-velocity regions. In applications, a less stringent condition than inequality (8) may sometimes be preferable.

ELIMINATION OF DOWNGOING ENERGY AT TIME ZERO

The conceptual basis for the migration of zero-offset data is the exploding reflector model (Loewenthal et al, 1976). According to this model, a zero-offset time section can be generated directly by halving all the acoustic velocities in the medium and placing explosive sources on the reflecting horizons. These explode at time zero with strengths proportional to the reflection coefficients. The need to replace physical reality by a model stems from the fact that a zero-offset section cannot be obtained from a single shot, but rather is composed from a series of shots. With the exploding reflector model, the aim of migration is to produce the pressure at time zero in all space (Loewenthal et al, 1976; Stolt, 1978).

The exploding reflector model applies directly to migration based on one-way wave equations since these propagate only upgoing energy. For the two-way wave equation, there is a nonuniqueness concerning downgoing energy. Consider the one-dimensional (1-D) time section in Figure 1a. When the velocity

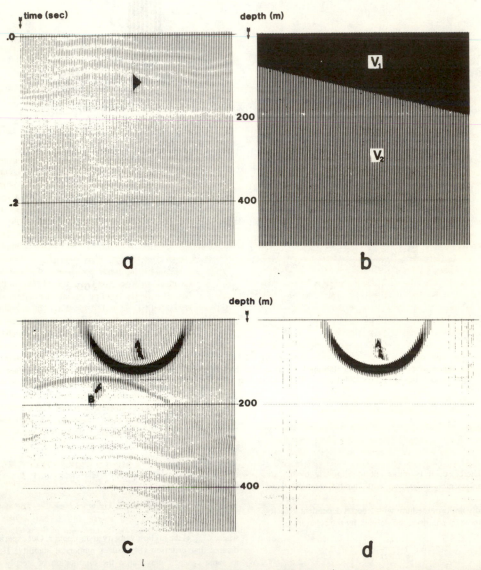

FIG. 2. (a) A time section containing a single spike. (b) Velocity model $V_1 = 2000$ m/sec; $V_2 = 1000$ m/sec. (c) Migrated section based on equation (3). (d) Migrated section after the elimination of downgoing energy at time $t = 0$.

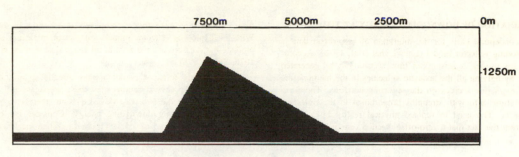

FIG. 3. Spatial configuration of the wedge model.

structure is as shown in Figure 1b, the migrated section from the solution of equation (3) becomes the section shown in Figure 1c. This section contains an upgoing event A which corresponds to the single event on the time section in Figure 1a. An additional event B containing downgoing energy only is also present. This event appears because of an inherent nonuniqueness in the conceptual model on which the migration is based. With the model, a surface recording alone cannot determine the amount of downgoing energy at time $t = 0$. In order to determine the amount of downgoing energy, a set of geophones also needs to be placed beneath the structure of interest. Obviously this option is not realizable in practice.

Downgoing energy at time $t = 0$ can be eliminated from the depth section by filtering out components with negative vertical wavenumbers. When this procedure is applied to the section of Figure 1c, the section shown in Figure 1d results. This section contains event A only.

Figure 2 shows the same elimination method for a two-dimensional (2-D) example. The time section in Figure 2a contains a single event which should give a depth section with a circular reflector (event A in Figure 2c). However, when the velocity is as in Figure 2b, an additional downgoing event B is produced on the depth section (Figure 2c). After the elimination procedure is applied, only event A remains (Figure 2d).

The nonuniqueness associated with downgoing energy becomes significant only in the presence of strong velocity contrasts. In many cases elimination of this energy is not necessary.

RELATION OF MIGRATION SCHEME TO THE PHASE-SHIFT METHOD

In a uniform or horizontally stratified region the migration scheme of this study is closely related to the phase-shift method (Gazdag, 1978). The point of departure is that the depth extrapolation is done here numerically instead of by phase shift.

For a homogeneous region with acoustic velocity C, equation (3) can be transformed with respect to x to give

$$\frac{\partial}{\partial z} \begin{bmatrix} \tilde{P} \\ \\ \dfrac{\partial \tilde{P}}{\partial z} \end{bmatrix} = \begin{bmatrix} 0 & 1 \\ \\ -\eta^2 & 0 \end{bmatrix} \begin{bmatrix} \tilde{P} \\ \\ \dfrac{\partial \tilde{P}}{\partial z} \end{bmatrix}, \qquad (9)$$

where $\eta^2 = \omega^2/C^2 - K_x^2$, and $\tilde{P}(K_x, z, \omega)$ and $(\partial \tilde{P}/\partial z)(K_x,$

$z, \omega)$ are, respectively, the doubly transformed pressure and vertical pressure gradient.

The solutions of equation (9) are given by

$$\tilde{P}(K_x, z, \omega) = A_+ \, e^{i\eta(z - z_0)} + A_- \, e^{-i\eta(z - z_0)}], \qquad (10)$$

and

$$\frac{\partial \tilde{P}}{\partial z} = i\eta[A_+ \, e^{i\eta(z - z_0)} A_- \, e^{i\eta(z - z_0)}], \qquad (11)$$

where A_+ and A_-, respectively, represent amplitudes of upgoing and downgoing waves and z_0 is a reference depth. Equations (10) and (11) can be used for depth extrapolation in a phase-shift migration. This migration can also be used for a horizontally stratified velocity model by using equations (10) and (11) within each layer and determining A_+ and A_- by the continuity conditions of $\tilde{P}$ and $\partial \tilde{P}/\partial z$ across the top interface of the layer. This type of phase-shift migration accounts for amplitude changes. When A_- is set equal to zero, only upgoing energy is considered and $\partial \tilde{P}/\partial z$ can no longer be made continuous. The migration then becomes the phase-shift method in Gazdag (1978). In this migration A_+ remains constant at all depths and the correct amplitudes of events are no longer restored (Appendix). The migration is then equivalent to migration with one-way wave equations.

When the migration method of this paper is compared with the phase-shift method for a stratified medium, it becomes apparent that the two methods are identical in the manner in which horizontal derivatives are calculated [the Fourier method calculates $\partial^2 P/\partial x^2$ in the (K_x, z, ω) domain by multiplication by $-K_x^2$]. The only difference between the two methods is that in this study, $\tilde{P}$ and $\partial \tilde{P}/\partial z$ are stepped down by a Runge-Kutta method, instead of by equations (10) and (11). Therefore, for sufficiently small Δz the two methods should give practically identical results.

EXAMPLE: A BURIED WEDGE STRUCTURE

In the example of a buried wedge structure, the input time section was obtained from the acoustic modeling tank at the Seismic Acoustics Laboratory at the Univ. of Houston. The same model was used in Kosloff and Baysal (1982) to compare forward modeling results. The scaled dimensions of the model are shown in Figure 3. The wedge structure was made of low-velocity room temperature vulcanized (RTV) rubber, whereas the base was made of high-velocity Plexiglas. The whole model was immersed in water which had a scaled velocity of 3950 m/sec. A zero-offset

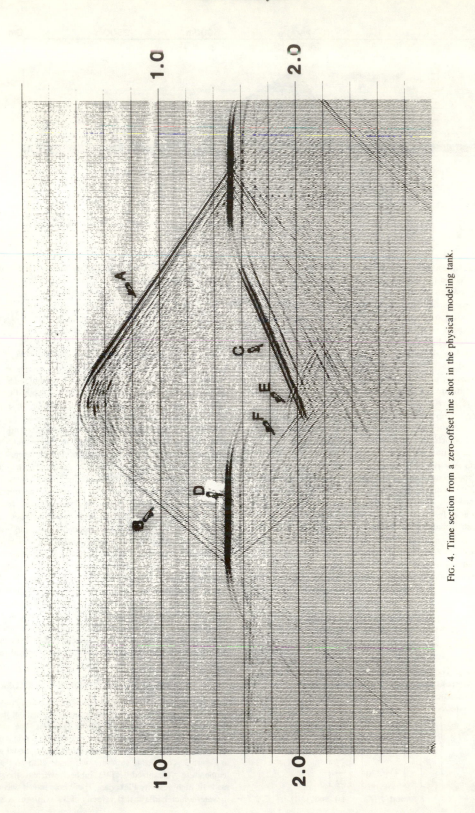

FIG. 4. Time section from a zero-offset line shot in the physical modeling tank.

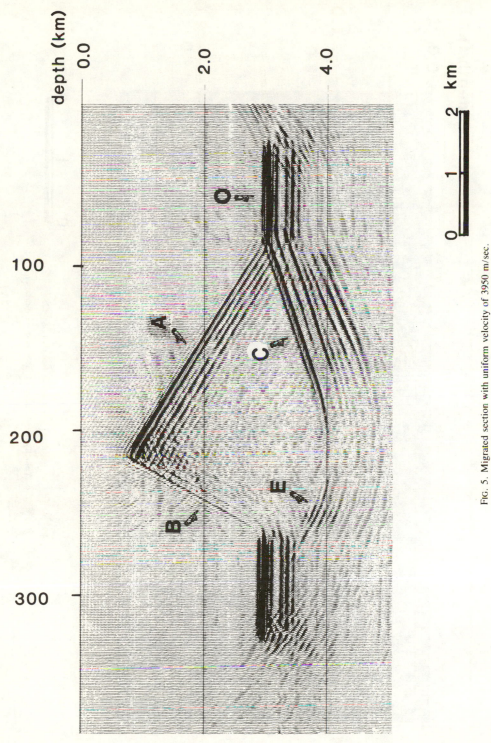

FIG. 5. Migrated section with uniform velocity of 3950 m/sec.

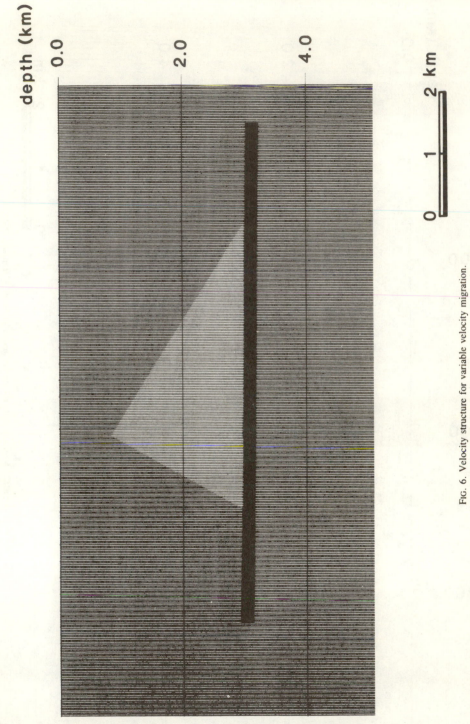

FIG. 6. Velocity structure for variable velocity migration.

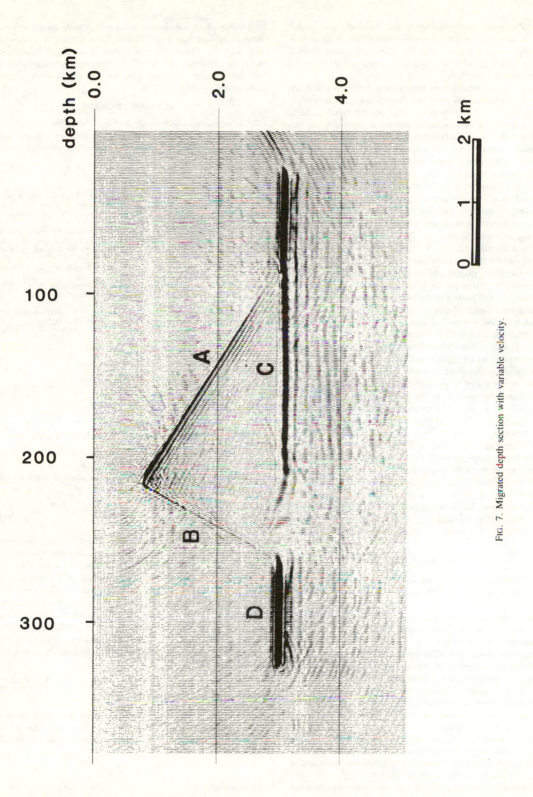

FIG. 7. Migrated depth section with variable velocity.

line was collected perpendicular to the symmetry axis of the wedge at a scaled height of approximately 800 m above the wedge tip, with a shot spacing of 26 m. Since this model includes steep dips and high velocity contrasts, it serves as a good test for the migration algorithm.

Figure 4 shows the observed time section. Events A and B are interpreted as reflections from the sloping sides of the wedge, whereas events C and D are reflections from the Plexiglas base. Events E and F are complicated events not accounted for by the exploding reflector model (Kosloff and Baysal, 1982).

In order to make the migration as objective as possible, no prior knowledge of the structure was assumed and at first the time section (Figure 4) was migrated with a uniform velocity equal to the scaled water velocity of 3950 m/sec. The migrated section is shown in Figure 5. In this figure events which traveled to the surface through water only were migrated to their respective proper positions including the 60 degree wedge interface B. On the other hand, the reflection C from the Plexiglas base underlying the wedge is undermigrated.

In the second stage, variable velocities were introduced. The velocity interfaces, except portions of the base under the wedge, were derived from the depth section of the constant velocity migration (Figure 5). The portion of the base under the RTV wedge was continued horizontally (Figures 5 and 6). The scaled velocities were taken as 3950 m/sec for water, 6000 m/sec for Plexiglas, and 2650 m/sec for the RTV (Figure 6).

The migrated section with variable velocity is shown in Figure 7. In this figure, the Plexiglas base C under the RTV wedge is defined in the correct location. However, part of the base is missing. This can be attributed to the fact that energy which propagates upward from this portion of the base encounters the steeply dipping side of the wedge at an angle beyond the critical angle for RTV and water. This type of propagation is not accounted for by the exploding reflector model on which the migration is based.

It is interesting to note that the migrated results are extremely sensitive to small changes in RTV velocity. In particular the Plexiglas base under the RTV wedge becomes misaligned whenever this velocity is perturbed, in the same manner as the misalignment in Figure 5. In fact, the RTV velocity of 2650 m/sec used for the migration is about 300 m/sec higher than the velocity which is usually quoted for this material. This sensitivity may suggest using migration as a means to determine velocities of physical modeling materials.

CONCLUSIONS

A migration scheme based on a direct integration in depth of the acoustic wave equation has been presented. After the problems of specification of surface boundary conditions and the removal of evanescent energy had been addressed, implementing this migration algorithm was not more complicated than implementing one-way equation schemes.

The present method may offer improvements over conventional finite-difference schemes with regards to the migration of steeply dipping structures, or migration in regions with high velocity contrasts. This is because the full acoustic wave equation is not based on any assumptions concerning the medium through which the waves propagate. Moreover, the numerical algorithm which was used is highly accurate because it uses the Fourier method for calculating horizontal derivatives and a fourth-order Runge-Kutta method for the depth extrapolation. Consequently, for a horizontally stratified medium it proved to be practically equivalent to the analytic phase-shift method (Gazdag, 1978).

The possibility of using the full acoustic wave equation for migration, which was demonstrated in this study for stacked sections, may gain added significance for migration of nonstacked time sections. There, the preservation of amplitude information becomes important and therefore a migration scheme based on the full acoustic wave equation may be necessary to achieve satisfactory results.

ACKNOWLEDGMENT

A large portion of this study was done while the authors were at the Seismic Acoustics Laboratory at the Univ. of Houston. Discussions with Norm Bleistein and Fred Hilterman were most helpful.

REFERENCES

Berkhout, A. J., and Van Wulfften Palthe, D. W., 1979, Migration in terms of spatial deconvolution: Geophys. Prosp., v. 27, p. 261–291.
Claerbout, J. F., 1976, Fundamentals of geophysical data processing: New York, McGraw-Hill Book Co., Inc.
Claerbout, J. F., and Doherty, S. M., 1972, Downward continuation of moveout corrected seismograms: Geophysics, v. 37, p. 741–768.
Gazdag, J., 1980, Wave equation migration with the accurate space derivative method: Geophys. Prosp., v. 28, p. 60–70.
Kosloff, D., and Baysal, E., 1982, Forward modeling by a Fourier method: Geophysics, v. 47, p. 1402–1412.
Loewenthal, D., Lu, L., Roberson, R., and Sherwood, J. W. C., 1976, The wave equation applied to migration: Geophys. Prosp., v. 24, p. 380–399.
Larner, K., Hatton, L., Gibson, B., and Hsu, I., 1981, Depth migration of imaged time sections: Geophysics, v. 46, p. 734–750.
Mathews, J., and Walker, R. L., 1964, Mathematical methods of physics: The Benjamin/Cummings Publ. Co.
Stolt, R. H., 1978, Migration by Fourier transform: Geophysics, v. 43, p. 23–48.

APPENDIX
WAVE AMPLITUDES FOR 1-D ONE-WAY WAVE EQUATIONS

This appendix shows that one-way wave equations fail to reproduce correct amplitudes for 1-D propagation.

A wide class of one-way equations are derived from a series expansion of the dispersion relation

$$\frac{\omega}{C} = \mp (K_x^2 + K_z^2)^{1/2} \qquad (A-1)$$

(e.g., Claerbout, 1976, p. 202), where K_z and K_x are, respectively, the vertical and horizontal wavenumbers, ω is the temporal frequency, and C is the acoustic velocity which is assumed to vary slowly in space. The 15 degree wave equation (Claerbout, 1976), for example, can be derived by retaining the first two terms in a Taylor series expansion of equation (A-1) and by replacing the dispersion relation by a differential equation and transforming the result to the moving coordinate system of Claerbout (1976).

In 1-D vertical propagation, K_x in equation (A-1) is set to zero and the one-way wave equation which corresponds to equation (A-1) becomes

$$\frac{1}{C(z)} \frac{\partial P}{\partial t} = \mp \frac{\partial P}{\partial z}. \qquad (A-2)$$

Equation (A-2) can also be derived directly from most one-way wave equations by setting all terms containing horizontal derivatives to zero. Equation (A-2) can be solved by using the variable separation

$$P = e^{i\omega t} U(z).$$

We obtain

$$P(z, t) = \exp\left[\mp i\omega \left(t - \int_0^z \frac{dy}{C(y)} \right) \right]. \quad (A-3)$$

Equation (A–3) is the zeroth order WKB solution to the full acoustic wave equation (Mathews and Walker, 1964). With this solution the amplitude is constant throughout the medium regardless of the values of $C(z)$. Conversely, it may be recalled that the first-order WKB solution to the acoustic wave equation is given by

$$P(z, t) = \frac{1}{(\omega/C)^{1/4}} \exp\left[\mp i\omega \left(t - \int_0^z \frac{dy}{C(y)} \right) \right] \quad (A-4)$$

(Mathews and Walker, 1964). This solution does contain amplitude modulation depending upon $C(z)$.

In the case of velocity jumps at interfaces, the one-way wave equation solution (A–3) gives the same wave amplitudes at both sides of the interfaces. On the other hand, for the acoustic wave equation, these amplitudes differ by the ratio of the transmission coefficients.

In conclusion, for both smooth and discontinuous velocity variation, the one-way wave equation does not produce the correct amplitude. These results also apply to wave propagation in two or three spatial dimensions.

Reprinted from *Geophysical Prospecting*, **31**, 413-420.

MIGRATION BY EXTRAPOLATION OF TIME-DEPENDENT BOUNDARY VALUES*

G.A. McMECHAN**

ABSTRACT

MCMECHAN, G.A. 1983, Migration by Extrapolation of Time-Dependent Boundary Values, Geophysical Prospecting 31, 413–420.

Migration of an observed zero-offset wavefield can be performed as the solution of a boundary value problem in which the data are extrapolated backward in time. This concept is implemented through a finite-difference solution of the two-dimensional acoustic wave equation. All depths are imaged simultaneously at time 0 (the imaging condition), and all dips (right up to vertical) are correctly migrated. Numerical examples illustrate this technique in both constant and variable velocity media.

INTRODUCTION

There are several schemes available for depth migration of zero-offset (or normal-moveout-corrected stacked) seismic sections. Among the common ones are Kirchhoff migration (French 1974, 1975), f–k migration (Stolt 1978), spatial deconvolution (Berkhout and Van Wulfften Palthe 1979, Berkhout 1980), and finite-difference methods (see Claerbout and Doherty 1972, Claerbout 1976). All involve some form of extrapolation based on the wave equation.

This paper describes a new finite-difference approach to migration that, for reasons that will become apparent, shall be referred to as *boundary value migration* (BVM). BVM is based on the reversibility of the wave equation; migration of a zero-offset section can be treated as the reverse of data modeling by the exploding reflector method. In this method, upward propagation of energy is initiated at all reflectors at time $t = 0$, the wave equation carries this energy through the model, and each recorder produces a response time history for all subsequent times.

The inverse problem of imaging reflectivity (i.e., migration) can be solved as a boundary value problem. The required boundary conditions over the recording surface are known at all times, and consist explicitly of the recorded observations.

* Received September 1981, revisions October 1982.
** Pacific Geoscience Center, Earth Physics Branch, Department of Energy, Mines and Resources, Sidney, BC V8L 4B2, Canada. Present address: Department of Computer Science, University of Victoria, PO Box 1700, Victoria, BC V8W 2Y2, Canada.

Migration is performed by driving the mesh at each recording point with the time reverse of the seismic trace recorded at that point. The wave equation propagates the energy downward into the medium; this extrapolation is continued backwards in time to $t = 0$, when all depths are imaged simultaneously.

The method is implemented with a two-dimensional, second-order-explicit, finite-difference solution of the two-way acoustic wave equation. Since no paraxial approximations are involved, even 90° dips are correctly migrated.

THEORY

In this section, the concept of migration as extrapolation of a data wavefield backwards in time (rather than the usual downward continuation in depth) is illustrated in the simple context of a point scatterer in a homogeneous half-space. In the following section, examples involving more complicated reflectors and heterogeneous velocity fields are presented.

The details of the formulation used for BVM by time extrapolation have been presented by McMechan (1982) in the context of imaging of earthquake sources. The examples in that paper are relevant here since migration also can be thought of as a source location problem in which sources are coincident with reflectors. Since the algorithm has been described by McMechan (1982), only a brief summary will be given here.

For use in illustrating the BVM technique, several synthetic zero-offset seismic data sections were generated by the exploding reflector method as referred to above. The finite-difference code used for this modeling is similar to that used for subsequent migration. The source function used in modeling was that presented by Alterman and Karal (1968). The absorbing boundary conditions described by Clayton and Engquist (1977) were used along the sides and bottom of the grid for both modeling and migration.

Energy is propagated through the finite-difference grid by the two-dimensional acoustic wave equation (see Claerbout 1976, Mitchell 1969):

$$U_{yy} + U_{zz} = \frac{1}{V^2(y, z)} U_{tt}, \tag{1}$$

where U is acoustic pressure, V is velocity, and subscripts denote partial derivatives with respect to y (the horizontal coordinate), z (the depth coordinate), or t (time). The model velocities V used in both modeling and migration are half the true velocities because total two-way traveltimes are twice the corresponding one-way times from the reflectors to the receivers. The acoustic wave equation is implemented in a second-order-explicit, finite-difference scheme. The characteristics of such schemes with respect to grid dispersion and numerical stability are discussed, among others, by Alford, Kelly and Boore (1974) and Mitchell (1969).

At a representative time step (at time t_i), three U wavefields are involved: $U(y, z, t_i)$ is the entire (y, z) wavefield at time t_i; $U(y, z, t_{i-1})$ is U at the previous time step $(t = t_{i-1})$; and $U(y, z, t_{i-2})$ is at t_{i-2}. The response $U(y_k, z_j, t_i)$ at the internal grid point (y_k, z_j) (located at the intersection of the jth horizontal mesh line with the kth vertical line) at time t_i is

$$U(y_k, z_j, t_i) = 2(1 - 2A^2)U(y_k, z_j, t_{i-1}) - U(y_k, z_j, t_{i-2})$$
$$+ A[U(y_{k+1}, z_j, t_{i-1}) + U(y_{k-1}, z_j, t_{i-1})$$
$$+ U(y_k, z_{j+1}, t_{i-1}) + U(y_k, z_{j-1}, t_{i-1})], \qquad (2)$$

where $A = V(y_k, z_j) \Delta t/h$, Δt is the time step between successive U-wavefields ($\Delta t = t_i - t_{i-1}$), and h is the distance between grid lines in both y- and z-directions. The condition for local stability of (2) is that $\Delta t < hV^{-1}2^{-1/2}$.

The BVM processing of the synthetic zero-offset section corresponding to a point reflector in a homogeneous half-space is shown in fig. 1. Figure 1(a) contains the data section to be migrated. The finite-difference grid over which the wave equation is to be solved is defined in the y–z plane (a vertical slice through the earth). The process of migration transfers data from the y–t plane to the y–z plane as follows. Each iteration (each time step) during migration consists of two operations: the wave equation carries all the previous energy away from the upper boundary of the y–z plane (toward its migrated position); then, values extracted from the seismogram profile along that slice of constant time that corresponds to the current iteration are inserted at each recording point as new boundary conditions. These operations are facilitated by having the horizontal mesh increment equal to the data trace separation, and the time digitization increment of the data equal to the time step in extrapolation. [Interpolation (see Larner, Gibson and Rothman 1981) may be required.] Consider a specific example: fig. 1(b) shows the y–z plane at time t_i. All contributions for times greater than t_i are already present and are migrating toward their final positions. The boundary values to be inserted at time t_i are found in the seismogram profile (fig. 1a) along the slice at time t_i. Extrapolation continues backward in time with new boundary values inserted at each time step. The imaging condition is time 0. At this time, all the energy is at its migrated position; the entire y–z plane is imaged simultaneously (fig. 1d). This is in contrast to schemes such as that of Claerbout (1976), in which the data are downward-continued in depth (rather than time) to image one depth slice at each (depth) iteration.

In summary, solution of the forward problem involves generating a response wavefield $U(y, z, t)$ from initial conditions $U(y, z, t_1)$ and $U(y, z, t_2)$ in a velocity model $V(y, z)$. The resulting seismograms at $z = 0$ are $U(y, 0, t)$. In a real survey situation, $U(y, 0, t)$ are the recorded data. (For simplicity, $z = 0$ is indicated as the recording datum, but any (y, z) points can be used). Migration of the data consists of successively solving (2) with the boundary values $U(y, 0, t_i)$, $U(y, 0, t_{i-1})$, and $U(y, 0, t_{i-2})$ obtained directly from the observations at each i.

Although the point reflector example in fig. 1 is a simple one, it contains all the elements necessary for the construction of more complicated examples (and, in fact, for migration of real data, although application will be left for another paper). Any number of arbitrarily shaped reflections can be treated as the superposition of the responses to point sources (Huygens' principle), and the spatial variation of velocity $[V(y, z)]$ is already part of the finite-difference solution (see the equation above).

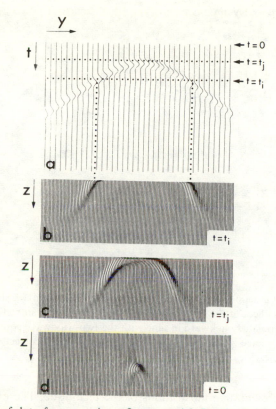

Fig. 1. Migration of data from a point reflector model. Panel (a) shows the synthetic zero-offset section computed for a point reflector model. Panels (b), (c), and (d) show the y–z plane (a vertical slice through the earth) at three iterations that are widely spaced in time. Time t_i is far from the image time, time t_j is intermediate, and $t = 0$ is the image condition. In (d), all the energy in the time section (a) has been successfully migrated back to the point reflector. The two small artifacts that remain faintly visible below the focus are due to the data truncation at the two edges of the profile. The dotted lines show the relationship between the seismograms and the boundary values in the y–z plane at $t = t_i$ (see text). For clarity, only every fourth seismogram that was computed is plotted in (a). In the migration, however, all the seismograms were used. This is also true of figs. 2, 3, and 4. Physical units are not included on the axes to emphasize that the method is independent of scale (e.g., very large-scale features can be imaged using long periods).

The following section contains three additional numerical examples of migration with the BVM method. In these examples, only the y–t plane and the imaged y–z plane will be shown; intermediate y–z plots such as fig. 1(b), (c) will not be included.

NUMERICAL EXAMPLES

In the previous section, the method of boundary value migration was presented through an example of a point reflector. This section contains three additional

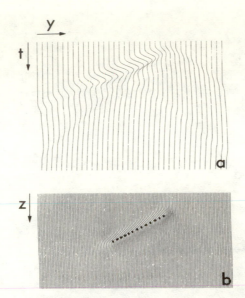

Fig. 2. Migration of a synthetic time section corresponding to a truncated, sloping reflector. Successive constant-time slices of the data section (a) are progressively introduced as boundary conditions and extrapolated backward in time with the two-way wave equation. The peak of the migrated reflector image in the y–z plane (b) coincides with the correct solution, which is shown as the dotted line. (See caption to fig. 1.)

examples that are representative illustrations of the technique. The examples are a truncated sloping reflector, a vertical reflector, and a sinusoidal reflector. In the first two, velocity is constant; in the third, it is a function of both y and z. Discussion of a broader range of applications is included.

Figure 2(a) contains the synthetic two-way time section for a truncated, sloping reflector. The diffractions generated at each end of the reflector are clear. Migration of this time section produces an image in the y–z plane as shown in fig. 2(b); the correct solution, which is a finite reflector dipping at 26.6°, is indicated by the dotted line.

Since the two-way wave equation is used in this migration algorithm, stable migration of very steep dips is possible. This is in contrast to other finite-difference algorithms that use an approximate (e.g. paraxial) one-way wave equation (see Claerbout 1976). The maximum dips that are correctly migrated by a paraxial approximation depend on the number of terms taken in the series expansion for the z extrapolation operator. Figure 3 illustrates migration of a vertical reflector (90° dip). Figure 3(a) contains the time section, and fig. 3(b) contains the migrated image. The dotted line in fig. 3(b) indicates the correct solution. Artifacts associated with spatial truncation of the data set, particularly at the right side, produce some broadening of the image.

Another potential advantage to the use of the two-way wave equation is that it can correctly migrate not only energy that moves monotonically downward in the

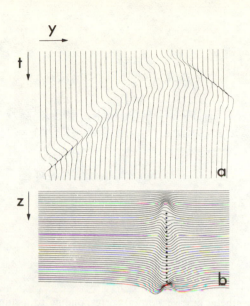

Fig. 3. Migration of a synthetic time section (a) corresponding to a vertical reflector. The peak of the migrated image in the y–z plane (b) coincides with the correct solution, which is shown as the dotted line. (See caption to fig. 1.)

z-direction during migration, but also energy whose migration path contains changes in the sign of $\partial z/\partial t$ (e.g., energy that leaves an exploding reflector downward). For example, it is possible to image the underside of an overhanging structure such as often occurs at the edge of a salt dome. For this energy to be recorded, it is necessary that there exists some deeper structure (such as a strong reflector, or a region of high velocity gradient) that turns the energy back toward the surface. Similarly, for this energy to be migrated properly, this deeper structure must be explicitly contained in the velocity field through which the migration is to be done. McMechan (1982) contains three examples of imaging of sources (for which one can directly substitute exploding reflectors) that use both up- and downgoing energy. It is of interest to note that this extension still conforms to the definition of Claerbout (1976) that a reflector exists at the spatial and temporal coincidence of up- and downgoing waves.

The final example of this paper, a sinusoidal reflector, is shown in fig. 4. This is a more complicated example than the previous ones, both in the shape of the reflector and in the velocity distribution. A linear velocity gradient was used in both y- and z-directions, with the velocity increasing from the upper left corner of the y–z grid to the lower right corner. The seismic time section (fig. 4a) was migrated to produce the sinusoidal image in the y–z plane (fig. 4b).

In all the examples shown above, migration by extrapolation of time-dependent boundary values has produced the correct solution. This migration technique, when applied to primary acoustic reflected waves, does not seem to have any problems or

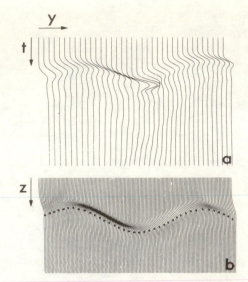

Fig. 4. Migration of a synthetic time section (a) corresponding to a sinusoidal reflector in a variable velocity medium. The peak of the migrated image in the y–z plane (b) coincides with the correct solution, which is shown as the dotted line. (See caption to fig. 1.)

inherent restrictions, other than those already associated with finite differences (see Alford et al. 1974), and these can generally be overcome by the use of a higher order difference scheme (only second order was used here), or by using higher data sampling rates or a finer computational mesh. The algorithm can also be used as it is for prestack migration by extrapolating data from each shot separately and then superimposing all the resultant y–z wavefields. The BVM algorithm should be directly applicable to real data in its present form; application is left for another paper. Finally, the concept of BVM is directly applicable in three, as well as two, space dimensions, requiring only a three-dimensional wave equation code for implementation.

SUMMARY

A new method of migrating seismic time sections in variable velocity media is presented and illustrated with four numerical examples. The data are introduced in successive time slices as boundary conditions to a finite-difference extrapolation backwards in time. At time 0 (the imaging condition), all depths are imaged simultaneously. Extrapolation through time is done with the two-way wave equation so all dips are correctly migrated. The method is easy to implement and has no inherent limitations other than those typically associated with finite differencing.

ACKNOWLEDGMENTS

The author gratefully acknowledges critical reviews by his colleagues in the Earth Physics Branch and by C. Yorath of the Geological Survey of Canada. Primary funding of this project was by the Department of Energy, Mines & Resources, Canada; additional support was provided by I. Barrodale through NSERC grant A5251. All computations were performed at the Computing Center of the University of Victoria. Contribution from the Earth Physics Branch No. 1025.

REFERENCES

ALFORD, R.M., KELLY, K.R. and BOORE, D.M. 1974, Accuracy of finite-difference modeling of the acoustic wave equation, Geophysics 39, 834–842.

ALTERMAN, Z. and KARAL, F.C. 1968, Propagation of elastic waves in layered media by finite difference methods, Bulletin of the Seismological Society of America 58, 367–398.

BERKHOUT, A.J. 1980, Seismic Migration, Elsevier, New York.

BERKHOUT, A.J. and VAN WULFFTEN PALTHE, D.W. 1979, Migration in terms of spatial deconvolution, Geophysical Prospecting 27, 261–291.

CLAERBOUT, J.F. 1976, Fundamentals of Geophysical Data Processing, McGraw-Hill, New York.

CLAERBOUT, J.F. and DOHERTY, S.M. 1972, Downward continuation of moveout corrected seismograms, Geophysics 37, 741–768.

CLAYTON, R.W. and ENGQUIST, B. 1977, Absorbing boundary conditions for acoustic and elastic wave equations, Bulletin of the Seismological Society of America 67, 1529–1540.

FRENCH, W.S. 1974, Two-dimensional and three-dimensional migration of model-experiment reflection profiles, Geophysics 39, 265–277.

FRENCH, W.S. 1975, Computer migration of oblique seismic reflection profiles, Geophysics 40, 961–980.

LARNER, K., GIBSON, B. and ROTHMAN, D. 1981, Trace interpolation and the design of seismic surveys (abstract), Geophysics 46, 407.

McMECHAN, G.A. 1982, Determination of source parameters by wavefield extrapolation, Geophysical Journal of the Royal Astronomical Society 71, 613–628.

MITCHELL, A.R. 1969, Computational Methods in Partial Differential Equations, J. Wiley & Sons, New York.

STOLT, R.H. 1978, Migration by Fourier transform, Geophysics 43, 23–48.

Reprinted from GEOPHYSICS, **48**, 1514-1524.

Reverse time migration

Edip Baysal*, Dan D. Kosloff‡, and John W. C. Sherwood§

ABSTRACT

Migration of stacked or zero-offset sections is based on deriving the wave amplitude in space from wave field observations at the surface. Conventionally this calculation has been carried out through a depth extrapolation.

We examine the alternative of carrying out the migration through a reverse time extrapolation. This approach may offer improvements over existing migration methods, especially in cases of steeply dipping structures with strong velocity contrasts. This migration method is tested using appropriate synthetic data sets.

INTRODUCTION

Migration of stacked or zero-offset data considered to consist of primary reflections only has usually been achieved through a downward continuation of the surface data (Claerbout and Doherty, 1972; Loewenthal et al, 1976; Stolt, 1978; Berkhout, 1980). The final migrated section is then given by the amplitude of the extrapolated field at time zero as a function of depth (Loewenthal et al, 1976; Judson et al, 1980). The velocity for the calculations should be taken as half the actual velocity in the medium (Loewenthal et al, 1976).

The imaging principle inherent in the migration of stacked sections permits a different approach to migration based on reverse time marching instead of a depth extrapolation. The stacked section is considered as a surface boundary condition for a reverse operation to the modeling type wave calculations that step forward in time (Kelly et al, 1976; Kosloff and Baysal, 1982). The calculations are carried out in reversed time from the time of the last sample on the time section until time zero when the amplitudes in all space are considered as the final migrated section. If the velocities for the migration are chosen correctly, the wave field at time zero should be coincident with the reflecting horizons in the medium.

Reverse time migration may offer a number of improvements over conventional depth extrapolation. In particular, the posing of the migration problem as an extrapolation in time instead of in depth avoids the problems associated with evanes-

cent energy (Kosloff and Baysal, 1983). Furthermore, this paper will show that it is possible to use wave equations containing no dip limitations for time stepping schemes and that a steep-dip depth migration can be achieved with ease.

In the following sections, we outline the main ingredients of reverse time migration and present a number of examples which shed light on its features.

DEPTH MIGRATION AS A REVERSE EXTRAPOLATION IN TIME

The basis for migration of stacked time sections is the "exploding reflector model" (Loewenthal et al, 1976). According to this model, an approximation to a stacked section can be obtained in a single experiment by replacing the subsurface with a medium containing half the actual velocities in the earth, and by initiating explosive sources at time zero on all the reflecting boundaries. With this model, the recorded surface time section approximates the stacked or zero-offset section which would be collected over the same region (Loewenthal et al, 1976).

The purpose of migration, based on the exploding reflector model, is to recover the amplitudes at time zero which give the location and strength of the reflectors. Let $P(x, z = 0, t)$ denote the surface recorded time section with x the horizontal midpoint coordinate along the seismic line and z the depth. The migrated section then becomes $P(x, z, t = 0)$. In the reverse time depth migration, it is assumed that $P(x, z, t) = 0$ for $t > T_L$ where T_L is the last recorded time sample. In other words, it is assumed that after this time, the energy has propagated away from the subsurface underneath the seismic line. The migration is then formulated as a wave propagation problem in which the waves are generated from the time reversed stacked or zero-offset section $P(x, z = 0, t)$ which is applied as a surface boundary condition.

There are a number of possibilities for choosing a wave equation for the migration. Since the subsurface recorded common-depth-point (CDP) stacked section is ideally free of multiple reflections, it seems appropriate to use a wave equation which avoids layer reflections. Thus, for the present study,

(*Text continued on p. 1519*)

Manuscript received by the Editor August 30, 1982; revised manuscript received December 14, 1982.
*Geophysical Development Corporation and Seismic Acoustics Laboratory, Houston, TX.
‡Geophysics Department, Tel-Aviv University, Tel-Aviv, Israel and Seismic Acoustics Laboratory of the University of Houston.
§Geophysical Development Corporation, 8401 Westheimer, Houston, TX 77063.

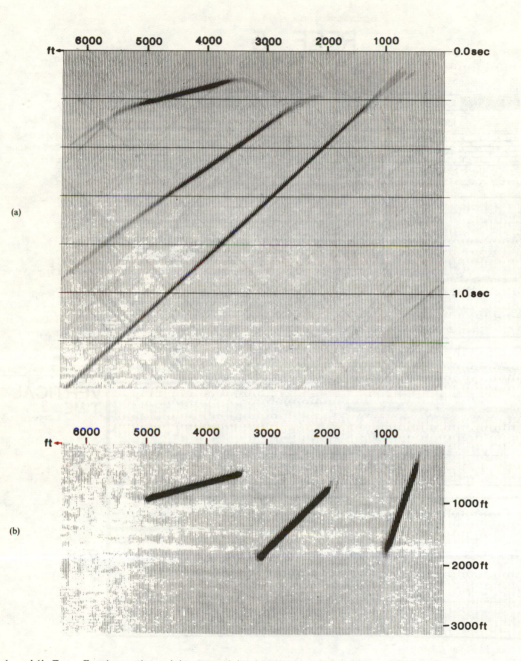

FIGS. 1a and 1b. Zero-offset time section and the reverse time migration result for a model consisting of reflector segments with dips of 15, 45, and 70 degrees.

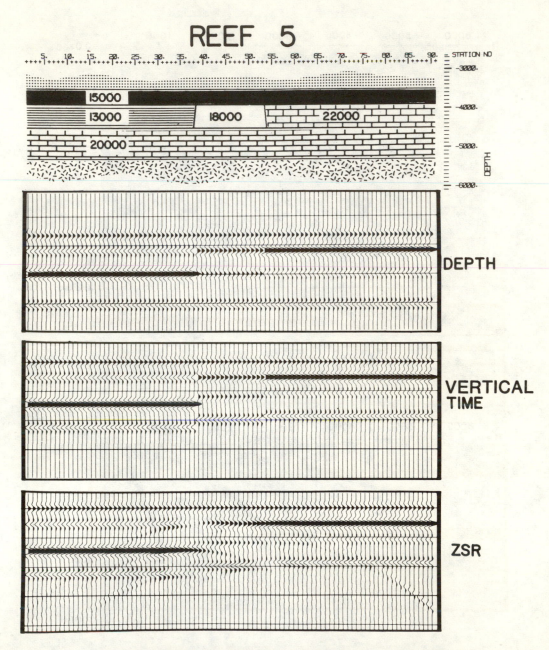

FIG. 2. Reef model, depth section, 1-D vertical time, and zero-offset 2-D seismic section.

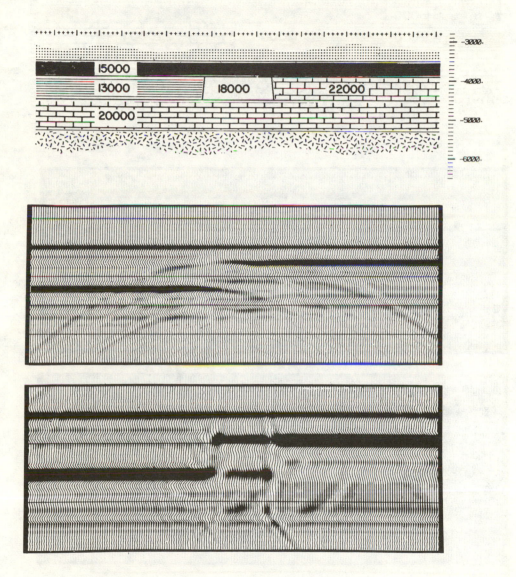

Fig. 3. Zero-offset seismic section of the reef model and the depth migration.

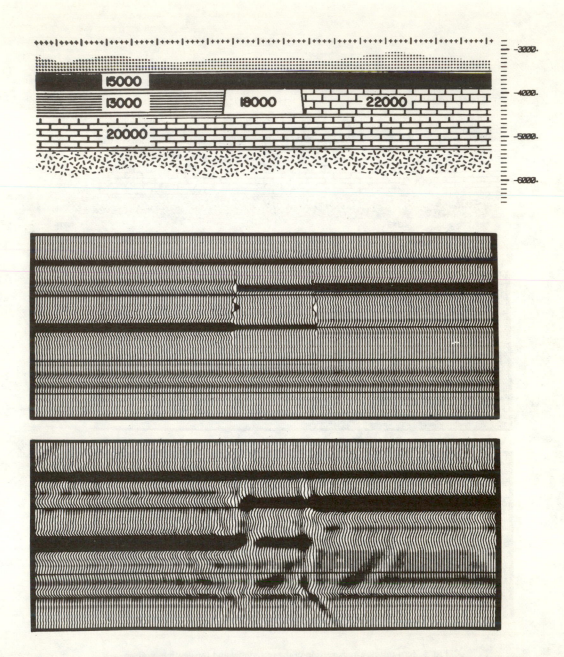

Fig. 4. Input model depth section for the reef model and the result of the depth migration.

we chose the 90-degree dip wave equation first presented by Gazdag (1981):

$$\pm \left[\frac{\partial^2}{\partial x^2} + \frac{\partial^2}{\partial z^2} \right]^{1/2} P = \frac{1}{c(x, z)} \frac{\partial P}{\partial t}. \quad (1)$$

In equation (1), $P(x, z, t)$ denotes the wave field (related either to the pressure or to the vertical velocity component), $c(x, z)$ is the velocity field, and x and z, respectively, are the horizontal and vertical coordinates. The square root derivative operator does not have an explicit representation in the spatial domain, but it can be handled in a natural manner using spatial Fourier transforms (Gazdag, 1980, 1981; Kosloff and Baysal, 1982, 1983). Using equation (1), a numerical estimate can be made of the time derivative of P at time T. $P(x, z, t = T)$ is first

2-D Fourier transformed to the wavenumber domain (k_x, k_z) using the fast Fourier transform algorithm. Subsequent multiplication by $[\text{sign} (k_z) i (k_x^2 + k_z^2)^{1/2}]$, 2-D Fourier inversion back to the (x, z) domain, and multiplication by the spatially varying velocity $c(x, z)$ yields the time derivative $\dot{P}(x, z, t = T)$. This is now approximated by the centered finite difference of $P(x, z, t = T - \Delta t)$ and $P(x, z, t = T + \Delta t)$:

$$[P(x, z, T + \Delta t) - P(x, z, T - \Delta t)]/2\Delta t = \dot{P}(x, z, T). \quad (2)$$

Hence, knowledge of the wave field $P(x, z, t)$ at times $(T + \Delta t)$ and T enables estimation of $P(x, z, t = T - \Delta t)$. Since the wave field is propagated back down toward the reflectors, it is physically impossible for this method to yield a sensible value for $P(x, z = 0, T - \Delta t)$. Instead these values at the $z = 0$ boundary

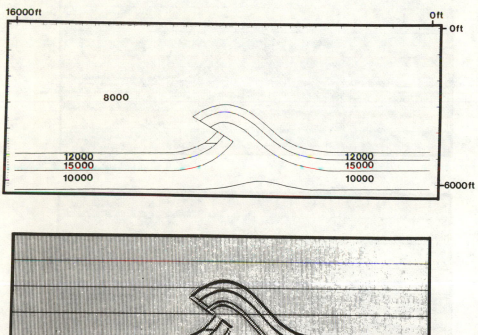

Fig. 5. Overthrust model and its depth section.

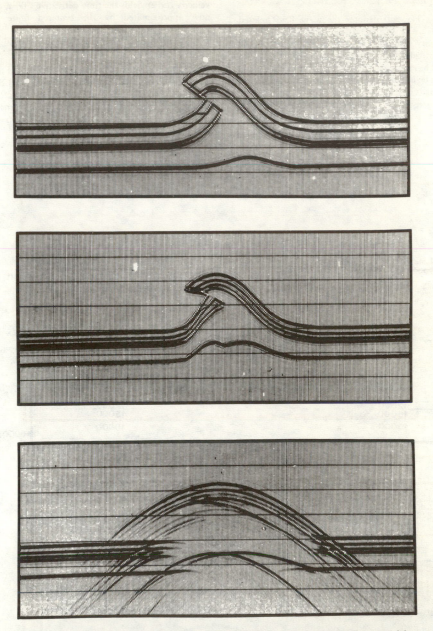

Fig. 6. Depth, 1-D vertical time, and zero-offset seismic sections for the overthrust model.

are provided from the stacked time section $P(x, z = 0, t)$, $0 \leq t \leq T_L$. The calculations proceed from time $t = T_L$ to time $t = 0$, the initial wave field being taken as zero at times $(T_L + \Delta t)$ and $(T_L + 2\Delta t)$.

This approach utilizing equations (1) and (2) appears eminently suitable for reverse time migration because it applies to dips reaching 90 degrees and it permits both vertical and lateral velocity variations. The method is also free of numerical dispersion and instability from exponentially growing evanescent waves.

EXAMPLES

The algorithm is demonstrated here with synthetic data. Rather than running the same program forward in time to generate input time sections, other algorithms were used to create the synthetic time sections.

The first example is a test for accuracy as a function of dip. The input model consists of three reflector segments with dips of 15, 45, and 70 degrees. The velocity of the medium is 8000 ft/sec. The time section resulting from f-k modeling (Stolt, 1978)

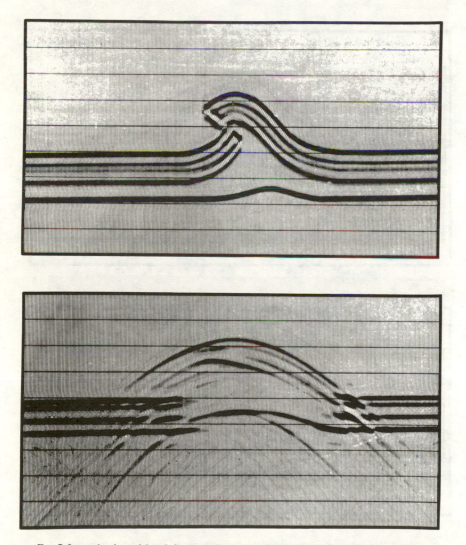

FIG. 7. Input depth model and the zero-offset seismic section filtered with a low-pass filter.

and the corresponding reverse time migration are shown in Figures 1a and 1b, respectively. It is apparent that the events have been migrated with crisp definition and no noticeable dispersion. The dips agree with those of the original model.

The second example is a stratigraphic model featuring a 600 ft high pinnacle reef that is approximately 1600 ft wide at the base. This reef model, its depth section, 1-D vertical time section, and zero-offset 2-D time section are shown in Figure 2. The zero-offset seismic section was obtained from a modified Kirchhoff modeling program (Larson and Hilterman, 1976). Note the dead zone that is on the left-hand side of the top of the reef reflection in the zero-offset section. This dead zone occurs because the reflection coefficient changes polarity laterally and, thus, for shotpoints near this reflection coefficient discontinuity the wavefront sees half of a positive reflecting boundary and half of a negative reflecting boundary. Notice also that the velocity pull-up is not as simple as the 1-D vertical time section suggests. The pull-up on the deeper reflector appears to be dipping.

In the migration the corrrect velocity field was used. Figure 3 shows the geologic model, its zero-offset seismic section, and the depth migration result. The wavelet used in the zero-offset seismic section was a 28 Hz Ricker wavelet. Prior to migration the input section was filtered with a low-pass filter with a

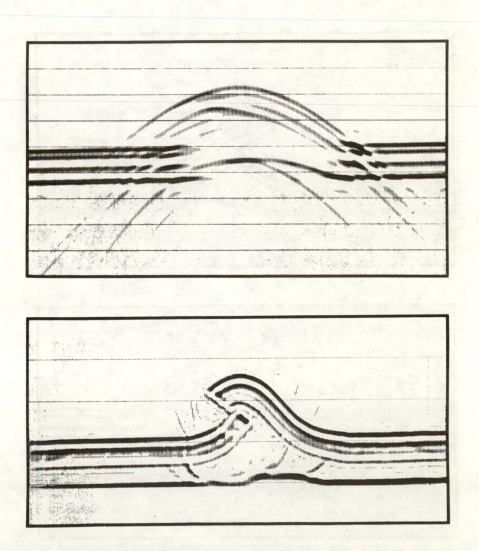

FIG. 8. Zero-offset seismic section of the overthrust model and its depth migration.

cut-off frequency of 20 Hz in order to shorten the calculation time (the time step size used in the migration was 1 msec). Thus in Figure 3 the zero-offset section has a sharper wavelet than the depth migration result. Also the vertical scale for the migration result is depth z, and therefore the wavelet observed in the the migrated section is a spatial wavelet. The trace spacing in the zero-offset seismic section was 50 ft, and in the migration a grid spacing of 50 ft ($\Delta x = \Delta z = 50$ ft) was used. The migrated depth section should be compared against the input depth model in order to examine the results. Figure 4 shows this comparison. The location of the reef is correctly presented in the depth migrated section. The change in the reflection coef-

ficient on both sides of the reef also conforms with the geologic model.

The third example is more representative of a structural model. Figure 5 shows the model and its depth section. The velocities used in the model range from 8000 to 15,000 ft/sec. The depth section, 1-D time section, and zero-offset seismic section are presented in Figure 6. The zero-offset seismic section presents a difficult case for interpretation, and it is obvious that the migration of this seismic section will be necessary for a sensible interpretation.

Trace spacing for this model was 50 ft. In the depth migration process a grid spacing of 50 ft ($\Delta x = \Delta z = 50$ ft) was used.

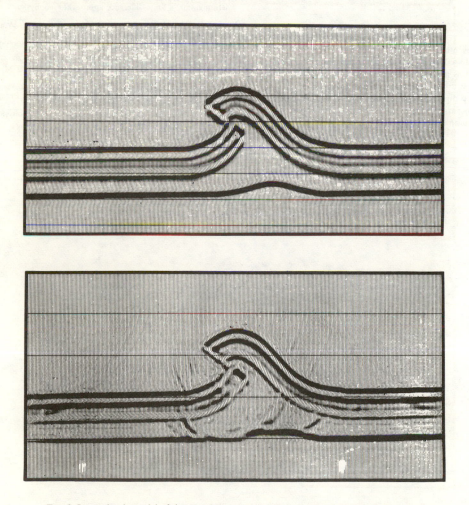

FIG. 9. Input depth model of the overthrust model and the result of the depth migration.

The zero-offset seismic section was filtered with a low-pass filter of 20 Hz cut-off frequency, which allowed a time sample rate of 1 msec to be used in the migration process. The filtered input data are shown in Figure 7.

The depth migration result is shown in Figure 8, together with the input zero-offset section. The vertical axis of the zero-offset section is time, whereas in the migration result the vertical scale represents depth in feet. Figure 8 indicates that migration of seismic data is a very important tool in interpretation. Since the input model was synthetic, the migration result can be compared against the input depth model. Figure 9 shows that the result of such a comparison is satisfactory. In the migration result the continuity of the folded sediments and the sharpness of the fault zone illustrate the accuracy of the algorithm. The deepest boundary of the geologic model under the overthrust zone is not accurately reconstructed. It is probable that the fault lies not with the depth migration but rather with the modeling program producing the zero-offset seismic section. This was a ray-tracing program (modified Kirchhoff modeling) which may be expected to be inaccurate with such a complex velocity model.

CONCLUSION

We have presented a migration method for stacked or zero-offset sections based on a reverse time extrapolation. Theoretical considerations and the synthetic examples presented indicate that reverse time depth migration can handle structures containing steep dips and strong velocity contrasts. In complicated areas this migration may offer a viable alternative to migration based on depth extrapolation.

ACKNOWLEDGMENTS

It is a pleasure to acknowledge the financial support and the provision of excellent computer facilities for two of the authors from the Seismic Acoustics Laboratory of the Univ. of Houston. Also we are very appreciative of the advice and assistance of Fred Hilterman in the construction of the synthetic models and for his enthusiastic participation in several lengthy discussions.

REFERENCES

Berkhout, A. J., 1980, Seismic migration: Elsevier Sci. Publ. Co., Amsterdam.
Claerbout, J., and Doherty, S., 1972, Downward continuation of moveout corrected seismograms: Geophysics, v. 37, p. 741–768.
Gazdag, J., 1980, Wave equation migration with the accurate space derivative method: Geophys. Prosp., v. 28, p. 60–70.
———— 1981, Modeling of the acoustic wave equation with transform methods: Geophysics, v. 46, p. 854–859.
Judson, D. R., Lin, J., Schultz, P. S., and Sherwood, J. W. C., 1980, Depth migration after stack: Geophysics, v. 45, p. 361–375.
Kelly, K. R., Ward, R. W., Treitel, S., and Alford, R. M., 1976, Synthetic seismograms: a finite-difference approach: Geophysics, v. 41, p. 2–27.
Kosloff, D. D., and Baysal, E., 1982, Forward modeling by a Fourier method: Geophysics, v. 47, p. 1402–1412.
———— 1983, Migration with the full acoustic wave equation: Geophysics, v. 48, p. 677–687.
Larson, D., and Hilterman, F. J., 1976, Diffractions: their generation and interpretation use: Presented at the 29th Annual Midwestern SEG Meeting, Dallas.
Loewenthal, D., Lu, L., Roberson, R., and Sherwood, J. W. C., 1976, The wave equation applied to migration: Geophys. Prosp., v. 24, p. 380–399.
Stolt, R., 1978, Migration by Fourier transform: Geophysics, v. 43, p. 23–48.